tenth edition

W9-ASE-403

Environmental Geology

Carla W. Montgomery

Professor Emerita
Northern Illinois University

Connect
Learn
Succeed™

ENVIRONMENTAL GEOLOGY, TENTH EDITION

Published by McGraw-Hill, a business unit of The McGraw-Hill Companies, Inc., 1221 Avenue of the Americas, New York, NY 10020. Copyright © 2014 by The McGraw-Hill Companies, Inc. All rights reserved. Printed in the United States of America. Previous editions © 2011, 2008, and 2006. No part of this publication may be reproduced or distributed in any form or by any means, or stored in a database or retrieval system, without the prior written consent of The McGraw-Hill Companies, Inc., including, but not limited to, in any network or other electronic storage or transmission, or broadcast for distance learning.

1 2 3 4 5 6 7 8 9 0 DOW/DOW 1 0 9 8 7 6 5 4 3

ISBN 978–0–07–352411–5
MHID 0–07–352411–5

Senior Vice President, Products & Markets: *Kurt L. Strand*
Vice President, General Manager, Products & Markets: *Marty Lange*
Vice President, Content Production & Technology Services: *Kimberly Meriwether David*
Managing Director: *Thomas Timp*
Brand Manager: *Michelle Vogler*
Director of Development: *Rose Koos*
Director of Digital Content: *Andrea Pellerito*
Marketing Manager: *Matt Garcia*
Director, Content Production: *Terri Schiesl*
Senior Project Manager: *Vicki Krug*
Buyer: *Nicole Baumgartner*
Designer: *Margarite Reynolds*
Interior Designer: *Linda Robertson*
Cover Designer: *Studio Montage, St. Louis, Missouri*
Cover Image: *© Pixtal/age Fotostock*
Lead Content Licensing Specialist: *Carrie K. Burger*
Photo Research: *Jerry Marshall/pictureresearching.com*
Senior Media Project Manager: *Tammy Juran*
Compositor: *Aptara®, Inc.*
Typeface: *10/12 Times LT Std Roman*
Printer: *R. R. Donnelley*

Library of Congress Cataloging-in-Publication Data

Cataloging-in-Publication Data has been requested from the Library of Congress.

The Internet addresses listed in the text were accurate at the time of publication. The inclusion of a website does not indicate an endorsement by the authors or McGraw-Hill, and McGraw-Hill does not guarantee the accuracy of the information presented at these sites.

In Dedication

Environmental Geology is affectionately dedicated to the memory of Ed Jaffe, whose confidence in an unknown author made the first edition possible.

–CWM–

Preface

About the Course

Environmental Geology Is Geology Applied to Living

The *environment* is the sum of all the features and conditions surrounding an organism that may influence it. An individual's physical environment encompasses rocks and soil, air and water, such factors as light and temperature, and other organisms. One's social environment might include a network of family and friends, a particular political system, and a set of social customs that affect one's behavior.

Geology is the study of the earth. Because the earth provides the basic physical environment in which we live, all of geology might in one sense be regarded as environmental geology. However, the term *environmental geology* is usually restricted to refer particularly to geology as it relates directly to human activities, and that is the focus of this book. Environmental geology is geology applied to living. We will examine how geologic processes and hazards influence human activities (and sometimes the reverse), the geologic aspects of pollution and waste-disposal problems, and several other topics.

Why Study Environmental Geology?

One reason for studying environmental geology might simply be curiosity about the way the earth works, about the *how* and *why* of natural phenomena. Another reason is that we are increasingly faced with environmental problems to be solved and decisions to be made, and in many cases, an understanding of one or more geologic processes is essential to finding an appropriate solution.

Of course, many environmental problems cannot be fully assessed and solved using geologic data alone. The problems vary widely in size and in complexity. In a specific instance, data from other branches of science (such as biology, chemistry, or ecology), as well as economics, politics, social priorities, and so on may have to be taken into account. Because a variety of considerations may influence the choice of a solution, there is frequently disagreement about which solution is "best." Our personal choices will often depend strongly on our beliefs about which considerations are most important.

About the Book

An introductory text cannot explore all aspects of environmental concerns. Here, the emphasis is on the physical constraints imposed on human activities by the geologic processes that have shaped and are still shaping our natural environment. In a real sense, these are the most basic, inescapable constraints; we cannot, for instance, use a resource that is not there, or build a secure home or a safe dam on land that is fundamentally unstable. Geology, then, is a logical place to start in developing an understanding of many environmental issues. The principal aim of this book is to present the reader with a broad overview of environmental geology. Because geology does not exist in a vacuum, however, the text introduces related considerations from outside geology to clarify other ramifications of the subjects discussed. Likewise, the present does not exist in isolation from the past and future; occasionally, the text looks both at how the earth developed into its present condition and where matters seem to be moving for the future. It is hoped that this knowledge will provide the reader with a useful foundation for discussing and evaluating specific environmental issues, as well as for developing ideas about how the problems should be solved.

Features Designed for the Student

This text is intended for an introductory-level college course. It does not assume any prior exposure to geology or college-level mathematics or science courses. The metric system is used throughout, except where other units are conventional within a discipline. (For the convenience of students not yet "fluent" in metric units, a conversion table is included on the inside back cover, and in some cases, metric equivalents in English units are included within the text.)

Each chapter opens with an introduction that sets the stage for the material to follow. In the course of the chapter, important terms and concepts are identified by boldface type, and these terms are collected as "Key Terms and Concepts" at the end of the chapter for quick review. The Glossary includes both these boldface terms and the additional, italicized terms that many chapters contain. Most chapters include actual case histories and specific real-world examples. Every chapter concludes with review questions and exercises, which allow students to test their comprehension and apply their knowledge. The "Exploring Further" section of each chapter includes a number of activities in which students can engage, some involving online data, and some, quantitative analysis. For example, they may be directed to examine real-time stream-gaging or landslide-monitoring data, or information on current or recent earthquake activity; they can manipulate historic climate data from NASA to examine trends by region or time period; they may calculate how big a wind farm or photovoltaic array would be required to replace a conventional power plant; they can even learn how to reduce sulfate pollution by buying SO_2 allowances.

Each chapter includes one or more case studies. Some involve a situation, problem, or application that might be encountered in everyday life. Others offer additional case histories or relevant examples. The tone is occasionally light, but the underlying issues are nonetheless real. (While some case studies were inspired by actual events, and include specific factual information, all of the characters quoted, and their interactions, are wholly fictitious.)

Additional online resources available on the website for each chapter are of two kinds. One is "NetNotes," a modest collection of Internet sites that provide additional information and/or images relevant to the chapter content. These should prove useful to both students and instructors. An effort has been made to concentrate on sites with material at an appropriate level for the book's intended audience and also on sites likely to be relatively stable in the very fluid world of the Internet (government agencies, educational institutions, or professional-association sites). The other is "Suggested Readings/References," some of which can also be accessed online. A previous appendix on maps and satellite imagery included in earlier editions has been moved to the text's website along with other readings formerly in the text.

New and Updated Content

Environmental geology is, by its very nature, a dynamic field in which new issues continue to arise and old ones to evolve. Every chapter has been updated with regard to data, examples, and illustrations.

Geology is a visual subject, and photographs, satellite imagery, diagrams, and graphs all enhance students' learning. Accordingly, this edition includes more than one hundred new photographs/images and forty new figures, with revisions having been made to dozens more.

Significant content additions and revisions to specific chapters include:

Chapter 1   Population data and projections have been updated; Case Study 1 includes new lunar data.

Chapter 2   The Libby, Montana, vermiculite case study has been updated and refined.

Chapter 3   The chapter has been reorganized for better flow, and discussion of compressive and tensile stress as related to tectonics and plate boundaries has been clarified.

Chapter 4   Discussion of waves, seismic waves, and seismographs has been enhanced. Much new material has been added on the recent earthquakes in Japan, Haiti, New Zealand, and elsewhere. Case Study 4.1 now includes extensive discussion of the Japanese tsunami, and the chapter notes application of the Japanese earthquake early warning system to this quake. Case Study 4.2 updates information on SAFOD results. The trial of the Italian seismologists who failed to predict the 2009 l'Aquila earthquake is noted.

Chapter 5   New material on the roles of fluid and of pressure reduction in promoting mantle melting has been added. The eruption of Eyjafjallajökull and its effects on air travel are discussed. The status of Redoubt as examined in Case Study 5.2 has been updated, and the possible role of seismic activity in triggering the eruption of Chaitén volcano noted.

Chapter 6   The 2011 Mississippi River flooding is discussed, including the role of deliberate breaching of levees.

Chapter 7   Material on the effects of Hurricanes Irene and Sandy has been added, particularly in Case Study 7.

Chapter 8   The 2010 landslide that buried Attabad, Pakistan, is discussed, together with its aftermath.

Chapter 9   Discussion of possible causes of past ice ages has been expanded, now including the potential role of the evolution of land plants; discussions of Milankovitch cycles and of desertification have been enhanced.

Chapter 10   Data on global temperature changes have been updated, as have data on changes in alpine glacier thickness; changes in the thickness and extent of Arctic sea ice cover are presented, and the breakup of the Wilkins ice shelf in Antarctica noted. Discussion of global-change impacts has been expanded and now includes ocean acidification. Climate-change vulnerability across Africa as identified by the U.N. Environment Programme is examined.

Chapter 11   Status of water levels in Lake Mead and of Lake Chad and the Aral Sea have been updated. Discussion of Darcy's Law has been clarified. Case Study 11 has been expanded, with information on radium in ground water nationally, and a note on bottled versus tap-water quality. New data on water withdrawals, nationally and by state, and on irrigation-water use by state, are presented; groundwater monitoring by satellite illustrates declining water levels.

Chapter 12   Data on soil erosion by wind and water, nationally and by region within the United States, have been updated, highlighting areas in which erosion rates exceed sustainable limits.

Chapter 13   The discussion of resources versus reserves is incorporated early in the chapter. Distribution of world reserves of a dozen key metals has been updated, along with data on U.S. per-capita consumption of select minerals and fuels. All tables of U.S. and world mineral production, consumption, and reserves have been updated. The former Case Study 13 (now 13.2) reflects recent commodity price rises. A new Case Study 13.1 has been added to focus on the rare-earth elements, their importance, and the current dominance of China in the world REE trade.

Chapter 14   All data on U.S. energy production and consumption by source have been updated. Information on shale gas, its distribution, and its significance for U.S. natural-gas reserves has been added, with concerns relating to fracking noted. The Deepwater Horizon oil spill is discussed; discussion of the Athabasca oil sands has been expanded.

Chapter 15 Extensive discussion of the Fukushima power-plant accident has been incorporated in Case Study 15.1. New data on world use of nuclear fission power and on U.S. solar and wind-energy potential across the country are presented. Iceland has been added as a new comparison case in figure 15.33, and data on the remaining countries' energy-source patterns updated.

Chapter 16 Data on the composition and fate of U.S. municipal wastes have been updated. A discussion of the challenge of estimating the effects of low-level radiation exposure has been added, and the status of Yucca Mountain and of world development of high-level nuclear-waste repositories updated.

Chapter 17 The significance of trace elements to health and the concept of the dose-response curve are incorporated early in the chapter. New data on nutrient loading in the Gulf of Mexico from the Mississippi River basin are presented. The growing problem of pharmaceuticals in wastewater is noted, and maps modeling nutrient and herbicide concentrations in ground water across the country are examined.

Chapter 18 Data on sources of U.S. air pollutants, air quality, and acid rain across the country have been updated. Trends in air pollution in selected major U.S. cities are shown, and particulate air pollution around the globe, including its varied sources, is considered. New data on ground-level ozone in the United States and stratospheric ozone globally are presented, and the recently recognized Arctic "ozone hole" is discussed.

Chapter 19 Seafloor imaging as it relates to resource rights under the Law of the Sea Treaty is illustrated. Discussion of the Montreal Protocol has been expanded to include the evolving problem of HFCs and HCFCs and ozone depletion. The Keystone XL pipeline has been added to Case Study 19.

Chapter 20 New data on land cover/use, recent population change, and population density for the United States are examined. In the engineering-geology section, discussions of the cases of the Leaning Tower of Pisa and of the St. Francis Dam have been expanded, and the case of the Taum Sauk dam failure added.

The online "NetNotes" have been checked, all URLs confirmed, corrected, or deleted as appropriate, and new entries have been added for every chapter. The "Suggested Readings/References" have likewise been updated, with some older materials removed and new items added in each chapter.

Organization

The book starts with some background information: a brief outline of earth's development to the present, and a look at one major reason why environmental problems today are so pressing—the large and rapidly growing human population. This is followed by a short discussion of the basic materials of geology—rocks and minerals—and some of their physical properties, which introduces a number of basic terms and concepts that are used in later chapters.

The next several chapters treat individual processes in detail. Some of these are large-scale processes, which may involve motions and forces in the earth hundreds of kilometers below the surface, and may lead to dramatic, often catastrophic events like earthquakes and volcanic eruptions. Other processes—such as the flow of rivers and glaciers or the blowing of the wind—occur only near the earth's surface, altering the landscape and occasionally causing their own special problems. In some cases, geologic processes can be modified, deliberately or accidentally; in others, human activities must be adjusted to natural realities. The section on surface processes concludes with a chapter on climate, which connects or affects a number of the surface processes described earlier.

A subject of increasing current concern is the availability of resources. A series of five chapters deals with water resources, soil, minerals, and energy, the rates at which they are being consumed, probable amounts remaining, and projections of future availability and use. In the case of energy resources, we consider both those sources extensively used in the past and new sources that may or may not successfully replace them in the future.

Increasing population and increasing resource consumption lead to an increasing volume of waste to be disposed of; thoughtless or inappropriate waste disposal, in turn, commonly creates increasing pollution. Three chapters examine the interrelated problems of air and water pollution and the strategies available for the disposal of various kinds of wastes.

The final two chapters deal with a more diverse assortment of subjects. Environmental problems spawn laws intended to solve them; chapter 19 looks briefly at a sampling of laws, policies, and international agreements related to geologic matters discussed earlier in the book, and some of the problems with such laws and accords. Chapter 20 examines geologic constraints on construction schemes and the broader issue of trying to determine the optimum use(s) for particular parcels of land—matters that become more pressing as population growth pushes more people to live in marginal places.

Relative to the length of time we have been on earth, humans have had a disproportionate impact on this planet. Appendix A explores the concept of geologic time and its measurement and looks at the rates of geologic and other processes by way of putting human activities in temporal perspective. Appendix B provides short reference keys to aid in rock and mineral identification, and the inside back cover includes units of measurement and conversion factors.

Of course, the complex interrelationships among geologic processes and features mean that any subdivision into chapter-sized pieces is somewhat arbitrary, and different instructors may prefer different sequences or groupings (streams and ground water together, for example). An effort has been made to design chapters so that they can be resequenced in such ways without great difficulty.

Supplements

McGraw-Hill offers various tools and technology products to support *Environmental Geology, 10th Edition*. Instructors can obtain teaching aids by calling the Customer Service Department at 800-338-3987, visiting the McGraw-Hill website at **www.mhhe.com**, or contacting their McGraw-Hill sales representative.

McGraw-Hill ConnectPlus®

McGraw-Hill ConnectPlus® (**www.mcgrawhillconnect.com**) is a web-based assignment and assessment platform that gives students the means to better connect with their coursework, with their instructors, and with the important concepts that they will need to know for success now and in the future. **ConnectPlus** provides students 24/7 online access to an interactive eBook—an online edition of the text—to aid them in successfully completing their work, wherever and whenever they choose.

With ConnectPlus Geology, instructors can deliver assignments, quizzes, and tests online. Instructors can edit existing questions and author entirely new problems; track individual student performance—by question, assignment; or in relation to the class overall—with detailed grade reports; integrate grade reports easily with Learning Management Systems, such as WebCT and Blackboard; and much more.

By choosing ConnectPlus, instructors are providing their students with a powerful tool for improving academic performance and truly mastering course material. ConnectPlus allows students to practice important skills at their own pace and on their own schedule. Importantly, students' assessment results and instructors' feedback are all saved online—so students can continually review their progress and plot their course to success.

The following resources are available in ConnectPlus:

- Auto-graded, interactive assessments
- Powerful reporting against learning outcomes and level of difficulty
- The full textbook as an integrated, dynamic eBook, which you can also assign
- Access to instructor resources, such as a test bank, PowerPoint slides, and an instructor's manual
- Digital files of all the textbook illustrations and photographs

McGraw-Hill Tegrity Campus™

Tegrity is a service that makes class time always available by automatically capturing every lecture in a searchable format for students to review when they study and complete assignments. With a simple one-click start and stop process, you capture all computer screens and corresponding audio. Students can replay any part of any class with easy-to-use browser-based viewing on a PC or Mac.

Educators know that the more students can see, hear, and experience class resources, the better they learn. With Tegrity, students quickly recall key moments by using Tegrity's unique search feature. This search helps students efficiently find what they need, when they need it across an entire semester of class recordings. Help turn all your students' study time into learning moments immediately supported by your lecture.

To learn more about Tegrity, watch a two-minute Flash demo at **tegritycampus.mhhe.com**.

Companion Website

A website is available to students and provides practice quizzes with immediate feedback and other resources, including links to relevant Internet sites. **www.mhhe.com/montgomery10e**

Customizable Textbooks: McGraw-Hill Create™

Create what you've only imagined. Introducing **McGraw-Hill Create™**, a new self-service website that allows you to create custom course materials—print and eBooks—by drawing upon McGraw-Hill's comprehensive, cross-disciplinary content. Add your own content quickly and easily. Tap into other rights-secured third party sources as well. Then, arrange the content in a way that makes the most sense for your course. Even personalize your book with your course name and information. Choose the best format for your course: color print, black and white print, or eBook. The eBook is now viewable on an iPad! And when you are finished customizing, you will receive a free PDF review copy in just minutes! Visit McGraw-Hill Create—**www.mcgrawhillcreate.com**—today and begin building your perfect book.

CourseSmart eBook

CourseSmart is a new way for faculty to find and review eBooks. It's also a great option for students who are interested in accessing their course materials digitally and saving money. *CourseSmart* offers thousands of the most commonly adopted textbooks across hundreds of courses. It is the only place for faculty to review and compare the full text of a textbook online, providing immediate access without the environmental impact of requesting a print exam copy. At *CourseSmart*, students can save up to 30% off the cost of a print book, reduce their impact on the environment, and gain access to powerful Web tools for learning including full text search, notes and highlighting, and email tools for sharing notes between classmates.

To review complimentary copies or to purchase an eBook, go to **www.coursesmart.com**.

McGraw-Hill Higher Education and Blackboard®

McGraw-Hill Higher Education and Blackboard have teamed up! What does this mean for you?

1. **Your life, simplified.** Now you and your students can access McGraw-Hill's ConnectPlus® and Create™ right from within your Blackboard course—all with one single sign-on. Say goodbye to the days of logging in to multiple applications.

2. **Deep integration of content and tools.** Not only do you get single sign-on with Connect and Create, you also get deep integration of McGraw-Hill content and content engines right in Blackboard. Whether you're choosing a book for your course or building Connect assignments, all the tools you need are right where you want them—inside of Blackboard.

3. **Seamless gradebooks.** Are you tired of keeping multiple gradebooks and manually synchronizing grades into Blackboard? We thought so. When a student completes an integrated Connect assignment, the grade for that assignment automatically (and instantly) feeds your Blackboard grade center.

4. **A solution for everyone.** Whether your institution is already using Blackboard or you just want to try Blackboard on your own, we have a solution for you. McGraw-Hill and Blackboard can now offer you easy access to industry-leading technology and content, whether your campus hosts it or we do. Be sure to ask your local McGraw-Hill representative for details. **www.domorenow.com**

Course Delivery Systems

With help from our partners WebCT, Blackboard, Top-Class, eCollege, and other course management systems, professors can take complete control of their course content. Course cartridges containing website content, online testing, and powerful student tracking features are readily available for use within these platforms.

Acknowledgments

A great many people have contributed to the development of one or another edition of this book. Portions of the manuscript of the first edition were read by Colin Booth, Lynn A. Brant, Arthur H. Brownlow, Ira A. Furlong, David Huntley, John F. Looney, Jr., Robert A. Matthews, and George H. Shaw, and the entire book was reviewed by Richard A. Marston and Donald J. Thompson. The second edition was enhanced through suggestions from Robert B. Furlong, Jeffrey J. Gryta, David Gust, Robert D. Hall, Stephen B. Harper, William N. Mode, Martin Reiter, and Laura L. Sanders; the third, with the assistance of Susan M. Cashman, Robert B. Furlong, Frank Hanna, William N. Mode, Paul Nelson, Laura L. Sanders, and Michael A. Velbel; the fourth, through the input of reviewers Herbert Adams, Randall Scott Babcock, Pascal de Caprariis, James Cotter, Dru Germanoski, Thomas E. Hendrix, Gordon Love, Steven Lund, Michael McKinney, Barbara Ruff, Paul Schroeder, Ali Tabidian, Clifford Thurber, and John Vitek. The fifth edition was improved thanks to reviews by Kevin Cole, Gilbert N. Hanson, John F. Hildenbrand, Ann E. Homes, Alvin S. Konigsberg, Barbara L. Ruff, Vernon P. Scott, Jim Stimson, Michael Whitsett, and Doreen Zaback; the sixth, by reviews from Ray Beiersdorfer, Ellin Beltz, William B. N. Berry, Paul Bierman, W. B. Clapham, Jr., Ralph K. Davis, Brian

E. Lock, Gregory Hancock, Syed E. Hasan, Scott W. Keyes, Jason W. Kelsey, John F. Looney Jr., Christine Massey, Steve Mattox, William N. Mode, William A. Newman, Clair R. Ossian, David L. Ozsvath, Alfred H. Pekarek, Paul H. Reitan, and Don Rimstidt. Improvements in the seventh edition were inspired by reviewers Thomas J. Algaeo, Ernest H. Carlson, Douglas Crowe, Richard A. Flory, Hari P. Garbharran, Daniel Horns, Ernst H. Kastning, Abraham Lerman, Mark Lord, Lee Ann Munk, June A. Oberdorfer, Assad I. Panah, James S. Reichard, Frederick J. Rich, Jennifer Rivers Coombs, Richard Sleezer, and Michael S. Smith, and the eighth edition benefited from suggestions by Richard Aurisano, Thomas B. Boving, Ernest H. Carlson, Elizabeth Catlos, Dennis DeMets, Hailiang Dong, Alexander Gates, Chad Heinzel, Edward Kohut, Richard McGehee, Marguerite Moloney, Lee Slater, and Dan Vaughn, and additional comments by Nathan Yee. The ninth edition was strengthed through the reviews of Christine Aide, James Bartholomew, Thomas Boving, Jim Constantopoulos, Mark Groszos, Duke Ophori, Bianca Pedersen, John Rockaway, Kevin Svitana, and Clifford H. Thurber. This tenth edition has, in turn, been improved as a result of the sharp eyes and thoughtful suggestions of reviewers Michael Caudill, Hocking College; Katherine Grote, University of Wisconsin–Eau Claire; Lee Slater, Rutgers–Newark; Alexis Templeton, University of Colorado; Adil Wadia, The University of Akron Wayne College; and Lee Widmer, Xavier University.

The input of all of the foregoing individuals, and of many other users who have informally offered additional advice, has substantially improved the text, and their help is most gratefully acknowledged. If, as one reviewer commented, the text "just keeps getting better," a large share of the credit certainly belongs to the reviewers. Any remaining shortcomings are, of course, my own responsibility.

M. Dalecheck, C. Edwards, I. Hopkins, and J. McGregor at the USGS Photo Library in Denver provided invaluable assistance with the photo research over the years. The encouragement of a number of my colleagues—particularly Colin Booth, Ron C. Flemal, Donald M. Davidson, Jr., R. Kaufmann, and Eugene C. Perry, Jr.—was a special help during the development of the first edition. The ongoing support and interest of fellow author, deanly colleague, and ecologist Jerrold H. Zar has been immensely helpful. Thanks are also due to the several thousand environmental geology students I have taught, many of whom in the early years suggested that I write a text, and whose classes collectively have provided a testing ground for many aspects of the presentations herein.

My family has been supportive of this undertaking from the inception of the first edition. A very special vote of appreciation goes to my husband Warren—ever-patient sounding board, occasional photographer and field assistant—in whose life this book has so often loomed so large.

Last, but assuredly not least, I express my deep gratitude to the entire McGraw-Hill book team for their enthusiasm, professionalism, and just plain hard work, without which successful completion of each subsequent edition of this book would have been impossible.

Carla W. Montgomery

Contents

SECTION II INTERNAL PROCESS

SECTION III SURFACE PROCESSES

CHAPTER 6

Streams and Flooding   120

CHAPTER 7

Coastal Zones and Processes   146

SECTION IV RESOURCES

CHAPTER **16**

Waste Disposal 364

CHAPTER **17**

Water Pollution 396

Planet and Population: An Overview

About five billion years ago, out of a swirling mass of gas and dust, evolved a system of varied planets hurtling around a nuclear-powered star—our solar system. One of these planets, and one only, gave rise to complex life-forms. Over time, a tremendous diversity of life-forms and ecological systems developed, while the planet, too, evolved and changed, its interior churning, its landmasses shifting, its surface constantly being reshaped. Within the last several million years, the diversity of life on earth has included humans, increasingly competing for space and survival on the planet's surface. With the control over one's surroundings made possible by the combination of intelligence and manual dexterity, humans have found most of the land on the planet inhabitable; they have learned to use not only plant and animal resources, but minerals, fuels, and other geologic materials; in

some respects, humans have even learned to modify natural processes that inconvenience or threaten them. As we have learned how to study our planet in systematic ways, we have developed an ever-increasing understanding of the complex nature of the processes shaping, and the problems posed by, our geological environment. **Environmental geology** explores the many and varied interactions between humans and that geologic environment.

As the human population grows, it becomes increasingly difficult for that population to survive on the resources and land remaining, to avoid those hazards that cannot be controlled, and to refrain from making irreversible and undesirable changes in environmental systems. The urgency of perfecting our understanding, not only of natural processes but also of our impact on the planet, is becoming more and more apparent, and has

Geology provides the ground we live on, the soil in which our crops are grown, many of our resources, and often, striking scenery. Here, the Orange River provides vital irrigation water for growing grapes in arid Namibia. The river also carves the landscape and carries sediment away.

Image by Jesse Allen, using data from NASA's EO-1 team and the U.S. Geological Survey; courtesy NASA.

motivated increased international cooperation and dialogue on environmental issues.

In 1992, more than 170 nations came together in Rio de Janeiro for the United Nations Conference on Environment and Development, to address such issues as global climate change, sustainable development, and environmental protection. Subsequent decades have seen further conferences, and some tangible progress, addressing these topics. (However, while nations may readily agree on *what* the problematic issues are, agreement on *solutions* is often much harder to achieve!)

Earth in Space and Time

The Early Solar System

In recent decades, scientists have been able to construct an ever-clearer picture of the origins of the solar system and, before that, of the universe itself. Most astronomers now accept some sort of "Big Bang" as the origin of today's universe. Just before it occurred, all matter and energy would have been compressed into an enormously dense, hot volume a few millimeters (much less than an inch) across. Then everything was flung violently apart across an ever-larger volume of space. The time of the Big Bang can be estimated in several ways. Perhaps the most direct is the back-calculation of the universe's expansion to its apparent beginning. Other methods depend on astrophysical models of creation of the elements or the rate of evolution of different types of stars. Most age estimates overlap in the range of 12 to 14 billion years.

Stars formed from the debris of the Big Bang, as locally high concentrations of mass were collected together by gravity, and some became large and dense enough that energy-releasing atomic reactions were set off deep within them. Stars are not permanent objects. They are constantly losing energy and mass as they burn their nuclear fuel. The mass of material that initially formed the star determines how rapidly the star burns; some stars burned out billions of years ago, while others are probably forming now from the original matter of the universe mixed with the debris of older stars.

Our sun and its system of circling planets, including the earth, are believed to have formed from a rotating cloud of gas and dust (small bits of rock and metal), some of the gas debris from older stars (figure 1.1). Most of the mass of the cloud coalesced to form the sun, which became a star and began to "shine," or release light energy, when its interior became so dense and hot from the crushing effects of its own gravity that nuclear reactions were triggered inside it. Meanwhile, dust condensed from the gases remaining in the flattened cloud disk rotating around the young sun. The dust clumped into planets, the formation of which was essentially complete over 4½ billion years ago.

The Planets

The compositions of the planets formed depended largely on how near they were to the hot sun. The planets formed nearest to the sun contained mainly metallic iron and a few minerals with very high melting temperatures, with little water or gas. Somewhat farther out, where temperatures were lower, the developing planets incorporated much larger amounts of lower-temperature minerals, including some that contain water locked

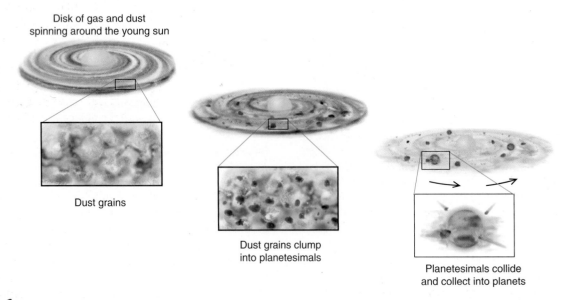

Disk of gas and dust spinning around the young sun

Dust grains

Dust grains clump into planetesimals

Planetesimals collide and collect into planets

Figure 1.1

Our solar system formed as dust condensed from the gaseous nebula, then clumped together to make planets.

Table 1.1 — Some Basic Data on the Planets

Planet	Mean Distance from Sun (millions of km)	Mean Temperature (°C)	Equatorial Diameter, Relative to Earth	Density* (g/cu. cm)	
Mercury	58	167	0.38	5.4	
Venus	108	464	0.95	5.2	Predominantly rocky/metal planets
Earth	150	15	1.00	5.5	
Mars	228	−65	0.53	3.9	
Jupiter	778	−110	11.19	1.3	
Saturn	1427	−140	9.41	0.7	Gaseous planets
Uranus	2870	−195	4.06	1.3	
Neptune	4479	−200	3.88	1.6	

Source: Data from NASA.

*No other planets have been extensively sampled to determine their compositions directly, though we have some data on their surfaces. Their approximate bulk compositions are inferred from the assumed starting composition of the solar nebula and the planets' densities. For example, the higher densities of the inner planets reflect a significant iron content and relatively little gas.

within their crystal structures. (This later made it possible for the earth to have liquid water at its surface.) Still farther from the sun, temperatures were so low that nearly all of the materials in the original gas cloud condensed—even materials like methane and ammonia, which are gases at normal earth surface temperatures and pressures.

The result was a series of planets with a variety of compositions, most quite different from that of earth. This is confirmed by observations and measurements of the planets. For example, the planetary densities listed in table 1.1 are consistent with a higher metal and rock content in the four planets closest to the sun and a much larger proportion of ice and gas in the planets farther from the sun (see also figure 1.2). These differences should be kept in mind when it is proposed that other planets could be mined for needed minerals. Both the basic chemistry of these other bodies and the kinds of ore-forming or other resource-forming processes that might occur on them would differ considerably from those on earth, and may not have led to products we would find useful. (This is leaving aside any questions of the economics or technical practicability of such mining activities!) In addition, our principal current energy sources required living organisms to form, and so far, no such life-forms have been found on other planets or moons. Venus—close to Earth in space, similar in size and density— shows marked differences: Its dense, cloudy atmosphere is thick with carbon dioxide, producing planetary surface temperatures hot enough to melt lead through runaway greenhouse-effect heating (see chapter 10). Mars would likewise be inhospitable: It is very cold, and we could not breathe its atmosphere. Though its surface features indicate the presence of liquid water in its past, there is none now, and only small amounts of water ice have been found. There is not so much as a blade of grass for vegetation; the brief flurry of excitement over possible evidence of life on Mars referred only to fossil microorganisms, and more-intensive investigations suggested that the tiny structures in question likely are inorganic, though the search for Martian microbes continues.

Earth, Then and Now

The earth has changed continuously since its formation, undergoing some particularly profound changes in its early history. The early earth was very different from what it is today, lacking the modern oceans and atmosphere and having a much different surface from its present one, probably more closely resembling the barren, cratered surface of the moon. Like other planets, Earth was formed by accretion, as gravity collected together the solid bits that had condensed from the solar nebula. Some water may have been contributed by gravitational capture of icy comets, though recent analyses of modern comets do not suggest that this was a major water source. The planet was heated by the impact of the colliding dust particles and meteorites as they came together to form the earth, and by the energy release from decay of the small amounts of several naturally radioactive elements that the earth contains. These heat sources combined to raise the earth's internal temperature enough that parts of it, perhaps eventually most of it, melted, although it was probably never molten all at once. Dense materials, like metallic iron, would have tended to sink toward the middle of the earth. As cooling progressed, lighter, low-density minerals crystallized and floated out toward the surface. The eventual result was an earth differentiated into several major compositional zones: the central **core;** the surrounding **mantle;** and a thin **crust** at the surface (see figure 1.3). The process was complete well before 4 billion years ago.

Although only the crust and a few bits of uppermost mantle that are carried up into the crust by volcanic activity can be sampled and analyzed directly, we nevertheless have a good deal of information on the composition of the earth's interior. First, scientists can estimate from analyses of stars the starting composition of the cloud from which the solar system formed. Geologists can also infer aspects of the earth's bulk composition from analyses of certain meteorites believed to have formed at the same time as, and under conditions similar to, the earth. Geophysical data demonstrate that the earth's

Figure 1.2

The planets of the solar system vary markedly in both composition and physical properties. For example, Mercury (A), as shown in this image from a 2008 *Messenger* spacecraft flyby, is rocky, iron-rich, dry, and pockmarked with craters. Mars (B) shares many surface features with Earth (volcanoes, canyons, dunes, slumps, stream channels, and more), but the surface is now dry and barren; (C) a 2008 panorama by the Mars rover *Spirit*. Jupiter (D) is a huge gas ball, with no solid surface at all, and dozens of moons of ice and rock that circle it to mimic the solar system in miniature. Note also the very different sizes of the planets (E). The Jovian planets—named for Jupiter—are gas giants; the terrestrial planets are more rocky, like Earth.

(A) NASA image courtesy Science Operations Center at Johns Hopkins University Applied Physics Laboratory; (B) courtesy NASA; (C) image by NASA/JPL/Cornell; (D) courtesy NSSDC Goddard Space Flight Center.

A

B

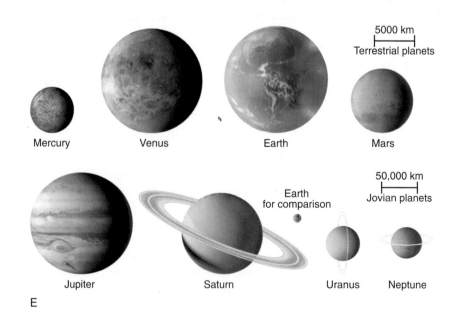

C

D

E

Mercury   Venus   Earth   Mars

5000 km
Terrestrial planets

Jupiter   Saturn   Uranus   Neptune

Earth for comparison

50,000 km
Jovian planets

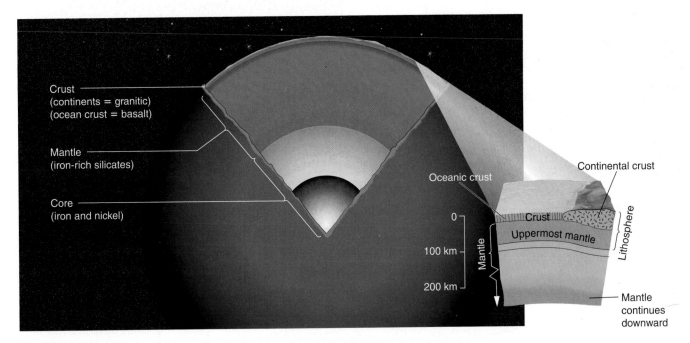

Figure 1.3

A chemically differentiated earth. The core consists mostly of iron; the outer part is molten. The mantle, the largest zone, is made up primarily of ferromagnesian silicates (see chapter 2) and, at great depths, of oxides of iron, magnesium, and silicon. The crust (not drawn to scale, but exaggerated vertically in order to be visible at this scale) forms a thin skin around the earth. Oceanic crust, which forms the sea floor, has a composition somewhat like that of the mantle, but is richer in silicon. Continental crust is both thicker and less dense. It rises above the oceans and contains more light minerals rich in calcium, sodium, potassium, and aluminum. The "plates" of plate tectonics (the lithosphere) comprise the crust and uppermost mantle. (100 km ≈ 60 miles)

interior is zoned and also provide information on the densities of the different layers within the earth, which further limits their possible compositions. These and other kinds of data indicate that the earth's core is made up mostly of iron, with some nickel and a few minor elements; the outer core is molten, the inner core solid. The mantle consists mainly of iron, magnesium, silicon, and oxygen combined in varying proportions in several different minerals. The earth's crust is much more varied in composition and very different chemically from the average composition of the earth (see table 1.2). Crust and uppermost mantle together form a somewhat brittle shell around the earth. As is evident from table 1.2, many of the metals we have come to prize as resources are relatively uncommon elements in the crust.

The heating and subsequent differentiation of the early earth led to another important result: formation of the atmosphere and oceans. Many minerals that had contained water or gases in their crystals released them during the heating and melting, and as the earth's surface cooled, the water could condense to form the oceans. Without this abundant surface water, which in the solar system is unique to earth, most life as we know it could not exist. The oceans filled basins, while the continents, buoyant because of their lower-density rocks and minerals, stood above the sea surface. At first, the continents were barren of life.

The earth's early atmosphere was quite different from the modern one, aside from the effects of modern pollution. The first atmosphere had little or no free oxygen in it. It probably consisted

Table 1.2	Most Common Chemical Elements in the Earth		
WHOLE EARTH		**CRUST**	
Element	Weight Percent	Element	Weight Percent
Iron	32.4	Oxygen	46.6
Oxygen	29.9	Silicon	27.7
Silicon	15.5	Aluminum	8.1
Magnesium	14.5	Iron	5.0
Sulfur	2.1	Calcium	3.6
Nickel	2.0	Sodium	2.8
Calcium	1.6	Potassium	2.6
Aluminum	1.3	Magnesium	2.1
(All others, total)	.7	(All others, total)	1.5

(Compositions cited are averages of several independent estimates.)

dominantly of nitrogen and carbon dioxide (the gas most commonly released from volcanoes, aside from water) with minor amounts of such gases as methane, ammonia, and various sulfur gases. Humans could not have survived in this early atmosphere. Oxygen-breathing life of any kind could not exist before the single-celled blue-green algae appeared in large numbers to modify

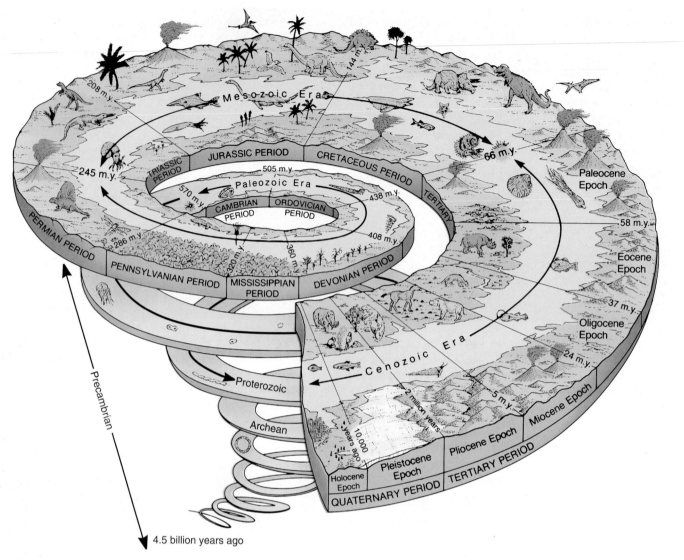

Figure 1.4

The "geologic spiral": Important plant and animal groups appear where they first occurred in significant numbers. If earth's whole history were equated to a 24-hour day, modern thinking humans (*Homo sapiens*) would have arrived on the scene just about ten seconds ago. For another way to look at these data, see table A.1 in appendix A.

Source: Modified after U.S. Geological Survey publication Geologic Time.

the atmosphere. Their remains are found in rocks as much as several billion years old. They manufacture food by photosynthesis, using sunlight for energy, consuming carbon dioxide, and releasing oxygen as a by-product. In time, enough oxygen accumulated that the atmosphere could support oxygen-breathing organisms.

Life on Earth

The rock record shows when different plant and animal groups appeared. Some are represented schematically in figure 1.4. The earliest creatures left very few remains because they had no hard skeletons, teeth, shells, or other hard parts that could be preserved in rocks. The first multicelled oxygen-breathing creatures probably developed about 1 billion

years ago, after oxygen in the atmosphere was well established. By about 550 million years ago, marine animals with shells had become widespread.

The development of organisms with hard parts—shells, bones, teeth, and so on—greatly increased the number of preserved animal remains in the rock record; consequently, biological developments since that time are far better understood. Dry land was still barren of large plants or animals half a billion years ago. In rocks about 500 million years old is the first evidence of animals with backbones—the fish—and soon thereafter, early land plants developed, before 400 million years ago. Insects appeared approximately 300 million years ago. Later, reptiles and amphibians moved onto the continents. The dinosaurs appeared about 200 million years ago and the first mammals at nearly the same time. Warm-blooded animals took to the air with the development of birds about 150 million years ago, and by 100 million years ago, both birds and mammals were well established.

Such information has current applications. Certain energy sources have been formed from plant or animal remains. Knowing the times at which particular groups of organisms appeared and flourished is helpful in assessing the probable amounts of these energy sources available and in concentrating the search for these fuels on rocks of appropriate ages.

On a timescale of billions of years, human beings have just arrived. The most primitive human-type remains are no more than 4 to 5 million years old, and modern, rational humans (*Homo sapiens*) developed only about half a million years ago. Half a million years may sound like a long time, and it is if compared to a single human lifetime. In a geologic sense, though, it is a very short time. If we equate the whole of earth's history to a 24-hour day, then shelled organisms appeared only about three hours ago; fish, about 2 hours and 40 minutes ago; land plants, two hours ago; birds, about 45 minutes ago—and *Homo sapiens* has been around for just the last ten *seconds*. Nevertheless, we humans have had an enormous impact on the earth, at least at its surface, an impact far out of proportion to the length of time we have occupied it. Our impact is likely to continue to increase rapidly as the population does likewise.

Geology, Past and Present

Two centuries ago, geology was mainly a descriptive science involving careful observation of natural processes and their products. The subject has become both more quantitative and more interdisciplinary through time. Modern geoscientists draw on the principles of chemistry to interpret the compositions of geologic materials, apply the laws of physics to explain these materials' physical properties and behavior, use the biological sciences to develop an understanding of ancient life-forms, and rely on engineering principles to design safe structures in the presence of geologic hazards. The emphasis on the "why," rather than just the "what," has also increased.

The Geologic Perspective

Geologic observations now are combined with laboratory experiments, careful measurements, and calculations to develop theories of how natural processes operate. Geology is especially challenging because of the disparity between the scientist's laboratory and nature's. In the research laboratory, conditions of temperature and pressure, as well as the flow of chemicals into or out of the system under study, can be carefully controlled. One then knows just what has gone into creating the product of the experiment. In nature, the geoscientist is often confronted only with the results of the "experiment" and must deduce the starting materials and processes involved.

Another complicating factor is time. The laboratory scientist must work on a timescale of hours, months, years, or, at most, decades. Natural geologic processes may take a million or a billion years to achieve a particular result, by stages too slow even to be detected in a human lifetime (table 1.3). This understanding may be one of the most significant contributions of early geoscientists: the recognition of the vast length of geologic history, sometimes described as "deep time." The qualitative and quantitative tools for sorting out geologic events and putting dates on them are outlined in appendix A. For now, it is useful to bear in mind that the immensity of geologic time can make it difficult to arrive at a full understanding of how geologic processes operated in the past from observations made on a human timescale. It dictates caution, too, as we try to project, from a few years' data on global changes associated with human activities, all of the long-range impacts we may be causing.

Also, the laboratory scientist may conduct a series of experiments on the same materials, but the experiments can be stopped and those materials examined after each stage. Over the vast spans of geologic time, a given mass of earth material may have been transformed a half-dozen times or more, under different conditions each time. The history of the rock that ultimately results may be very difficult to decipher from the end product alone.

Table 1.3	Some Representative Geologic-Process Rates

Process	Occurs Over a Time Span of About This Magnitude
Rising and falling of tides	1 day
"Drift" of a continent by 2–3 centimeters (about 1 inch)	1 year
Accumulation of energy between large earthquakes on a major fault zone	10–100 years
Rebound (rising) by 1 meter of a continent depressed by ice sheets during the Ice Age	100 years
Flow of heat through 1 meter of rock	1000 years
Deposition of 1 centimeter of fine sediment on the deep-sea floor	1000–10,000 years
Ice sheet advance and retreat during an ice age	10,000–100,000 years
Life span of a small volcano	100,000 years
Life span of a large volcanic center	1–10 million years
Creation of an ocean basin such as the Atlantic	100 million years
Duration of a major mountain-building episode	100 million years
History of life on earth	Over 3 billion years

Geology and the Scientific Method

The **scientific method** is a means of discovering basic scientific principles. One begins with a set of observations and/or a body of data, based on measurements of natural phenomena or on experiments. One or more *hypotheses* are formulated to explain the observations or data. A **hypothesis** can take many forms, ranging from a general conceptual framework or model describing the functioning of a natural system, to a very precise mathematical formula relating several kinds of numerical data. What all hypotheses have in common is that they must all be susceptible to testing and, particularly, to *falsification*. The idea is not simply to look for evidence to support a hypothesis, but to examine relevant evidence with the understanding that it may show the hypothesis to be wrong.

In the classical conception of the scientific method, one uses a hypothesis to make a set of predictions. Then one devises and conducts experiments to test each hypothesis, to determine whether experimental results agree with predictions based on the hypothesis. If they do, the hypothesis gains credibility. If not, if the results are unexpected, the hypothesis must be modified to account for the new data as well as the old or, perhaps, discarded altogether. Several cycles of modifying and retesting hypotheses may be required before a hypothesis that is consistent with all the observations and experiments that one can conceive is achieved. A hypothesis that is repeatedly supported by new experiments advances in time to the status of a **theory,** a generally accepted explanation for a set of data or observations.

Much confusion can arise from the fact that in casual conversation, people often use the term *theory* for what might better be called a hypothesis, or even just an educated guess. ("So, what's your theory?" one character in a TV mystery show may ask another, even when they've barely looked at the first evidence.) Thus people may assume that a scientist describing a theory is simply telling a plausible story to explain some data. A scientific theory, however, is a very well-tested model with a very substantial and convincing body of evidence that supports it. A hypothesis may be advanced by just one individual; a theory has survived the challenge of extensive testing to merit acceptance by many, often most, experts in a field. The Big Bang theory is not just a creative idea. It accounts for the decades-old observation that all the objects we can observe in the universe seem to be moving apart. If it is correct, the universe's origin was very hot; scientists have detected the cosmic microwave background radiation consistent with this. And astrophysicists' calculations predict that the predominant elements that the Big Bang would produce would be hydrogen and helium—which indeed overwhelmingly dominate the observed composition of our universe.

The classical scientific method is not strictly applicable to many geologic phenomena because of the difficulty of experimenting with natural systems, given the time and scale considerations noted earlier. For example, one may be able to conduct experiments on a single rock, but not to construct a whole volcano in the laboratory. In such cases, hypotheses are often tested entirely through further observations or theoretical calculations

and modified as necessary until they accommodate all the relevant observations (or are discarded when they cannot be reconciled with new data). This broader conception of the scientific method is well illustrated by the development of the theory of plate tectonics, discussed in chapter 3. "Continental drift" was once seen as a wildly implausible idea, advanced by an eccentric few, but in the latter half of the twentieth century, many kinds of evidence were found to be explained consistently and well by movement of plates—including continents—over earth's surface. Still, the details of plate tectonics continue to be refined by further studies. Even a well-established theory may ultimately be proved incorrect. (Plate tectonics in fact supplanted a very different theory about how mountain ranges form.) In the case of geology, complete rejection of an older theory has most often been caused by the development of new analytical or observational techniques, which make available wholly new kinds of data that were unknown at the time the original theory was formulated.

The Motivation to Find Answers

In spite of the difficulties inherent in trying to explain geologic phenomena, the search for explanations goes on, spurred not only by the basic quest for knowledge, but also by the practical problems posed by geologic hazards, the need for resources, and concerns about possible global-scale human impacts, such as ozone destruction and global warming.

The hazards may create the most dramatic scenes and headlines, the most abrupt consequences: The 1989 Loma Prieta (California) earthquake caused more than $5 billion in damage; the 1995 Kobe (Japan) earthquake (figure 1.5), similar in size to Loma Prieta, caused over 5200 deaths and about $100 billion in property damage; the 2004 Sumatran earthquake claimed nearly 300,000 lives; the 2011 quake offshore from

Figure 1.5

Overturned section of Hanshin Expressway, eastern Kobe, Japan, after 1995 earthquake. This freeway, elevated to save space, was built in the 1960s to then-current seismic design standards.

Photograph by Christopher Rojahn, Applied Technology Council.

Figure 1.6

Ash pours from Mount St. Helens, May 1980.

Photograph by Peter Lipman, courtesy USGS.

Figure 1.7

A major river like the Mississippi floods when a large part of the area that it drains is waterlogged by more rain or snowmelt than can be carried away in the channel. Such floods—like that in summer 1993, shown here drenching Jefferson City, Missouri—can be correspondingly long-lasting. Over millennia, the stream builds a floodplain into which the excess water naturally flows; we build there at our own risk.

Photograph by Mike Wright, courtesy Missouri Department of Transportation.

Honshu, Japan, killed over 15,000 people and caused an estimated $300 billion in damages. The 18 May 1980 eruption of Mount St. Helens (figure 1.6) took even the scientists monitoring the volcano by surprise, and the 1991 eruption of Mount Pinatubo in the Philippines not only devastated local residents but caught the attention of the world through a marked decline in 1992 summer temperatures. Efforts are underway to provide early warnings of such hazards as earthquakes, volcanic eruptions, and landslides so as to save lives, if not property. Likewise, improved understanding of stream dynamics and more prudent land use can together reduce the damages from flooding (figure 1.7), which amount in the United States to over $1 billion a year and the loss of dozens of lives annually. Landslides and other slope and ground failures (figure 1.8) take a similar toll in property damage, which could be reduced by more attention to slope stability and improved engineering practices. It is not only the more dramatic hazards that are costly: on average, the cost of structural damage from unstable soils each year approximately equals the combined costs of landslides, earthquakes, and flood damages in this country.

Our demand for resources of all kinds continues to grow and so do the consequences of resource use. In the United States, average per-capita water use is 1500 gallons per day; in many places, groundwater supplies upon which we have come to rely heavily are being measurably depleted. Worldwide, water-resource disputes between nations are increasing in number. As we mine more extensively for mineral resources, we face the problem of how to minimize associated damage to the mined lands (figure 1.9). The grounding of the *Exxon Valdez* in 1989, dumping 11 million gallons of oil into Prince William Sound, Alaska, and the massive spill from the 2010 explosion of the *Deepwater Horizon* drilling platform in the Gulf of Mexico were reminders of the negative consequences of petroleum exploration, just as the 1991 war in Kuwait, and the later invasion of Iraq, were reminders of U.S. dependence on imported oil.

Figure 1.8

Slope failure on a California hillside undercuts homes.

Photograph by J. T. McGill, USGS Photo Library, Denver, CO.

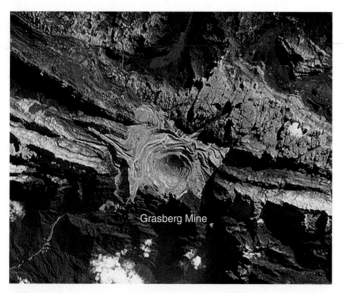

Figure 1.9

The Grasberg Mine in Irian Jaya, Indonesia, is one of the world's largest gold- and copper-mining operations. The surface pit is nearly 4 km (2½ miles) across; note the sharp contrast with surrounding topography. Slopes oversteepened by mining have produced deadly landslides, and local residents worry about copper and acid contamination in runoff water.

Image courtesy of Earth Sciences and Image Analysis Laboratory, NASA Johnson Space Center.

Figure 1.10

Bit by bit, lava flows like this one on Kilauea have built the Hawaiian Islands.

As we consume more resources, we create more waste. In the United States, total waste generation is estimated at close to 300 million tons per year—or more than a ton per person. Careless waste disposal, in turn, leads to pollution. The Environmental Protection Agency continues to identify toxic-waste disposal sites in urgent need of cleanup; by 2000, over 1500 so-called priority sites had been identified. Cleanup costs per site have risen to over $30 million, and the projected total costs to remediate these sites alone is over $1 trillion. As fossil fuels are burned, carbon dioxide in the atmosphere rises, and modelers of global climate strive to understand what that may do to global temperatures, weather, and agriculture decades in the future.

These are just a few of the kinds of issues that geologists play a key role in addressing.

Wheels Within Wheels: Earth Cycles and Systems

The earth is a dynamic, constantly changing planet—its crust shifting to build mountains; lava spewing out of its warm interior; ice and water and windblown sand and gravity reshaping its surface, over and over. Some changes proceed in one direction only: for example, the earth has been cooling progressively since its formation, though considerable heat remains in its interior. Many of the processes, however, are cyclic in nature.

Consider, for example, such basic materials as water or rocks. Streams drain into oceans and would soon run dry if not replenished; but water evaporates from oceans, to make the rain and snow that feed the streams to keep them flowing. This describes just a part of the *hydrologic* (water) *cycle,* explored more fully in chapters 6 and 11. Rocks, despite their appearance of permanence in the short term of a human life, participate in the *rock cycle* (chapters 2 and 3). The kinds of evolutionary paths rocks may follow through this cycle are many, but consider this illustration: A volcano's lava (figure 1.10) hardens into rock; the rock is weathered into sand and dissolved chemicals; the debris, deposited in an ocean basin, is solidified into a new rock of quite different type; and some of that new rock may be carried into the mantle via plate tectonics, to be melted into a new lava. The time frame over which this process occurs is generally much longer than that over which water cycles through atmosphere and oceans, but the principle is similar. The Appalachian or Rocky Mountains as we see them today are not as they formed, tens or hundreds of millions of years ago; they are much eroded from their original height by water and ice, and, in turn, contain rocks formed in water-filled basins and deserts from material eroded from more-ancient mountains before them (figure 1.11).

Chemicals, too, cycle through the environment. The carbon dioxide that we exhale into the atmosphere is taken up by plants through photosynthesis, and when we eat those plants for food energy, we release CO_2 again. The same exhaled CO_2 may also dissolve in rainwater to make carbonic acid that dissolves continental rock; the weathering products may wash into the ocean, where dissolved carbonate then contributes to the formation of carbonate shells and carbonate rocks in the ocean basins; those rocks may later be exposed and weathered by rain, releasing CO_2 back into the atmosphere or dissolved carbonate into streams that carry it back to the ocean. The cycling of chemicals and materials in the environment may be complex, as we will see in later chapters.

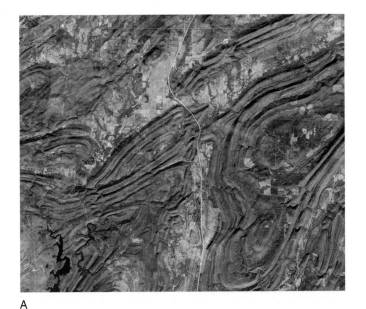

A

B

Figure 1.11

Rocks tell a story of constant change. (A) The folds of the Ouachita Mountains of Oklahoma formed deep in the crust when Africa and North America converged hundreds of millions of years ago; now they are eroding and crosscut by rivers. (B) The sandstones of Zion National Park preserve ancient windswept dunes, made of sand eroded from older rocks, deeply buried and solidified into new rock, then uplifted to erode again.

(A) Image by Jesse Allen and Robert Simmon using data provided by NASA/GSFC/METI/ERSDAC/JAROS and the U.S./Japan ASTER Science Team.

Furthermore, these processes and cycles are often interrelated, and seemingly local actions can have distant consequences. We dam a river to create a reservoir as a source of irrigation water and hydroelectric power, inadvertently trapping stream-borne sediment at the same time; downstream, patterns of erosion and deposition in the stream channel change, and at the coast, where the stream pours into the ocean, coastal erosion of beaches increases because a part of their sediment supply, from the stream, has been cut off. The volcano that erupts the lava to make the volcanic rock also releases water vapor into the atmosphere, and sulfur-rich gases and dust that influence the amount of sunlight reaching earth's surface to heat it, which, in turn, can alter the extent of evaporation and resultant rainfall, which will affect the intensity of landscape erosion and weathering of rocks by water. . . . So although we divide the great variety and complexity of geologic processes and phenomena into more manageable, chapter-sized units for purposes of discussion, it is important to keep such interrelationships in mind. And superimposed on, influenced by, and subject to all these natural processes are humans and human activities.

Nature and Rate of Population Growth

Animal populations, as well as primitive human populations, are generally quite limited both in the areas that they can occupy and in the extent to which they can grow. They must live near food and water sources. The climate must be one to which they can adapt. Predators, accidents, and disease take a toll. If the population grows too large, disease and competition for food are particularly likely to cut it back to sustainable levels.

The human population grew relatively slowly for hundreds of thousands of years. Not until the middle of the nineteenth century did the world population reach 1 billion. However, by then, a number of factors had combined to accelerate the rate of population increase. The second, third, and fourth billion were reached far more quickly; the world population is now over 7 billion and is expected to rise to over 9 billion by 2050 (figure 1.12).

Humans are no longer constrained to live only where conditions are ideal. We can build habitable quarters even in extreme climates, using heaters or air conditioners to bring the temperature to a level we can tolerate. In addition, people need not live where food can be grown or harvested or where there is abundant fresh water: The food and water can be transported, instead, to where the people choose to live.

Growth Rates: Causes and Consequences

Population growth occurs when new individuals are added to the population faster than existing individuals are removed from it. On a global scale, the population increases when its birthrate exceeds its death rate. In assessing an individual nation's or region's rate of population change, immigration and emigration must also be taken into account. Improvements in nutrition and health care typically increase life expectancies, decrease mortality rates, and thus increase the rate of population growth.

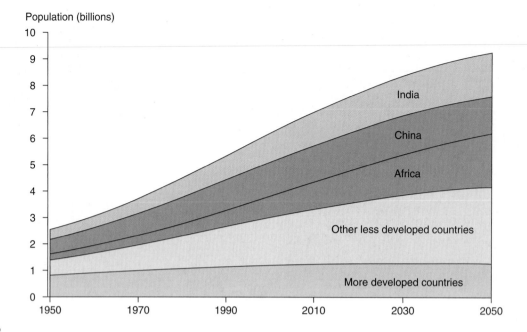

Figure 1.12

World population, currently over 7 billion, is projected to reach over 9 billion by 2050. Most of that population increase will occur in less-developed countries. It is important to realize, too, that in some large developing nations (notably China and India) the middle class is growing very rapidly, which has serious implications for resource demand, as will be explored in later chapters.

Source: United Nations, Department of Economic and Social Affairs, Population Division (2007). World Population Prospects: The 2006 Revision, Highlights, Working Paper No. ESA/P/WP. 202.

Increased use of birth-control or family-planning methods reduces birthrates and, therefore, also reduces the rate of population growth; in fact, a population can begin to decrease if birthrates are severely restricted.

The sharply rising rate of population growth over the past few centuries can be viewed another way. It took until about A.D. 1830 for the world's population to reach 1 billion. The population climbed to 2 billion in the next 130 years, and to 3 billion in just 30 more years, as ever more people contributed to the population growth and individuals lived longer. The last billion people have been added to the world's population in just a dozen years.

There are wide differences in growth rates among regions (table 1.4; figures 1.12 and 1.13). The reasons for this are many. Religious or social values may cause larger or smaller families to be regarded as desirable in particular regions or cultures. High levels of economic development are commonly associated with reduced rates of population growth; conversely, low levels of development are often associated with rapid population growth. The impact of improved education, which may accompany economic development, can vary: It may lead to better nutrition, prenatal and child care, and thus to increased growth rates, but it may also lead to increased or more effective practice of family-planning methods, thereby reducing growth rates.

A few governments of nations with large and rapidly growing populations have considered encouraging or mandating

Table 1.4	World and Regional Population Growth and Projections (in millions)						
Year	World	North America	Latin America and Caribbean	Africa	Europe	Asia	Oceania
1950	2520	172	167	221	547	1402	13
Mid-2011	6987	346	596	1051	740	4216	37
2050 (projected)	9587	470	746	2300	725	5284	62
Growth rate (%/year)	1.2	0.5	1.2	2.4	0	1.1	1.2
Doubling time (years)	58	140	58	29	*	64	58

*Not applicable; population declining.

Source: United Nations World Population Estimates and Projections (1950 data; http://www.popin.org/pop1998/) and 2011 World Population Data Sheet, Population Reference Bureau, 2011.

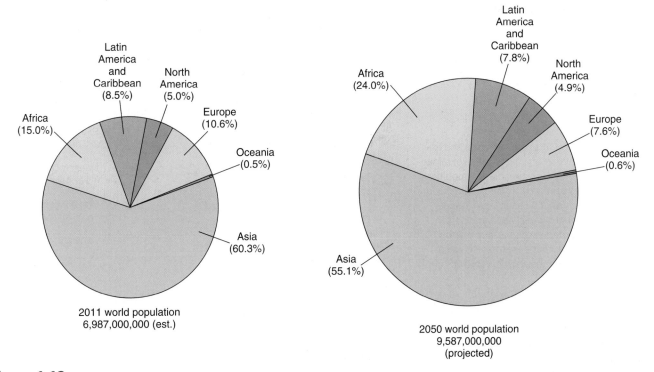

Figure 1.13

Population distribution by region in mid-2011 with projections to the year 2050. Size of circle reflects relative total population. The most dramatic changes in proportion are the relative growth of the population of Africa and decline of population in Europe. Data from table 1.4.

family planning. India and the People's Republic of China have taken active measures, with varying results: China's population growth rate is just 0.5% per year, but India's remains a relatively high 1.5% per year. At those rates, the population of India will surpass that of China by 2020, at which time both countries' populations will exceed 1.4 billion.

A relatively new factor that strongly affects population growth in some less-developed nations is AIDS. In more-developed countries, typically less than one-half of 1% of the population aged 15 to 49 is afflicted; the prevalence in some African nations is over 20%. In Swaziland, where AIDS prevalence in this age group is about 26%, life expectancy is down to 49 years (world average life expectancy is 70 years). How population in such countries will change over time depends greatly on how effectively the AIDS epidemic is controlled.

Even when the population growth rate is constant, the number of *individuals* added per unit of time increases over time. This is called **exponential growth,** a concept to which we will return when discussing resources in Section IV. The effect of exponential growth is similar to interest compounding. If one invests $100 at 10% per year and withdraws the interest each year, one earns $10/year, and after 10 years, one has collected $100 in interest. If one invests $100 at 10% per year, compounded annually, then, after one year, $10 interest is credited, for a new balance of $110. But if the interest is not withdrawn, then at the end of the second year, the interest is $11 (10% of $110), and the new balance is $121. By the end of the tenth year

(assuming no withdrawals), the interest for the year is $23.58, but the interest *rate* has not increased. And the balance is $259.37 so, subtracting the original investment of $100, this means total interest of $159.37 rather than $100. Similarly with a population of 1 million growing at 5% per year: In the first year, 50,000 persons are added; in the tenth year, the population grows by 77,566 persons. The result is that a graph of population versus time steepens over time, even at a constant growth rate. If the growth rate itself also increases, the curve rises still more sharply.

For many mineral and fuel resources, consumption has been growing very rapidly, even more rapidly than the population. The effects of exponential increases in resource demand are like the effects of exponential population growth (figure 1.14). If demand increases by 2% per year, it will double not in 50 years, but in 35. A demand increase of 5% per year leads to a doubling in demand in 14 years and a tenfold increase in demand in 47 years! In other words, a prediction of how soon mineral or fuel supplies will be used up is very sensitive to the assumed rate of change of demand. Even if population is no longer growing exponentially, consumption of many resources is.

Growth Rate and Doubling Time

Another way to look at the rapidity of world population growth is to consider the expected **doubling time,** the length of time required for a population to double in size. Doubling time (*D*)

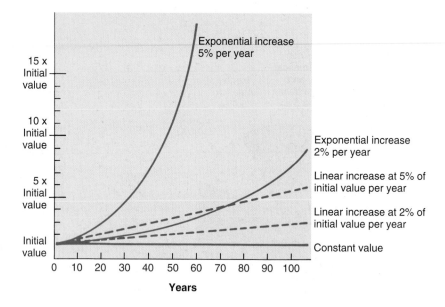

Figure 1.14

Graphical comparison of the effects of linear and exponential growth, whether on consumption of minerals, fuels, water, and other consumable commodities, or population. With linear growth, one adds a fixed percentage of the *initial* value each year (dashed lines). With exponential growth, the same percentage increase is computed each year, but year by year the value on which that percentage is calculated increases, so the annual increment keeps getting larger.

in years may be estimated from growth rate *(G)*, expressed in percent per year, using the simple relationship $D = 70/G$, which is derived from the equation for exponential growth (see "Exploring Further" question 2 at the chapter's end). The higher the growth rate, the shorter the doubling time of the population (see again table 1.4). By region, the most rapidly growing segment of the population today is that of Africa. Its population, estimated 1.05 billion in 2011, is growing at about 2.4% per year. The largest segment of the population, that of Asia, is increasing at 1.1% per year, and since the over 4 billion people there represent well over half of the world's total population, this leads to a relatively high global average growth rate. Europe's population has begun to decline slightly, but Europe contains only about 11% of the world's population. Thus, the fastest growth in general is taking place in the largest segments of the population.

The average worldwide population growth rate is about 1.2% per year. This may sound moderate, but it corresponds to a relatively short doubling time of about 58 years. At that, the present population growth rate actually represents a decline from nearly 2% per year in the mid-1960s. However, a decreasing *growth rate* is not at all the same as a decreasing population. Depending upon projected fertility rates, estimates of world population in the year 2050 can vary by several billion. Using a medium fertility rate, the Population Reference Bureau projects a 2050 world population of almost 9.6 billion. Figure 1.13 illustrates how those people will be distributed by region, considering differential growth rates from place to place.

Even breaking the world down into regions of continental scale masks a number of dramatic individual-country cases. Discussion of these, and of their political, economic, and cultural implications, is well beyond the scope of this chapter, but consider the following: While the population of Europe is nearly stable, declining only slightly overall, in many parts of northern and eastern Europe, sharp declines are occurring. By 2050, the populations of Russia, Ukraine, and Bulgaria are

expected to drop by 12, 20, and 24 percent, respectively. Conversely, parts of the Middle East are experiencing explosive population growth, with projected increases by 2050 of 44% in Israel, 60% in Saudi Arabia, 111% in Jordan, 133% in the Palestinian Territory, 150% in Syria, and 155% in Iraq. The demographics differ widely between countries, too. Globally, 27% of the population is under 15, and only 8% above age 65. But in Japan, only 13% of the population is below age 15, with 23% age 65 or older; in Afghanistan, 44% of people are under 15 and only 2% age 65 or over. Thus, different nations face different challenges. Where rapid population growth meets scarcity of resources, problems arise.

Impacts of the Human Population

The problems posed by a rapidly growing world population have historically been discussed most often in the context of food: that is, how to produce sufficient food and distribute it effectively to prevent the starvation of millions in overcrowded countries or in countries with minimal agricultural development. This is a particularly visible and immediate problem, but it is also only one facet of the population challenge.

Farmland and Food Supply

Whether or not the earth can support 7 billion people, or 9 or 11 billion, is uncertain. In part, it depends on the quality of life, the level of technological development, and other standards that societies wish to maintain. Yet even when considering the most basic factors, such as food, it is unclear just how many people the earth can sustain. Projections about the adequacy of food production, for example, require far more information than just the number of people to be fed and the amount of available land. The total arable land (land suitable for cultivation) in the world has been estimated at 7.9 billion acres, or about 1.1 acres per person of the present population. The major limitation on this figure is availability of water, either as rainfall or through irrigation. Further considerations relating to the nature of the soil include the soil's fertility, water-holding capacity, and general suitability for farming. Soil character varies tremendously, and its productivity is similarly variable. Moreover, farmland can deteriorate through loss of nutrients and by wholesale erosion of topsoil, and this degradation must be considered when making production predictions.

There is also the question of what crops can or should be grown. Today, this is often a matter of preference or personal taste, particularly in farmland-rich (and energy- and water-rich) nations. The world's people are not now always being fed in the most resource-efficient ways. To produce one ton of corn requires about 250,000 gallons of water; a ton of wheat, 375,000 gallons; a ton of rice, 1,000,000 gallons; a ton of beef, 7,500,000 gallons. Some new high-yield crop varieties may require irrigation, whereas native varieties did not. The total irrigated acreage in the world has more than doubled in three decades. However, water resources are dwindling in many places; the water costs of food production must increasingly be taken into account.

Genetic engineering is now making important contributions to food production, as varieties are selectively developed for high yield, disease resistance, and other desirable qualities. These advances have led some to declare that fears of global food shortages are no longer warranted, even if the population grows by several billion more. However, two concerns remain: One, poor nations already struggling to feed their people may be least able to afford the higher-priced designer seed or specially developed animal strains to benefit from these advances. Second, as many small farms using many, genetically diverse strains of food crops are replaced by vast areas planted with a single, new, superior variety, there is the potential for devastating losses if a new pest or disease to which that one strain is vulnerable enters the picture.

Food production as practiced in the United States is also a very energy-intensive business. The farming is heavily mechanized, and much of the resulting food is extensively processed, stored, and prepared in various ways requiring considerable energy. The products are elaborately packaged and often transported long distances to the consumer. Exporting the same production methods to poor, heavily populated countries short on energy and the capital to buy fuel, as well as food, may be neither possible nor practical. Even if possible, it would substantially increase the world's energy demands.

Population and Nonfood Resources

Food is at least a renewable resource. Within a human life span, many crops can be planted and harvested from the same land and many generations of food animals raised. By contrast, the supplies of many of the resources considered in later chapters—minerals, fuels, even land itself—are finite. There is only so much oil to burn, rich ore to exploit, and suitable land on which to live and grow food. When these resources are exhausted, alternatives will have to be found or people will have to do without.

The earth's supply of many such materials is severely limited, especially considering the rates at which these resources are presently being used. Many could be effectively exhausted within decades, yet most people in the world are consuming very little in the way of minerals or energy. Current consumption is strongly concentrated in a few highly industrialized societies. Per-capita consumption of most mineral and energy resources is higher in the United States than in any other nation. For the world population to maintain a standard of living comparable to that of the United States, mineral production would have to increase about fourfold, on the average. There are neither the recognized resources nor the production capability to maintain that level of consumption for long, and the problem becomes more acute as the population grows.

Some scholars believe that we are already on the verge of exceeding the earth's **carrying capacity,** its ability to sustain its population at a basic, healthy, moderately comfortable standard of living. Estimates of sustainable world population made over the last few decades range from under 7 billion—and remember, we are already there—to over 100 billion persons. The wide range is attributable to considerable variations in model assumptions, including standard of living and achievable productivity of farmland. Certainly, given global resource availability and technology, even the present world population could not enjoy the kind of high-consumption lifestyle to which the average resident of the United States has become accustomed.

It is true that, up to a point, the increased demand for minerals, fuels, and other materials associated with an increase in population tends to raise prices and promote exploration for these materials. The short-term result can be an apparent increase in the resources' availability, as more exploration leads to discoveries of more oil fields, ore bodies, and so on. However, the quantity of each of these resources is finite. The larger and more rapidly growing the world population, and the faster its level of development and standard of living rise, the more rapidly limited resources are consumed, and the sooner those resources will be exhausted.

Land is clearly a basic resource. Seven, 9, or 100 billion people must be *put* somewhere. Already, the global average population density is about 51 persons per square kilometer of land surface (130 persons per square mile), and that is counting *all* the land surface, including jungles, deserts, and mountain ranges, leaving out only the Antarctic continent. The ratio of people to readily inhabitable land is plainly much higher. Land is also needed for manufacturing, energy production, transportation, and a variety of other uses. Large numbers of people consuming vast quantities of materials generate vast quantities of wastes. Some of these wastes can be recovered and recycled, but others cannot. It is essential to find places to put the latter, and ways to isolate the harmful materials from contact with the growing population. This effort claims still more land and, often, resources. All of this is why land-use planning—making the most of every bit of land available—is becoming increasingly important. At present, it is too often true that the ever-growing population settles in areas that are demonstrably unsafe or in which the possible problems are imperfectly known (figures 1.15 and 1.16).

Figure 1.16

Even where safer land is abundant, people may choose to settle in hazardous—but scenic—places, such as barrier islands. South Dade County, Florida.

Photograph © Alan Schein Photography/Corbis.

Figure 1.15

The landslide hazard to these structures sitting at the foot of steep slopes in Rio de Janeiro is obvious, but space for building here is limited.

Photograph © Will and Deni McIntyre/Getty Images.

Uneven Distribution of People and Resources

Even if global carrying capacity were ample in principle, that of an individual region may not be. None of the resources—livable land, arable land, energy, minerals, or water—is uniformly distributed over the earth. Neither is the population (see figure 1.17). In 2011, the population density in Singapore was about 19,370 persons per square mile; in Australia and Canada, 8.

Many of the most densely populated countries are resource-poor. In some cases, a few countries control the major share of one resource. Oil is a well-known example, but there are many others. Thus, economic and political complications enter into the question of resource adequacy. Just because one nation controls enough of some commodity to supply all the world's needs does not necessarily mean that the country will choose to share its resource wealth or to distribute it at modest cost to other nations. Some resources, like land, are simply not transportable and therefore cannot readily be shared. Some of the complexities of global resource distribution will be highlighted in subsequent chapters.

Disruption of Natural Systems

Natural systems tend toward a balance or equilibrium among opposing factors or forces. When one factor changes, compensating changes occur in response. If the disruption of the system is relatively small and temporary, the system may, in time, return to its original condition, and evidence of the disturbance will disappear. For example, a coastal storm may wash away beach vegetation and destroy colonies of marine organisms living in a tidal flat, but when the storm has passed, new organisms will start to move back into the area, and new grasses will take root in the dunes. The violent eruption of a volcano like Mount Pinatubo may spew ash and gases high into the atmosphere, partially blocking sunlight and causing the earth to cool, but within a few years, the ash will have settled back to the ground, and normal temperatures will be restored. Dead leaves falling into a lake provide food for the microorganisms that within weeks or months will break the leaves down and eliminate them.

This is not to say that permanent changes never occur in natural systems. The size of a river channel reflects the maximum amount of water it normally carries. If long-term climatic or other conditions change so that the volume of water regularly reaching the stream increases, the larger quantity of water will, in time, carve out a correspondingly larger channel to accommodate it. The soil carried downhill by a landslide certainly does not begin moving back upslope after the landslide is over;

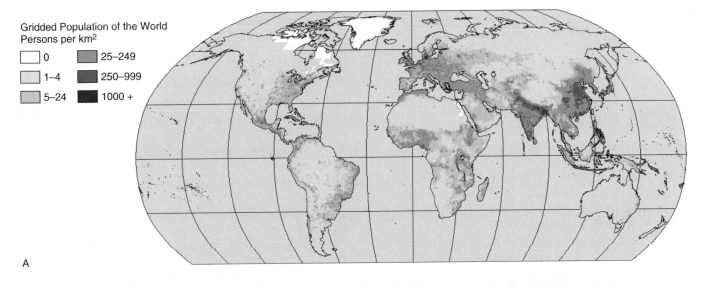

Gridded Population of the World
Persons per km²

☐ 0	☐ 25–249
☐ 1–4	☐ 250–999
☐ 5–24	■ 1000 +

A

B

Figure 1.17

(A) Global population density, 2000; the darker the shading, the higher the population density. Comparison with the distribution of lights at night in 2002 (B) shows that overall population density on the one hand, and urbanization/development on the other, are often, but not always, closely correlated.

(A) Source: Center for International Earth Science Information Network (CIESIN), Columbia University; and Centro Internacional de Agricultura Tropical (CIAT), Gridded Population of the World (GPW), Version 3. Palisades, NY: CIESIN, Columbia University. Available at: http://sedac.ciesin.columbia.edu/gpw.

(B) Image courtesy C. Mayhew & R. Simmon (NASA/GSFC), NOAA/NGDC, DMSP Digital Archive.

the face of the land is irreversibly changed. Even so, a hillside forest uprooted and destroyed by the slide may, within decades, be replaced by new growth in the soil newly deposited at the bottom of the hill.

Human activities can cause or accelerate permanent changes in natural systems. The impact of humans on the global environment is broadly proportional to the size of the population, as well as to the level of technological development achieved. This can be illustrated especially easily in the context of pollution. The smoke from one campfire pollutes only the air in the immediate vicinity; by the time that smoke is dispersed through the atmosphere, its global impact is negligible. The collective smoke from a century and a half of

increasingly industrialized society, on the other hand, has caused measurable increases in several atmospheric pollutants worldwide, and these pollutants continue to pour into the air from many sources. It was once assumed that the seemingly vast oceans would be an inexhaustible "sink" for any extra CO_2 that we might generate by burning fossil fuels, but decades of steadily climbing atmospheric CO_2 levels have proven that in this sense, at least, the oceans are not as large as we thought. Likewise, seven people carelessly dumping wastes into the ocean would not appreciably pollute that huge volume of water. The prospect of 7 *billion* people doing the same thing, however, is quite another matter. And every hour, now, world population increases by more than *9500 people.*

Earth's Moon

Scientists have long strived to explain the origin of earth's large and prominent satellite. Through much of the twentieth century, several different models competed for acceptance; within the last few decades a new theory of lunar origin has been developed. While a complete discussion of the merits and shortcomings of these is beyond the scope of this book, they provide good examples of how objective evidence can provide support for, or indicate weaknesses in, hypotheses and theories.

Any acceptable theory of lunar origin has to explain a number of facts. The moon orbits the earth in an unusual orientation (figure 1), neither circling around earth's equator nor staying in the plane in which the planets' orbits around the sun lie (the ecliptic plane). Its density is much lower than that of earth, meaning that it contains a much lower proportion of iron. Otherwise, it is broadly similar in composition to the earth's mantle. However, analysis of samples from the *Apollo* missions revealed that relative to earth, the moon is depleted not only in most gases, but also in volatile (easily vaporized) metals such as lead and rubidium, indicating a hot history for lunar material.

The older "sister-planet" model proposed that the earth and moon accreted close together during solar system formation, and that is how the moon comes to be orbiting the earth. But in that case, why is the moon not orbiting in the ecliptic plane, and why the significant chemical differences between the two bodies?

The "fission hypothesis" postulated that the moon was spun off from a rapidly rotating early earth after earth's core had been differentiated, so the moon formed mainly from earth's mantle. That would account for the moon's lower density and relatively lower iron content. But analyses of the lunar samples revealed the many additional chemical differences between the moon and earth's mantle. Furthermore, calculations show that a moon formed this way should be orbiting in the equatorial plane, and that far more angular momentum would be required to make it happen than is present in the earth-moon system.

A third suggestion was that the moon formed elsewhere in the solar system and then passed close enough to earth to be "captured" into orbit by gravity. A major flaw in this idea involves the dynamics necessary for capture. The moon is a relatively large satellite for a body the size of earth. For earth's gravity to snare it, the rate at which the moon came by would have to be very, very slow. But earth is hurtling around the sun at about 107,000 km/hr (66,700 mph), so the probability of the moon happening by at just the right distance and velocity for capture to occur is extremely low. Nor does capture account for a hot lunar origin.

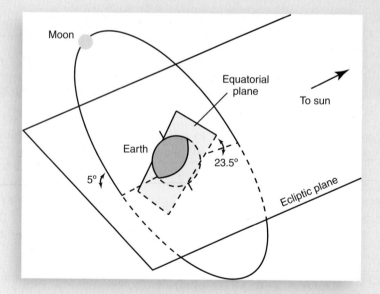

Figure 1

Summary

The solar system formed over 4½ billion years ago. The earth is unique among the planets in its chemical composition, abundant surface water, and oxygen-rich atmosphere. The earth passed through a major period of internal differentiation early in its history, which led to the formation of the atmosphere and the oceans. Earth's surface features have continued to change throughout the last 4^+ billion years, through a series of processes that are often cyclical in nature, and commonly interrelated. The oldest rocks in which remains of simple organisms are recognized are more than 3 billion years old. The earliest plants

So how to explain the moon? The generally accepted theory for the past two decades (sometimes informally described as the "Big Whack") involves collision between the earth and a body about the size of Mars, whose orbit in the early solar system put it on course to intercept earth. The tremendous energy of the collision would have destroyed the impactor and caused extensive heating and melting on earth, ejecting quantities of vaporized minerals into space around earth. If this impact occurred after core differentiation, that material would have come mainly from the mantles of Earth and the impactor. The orbiting material would have condensed and settled into a rotating disk of dust that later accreted to form the moon.

This theory, exotic as it sounds, does a better job of accounting for all the necessary facts. It provides for a (very) hot origin for the material that became the moon, explaining the loss of volatiles. With mantle material primarily involved, a resulting lunar composition similar to that of earth's mantle is reasonable, and contributions from the impactor would introduce some differences as well. An off-center hit by the impactor could easily produce a dust disk (and eventual lunar orbit) oriented at an angle to both the ecliptic plane and earth's equatorial plane. We know that the moon was extensively cratered early in its history, and accretion of debris from the collision could account for this. And computer models designed to test the mechanical feasibility of this theory have shown that indeed, it is physically plausible. So the "Big Whack" is likely to remain the prevailing theory of lunar origin—until and unless new evidence is found that does not fit.

Though humans have not been back to the moon for four decades, new information about it can still be generated. Since *Apollo*, computers have become much more powerful, and chemical-analysis techniques more sensitive and sophisticated. Reprocessing of *Apollo*-era seismic data has indicated that the moon has an iron-rich core like the Earth's, though proportionately much smaller. Reanalysis of lunar samples has provided more-detailed information on the moon's composition, including evidence for water in the lunar mantle. Such data will provide further tests of the "Big Whack" theory.

Interest in returning humans to the moon, to study and even to live, has recently been growing. However, the environment is a daunting one (figure 2). The moon has essentially no atmosphere, to breathe or to trap heat to moderate surface temperatures, so in sunlight the surface soars to about 120°C (250°F) and during the lunar night plunges to about −175°C (−280°F). There is no vegetation or other life. In 2009, an ingenious experiment demonstrated that some water ice is buried in the lunar soil. A spent rocket was deliberately crashed into a shaded area within a crater; an accompanying spacecraft analyzed the resultant debris and identified ice and water vapor in it. However, how much ice is there, and how widespread, is unknown. To bring water to the moon for human use would cost an estimated $7000–$70,000 *per gallon*. Many other raw materials would certainly have to be shipped there. Furthermore (as the *Apollo* astronauts discovered) the moon presents another special challenge: a surface blanket of abrasive rock dust, the result of pounding by many meteorites over its history—dust that can abrade and foul equipment and, if breathed, injure lungs. So at least for now, any lunar colony is likely to be very small, and necessarily contained in a carefully controlled environment.

Figure 2

The lunar surface is not an environment in which humans could live outside protective structures, with high energy and resource costs. Geologist/astronaut Harrison Schmitt, lunar module commander of the Apollo 17 mission.

Photograph courtesy NASA.

were responsible for the development of free oxygen in the atmosphere, which, in turn, made it possible for oxygen-breathing animals to survive. Human-type remains are unknown in rocks over 4 to 5 million years of age. In a geologic sense, therefore, human beings are quite a new addition to the earth's cast of characters, but they have had a very large impact. Geology, in turn, can have an equally large impact on us.

The world population, now over 7 billion, might well be over 9 billion by the year 2050. Even our present population cannot entirely be supported at the level customary in the more developed countries, given the limitations of land and resources. Extraterrestrial resources cannot realistically be expected to contribute substantially to a solution of this problem.

Key Terms and Concepts

carrying capacity 15
core 3
crust 3

doubling time 13
environmental
    geology 1

exponential growth 13
hypothesis 8
mantle 3

scientific method 8
theory 8

Exercises

Questions for Review

1. Describe the process by which the solar system is believed to have formed, and explain why it led to planets of different compositions, even though the planets formed simultaneously.

2. How old is the solar system? How recently have human beings come to influence the physical environment?

3. Explain how the newly formed earth differed from the earth we know today.

4. What kinds of information are used to determine the internal composition of the earth?

5. How were the earth's atmosphere and oceans formed?

6. What are the differences among facts, scientific hypotheses, and scientific theories?

7. Many earth materials are transformed through processes that are cyclical in nature. Describe one example.

8. The size of the earth's human population directly affects the severity of many environmental problems. Explain this idea in the context of (a) resources and (b) pollution.

9. If earth's population has already exceeded the planet's carrying capacity, what are the implications for achieving a comfortable standard of living worldwide? Explain.

10. What is the world's present population, to the nearest billion? How do recent population growth rates (over the last few centuries) compare with earlier times? Why?

11. Explain the concept of doubling time. How has population doubling time been changing through history? What is the approximate doubling time of the world's population at present?

12. What regions of the world currently have the fastest rates of population growth? The slowest?

13. Describe any one of the older theories of lunar origin, and note at least one fact about the moon that it fails to explain.

Exploring Further: Working with the Numbers

1. The urgency of population problems can be emphasized by calculating such "population absurdities" as the time at which there will be one person per square meter or per square foot of land or the time at which the weight of people will exceed the weight of the earth. Try calculating these "population absurdities" by using the world population projections from table 1.4 and the following data concerning the earth:

   Mass 5976×10^{21} kg*

   Land surface area (approx.) 149,000,000 sq. km

   Average weight of human body (approx.) 75 kg

2. Derive the relation between doubling time and growth rate, starting from the exponential-growth relation

   $$N = N_0 e^{(G/100) \cdot t}$$

   where N = the growing quantity (number of people, tons of iron ore consumed annually, dollars in a bank account, or whatever); N_0 is the initial quantity at the start of the time period of interest; G is growth rate expressed in percent per year, and "percent" means "parts per hundred"; and t is the time in years over which growth occurs. Keep in mind that doubling time, D, is the length of time required for N to double.

3. Select a single country or small set of related countries; examine recent and projected population growth rates in detail, including the factors contributing to the growth rates and trends in those rates. Compare with similar information for the United States or Canada. A useful starting point may be the latest World Population Data Sheet from the Population Reference Bureau (www.prb.org).

---

* This number is so large that it has been expressed in scientific notation, in terms of powers of 10. It is equal to 5976 with twenty-one zeroes after it. For comparison, the land surface area could also have been written as 149×10^6 sq. km, or $149,000 \times 10^3$ sq. km, and so on.

Rocks and Minerals— A First Look

It is difficult to talk for long about geology without discussing the rocks and minerals of which the earth is made. Considering that most common rocks and minerals are composed of a small subset of the chemical elements, they are remarkably diverse in color, texture, and other physical properties. The differences in the physical properties of rocks, minerals, and soils determine their suitability for different purposes—extraction of water or of metals, construction, manufacturing, waste disposal, agriculture, and other uses. For this reason, it is helpful to understand something of the nature of these materials. Also, each rock contains clues to the kinds of processes that formed it and to the geologic setting where it is likely to be found. For example, we will see that the nature of a volcano's rocks may indicate what hazards it presents to us; our search for new ores or fuels is often guided by an understanding of the specialized geologic environments in which they occur.

Minerals can be both beautiful and useful. These delicate flakes are crystals of wulfenite, an oxide of lead and molybdenum sometimes used as a molybdenum ore.

Atoms, Elements, Isotopes, Ions, and Compounds

Atomic Structure

All natural and most synthetic substances on earth are made from the ninety naturally occurring chemical elements. An **atom** is the smallest particle into which an element can be divided while still retaining the chemical characteristics of that element (see figure 2.1). The **nucleus,** at the center of the atom, contains one or more particles with a positive electrical charge (**protons**) and usually some particles of similar mass that have no charge (**neutrons**). Circling the nucleus are the negatively charged **electrons.** Protons and neutrons are similar in mass, and together they account for most of the mass of an atom. The much lighter electrons are sometimes represented as a "cloud" around the nucleus, as in figure 2.1, and are sometimes shown as particles, as in figure 2.3. The −1 charge of one electron exactly balances the +1 charge of a single proton.

The number of protons in the nucleus determines what chemical element that atom is. Every atom of hydrogen contains one proton in its nucleus; every oxygen atom contains eight protons; every carbon atom, six; every iron atom, twenty-six; and so on. The characteristic number of protons is the **atomic number** of the element.

Elements and Isotopes

With the exception of the simplest hydrogen atoms, all nuclei contain neutrons, and the number of neutrons is similar to or somewhat greater than the number of protons. The number of neutrons in atoms of one element is not always the same. The sum of the number of protons and the number of neutrons in a nucleus is the atom's **atomic mass number.** Atoms of a given element with different atomic mass numbers—in other words, atoms with the same number of protons but different numbers of neutrons—are distinct **isotopes** of that element. Some elements have only a single isotope, while others may have ten or more. (The reasons for these phenomena involve principles of nuclear physics and the nature of the processes by which the elements are produced in the interiors of stars, and we will not go into them here!)

For most applications, we are concerned only with the elements involved, not with specific isotopes. When a particular isotope is to be designated, this is done by naming the element (which, by definition, specifies the atomic number, or number of protons) and the atomic mass number (protons plus neutrons). Carbon, for example, has three natural isotopes. By far the most abundant is carbon-12, the isotope with six neutrons in the nucleus in addition to the six protons common to all carbon atoms. The two rarer isotopes are carbon-13 (six protons plus seven neutrons) and carbon-14 (six protons plus eight neutrons). Chemically, all behave alike. The human body cannot, for instance, distinguish between sugar containing carbon-12 and sugar containing carbon-13.

Other differences between isotopes may, however, make a particular isotope useful for some special purpose. Some isotopes are *radioactive,* meaning that over time, their nuclei will decay (break down) into nuclei of other elements, releasing energy. Each such radioactive isotope will decay at its own specific rate, which allows us to use such isotopes to date geologic materials and events, as described in appendix A. A familiar example is carbon-14, used to date materials containing carbon, including archeological remains such as cloth, charcoal, and bones. Differences in the properties of two uranium isotopes are important in understanding nuclear power options: only one of the two common uranium isotopes is suitable for use as reactor fuel, and must be extracted and concentrated from natural uranium as it occurs in uranium ore. The fact that radioactive elements will inexorably decay—releasing energy—at their own fixed, constant rates is part of what makes radioactive-waste disposal such a challenging problem, because no chemical or physical treatment can make those waste isotopes nonradioactive and inert.

Ions

In an electrically neutral atom, the number of protons and the number of electrons are the same; the negative charge of one electron just equals the positive charge of one proton. Most atoms, however, can gain or lose some electrons. When this happens, the atom has a positive or negative electrical charge and is called an **ion.** If it loses electrons, it becomes positively charged, since the number of protons then exceeds the number of electrons. If the atom gains electrons, the ion has a negative electrical charge. Positively and negatively charged ions are called, respectively, **cations** and **anions.** Both solids and liquids are, overall, electrically neutral, with the total positive and negative charges of cations and anions balanced. Moreover, free ions do not exist in solids; cations and anions are bonded together. In a solution, however, individual ions may exist and move independently. Many minerals break down into ions as they dissolve in water. Individual ions may then be taken up by plants as nutrients or react with other materials. The concentration of hydrogen ions (pH) determines a solution's acidity.

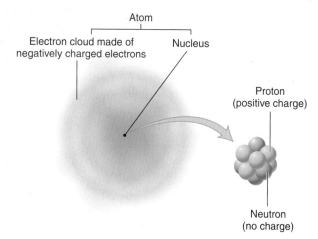

Figure 2.1

Schematic drawing of atomic structure (greatly enlarged and simplified). The nucleus is actually only about 1/1000th the overall size of the atom.

Figure 2.2 The periodic table.

Legend:
- 12 — Atomic number
- Mg — Chemical symbol
- 24.31 — Atomic weight (approximate when in parentheses)

* Elements heavier than uranium synthesized experimentally

1a	IIa	IIIb	IVb	Vb	VIb	VIIb	VIII			Ib	IIb	IIIa	IVa	Va	VIa	VIIa	0
1 H 1.008																	2 He 4.00
3 Li 6.94	4 Be 9.01											5 B 10.81	6 C 12.01	7 N 14.00	8 O 15.99	9 F 18.99	10 Ne 20.18
11 Na 22.99	12 Mg 24.31											13 Al 26.98	14 Si 28.09	15 P 30.97	16 S 32.06	17 Cl 35.45	18 Ar 39.95
19 K 39.10	20 Ca 40.08	21 Sc 44.96	22 Ti 47.90	23 V 50.94	24 Cr 51.99	25 Mn 54.94	26 Fe 55.85	27 Co 58.93	28 Ni 58.71	29 Cu 63.54	30 Zn 65.41	31 Ga 69.72	32 Ge 72.59	33 As 74.92	34 Se 78.96	35 Br 79.91	36 Kr 83.80
37 Rb 85.47	38 Sr 87.62	39 Y 88.91	40 Zr 91.22	41 Nb 92.91	42 Mo 95.94	43 Tc (99)	44 Ru 101.97	45 Rh 102.91	46 Pd 106.4	47 Ag 107.87	48 Cd 112.40	49 In 114.82	50 Sn 118.69	51 Sb 121.75	52 Te 127.60	53 I 126.90	54 Xe 131.30
55 Cs 132.91	56 Ba 137.34	57–71 see below	72 Hf 178.49	73 Ta 180.95	74 W 183.85	75 Re 186.2	76 Os 190.2	77 Ir 192.2	78 Pt 195.09	79 Au 196.97	80 Hg 200.59	81 Tl 204.37	82 Pb 207.19	83 Bi 208.98	84 Po (210)	85 At (210)	86 Rn (222)
87 Fr (223)	88 Ra (226)	89–103 see below	104* Rf (263)	105* Ha (262)	106* Sg (266)												

57 La 138.91	58 Ce 140.12	59 Pr 140.91	60 Nd 144.24	61 Pm (147)	62 Sm 150.35	63 Eu 151.96	64 Gd 157.25	65 Tb 158.92	66 Dy 162.50	67 Ho 164.93	68 Er 167.26	69 Tm 168.93	70 Yb 173.04	71 Lu 174.97
89 Ac (227)	90 Th 232.04	91 Pa (231)	92 U 238.03	93* Np (237)	94* Pu (242)	95* Am (243)	96* Cm (247)	97* Bk (247)	98* Cf (251)	99* Es (254)	100* Fm (253)	101* Md (256)	102* No (254)	103* Lw (257)

The Periodic Table

Some idea of the probable chemical behavior of elements can be gained from a knowledge of the **periodic table** (figure 2.2). The Russian scientist Dmitri Mendeleyev first observed that certain groups of elements showed similar chemical properties, which seemed to be related in a regular way to their atomic numbers. At the time (1869) that Mendeleyev published the first periodic table, in which elements were arranged so as to reflect these similarities of behavior, not all elements had even been discovered, so there were some gaps. The addition of elements identified later confirmed the basic concept. In fact, some of the missing elements were found more easily because their properties could to some extent be anticipated from their expected positions in the periodic table.

We now can relate the periodicity of chemical behavior to the electronic structures of the elements. Electrons around an atom occur in shells of different energies, each of which can hold a fixed maximum number of electrons. Those elements in the first column of the periodic table, known as the alkali metals, have one electron in the outermost shell of the neutral atom. Thus, they all tend to form cations of +1 charge by losing that odd electron. Outermost electron shells become increasingly full from left to right across a row. The next-to-last column, the halogens, are those elements lacking only one electron in the outermost shell, and they thus tend to gain one electron to form anions of charge −1. In the right-hand column are the inert gases, whose neutral atoms contain all fully filled electron shells.

The elements that occur naturally on earth have now been known for decades. Additional new elements must be created, not simply discovered, because these very heavy elements, with atomic numbers above 92 (uranium), are too unstable to have lasted from the early days of the solar system to the present. Some, like plutonium, are by-products of nuclear-reactor operation; others are created by nuclear physicists who, in the process, learn more about atomic structure and stability.

Compounds

A **compound** is a chemical combination of two or more chemical elements, bonded together in particular proportions, that has a distinct set of physical properties (often very different from those of any of the individual elements in it). In minerals, most bonds are *ionic* or *covalent,* or a mix of the two. In **ionic bonding,** the bond is based on the electrical attraction between oppositely charged ions. Bonds between atoms may also form if the atoms share electrons. This is **covalent bonding.** Table salt (sodium chloride) provides a common example of

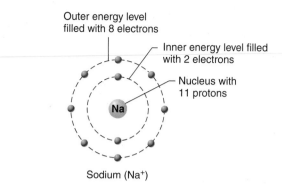

Sodium (Na⁺)

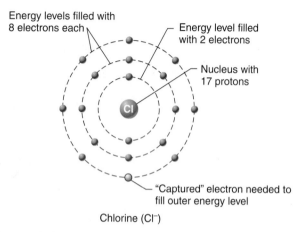

Chlorine (Cl⁻)

Figure 2.3

Sodium, with 11 protons and electrons, has two filled shells and one "leftover" electron in its outermost shell. Chlorine can accept that odd electron, filling its own outermost shell exactly. The resulting oppositely charged ions attract and bond.

ionic bonding (figure 2.3). Sodium, an alkali metal, loses its outermost electron to chlorine, a halogen. The two elements now have filled electron shells, but sodium is left with a +1 net charge, chlorine −1. The ions bond ionically to form sodium chloride. Sodium is a silver metal, and chlorine is a greenish gas that is poisonous in large doses. When equal numbers of sodium and chlorine atoms combine to make table salt, or sodium chloride, the resulting compound forms colorless crystals that do not resemble either of the component elements.

Minerals—General

Minerals Defined

A **mineral** is a naturally occurring, inorganic, solid element or compound with a definite chemical composition and a regular internal crystal structure. *Naturally occurring,* as distinguished from synthetic, means that minerals do not include the thousands of chemical substances invented by humans. *Inorganic,* in this context, means not produced solely by living organisms or by biological processes. That minerals must be *solid* means that the ice of a glacier is a mineral, but liquid water is not. Chemically, minerals may consist either of one element—like diamonds, which

are pure carbon—or they may be compounds of two or more elements. Some mineral compositions are very complex, consisting of ten elements or more. Minerals have a definite chemical composition or a compositional range within which they fall. The presence of certain elements in certain proportions is one of the identifying characteristics of each mineral. Finally, minerals are crystalline, at least on the microscopic scale. **Crystalline** materials are solids in which the atoms or ions are arranged in regular, repeating patterns (figure 2.4). These patterns may not be apparent to the naked eye, but most solid compounds are crystalline, and their crystal structures can be recognized and studied using X rays and other techniques. Examples of noncrystalline solids include glass (discussed later in the chapter) and plastic.

Identifying Characteristics of Minerals

The two fundamental characteristics of a mineral that together distinguish it from all other minerals are its chemical composition and its crystal structure. No two minerals are identical in both respects, though they may be the same in one. For example, diamond and graphite (the "lead" in a lead pencil) are chemically the same—both are made up of pure carbon. Their physical properties, however, are vastly different because of the differences in their internal crystalline structures. In a diamond, each carbon atom is firmly bonded to every adjacent carbon atom in every direction by covalent bonds. In graphite, the carbon atoms are bonded strongly in two dimensions into sheets, but the sheets are only weakly held together in the third dimension. Diamond is clear, colorless, and very hard, and a jeweler can cut it into beautiful precious gemstones. Graphite is black, opaque, and soft, and its sheets of carbon atoms tend to slide apart as the weak bonds between them are broken.

A mineral's composition and crystal structure can usually be determined only by using sophisticated laboratory equipment. When a mineral has formed large crystals with well-developed shapes, a trained mineralogist may be able to infer some characteristics of its internal atomic arrangement because crystals' shapes are controlled by and related to this atomic structure, but most mineral samples do not show large symmetric crystal forms by which they can be recognized with the naked eye. Moreover, many minerals share similar external forms, and the same mineral may show different external forms, though it will always have the same internal structure (figure 2.5). No one can look at a mineral and know its chemical composition without first recognizing what mineral it is. Thus, when scientific instruments are not at hand, mineral identification must be based on a variety of other physical properties that in some way reflect the mineral's composition and structure. These other properties are often what make the mineral commercially valuable. However, they are rarely unique to one mineral and often are deceptive. A few examples of such properties follow.

Other Physical Properties of Minerals

Perhaps surprisingly, color is often not a reliable guide to mineral identification. While some minerals always appear the same color, many vary from specimen to specimen. Variation in

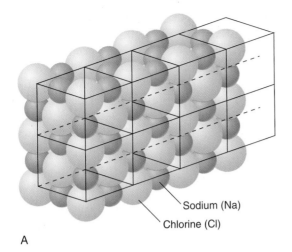

A

B

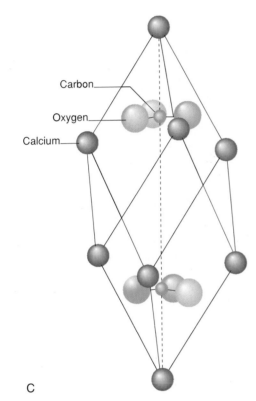

Carbon

Oxygen

Calcium

C

D

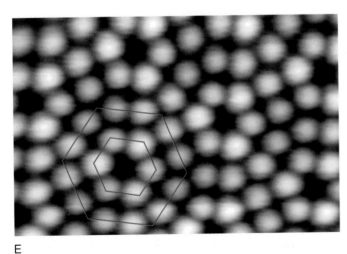

E

Figure 2.4

(A) Sodium and chloride ions are arranged alternately in the halite structure. Lines show the cubes that make up the repeating unit of the crystal structure; the resultant cubic crystal form is shown in (B). (C) The crystal structure of calcite (calcium carbonate, $CaCO_3$) is a bit more complex. Here, the atoms have been spread apart so that the structure is easier to see; again, lines show the shape of the repeating structural unit of the crystal, which may be reflected in the external form of calcite crystals (D). (E) Scanning tunneling microscope image of individual atoms in crystalline silicon. The diameter of each atom is about 0.00000002 inches. Note the regular hexagonal arrangement of atoms in this plane (lines added to highlight this).

(B) Photograph ©The McGraw-Hill Companies, Inc./Doug Sherman, photographer. (E) Image courtesy Jennifer MacLeod and Alastair McLean, Queen's University, Canada.

A

B

C

D

Figure 2.5

Many minerals may share the same external crystal form: (A) galena (PbS) and (B) fluorite (CaF_2) form cubes, as do halite (figure 2.4B) and pyrite (C). However, these minerals may show other forms; (D), for example, is a distinctive form of pyrite called a pyritohedron.

Photographs (A) and (C) © The McGraw-Hill Companies Inc./Doug Sherman, photographer.

color is usually due to the presence of small amounts of chemical impurities in the mineral that have nothing to do with the mineral's basic characteristic composition, and such variation is especially common when the pure mineral is light-colored or colorless. The very common mineral quartz, for instance, is colorless in its pure form. However, quartz also occurs in other colors, among them rose pink, golden yellow, smoky brown, purple (amethyst), and milky white. Clearly, quartz cannot always be recognized by its color, or lack of it.

Another example is the mineral corundum (figure 2.6A), a simple compound of aluminum and oxygen. In pure form, it too is colorless, and quite hard, which makes it a good abrasive. It is often used for the grit on sandpaper. Yet a little color from trace impurities not only disguises corundum, it can transform this utilitarian material into highly prized gems: Traces of chromium produce the deep bluish-red gem we call ruby, while sapphire is just corundum tinted blue by iron and titanium. The color of a mineral can vary within a single crystal (figure 2.6B). Even when the color shown by a mineral sample is the true color of the pure mineral, it is probably not unique. There are approximately 4400 known minerals, so there are usually many of any one particular color. (Interestingly, *streak,* the color of the powdered mineral as revealed when the mineral is scraped across a piece of unglazed tile, may be quite different from the color of the bulk sample, and more consistent for a single mineral. However, a tile is not always handy, and some samples are too valuable to treat this way.)

A

B

Figure 2.6

(A) The many colors of these corundum gemstones illustrate why color is a poor guide in mineral identification. See also figure 2.7B. (B) If chemical conditions change as a crystal grows, different parts may be different colors, as in this tourmaline.

Hardness, the ability to resist scratching, is another easily measured physical property that can help to identify a mineral, although it usually does not uniquely identify the mineral. Classically, hardness is measured on the Mohs hardness scale (table 2.1), in which ten common minerals are arranged in order of hardness. Unknown minerals are assigned a hardness on the basis of which minerals they can scratch and which minerals scratch them. A mineral that scratches gypsum and is scratched by calcite is assigned a hardness of 2½ (the hardness of an average fingernail). Because a diamond is the hardest natural substance known on earth, and corundum the second-hardest mineral, these minerals might be identifiable from their hardnesses. Among the thousands of "softer" (more readily scratched) minerals, however, there are many of any particular hardness, just as there are many of any particular color.

Not only is the external form of crystals related to their internal structure; so is **cleavage,** a distinctive way some minerals may break up when struck. Some simply crumble or shatter into irregular fragments (fracture). Others, however, break cleanly in certain preferred directions that correspond to planes of weak bonding in the crystal, producing planar cleavage faces.

Table 2.1 The Mohs Hardness Scale

Mineral	Assigned Hardness
talc	1
gypsum	2
calcite	3
fluorite	4
apatite	5
orthoclase	6
quartz	7
topaz	8
corundum	9
diamond	10

For comparison, the approximate hardnesses of some common objects, measured on the same scale, are: fingernail, 2½; copper penny, 3; glass, 5 to 6; pocketknife blade, 5 to 6. These materials can be used to estimate an unknown mineral's hardness when samples of these reference minerals are not available.

There may be only one direction in which a mineral shows good cleavage (as with mica, discussed later in the chapter), or there may be two or three directions of good cleavage. Cleavage surfaces are characteristically shiny (figure 2.7).

A number of other physical properties may individually be common to many minerals. *Luster* describes the appearance of the surfaces—glassy, metallic, pearly, etc. Some minerals are noticeably denser than most; a few are magnetic. Usually it is only by considering a whole set of such nonunique properties as color, hardness, cleavage, density, and others together that a mineral can be identified without complex instruments. For instance, there are many colorless minerals; but if a sample of such a mineral has a hardness of only 3, cleaves into rhombohedral shapes, and fizzes when weak acid is dripped on it, it is probably calcite (the fizzing is due to release of carbon dioxide, CO_2, as the calcite reacts with the acid).

Unique or not, the physical properties arising from minerals' compositions and crystal structures are often what give minerals value from a human perspective—the slickness of talc (main ingredient of talcum powder), the malleability and conductivity of copper, the durability of diamond, and the rich colors of tourmaline gemstones are all examples. Some minerals have several useful properties: table salt (halite), a necessary nutrient, also imparts a taste we find pleasant, dissolves readily to flavor liquids but is soft enough not to damage our teeth if we munch on crystals of it sprinkled on solid food, and will serve as a food preservative in high concentrations, among other helpful qualities.

Types of Minerals

As was indicated earlier, minerals can be grouped or subdivided on the basis of their two fundamental characteristics—composition and crystal structure. Compositionally, classification is typically on the basis of ions or ion groups that a set of

C

Figure 2.7

Another relationship between structures and physical properties is cleavage. Because of their internal crystalline structures, many minerals break apart preferentially in certain directions. (A) Halite has the cubic structure shown in figure 2.4A, and breaks into cubic or rectangular pieces, cleaving parallel to the crystal faces. (B) Fluorite also forms cubic crystals, but cleaves into octahedral fragments, breaking along different planes of its internal structure. (Note the variety of colors, too.) (C) Calcite cleaves into rhombohedra; compare with figure 2.4C.

Photograph (A) © The McGraw-Hill Companies Inc./Bob Coyle, photographer; (B) © The McGraw-Hill Companies, Inc./Charles D. Winters, photographer.

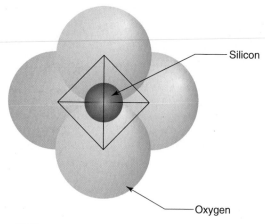

Figure 2.8

The basic silica tetrahedron, building block of all silicate minerals. (In figure 2.9, this group of atoms is represented only by the tetrahedron.)

minerals have in common. In this section, we will briefly review some of the basic mineral categories. A comprehensive survey of minerals is well beyond the scope of this book, and the interested reader should refer to standard mineralogy texts for more information. A summary of physical properties of selected minerals is found in appendix B.

Silicates

In chapter 1, we noted that the two most common elements in the earth's crust are oxygen and silicon. It comes as no surprise, therefore, that by far the largest compositional group of minerals is the **silicate** group, all of which are compounds containing silicon and oxygen, and most of which contain other elements as well. Because this group of minerals is so large, it is subdivided on the basis of crystal structure, by the ways in which the silicon and oxygen atoms are linked together. The basic building block of all silicates is a tetrahedral arrangement of four oxygen atoms (anions) around the much smaller silicon cation (figure 2.8). In different silicate minerals, these *silica tetrahedra* may be linked into chains, sheets, or three-dimensional frameworks by the sharing of oxygen atoms. Some of the physical properties of silicates and other minerals are closely related to their crystal structures (see figure 2.9). In general, we need not go into the structural classes of the silicates in detail. It is more useful to mention briefly a few of the more common, geologically important silicate minerals.

While not the most common, *quartz* is probably the best-known silicate. Compositionally, it is the simplest, containing only silicon and oxygen. It is a framework silicate, with silica tetrahedra linked in three dimensions, which helps make it relatively hard and weathering-resistant. Quartz is found in a variety of rocks and soils. Commercially, the most common use of pure quartz is in the manufacture of

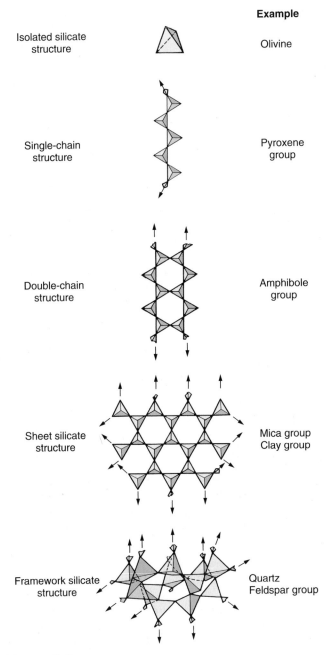

	Example
Isolated silicate structure	Olivine
Single-chain structure	Pyroxene group
Double-chain structure	Amphibole group
Sheet silicate structure	Mica group Clay group
Framework silicate structure	Quartz Feldspar group

Figure 2.9

Silica tetrahedra link together by sharing oxygen atoms (where the corners of the tetrahedra meet) to form a variety of structures. (Olivine and the pyroxenes and amphiboles are among the ferromagnesian silicates, described in the text.) Other structural arrangements (such as stacked rings of tetrahedra) exist, but they are less common.

glass, which also consists mostly of silicon and oxygen. Quartz-rich sand and gravel are used in very large quantities in construction.

The most abundant group of minerals in the crust is a set of chemically similar minerals known collectively as the *feldspars*. They are composed of silicon, oxygen, aluminum, and either sodium, potassium, or calcium, or some combination of these three. Again, logically enough, these common minerals are made from elements abundant in the crust. They are used extensively in the manufacture of ceramics.

Iron and magnesium are also among the more common elements in the crust and are therefore found in many silicate minerals. **Ferromagnesian** is the general term used to describe those silicates—usually dark-colored (black, brown, or green)— that contain iron and/or magnesium, with or without additional elements.

Most ferromagnesian minerals weather relatively readily. Rocks containing a high proportion of ferromagnesian minerals, then, also tend to weather easily, which is an important consideration in construction. Individual ferromagnesian minerals may be important in particular contexts. *Olivine,* a simple ferromagnesian mineral, is a major constituent of earth's mantle; gem-quality olivines from mantle-derived volcanic rocks are the semiprecious gem *peridot.*

Like the feldspars, the *micas* are another group of several silicate minerals with similar physical properties, compositions, and crystal structures. Micas are sheet silicates, built on an atomic scale of stacked-up sheets of linked silicon and oxygen atoms. Because the bonds between sheets are relatively weak, the sheets can easily be broken apart (figure 2.10C).

Clays are another family within the sheet silicates; in clays, the sheets tend to slide past each other, a characteristic that contributes to the slippery feel of many clays and related minerals. Clays are somewhat unusual among the silicates in that their structures can absorb or lose water, depending on how wet conditions are. Absorbed water may increase the slippery tendencies of the clays. Also, some clays expand as they soak up water and shrink as they dry out. A soil rich in these "expansive clays" is a very unstable base for a building, as we will see in later chapters. On the other hand, clays also have important uses, especially in making ceramics and in building materials. Other clays are useful as lubricants in the muds used to cool the drill bits in oil-drilling rigs.

A sampling of the variety of silicates is shown in figure 2.10.

Nonsilicates

Just as the silicates, by definition, all contain silicon plus oxygen as part of their chemical compositions, each nonsilicate mineral group is defined by some chemical constituent or characteristic that all members of the group have in common. Most often, the common component is the same negatively charged ion or group of atoms. Discussion of some of the nonsilicate mineral groups with examples of common or familiar members of each follows. See also table 2.2.

The **carbonates** all contain carbon and oxygen combined in the proportions of one atom of carbon to three atoms of oxygen (written CO_3). The carbonate minerals all dissolve relatively easily, particularly in acids, and the oceans contain a great deal of dissolved carbonate. Geologically, the most important, most abundant carbonate mineral is calcite, which is

A

B

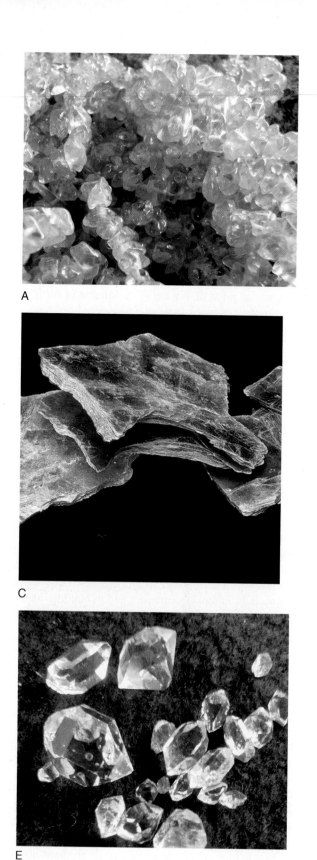

C

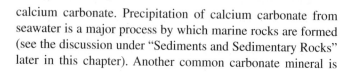

D

Figure 2.10

A collection of silicates: (A) Crystals of olivine that have been tumbled for use as a semiprecious gem, peridot; (B) tremolite, a variety of amphibole; (C) mica, showing cleavage between sheets of tetrahedra; (D) potassium feldspar; and (E) quartz.

Photograph (C) © The McGraw-Hill Companies, Inc./Bob Coyle, photographer; (D) © Doug Sherman/Geofile.

E

calcium carbonate. Precipitation of calcium carbonate from seawater is a major process by which marine rocks are formed (see the discussion under "Sediments and Sedimentary Rocks" later in this chapter). Another common carbonate mineral is dolomite, which contains both calcium and magnesium in approximately equal proportions. Carbonates may contain many other elements—iron, manganese, or lead, for example. The limestone and marble we use extensively for building and sculpture consist mainly of carbonates, generally calcite; calcite is also an important ingredient in cement.

The **sulfates** all contain sulfur and oxygen in the ratio of 1:4 (SO_4). A calcium sulfate—gypsum—is the most important, for it is both relatively abundant and commercially useful,

Table 2.2   Some Nonsilicate Mineral Groups*

Group	Compositional Characteristic	Examples
carbonates	metal(s) plus carbonate (1 carbon + 3 oxygen ions, CO_3)	calcite (calcium carbonate, $CaCO_3$)
		dolomite (calcium-magnesium carbonate, $CaMg(CO_3)_2$)
sulfates	metal(s) plus sulfate (1 sulfur + 4 oxygen ions, SO_4)	gypsum (calcium sulfate, with water, $CaSO_4 \cdot 2H_2O$)
		barite (barium sulfate, $BaSO_4$)
sulfides	metal(s) plus sulfur, without oxygen	pyrite (iron disulfide, FeS_2)
		galena (lead sulfide, PbS)
		cinnabar (mercury sulfide, HgS)
oxides	metal(s) plus oxygen	magnetite (iron oxide, Fe_3O_4)
		hematite (ferric iron oxide, Fe_2O_3)
		corundum (aluminum oxide, Al_2O_3)
		spinel (magnesium-aluminum oxide, $MgAl_2O_4$)
hydroxides	metal(s) plus hydroxyl (1 oxygen + 1 hydrogen ion, OH)	gibbsite (aluminum hydroxide, $Al(OH)_3$; found in aluminum ore)
		brucite (magnesium hydroxide, $Mg(OH)_2$; one ore of magnesium)
halides	metal(s) plus halogen element (fluorine, chlorine, bromine, or iodine)	halite (sodium chloride, $NaCl$)
		fluorite (calcium fluoride, CaF_2)
native elements	mineral consists of a single chemical element	gold (Au), silver (Ag), copper (Cu), sulfur (S), graphite (carbon, C)

*Other groups exist, and some complex minerals contain components of several groups (carbonate and hydroxyl groups, for example).

particularly as the major constituent in plaster of paris. Sulfates of many other elements, including barium, lead, and strontium, are also found.

When sulfur is present without oxygen, the resultant minerals are called **sulfides.** A common and well-known sulfide mineral is the iron sulfide *pyrite.* Pyrite (figure 2.5C, D) has also been called "fool's gold" because its metallic golden color often deceived early gold miners and prospectors into thinking they had struck it rich. Pyrite is not a commercial source of iron because there are richer ores of this metal. Nonetheless, sulfides comprise many economically important metallic ore minerals. An example that may be familiar is the lead sulfide mineral *galena,* which often forms in silver-colored cubes (figure 2.5A). The rich lead ore deposits near Galena, Illinois, gave the town its name. Sulfides of copper, zinc, and numerous other metals may also form valuable ore deposits (see chapter 13). Sulfides may also be problematic: when pyrite associated with coal is exposed by strip-mining and weathered, the result is sulfuric acid in runoff water from the mine.

Minerals containing just one or more metals combined with oxygen, and lacking the other elements necessary to classify them as silicates, sulfates, carbonates, and so forth, are the **oxides.** Iron combines with oxygen in different proportions to form more than one oxide mineral. One of these, magnetite, is, as its name suggests, magnetic, which is relatively unusual among minerals. Magnetic rocks rich in magnetite were known as lodestone in ancient times and were used as navigational aids like today's more compact compasses. Another iron oxide, hematite, may sometimes be silvery black but often has a red color and gives a reddish tint to many soils. Iron oxides on Mars's surface are responsible for that planet's orange hue. Many other oxide minerals also exist, including corundum, the aluminum oxide mineral mentioned earlier.

Native elements, as shown in table 2.2, are even simpler chemically than the other nonsilicates. Native elements are minerals that consist of a single chemical element, and the minerals' names are usually the same as the corresponding elements. Not all elements can be found, even rarely, as native elements. However, some of our most highly prized materials, such as gold, silver, and platinum, often occur as native elements. Diamond and graphite are both examples of native carbon. Sulfur may occur as a native element, either with or without associated sulfide minerals. Some of the richest copper ores contain native copper. Other metals that may occur as native elements include tin, iron, and antimony.

Interestingly, several of the factors that make the Earth unique in the solar system also increase its mineralogic diversity. Earth is large enough still to retain sufficient internal heat to keep churning and reprocessing crust and mantle. Abundant surface water not only allows for hydrous minerals and supports life on Earth; organisms, in turn, modify the chemistry of atmosphere and oceans, which creates additional mineral possibilities. Altogether, it is estimated that Mars is likely to have only about one-tenth as many different minerals as are found on the Earth, while other rocky planets and Earth's moon would have still fewer. Many of Earth's thousands of minerals are rare curiosities, but many others have become vital resources.

may become more rounded and thus not pack together very tightly or interlock as do the mineral grains in an igneous rock. Many clastic sedimentary rocks are therefore not particularly strong structurally, unless they have been extensively cemented.

Chemical sedimentary rocks form not from mechanical breakup and transport of fragments, but from crystals formed by precipitation or growth from solution. A common example is *limestone,* composed mostly of calcite (calcium carbonate). The chemical sediment that makes limestone may be deposited from fresh or salt water; under favorable chemical conditions, thick limestone beds, perhaps hundreds of meters thick, may form. Another example of a chemical sedimentary rock is *rock salt,* made up of the mineral halite, which is the mineral name for ordinary table salt (sodium chloride). A salt deposit may form when a body of salt water is isolated from an ocean and dries up.

Some chemical sediments have a large biological contribution. For example, many organisms living in water have shells or skeletons made of calcium carbonate or of silica (SiO_2, chemically equivalent to quartz). The materials of these shells or skeletons are drawn from the water in which the organisms grow. In areas where great numbers of such creatures live and die, the "hard parts"—the shells or skeletons—may pile up on the bottom, eventually to be buried and lithified. A sequence of sedimentary rocks may include layers of **organic sediments,** carbon-rich remains of living organisms; *coal* is an important example, derived from the remains of land plants that flourished and died in swamps.

Gravity plays a role in the formation of all sedimentary rocks. Mechanically broken-up bits of materials accumulate when the wind or water is moving too weakly to overcome gravity and move the sediments; repeated cycles of transport and deposition can pile up, layer by layer, a great thickness of sediment. Minerals crystallized from solution, or the shells of dead organisms, tend to settle out of the water under the force of gravity, and again, in time, layer on layer of sediment can build up. Layering, then, is a very common feature of sedimentary rocks and is frequently one way in which their sedimentary origins can be identified. Figure 2.12 shows several kinds of sedimentary rocks.

Sedimentary rocks can yield information about the settings in which the sediments were deposited. For example, the energetic surf of a beach washes away fine muds and leaves coarser sands and gravels; sandstone may, in turn, reflect the presence of an ancient beach. Distribution of glacier-deposited sediments of various ages contributes to our understanding not only of global climate change but also of plate tectonics (see chapter 3).

Metamorphic Rocks

The name *metamorphic* comes from the Greek for "changed form." A **metamorphic** rock is one that has formed from another, preexisting rock that was subjected to heat and/or pressure. The temperatures required to form metamorphic rocks are not as high as the temperatures at which the rocks would melt. Significant changes can occur in a solid rock at temperatures well below melting. Heat and pressure commonly cause the minerals in the rock to recrystallize. The original minerals may form larger crystals that perhaps interlock more tightly than before. Also, some

minerals may break down completely, while new minerals form under the new temperature and pressure conditions. Pressure may cause the rock to become deformed—compressed, stretched, folded, or compacted. All of this occurs while the rock is still solid; it does not melt during metamorphism.

The sources of the elevated pressures and temperatures of metamorphism are many. An important source of pressure is simply burial under many kilometers of overlying rock. The weight of the overlying rock can put great pressure on the rocks below. One source of elevated temperatures is the fact that temperatures increase with depth in the earth. In general, in most places, crustal temperatures increase at the rate of about 30°C per kilometer of depth (close to 60°F per mile)—which is one reason deep mines have to be air-conditioned! Deep in the crust, rocks are subjected to enough heat and pressure to show the deformation and recrystallization characteristic of metamorphism. Another heat source is a cooling magma. When hot magma formed at depth rises to shallower levels in the crust, it heats the adjacent, cooler rocks, and they may be metamorphosed; this is **contact metamorphism.** Metamorphism can also result from the stresses and heating to which rocks are subject during mountain-building or plate-tectonic movement. Such metamorphism on a large scale, not localized around a magma body, is **regional metamorphism.**

Any kind of preexisting rock can be metamorphosed. Some names of metamorphic rocks suggest what the earlier rock may have been. *Metaconglomerate* and *metavolcanic* describe, respectively, a metamorphosed conglomerate and a metamorphosed volcanic rock. *Quartzite* is a quartz-rich metamorphic rock, often formed from a very quartz-rich sandstone. The quartz crystals are typically much more tightly interlocked in the quartzite, and the quartzite is a more compact, denser, and stronger rock than the original sandstone. *Marble* is metamorphosed limestone in which the individual calcite grains have recrystallized and become tightly interlocking. The remaining sedimentary layering that the limestone once showed may be folded and deformed in the process, if not completely obliterated by the recrystallization.

Some metamorphic-rock names indicate only the rock's current composition, with no particular implication of what it was before. A common example is *amphibolite,* which can be used for any metamorphic rock rich in amphibole. It might have been derived from a sedimentary, metamorphic, or igneous rock of appropriate chemical composition; the presence of abundant metamorphic amphibole indicates moderately intense metamorphism and the fact that the rock is rich in iron and magnesium, but not the previous rock type.

Other metamorphic rock names describe the characteristic texture of the rock, regardless of its composition. Sometimes the pressure of metamorphism is not uniform in all directions; rocks may be compressed or stretched in a particular direction (*directed stress*). When you stamp on an aluminum can before tossing it in a recycling bin, you are technically subjecting it to a directed stress—vertical compression—and it flattens in that direction in response. In a rock subjected to directed stress, minerals that form elongated or platy crystals may line up parallel to each other. The resultant texture is described as **foliation,** from the Latin for "leaf" (as in the parallel leaves, or pages, of a book).

Figure 2.12

Sedimentary rocks, formed at low temperatures. (A) Limestone. (B) The fossils preserved in this limestone are *crinoids,* ancient echinoderms related to modern sea urchins and sea stars. (C) Shale, made of many fine layers of tiny clay particles. (D) Sandstone (note the rounding of larger grains). (E) Conglomerate, a coarser-grained rock similar to sandstone; many of the fragments here are rocks, not individual mineral grains. (F) Coal seams (dark layers) in a sequence of sedimentary rocks exposed in a sea cliff in southern Alaska.

Photograph (A) by I. J. Witkind, USGS Photo Library, Denver, CO.

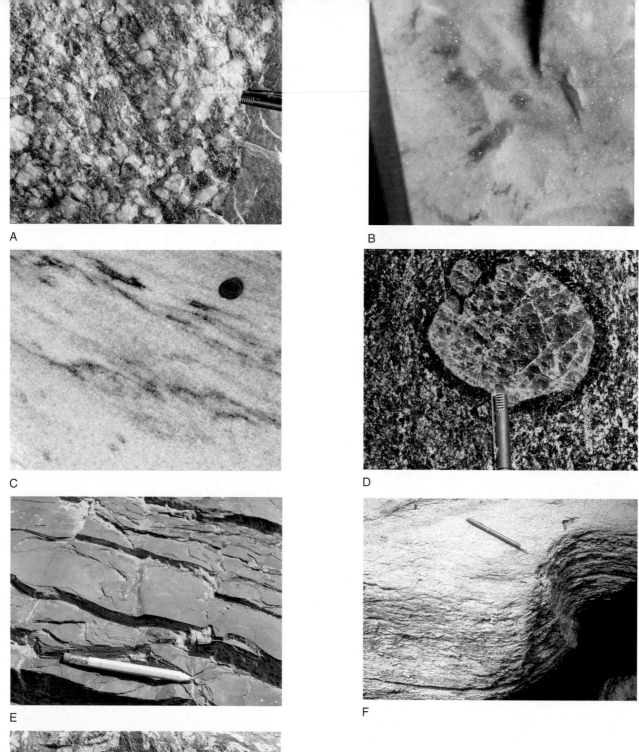

A

B

C

D

E

F

G

Figure 2.13

Metamorphic rocks have undergone mineralogical, chemical, and/
or structural change. (A) Metaconglomerate; note that it has
become so cohesive that the broken surface cuts through the
pebbles rather than breaking around them (as in figure 2.12E).
(B) Quartzite, metamorphosed sandstone; this sample shows the
glittering, "sugary" appearance due to recrystallizing of the quartz.
(C) Marble, metamorphosed limestone; look closely to see the
coarse, interlocking crystals of recrystallized calcium carbonate.
(D) Amphibolite (the dark crystals are amphibole). This sample also
contains large crystals of garnet, a common metamorphic mineral,
so it could be called a garnet amphibolite. (E) Slate, showing a tendency to split along parallel foliation planes. (F) Schist. (G) Gneiss.

Slate is a metamorphosed shale that has developed foliation under stress. The resulting rock tends to break along the foliation planes, parallel to the alignment of those minerals, and this characteristic makes it easy to break up into slabs for flagstones. The same characteristic is observed in *schist,* a coarser-grained, mica-rich metamorphic rock in which the mica flakes are similarly oriented. The presence of foliation can cause planes of structural weakness in the rock, affecting how it weathers and whether it is prone to slope failure or landslides. In other metamorphic rocks, different minerals may be concentrated in irregular bands, often with darker, ferromagnesian-rich bands alternating with light bands rich in feldspar and quartz. Such a rock is called a *gneiss* (pronounced "nice"). Because such terms as *schist* and *gneiss* are purely textural, the rock name can be modified by adding key features of the rock composition: "biotite-garnet schist," "granitic gneiss," and so on. Several examples of metamorphic rocks are illustrated in figure 2.13.

The Rock Cycle

It should be evident from the descriptions of the major rock types and how they form that rocks of any type can be transformed into rocks of another type or into another distinct rock of the same general type through the appropriate geologic processes. A sandstone may be weathered until it breaks up; its fragments may then be transported, redeposited, and lithified to form another sedimentary rock. It might instead be deeply buried, heated, and compressed, which could transform it into the metamorphic rock quartzite; or it could be heated until some or all of it melted, and from that melt, an igneous rock could be formed. Likewise, a schist could be broken up into small bits, forming a sediment that might eventually become sedimentary rock; more-intense metamorphism could transform it into a gneiss; or extremely high temperatures could melt it to produce a magma from which a granite could crystallize. Crustal rocks can be carried into the mantle and melted; fresh magma cools and crystallizes to form new rock; erosion and weathering processes constantly chip away at the surface. Note that the appearance (texture) of a rock can offer a good first clue to the conditions under which it (last) formed. A more comprehensive view of the links among different rock types is shown in generalized form in figure 2.14. In chapter 3, we will look at the rock cycle again, but in a somewhat different way, in the context of plate tectonics. Most interactions of people with the rock cycle involve the sedimentary and volcanic components of the cycle.

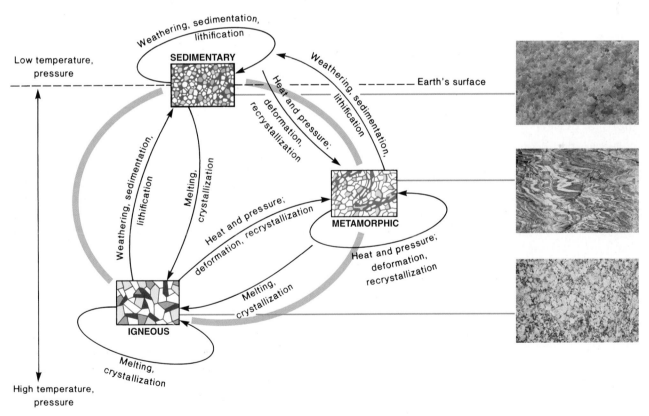

Figure 2.14

The rock cycle—a schematic view. Basically, a variety of geologic processes can transform any rock into a new rock of the same or a different class. The geologic environment is not static; it is constantly changing. The full picture, too, is more complex: any type of rock may be found at the earth's surface, and weathered; though the melts that form volcanic rocks are created at depth, the melts crystallize into rock near the surface; and so on.

Igneous rock photograph by N. K. Huber, USGS Photo Library, Denver, CO; other photographs © The McGraw-Hill Companies Inc./Doug Sherman, photographer.

Summary

The smallest possible unit of a chemical element is an atom. Isotopes are atoms of the same element that differ in atomic mass number; chemically, they are indistinguishable. Atoms may become electrically charged ions through the gain or loss of electrons. When two or more elements combine chemically in fixed proportions, they form a compound.

Minerals are naturally occurring inorganic solids, each of which is characterized by a particular composition and internal crystalline structure. They may be compounds or single elements. By far the most abundant minerals in the earth's crust and mantle are the silicates. These can be subdivided into groups on the basis of their crystal structures, by the ways in which the silicon and oxygen atoms are arranged. The nonsilicate minerals are generally grouped on the basis of common chemical characteristics.

Rocks are cohesive solids formed from rock or mineral grains or glass. The way in which rocks form determines how they are classified into one of three major groups: igneous rocks, formed from magma; sedimentary rocks, formed from low-temperature accumulations of particles or by precipitation from solution; and metamorphic rocks, formed from preexisting rocks through the application of heat and pressure. Through time, geologic processes acting on older rocks change them into new and different ones so that, in a sense, all kinds of rocks are interrelated. This concept is the essence of the rock cycle.

Key Terms and Concepts

anion 22	covalent bonding 23	lithification 34	proton 22
atom 22	crystalline 24	magma 34	regional metamorphism 36
atomic mass number 22	electron 22	metamorphic 36	rock 32
atomic number 22	ferromagnesian 29	mineral 24	rock cycle 33
carbonate 29	foliation 36	native element 31	sediment 34
cation 22	glass 34	neutron 22	sedimentary 34
chemical sedimentary rock 36	igneous 34	nucleus 22	silicate 28
clastic sedimentary rock 34	ion 22	organic sediments 36	sulfate 30
cleavage 27	ionic bonding 23	oxide 31	sulfide 31
compound 23	isotope 22	periodic table 23	volcanic 34
contact metamorphism 36	lava 34	plutonic 34	

Exercises

Questions for Review

1. Briefly define the following terms: *ion, isotope, compound, mineral,* and *rock.*

2. What two properties uniquely define a particular mineral?

3. Give the distinctive chemical characteristics of each of the following mineral groups: silicates, carbonates, sulfides, oxides, and native elements.

4. What is an igneous rock? How do volcanic and plutonic rocks differ in texture? Why?

5. What are the two principal classes of sedimentary rocks?

6. Describe how a granite might be transformed into a sedimentary rock.

7. Name several possible sources of the heat or pressure that can cause metamorphism. What kinds of physical changes occur in the rock as a result?

8. What is the rock cycle?

Exploring Further

1. The variety of silicate formulas arises partly from the different ways the silica tetrahedra are linked. Consider a single tetrahedron: If the silicon cation has a +4 charge, and the oxygen anion −2, there must be a net negative charge of −4 on the tetrahedron as a unit. Yet minerals must be electrically neutral. In quartz, this is accomplished by the sharing of all four oxygen atoms between tetrahedra, so only half the negative charge of each must be balanced by Si^{+4}, and there is a net of two oxygen atoms per Si (SiO_2). But in olivine, the tetrahedra share no oxygen atoms. If the magnesium (Mg) cation has a +2 charge, explain why the formula for a Mg-rich olivine is Mg_2SiO_4. Now consider a pyroxene, in which tetrahedra are linked in one dimension into chains, by the sharing of two of the four oxygen atoms. Again using Mg^{+2} for charge-balancing, show that the formula for a Mg-rich pyroxene would be $MgSiO_3$. Finally, consider the feldspars. They are framework silicates like quartz, but Al^{+3} substitutes for Si^{+4} in some of their tetrahedra. Explain how this allows feldspars to contain some sodium, potassium, or calcium cations and still remain electrically neutral.

2. Crystal skulls such as that featured in the movie *Indiana Jones and the Kingdom of the Crystal Skull* have been known to collectors and museums for over a century, but not everyone has believed in the alleged antiquity, or Mayan origin, of these artifacts. Think about how you might test such a skull for authenticity; then see the September 2008 issue of *Earth* for how experts approached this problem.

3. What kinds of rocks underlie your region of the country? (Your local geological survey could assist in providing the information.) Many states have a state mineral and/or rock. Does yours? If so, what is it, and why was it chosen?

Plate Tectonics

Several centuries ago, observers looking at global maps noticed the similarity in outline of the eastern coast of South America and the western coast of Africa (figure 3.1). In 1855, Antonio Snider went so far as to publish a sketch showing how the two continents could fit together, jigsaw-puzzle fashion. Such reconstructions gave rise to the bold suggestion that perhaps these continents had once been part of the same landmass, which had later broken up.

Climatologist Alfred Wegener was struck not only by the matching coastlines, but by geologic evidence from the continents. Continental rocks form under a wide range of conditions and yield varied kinds of information. Sedimentary rocks, for example, may preserve evidence of the ancient climate of the time and place in which the sediments were deposited. Such evidence shows that the climate in many places has varied widely

through time. We find evidence of extensive glaciation in places now located in the tropics, in parts of Australia, southern Africa, and South America (figure 3.2). There are desert sand deposits in the rocks of regions that now have moist, temperate climates and the remains of jungle plants in now-cool places. There are coal deposits in Antarctica, even though coal deposits form from the remains of a lush growth of land plants. These observations, some of which were first made centuries ago, cannot be explained as the result of global climatic changes, for the continents do not all show the same warming or cooling trends at the same time. However, climate is to a great extent a function of latitude, which strongly influences surface temperatures: Conditions are generally warmer near the equator and colder near the poles. Dramatic shifts in an individual continent's climate might result from changes in its latitude in the course of continental drift.

Better understanding of plate tectonics leads to better understanding of the where and why of earthquakes. Damage in Port-au-Prince after the 2010 Haiti earthquake.

Image by Marko Kokic, IFRC, courtesy NOAA/National Geophysical Data Center.

Figure 3.4

Global relief map of the world, including seafloor topography. Note ridges and trenches on the sea floor.

Source: Image courtesy of NOAA/National Geophysical Data Center.

A hot magma is therefore not magnetic, but as it cools and so-lidifies, and iron-bearing magnetic minerals (particularly magne-tite) crystallize from it, those magnetic crystals tend to line up in the same direction. Like tiny compass needles, they align them-selves parallel to the lines of force of the earth's magnetic field, which run north-south, and they point to the magnetic north pole (figure 3.5). They retain their internal magnetic orientation unless they are heated again. This is the basis for the study of **paleomag-netism,** "fossil magnetism" in rocks.

Magnetic north, however, has not always coincided with its present position. In the early 1900s, scientists investigating the direction of magnetization of a sequence of volcanic rocks in France discovered some flows that appeared to be magne-tized in the opposite direction from the rest: their magnetic min-erals pointed south instead of north. Confirmation of this discovery in many places around the world led to the suggestion in the late 1920s that the earth's magnetic field had "flipped," or reversed polarity; that is, that the north and south poles had switched places. During the time those surprising rocks had crystallized, a compass needle would have pointed to the mag-netic south pole, not north.

Today, the phenomenon of magnetic reversals is well documented. Rocks crystallizing at times when the earth's field was in the same orientation as it is at present are said to be nor-mally magnetized; rocks crystallizing when the field was ori-ented the opposite way are described as reversely magnetized. Over the history of the earth, the magnetic field has reversed many times, at variable intervals. Sometimes, the polarity re-mained constant for millions of years before reversing, while other reversals are separated by only a few tens of thousands of years. Through the combined use of magnetic measurements and age determinations on the magnetized rocks, geologists have been able to reconstruct the reversal history of the earth's magnetic field in some detail.

The explanation for magnetic reversals must be related to the origin of the magnetic field. The outer core is a metallic fluid, consisting mainly of iron. Motions in an electrically conducting fluid can generate a magnetic field, and this is believed to be the origin of the earth's field. (The simple pres-ence of iron in the core is not enough to account for the magnetic field, as core temperatures—over 3500°C—are far above the Curie temperature of iron, 770°C.) Perturbations or

Paleomagnetism and Seafloor Spreading

The ocean floor is made up largely of basalt, a volcanic rock rich in ferromagnesian minerals. During the 1950s, the first large-scale surveys of the magnetic properties of the sea floor produced an entirely unexpected result. As they tracked across a ridge, they recorded alternately stronger and weaker magnetism below.

At first, this seemed so incredible that it was assumed that the instruments or measurements were faulty. However, other studies consistently obtained the same results. For several years, geoscientists strove to find a convincing explanation for these startling observations.

Then, in 1963, an elegant explanation was proposed by the team of F. J. Vine and D. H. Matthews, and independently by L. W. Morley. The magnetic stripes could be explained as a result of alternating bands of normally and reversely magnetized rocks on the sea floor. Adding the magnetism of normally magnetized rocks to earth's current field produced a little stronger magnetization; reversely magnetized rocks would counter the earth's (much stronger) field somewhat, resulting in apparently lower net magnetic strength.

These bands or "stripes" of normally and reversely magnetized rocks were also parallel to, and symmetrically arranged on either side of, the seafloor ridges. Thus was born the concept of **seafloor spreading,** the parting of seafloor rocks at the ocean ridges. If the oceanic plates split and move apart, a rift begins to open, and mantle material from below can flow upward. The resulting lowering of pressure allows extensive melting to occur in this material (as will be explored in chapter 5), and magma rises up into the rift to form new sea floor. As the magma cools and solidifies to form new basaltic rock, that rock becomes magnetized in the prevailing direction of the earth's magnetic field. If the plates continue to move apart, the new rock will also split and part, making way for more magma to form still younger rock, and so on.

If the polarity of the earth's magnetic field reverses during the course of seafloor spreading, the rocks formed after a reversal are polarized oppositely from those formed before it. The ocean floor is a continuous sequence of basalts formed over tens or hundreds of millions of years, during which time there have been dozens of polarity reversals. The basalts of the sea floor have acted as a sort of magnetic tape recorder throughout that time, preserving a record of polarity reversals in the alternating bands of normally and reversely magnetized rocks. The process is illustrated schematically in figure 3.6.

Age of the Ocean Floor

The ages of seafloor basalts themselves lend further support to this model of seafloor spreading. Specially designed research ships can sample sediment from the deep-sea floor and drill into the basalt beneath.

The time at which an igneous rock, such as basalt, crystallized from its magma can be determined by methods described in appendix A. When this is done for many samples of seafloor basalt, a pattern emerges. The rocks of the sea floor are youngest close to the ocean ridges and become progressively older the farther away they are from the ridges on either side (see figure 3.7).

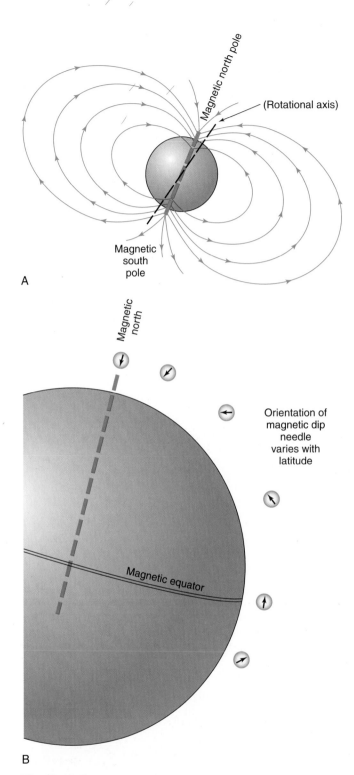

Figure 3.5

(A) Lines of force of earth's magnetic field. (B) A magnetized needle—or magnetic mineral's magnetism—will dip (be deflected from the horizontal) more steeply nearer the poles, just as the lines of force of the earth's field do.

changes in the fluid motions, then, could account for reversals of the field. The details of the reversal process remain to be determined.

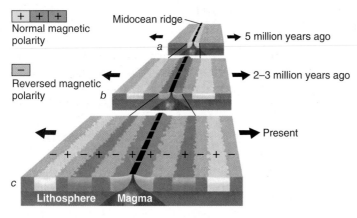

+ + +
Normal magnetic polarity

−
Reversed magnetic polarity

Midocean ridge
a → 5 million years ago
b → 2–3 million years ago
→ Present
− + − + − + − + − + − + − + − +
c
Lithosphere   Magma

Figure 3.6

Seafloor spreading leaves a record of bands of normally and reversely magnetized rocks on the sea floor. Different colors of normally polarized rocks are used here to emphasize their different crystallization ages.

After USGS publication This Dynamic Earth *by W. J. Kious and R. I. Tilling.*

Like the magnetic stripes, the age pattern is symmetric across each ridge. As seafloor spreading progresses, previously formed rocks are continually spread apart and moved farther from the ridge, while fresh magma rises to form new sea floor at the ridge. The oldest rocks recovered from the sea floor, well away from active ridges, are about 200 million years old.

Ages of sediments from the ocean basins reinforce this age pattern. Logically, sea floor must form before sediment can accumulate on it, and unless the sediment is disturbed, younger sediments then accumulate on older ones. When cores of oceanic sediment are collected and the ages of the oldest (deepest) sediments examined, it is found that those deepest sediments are older at greater distances from the seafloor ridges, and that only quite young sediments have been deposited close to the ridges.

Polar-Wander Curves

Evidence for plate movements does not come only from the sea floor. For reasons outlined later in the chapter, much older rocks are preserved on the continents than in the ocean—some continental samples are over 4 billion years old—so longer periods of earth history can be investigated through continental rocks. Studies of paleomagnetic orientations of continental rocks can span many hundreds of millions of years and yield quite complex data.

The lines of force of earth's magnetic field not only run north-south, they vary in dip with latitude: vertical at the magnetic poles, horizontal at the equator, at varying intermediate

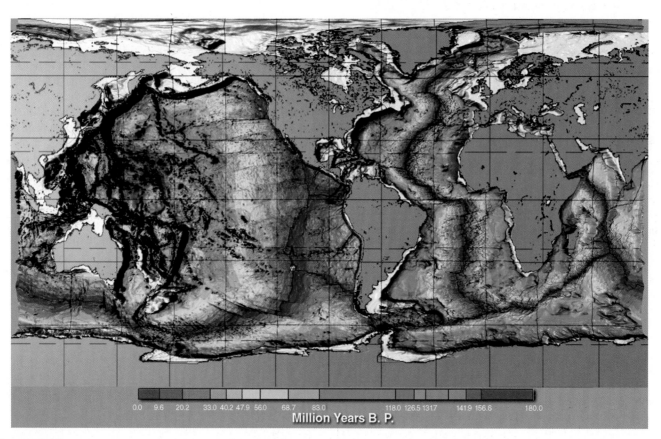

0.0 9.6 20.2 33.0 40.2 47.9 56.0 68.7 83.0 118.0 126.5 131.7 141.9 156.6 180.0
Million Years B. P.

Figure 3.7

Age distribution of the sea floor superimposed on a shaded relief map. ("B.P." means "before present.") Note relative spreading rates of Mid-Atlantic Ridge (slower) and East Pacific Rise (faster), shown by wider color bands in the Pacific.

Source: Marine Geology and Geophysics Division of the NOAA National Geophysical Data Center.

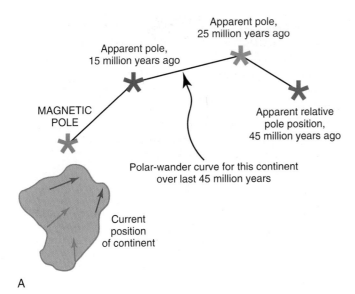

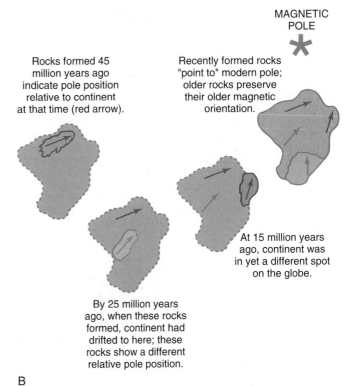

Figure 3.8

(A) Assuming a stationary continent, the shifting relative pole positions would suggest "polar wander." In fact, "polar wander curves" actually reflect wandering continents, attached to moving plates: (B) As rocks crystallize, their magnetic minerals align with the contemporary magnetic field. But continental movement changes the relative position of continent and magnetic pole over time.

dips in between, as was shown in figure 3.5B. When the orientation of magnetic minerals in a rock is determined in three dimensions, one can thus determine not only the direction of magnetic north, but (magnetic) latitude as well, or how far removed that region was from magnetic north at the time the rock's magnetism assumed its orientation.

Magnetized rocks of different ages on a single continent may point to very different apparent magnetic pole positions. The magnetic north and south poles may not simply be reversed, but may be rotated or tilted from the present magnetic north and south. When the directions of magnetization and latitudes of many rocks of various ages from one continent are determined and plotted on a map, it appears that the magnetic poles have meandered far over the surface of the earth—*if* the position of the continent is assumed to have been fixed on the earth throughout time. The resulting curve, showing the apparent movement of the magnetic pole relative to the continent as a function of time, is called the **polar-wander curve** for that continent (figure 3.8). We know now, however, that it isn't the poles that have "wandered" so much.

From their discovery, polar-wander curves were puzzling because there are good geophysical reasons to believe that the earth's magnetic poles should remain close to the geographic (rotational) poles. In particular, the fluid motions in the outer core that cause the magnetic field could be expected to be strongly influenced by the earth's rotation. Modern measurements do indicate some shifting in magnetic north relative to the north pole (geographic north), but not nearly as much as indicated by some polar-wander curves.

Additionally, the polar-wander curves for different continents do not match. Rocks of exactly the same age from two dif-

ferent continents may seem to point to two very different magnetic poles (figure 3.9). This confusion can be eliminated, however, if it is assumed that the magnetic poles have always remained close

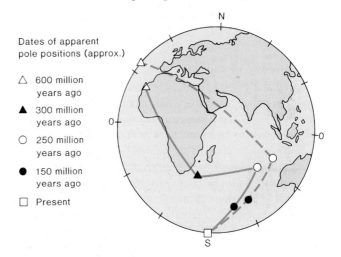

Figure 3.9

Examples of polar-wander curves. The apparent position of the magnetic pole relative to the continent is plotted for rocks of different ages, and these data points are connected to form the curve for that landmass. The present positions of the continents are shown for reference. Polar-wander curves for Africa (solid line) and Arabia (dashed line) suggest that these landmasses moved quite independently up until about 250 million years ago, when the polar-wander curves converge.

After M. W. McElhinny, Paleomagnetism and Plate Tectonics. *Copyright © 1973 by Cambridge University Press, New York, NY. Reprinted by permission.*

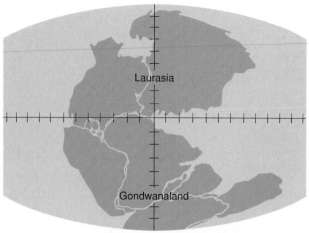

A 200 million years ago

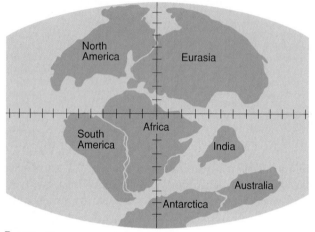

B 100 million years ago

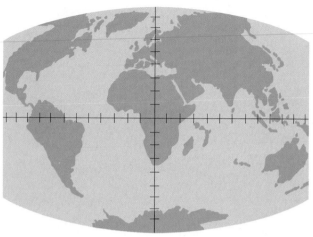

C Today

Figure 3.10

Reconstructed plate movements during the last 200 million years. *Laurasia* and *Gondwanaland* are names given to the northern and southern portions of Pangaea, respectively. Although these changes occur very slowly in terms of a human lifetime, they illustrate the magnitude of some of the natural forces to which we must adjust. (A) 200 million years ago. (B) 100 million years ago. (C) Today.

After R. S. Dietz and J. C. Holden, "Reconstruction of Pangaea," Journal of Geophysical Research, *75:4939–4956, 1970, copyright by the American Geophysical Union.*

to the geographic poles but that the *continents* have moved and rotated. The polar-wander curves then provide a way to map the directions in which the continents have moved through time, relative to the stationary poles and relative to each other.

Efforts to reconstruct the ancient locations and arrangements of the continents have been relatively successful, at least for the not-too-distant geologic past. It has been shown, for example, that a little more than 200 million years ago, there was indeed a single, great supercontinent, the one that Wegener envisioned and named Pangaea. The present seafloor spreading ridges are the lithospheric scars of the subsequent breakup of Pangaea (figure 3.10). However, they are not the only kind of boundary found between plates, as we will see later in the chapter.

Plate Tectonics—Underlying Concepts

As already noted, a major obstacle to accepting the concept of continental drift was imagining solid continents moving over solid earth. However, the earth is not rigidly solid from the surface to the center of the core. In fact, a plastic zone lies relatively close to

the surface. A thin shell of relatively rigid rock can move over this plastic layer below. The existence of plates, and the occurrence of earthquakes in them, reflect the way rocks respond to stress.

Stress and Strain in Geologic Materials

An object is under **stress** when force is being applied to it. **Compressive stress** squeezes or compresses the object; squeezing a sponge subjects it to compressive stress, and rocks deep in the crust are under compressive stress from the weight of rocks above them. **Tensile stress** stretches an object or pulls it apart; a stretched guitar or violin string being tuned is under tensile stress, as are rocks at a seafloor spreading ridge as plates move apart. A **shearing stress** is one that tends to cause different parts of the object to move in different directions across a plane or to slide past one another, as when a deck of cards is spread out on a tabletop by a sideways sweep of the hand.

Strain is deformation resulting from stress. It may be either temporary or permanent, depending on the amount and type of stress and on the physical properties of the material. If **elastic deformation** occurs, the amount of deformation is proportional to the stress applied (straight-line segments in

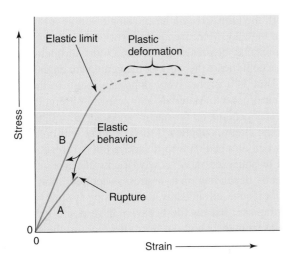

Figure 3.11

A stress-strain diagram. Elastic materials deform in proportion to the stress applied (straight-line segments). A very brittle material (curve A) may rupture before the elastic limit is reached. Other materials subjected to stresses above their elastic limits deform plastically until rupture (curve B).

figure 3.11), and the material returns to its original size and shape when the stress is removed. A gently stretched rubber band or a tennis ball that is squeezed or hit by a racket shows elastic behavior: release the stress and the object returns to its original dimensions. Rocks, too, may behave elastically, although much greater stress is needed to produce detectable strain. Once the **elastic limit** of a material is reached, the material may go through a phase of **plastic deformation** with increased stress (dashed section of line B in figure 3.11). During this stage, relatively small added stresses yield large corresponding strains, and the changes are permanent: the material does not return to its original size and shape after removal of the stress. A glassblower, an artist shaping clay, a carpenter fitting caulk into cracks around a window, and a blacksmith shaping a bar of hot iron into a horseshoe are all making use of the plastic behavior of materials. In rocks, folds result from plastic deformation (figure 3.12). A **ductile** material is one that can undergo extensive plastic deformation without breaking.

If stress is increased sufficiently, most solids eventually break, or **rupture.** In **brittle** materials, rupture may occur before there is any plastic deformation. Brittle behavior is characteristic of most rocks at the low temperatures and pressures near earth's surface. It leads to faults and fractures (figure 3.13), which will be explored further in chapter 4. Graphically, brittle behavior is illustrated by line A in figure 3.11.

Different types of rocks may tend to be more or less brittle or ductile, but other factors influence their behavior as well. One is temperature: All else being equal, rocks tend to behave more plastically at higher temperatures than at lower ones. The effect of temperature can be seen in the behavior of cold and hot glass. A rod of cold glass is brittle and snaps under stress before appreciable plastic deformation occurs, while a warmed glass rod may be bent and twisted without breaking. Pressure is an-

A

B

Figure 3.12

(A) Folding such as this occurs while rocks are deeper in the crust, at elevated temperatures and pressures. (B) The tightly folded marbles and slates of this roadcut were originally flat-lying layers of limestone and shale, later metamorphosed and deformed. The fracture across the bottom of this picture could have been created during blasting of the roadcut, when the rocks were at the surface, colder and more brittle.

(A) Photograph by M. R. Mudge, USGS Photo Library, Denver, CO.

other factor influencing rock behavior. *Confining pressure* is that uniform pressure which surrounds a rock at depth. Higher confining pressure also tends to promote more-plastic, less-brittle behavior in rocks. Confining pressure increases with depth in the earth, as the weight of overlying rock increases;

A

B

Figure 3.13

(A) This large fault in Cook Inlet, Alaska, formed during a collision of continental landmasses. (B) This rock suffered brittle failure under stress, which produced a much smaller fault.

(A) Photograph by N. J. Silberling, (B) Photograph by W. B. Hamilton, both courtesy USGS Photo Library, Denver, CO.

temperature likewise generally increases with depth. Thus, rocks such as the gneiss of figure 2.13G, metamorphosed deep in the crust, commonly show the folds of plastic deformation.

Rocks also respond differently to different types of stress; most are far stronger under compression than under tension. Natural samples may be weakened by fracturing or weathering.

In short, the term *strength* has no single simple meaning when applied to rocks. Moreover, several rock types of quite different properties may be present at the same site. Considerations such as these complicate the work of the geological engineer.

Time is a further, very important factor in the physical behavior of a rock. Materials may respond differently to given stresses, depending on the rate of stress, the period of time over which the stress is applied. A ball of putty will bounce elastically when dropped on a hard surface but deform plastically if pulled or squeezed more slowly. Rocks can likewise respond differently depending on how stress is applied, showing elastic behavior if stressed suddenly, as by passing seismic waves from an earthquake, but ductile behavior in response to prolonged stress, as from the weight of overlying rocks or the slow movements of plates over the longer span of geologic time.

Lithosphere and Asthenosphere

The earth's crust and uppermost mantle are somewhat brittle and elastic. Together they make up the outer solid layer of the earth called the **lithosphere,** from the Greek word *lithos,* meaning "rock." The lithosphere varies in thickness from place to place on the earth. It is thinnest underneath the oceans, where it extends to a depth of about 50 kilometers (about 30 miles). The lithosphere under the continents is both thicker on average than is oceanic lithosphere, and more variable in thickness, extending in places to about 250 kilometers (over 150 miles).

The layer below the lithosphere is the **asthenosphere,** which derives its name from the Greek word *asthenes,* meaning "without strength." The asthenosphere extends to an average depth of about 300 kilometers (close to 200 miles) in the mantle. Its lack of strength or rigidity results from a combination of high temperatures and moderate confining pressures that allows the rock to flow plastically under stress. Below the asthenosphere, as pressures increase faster than temperatures with depth, the mantle again becomes more rigid and elastic.

The asthenosphere was discovered by studying the behavior of seismic waves from earthquakes. Its presence makes the concept of continental drift more plausible. The continents need not scrape across or plow through solid rock; instead, they can be pictured as sliding over a softened, deformable layer underneath the lithospheric plates. The relationships among crust, mantle, lithosphere, and asthenosphere are illustrated in figure 3.14.

Recognition of the existence of the plastic asthenosphere made plate motions more plausible, but it did not prove that they had occurred. For most scientists to accept the concept of plate tectonics required the gathering of much additional information, such as that described in the last section.

Locating Plate Boundaries

The distribution of earthquakes and volcanic eruptions indicates that these phenomena are far from uniformly distributed over the earth (figures 4.6, 5.7). They are, for the most part, concentrated in belts or linear chains. This is consistent with the idea that the rigid shell of lithosphere is cracked in places, broken up into pieces, or plates. The volcanoes and earthquakes are concentrated

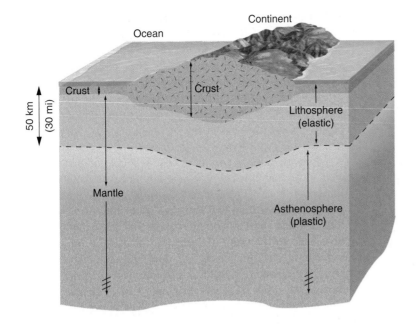

Figure 3.14

The outer zones of the earth (not to scale). The terms *crust* and *mantle* have compositional implications (see chapter 1); *lithosphere* and *asthenosphere* describe physical properties (the former, elastic and more rigid; the latter, plastic). The lithosphere includes the crust and uppermost mantle. The asthenosphere lies entirely within the upper mantle. Below it, the rest of the mantle is more rigidly solid again.

at the boundaries of these lithospheric plates, where plates jostle or scrape against each other. (The effect is somewhat like ice floes on an arctic sea: most of the grinding and crushing of ice and the spurting-up of water from below occur at the edges of the blocks of ice, while their solid central portions are relatively undisturbed.) Fewer than a dozen very large lithospheric plates have been identified (figure 3.15); as research continues, many smaller ones have been recognized in addition.

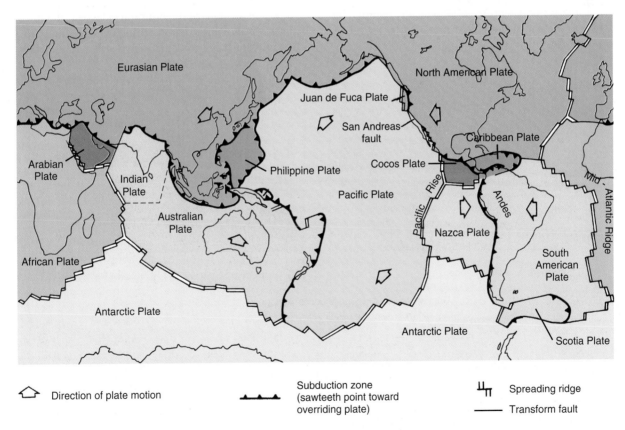

Figure 3.15

Principal world lithospheric plates. Arrows denote approximate relative directions of plate movement. The types of plate boundaries are discussed later in the chapter. (The nature of the recently recognized boundary between the Indian and Australian plates is still being determined.)

Source: After W. Hamilton, U.S. Geological Survey.

Types of Plate Boundaries

Different things happen at the boundaries between two lithospheric plates, depending in part on the relative motions of the plates, and in part on whether continental or oceanic lithosphere is at the edge of each plate where they meet.

Divergent Plate Boundaries

At a **divergent plate boundary,** lithospheric plates move apart, subject to tensional stress. The release of pressure facilitates some melting in the asthenosphere, and magma wells up from the asthenosphere, its passage made easier by deep fractures formed by the tensional stresses. A great deal of volcanic activity thus occurs at divergent plate boundaries. In addition, the pulling-apart of the plates of lithosphere results in earthquakes along these boundaries.

Seafloor spreading ridges are the most common type of divergent boundary worldwide, and we have already noted the formation of new oceanic lithosphere at these ridges. (Because lithosphere is created here, these are sometimes described as *constructive* plate boundaries.) As seawater circulates through this fresh, hot lithosphere, it is heated and reacts with the rock, becoming metal-rich. As it gushes back out of the sea floor, cooling and reacting with cold seawater, it may precipitate potentially valuable mineral deposits. Economically unprofitable to mine now, they may nevertheless prove to be useful resources for the future.

Continents can be rifted apart, too, and in fact most ocean basins are believed to have originated through continental rifting, as shown in figure 3.16. The process may be initiated either by tensional forces pulling the plates apart, or by rising hot asthenosphere along the rift zone. In the early stages of continental rifting, volcanoes may erupt along the rift, or great flows of basaltic lava may pour out through the fissures in the continent. If the rifting continues, a new ocean basin will eventually form between the pieces of the continent. This is happening now in northeast Africa, where three rift zones meet in what is called a *triple junction.* The Afar depression in Ethiopia (figure 3.17) is, in fact, a rift valley; note the valley's parallel edges. Along two branches—which have formed the Red Sea and the Gulf of Aden—continental separation has proceeded to the point of formation of oceanic lithosphere. More-limited continental rifting, now believed to have halted, created the zone of weakness in central North America that includes the New Madrid fault zone, still a seismic hazard, which will be discussed in chapter 4. The upper Rio Grande River flows south through New Mexico along the depression formed by the Rio Grande Rift, which one day (in a few million years) may become a new ocean basin.

Convergent Plate Boundaries

At a **convergent plate boundary,** as the name indicates, plates are moving toward each other, subjecting the rocks in the collision zone to compressive stress. The details of just

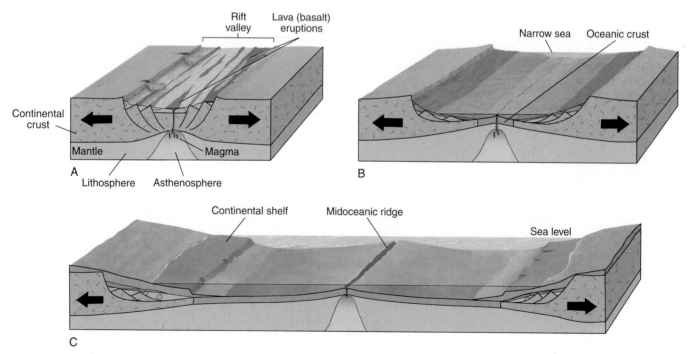

Figure 3.16

As continental rifting begins, crust is stretched, thinned, and fractured (A). Eventually, the continental pieces are fully separated, and oceanic crust is formed between them (B). The ocean basin widens as divergence continues (C).

Figure 3.17

Continental crust has been rifted, and blocks of crust have dropped into the gap to form the Afar depression. North-south cracks at bottom of image are also due to tension and rifting.

Image courtesy NASA and U.S./Japan ASTER Science Team.

oceanic lithosphere is less buoyant and more easily forced down into the asthenosphere as plates move together.

Most commonly, oceanic lithosphere is at the leading edge of one or both of the converging plates. One plate of oceanic lithosphere may be pushed under the other plate and descend into the asthenosphere. This type of plate boundary, where one plate is carried down below (subducted beneath) another, is called a **subduction zone** (see figure 3.18A,B). The nature of subduction zones can be demonstrated in many ways, a key one of which involves earthquake depths, as described in chapter 4. As the down-going slab is subjected to higher pressures deeper in the mantle, the rocks may be metamorphosed into denser ones, and this, in turn, will promote more subduction as gravity pulls on the denser lithosphere.

The subduction zones of the world balance the seafloor equation. If new oceanic lithosphere is constantly being created at spreading ridges, an equal amount must be destroyed somewhere, or the earth would simply keep getting bigger. This "excess" sea floor is consumed in subduction zones, which are sometimes also described as *destructive* plate boundaries. The subducted plate is heated by the hot asthenosphere. Fluids are released from it into the overlying mantle; some of it may become hot enough to melt; the dense, cold residue eventually breaks off and sinks deeper into the mantle. Meanwhile, at the spreading ridges, other melts rise, cool, and crystallize to make new sea floor. So, in a sense, the oceanic lithosphere is constantly being recycled, which explains why few very ancient seafloor rocks are known. Subduction also accounts for the fact that we do not always see the whole symmetric seafloor-age pattern around a ridge.

what happens depend on what sort of lithosphere is at the leading edge of each plate; one may have ocean-ocean, ocean-continent, or continent-continent convergence. Continental crust is relatively low in density, so continental lithosphere is therefore buoyant with respect to the dense, iron-rich mantle, and it tends to "float" on the asthenosphere. Oceanic crust is more similar in density to the underlying asthenosphere, so

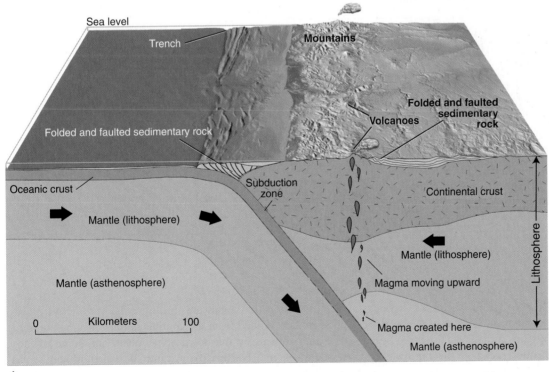

Figure 3.18

(A) At ocean-continent convergence, sea floor is consumed, and volcanoes form on the overriding continent. Here the trench is partially filled by sediment. (B) Volcanic activity at ocean-ocean convergence creates a string of volcanic islands. (C) Sooner or later, a continental mass on the subducting plate meets a continent on the overriding plate, and the resulting collision creates a great thickness of continental lithosphere, as in the Himalayas (D).

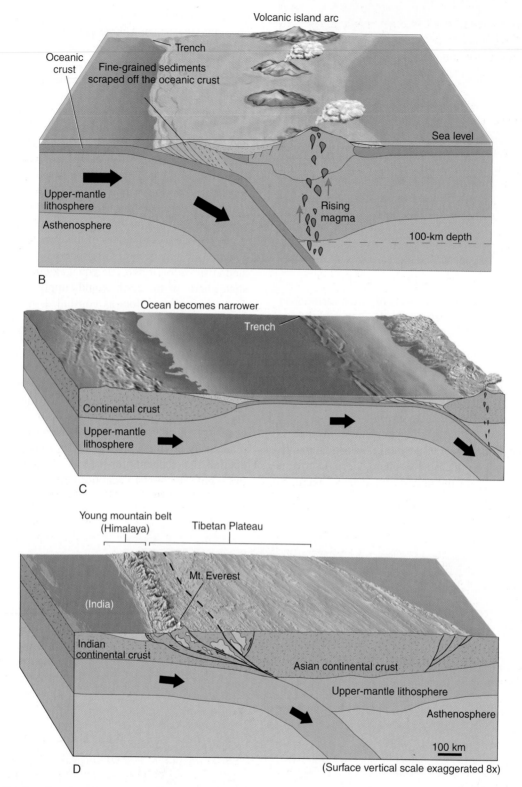

Figure 3.18 *Continued*

Consider the East Pacific Rise (ridge) in figure 3.7. Subduction beneath the Americas has consumed much of the sea floor east of the rise, so there is no longer older sea floor east of the rise to correspond to the older rocks of the northwestern Pacific Ocean floor.

Subduction zones are, geologically, very active places. Sediments eroded from the continents may accumulate in the trench formed by the down-going plate, and some of these sediments may be carried down into the asthenosphere to contribute to melts being produced there, as described further in chapter 5. Volcanoes form where molten material rises up through the overlying plate to the surface. At an ocean-ocean convergence, the result is commonly a line of volcanic islands, an **island arc** (see figure 3.18B). The great stresses involved in convergence and subduction give rise to numerous earthquakes. The bulk of mountain building, and its associated volcanic and earthquake activity, is related to subduction zones. Parts of the world near or above modern subduction zones are therefore prone to both volcanic and earthquake activity. These include the Andes region of South America, western Central America, parts of the northwest United States and Canada, the Aleutian Islands, China, Japan, and much of the rim of the Pacific Ocean basin, sometimes described as the "Ring of Fire" for its volcanism.

Only rarely is a bit of sea floor caught up in a continent during convergence and preserved. Most often, the sea floor in a zone of convergence is subducted and destroyed. The buoyant continents are not so easily reworked in this way; hence, very old rocks may be preserved on them. Because all continents are part of moving plates, sooner or later they all are inevitably transported to a convergent boundary, as leading oceanic lithosphere is consumed.

If there is also continental lithosphere on the plate being subducted at an ocean-continent convergent boundary, consumption of the subducting plate will eventually bring the continental masses together (figure 3.18C). The two landmasses collide, crumple, and deform. One may partially override the other, but the buoyancy of continental lithosphere ensures that neither sinks deep into the mantle, and a very large thickness of continent may result. Earthquakes are frequent during continent-continent collision as a consequence of the large stresses involved in the process. The extreme height of the Himalaya Mountains is attributed to just this sort of collision. India was not always a part of the Asian continent. Paleomagnetic evidence indicates that it drifted northward over tens of millions of years until it "ran into" Asia (recall figure 3.10), and the Himalayas were built up in the process (figure 3.18D). Earlier, the ancestral Appalachian Mountains were built in the same way, as Africa and North America converged prior to the breakup of Pangaea. In fact, many major mountain ranges worldwide represent sites of sustained plate convergence in the past, and much of the western portion of North America consists of bits of continental lithosphere "pasted onto" the continent in this way (figure 3.19). This process accounts for the juxtaposition of rocks that are quite different in age and geology.

Transform Boundaries

The actual structure of a spreading ridge is more complex than a single, straight crack. A close look at a mid-ocean spreading ridge reveals that it is not a continuous rift thousands of kilometers long. Rather, ridges consist of many short segments slightly offset from one another. The offset is a special kind of fault, or break in the lithosphere, known as a **transform fault** (see figure 3.20A). The opposite sides of a transform fault belong to two different plates, and these are moving in opposite directions in a shearing motion. As the plates scrape past each other, earthquakes occur along the transform fault. Transform faults may also occur between a trench (subduction zone) and a spreading ridge, or between two trenches (figure 3.20B), but these are less common.

The famous San Andreas fault in California is an example of a transform fault. The East Pacific Rise disappears under the edge of the continent in the Gulf of California. Along the northwest coast of the United States, subduction is consuming the small Juan de Fuca Plate (figure 3.15).The San Andreas is the transform fault between the subduction zone and the spreading ridge. Most of North America is part of the North American Plate. The thin strip of California on the west side of the San Andreas fault, however, is moving northwest with the Pacific Plate. The stress resulting from the shearing displacement across the fault leads to ongoing earthquake activity.

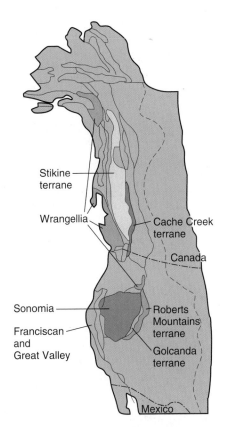

Figure 3.19

A *terrane* is a region of rocks that share a common history, distinguished from nearby, but genetically unrelated, rocks. Geologists studying western North America have identified and named many different terranes, some of which are shown here. A terrane that has been shown—often by paleomagnetic evidence—to have been transported and added onto a continent from some distance away is called an *accreted terrane*.

After U.S. Geological Survey Open-File Map 83-716.

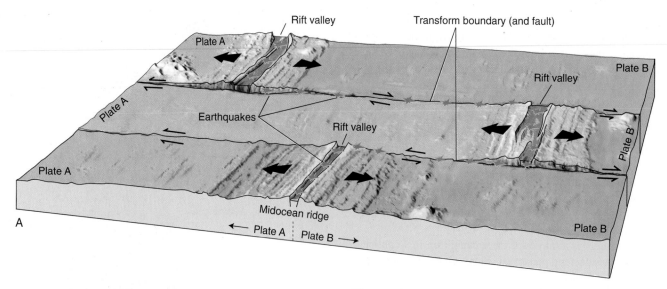

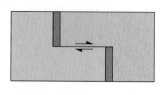
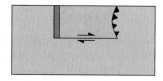
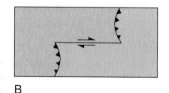

B

Figure 3.20

(A) Seafloor spreading-ridge segments are offset by transform faults. Red asterisks show where earthquakes occur in the ridge system. (B) Types of transform boundaries. Red segments are spreading ridges; "sawteeth" point to the overriding plate at a subduction boundary.

Past Motions, Present Velocities

Rates and directions of plate movement have been determined in a variety of ways. As previously discussed, polar-wander curves from continental rocks can be used to determine how the continents have shifted. Seafloor spreading is another way of determining plate movement. The direction of seafloor spreading is usually obvious: away from the ridge. Rates of seafloor spreading can be found very simply by dating rocks at different distances from the spreading ridge and dividing the distance moved by the rock's age (the time it has taken to move that distance from the ridge at which it formed). For example, if a 10-million-year-old piece of sea floor is collected at a distance of 100 kilometers from the ridge, this represents an average rate of movement over that time of 100 kilometers/ 10 million years. If we convert that to units that are easier to visualize, it works out to about 1 centimeter (a little less than half an inch) per year.

Another way to monitor rates and directions of plate movement is by using mantle **hot spots.** These are isolated areas of volcanic activity usually not associated with plate boundaries. They are attributed to columns of warm mantle material *(plumes),* perhaps originating at the base of the mantle (figure 3.21), that rise up through colder, denser mantle. Reduction in pressure as a plume rises can lead to partial melting, and the resultant magma can work its way up through the overlying plate to create a volcano. If we assume that mantle hot spots remain fixed in position while the lithospheric plates move over them, the result should be a trail of volcanoes of differing ages with the youngest closest to the hot spot.

How Far, How Fast, How Long, How Come?

Geologic and topographic information together allow us to identify the locations and nature of the major plate boundaries shown in figure 3.15. Ongoing questions include the reasons that plates move, and for how much of geologic history plate-tectonic processes have operated. Recent studies have suggested that features characteristic of modern plate-tectonic processes can be documented in rocks at least three billion years old. Long before Pangaea, then, supercontinents were evidently forming and breaking up as a consequence of plate-tectonic activity.

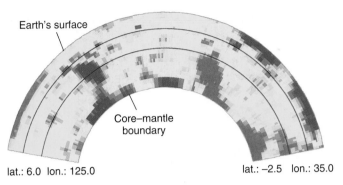

Figure 3.21

Seismic tomography is a technique that uses seismic-wave velocities to locate areas of colder rock (shown here in blue) and warmer rock (red). The deep cold rocks in this cross section may be sunken slabs of cold subducted lithosphere; the warmer columns at right, mantle plumes.

Image courtesy of Stephen Grand, University of Texas at Austin.

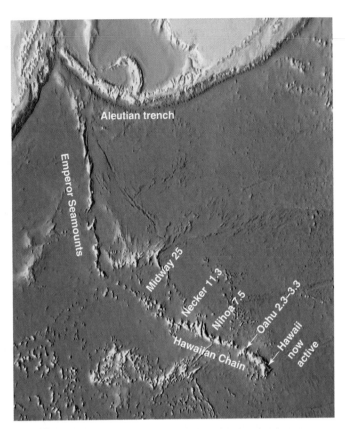

Figure 3.22

The Hawaiian Islands and other volcanoes in a chain formed over a hot spot. Numbers indicate ages of volcanic eruptions, in millions of years. Movement of the Pacific Plate has carried the older volcanoes far from the hot spot, now under the active volcanic island of Hawaii.

Image courtesy of NOAA.

A good example can be seen in the north Pacific Ocean (see figure 3.22). A topographic map shows a V-shaped chain of volcanic islands and submerged volcanoes. When rocks from these volcanoes are dated, they show a progression of ages, from about 75 million years at the northwestern end of the chain, to about 40 million years at the bend, through progressively younger islands to the still-active volcanoes of the island of Hawaii, located at the eastern end of the Hawaiian Island group. From the distances and age differences between pairs of points, we can again determine the rate of plate motion. For instance, Midway Island and Hawaii are about 2700 kilometers (1700 miles) apart. The volcanoes of Midway were active about 25 million years ago. Over the last 25 million years, then, the Pacific Plate has moved over the mantle hot spot at an average rate of 2700 kilometers/25 million years, or about 11 centimeters (4.3 inches) per year. The orientation of the volcanic chain shows the direction of plate movement—west-northwest. The kink in the chain at about 40 million years ago indicates that the direction of movement of the Pacific Plate changed at that time.

Satellite-based technology now allows direct measurement of modern plate movement (figure 3.23). Looking at many such determinations from all over the world, geologists find that average rates of plate motion are 2 to 3 centimeters (about 1 inch) per year, though as the figure shows, they can vary quite a bit. This seemingly trivial amount of motion does add up through geologic time. Movement of 2 centimeters per year for 100 million years means a shift of 2000 kilometers, or about 1250 miles!

Why Do Plates Move?

A driving force for plate tectonics has not been definitely identified. Several proposed mechanisms are shown in figure 3.24.

For many years, the most widely accepted explanation was that the plates were moved by large **convection cells** slowly churning in the plastic asthenosphere. According to this model, hot material rises at the spreading ridges; some magma escapes to form new lithosphere, but the rest of the rising asthenospheric material spreads out sideways beneath the lithosphere, slowly cooling in the process. As it flows outward, it drags the overlying lithosphere outward with it, thus continuing to open the ridges. When it has cooled somewhat, the flowing material is dense enough to sink back deeper into the asthenosphere—for example, under subduction zones. However, the existence of convection cells in the asthenosphere has not been proven definitively, even with sophisticated modern tools such as seismic tomography. There is some question, too, whether flowing asthenosphere could exercise enough drag on the lithosphere above to propel it laterally and force it into convergence.

One alternative explanation, for which considerable evidence has accumulated, is that the weight of the dense, downgoing slab of lithosphere in the subduction zone pulls the rest of the trailing plate along with it, opening up the spreading ridges

Table 4.1   **Continued**

Year	Location	Magnitude*	Deaths	Damages[†]
2001	United States: Washington State	6.8	0	$1–4 billion
2002	Afghanistan (2 events)	7.4, 6.1	1150	
2002	United States: Central Alaska	7.9	0	
2004	Northern Sumatra	9.1	283,000[+]	
2005	Northern Sumatra	8.6	1300[+]	
2005	Pakistan	7.6	86,000[+]	
2006	Indonesia: Java	6.3	5750	$3.1 billion
2008	China: Sichuan	7.9	87,650	$86 billion
2008	Pakistan	6.4	166	
2009	Samoa Islands	8.1	192	
2010	Haiti	7.0	316,000	$7.8 billion
2010	Offshore Chile	8.8	523	$30 billion
2011	Christchurch, New Zealand	6.1	363	$20 billion
2011	Offshore Honshu, Japan	9.0	20,900	$309 billion

Note: This table is not intended to be a complete or systematic compilation but to give examples of major damaging earthquakes in history.

Primary source (pre-1988 entries): Catalog of Significant Earthquakes 2000 B.C.–1979 *and supplement to 1985, World Data Center A, Report SE-27. Later data from National Earthquake Information Center and National Geophysical Data Center.*

*Where actual magnitude is not available, maximum Mercalli intensity is sometimes given in parentheses (see table 4.3). Earlier magnitudes are Richter magnitudes; for most recent events, and 1960 Chilean and 1964 Alaskan quakes, moment magnitudes are given.

[†]Estimated in 1979 dollars for pre-1980 entries. In many cases, precise damages are unknown.

Earthquakes—Terms and Principles

Basic Terms

Major earthquakes dramatically demonstrate that the earth is a dynamic, changing system. Earthquakes, in general, represent a release of built-up stress in the lithosphere. They occur along **faults,** planar breaks in rock along which there is displacement of one side relative to the other. Sometimes, the stress produces new faults or breaks; sometimes, it causes slipping along old, existing faults. When movement along faults occurs gradually and relatively smoothly, it is called **creep** (figure 4.1). Creep— sometimes termed *aseismic slip,* meaning fault displacement without significant earthquake activity—can be inconvenient but rarely causes serious damage or loss of life.

When friction between rocks on either side of a fault prevents the rocks from slipping easily or when the rock under stress is not already fractured, some elastic deformation will occur before failure. When the stress at last exceeds the rupture strength of the rock (or the friction along a preexisting fault), a sudden movement occurs to release the stress. This is an **earthquake,** or *seismic slip.* With the sudden displacement and associated stress release, the rocks snap back elastically to their previous dimensions; this behavior is called **elastic rebound** (figure 4.2; recall the discussion of elastic behavior from chapter 3).

Faults come in all sizes, from microscopically small to thousands of kilometers long. Likewise, earthquakes come in all sizes, from tremors so small that even sensitive instruments can barely detect them, to massive shocks that can level cities. (Indeed, the "aseismic" movement of creep is actually characterized by many microearthquakes, so small that they are typically not felt at all.) The amount of damage associated with an earthquake is partly a function of the amount of accumulated energy released as the earthquake occurs.

The point on a fault at which the first movement or break occurs during an earthquake is called the earthquake's **focus,** or *hypocenter* (figure 4.3). In the case of a large earthquake, for example, a section of fault many kilometers long may slip, but there is always a point at which the first movement occurred, and this point is the focus. Earthquakes are sometimes characterized by their focal depth, with those having a focus at 0–70 km depth described as "shallow"; those with focal depth in the 70–350 km range, "intermediate"; and those with focus at 350–700 km, "deep." (Note, therefore, that even the so-called deep-focus earthquakes are confined to the upper mantle.) The point on the earth's surface directly above the focus is called the **epicenter.** When news accounts tell where an earthquake occurred, they report the location of the epicenter.

A

B

Figure 4.1

(A) Walls of a culvert at the Almaden Winery that straddles the San Andreas fault are being offset by about 1.5 cm/yr by creep along the fault. Note especially the offset by subject's left foot. (B) Curbstone offset by creep along the Hayward fault in Hayward, California, from 1974 (top) to 1993 (bottom).

Photograph (A) by Joe Dellinger, (B) by Sue Hirschfeld, both courtesy NOAA/National Geophysical Data Center.

Types of Faults

Faults are sometimes described in terms of the nature of the displacement—how the rocks on either side move relative to each other—which commonly reflects the nature of the stresses involved. Certain types of faults are characteristic of particular

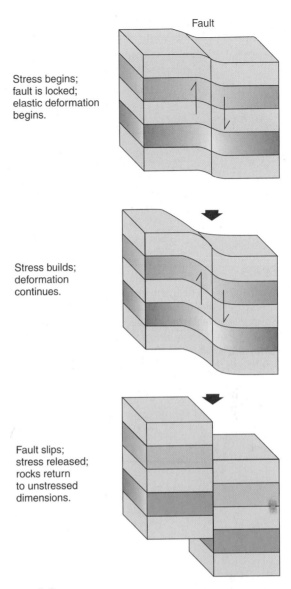

Fault

Stress begins; fault is locked; elastic deformation begins.

Stress builds; deformation continues.

Fault slips; stress released; rocks return to unstressed dimensions.

Figure 4.2

The phenomenon of elastic rebound: After the fault slips, the rocks spring back elastically to an undeformed condition. A rubber band stretched to the point of breaking shows similar behavior: after rupture, the fragments return to their original dimensions.

plate-tectonic settings. In describing the orientation of a fault (or other planar feature, such as a layer of sedimentary rock), geologists use two measures (figure 4.4). The *strike* is the compass orientation of the line of intersection of the plane of interest with the earth's surface (e.g., the fault trace in figure 4.3). The *dip* of the fault is the angle the plane makes with the horizontal, a measure of the steepness of slope of the plane.

A **strike-slip fault,** then, is one along which the displacement is parallel to the strike (horizontal). A transform fault, such as shown in figure 3.20, is a type of strike-slip fault, and reflects stresses acting horizontally. The San Andreas is a strike-slip fault. A **dip-slip fault** is one in which

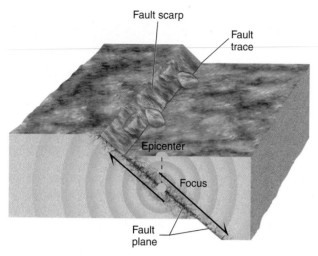

Figure 4.3

Simplified diagram of a fault, showing the focus, or hypocenter (point of first break along the fault), and the epicenter (point on the surface directly above the focus). Seismic waves dissipate the energy released by the earthquake as they travel away from the fault zone. Fault scarp is cliff formed along the fault plane at ground surface; fault trace is the line along which the fault plane intersects the ground surface.

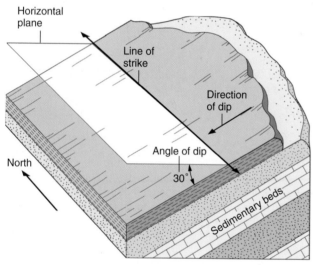

Figure 4.4

Strike and dip of a plane. Note that the dip direction is the direction in which water would run down the sloping plane. This particular plane (bed) would be described as having a north–south strike and dipping 30° west.

the displacement is vertical, up or down in the direction of dip. A dip-slip fault in which the block above the fault has moved down relative to the block below (as in figure 4.3) is called a *normal fault*. Rift valleys, whether along seafloor spreading ridges or on continents, are commonly bounded by steeply sloping normal faults, resulting from the tensional

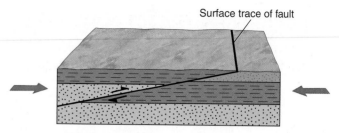

Figure 4.5

Along a thrust fault, compressional stress (indicated by red arrows) pushes one block up and over the other.

stress of rifting (recall figures 3.16 and 3.17). A dip-slip fault in which the block above has moved up relative to the block below the fault is a *reverse fault*, and indicates compressional stress rather than tension. Convergent plate boundaries are often characterized by **thrust faults,** which are just reverse faults with relatively shallowly dipping fault planes (figure 4.5)

Earthquake Locations

As shown in figure 4.6, the locations of major earthquake epicenters are concentrated in linear belts. These belts correspond to plate boundaries; recall figure 3.15. Not all earthquakes occur at plate boundaries, but most do. These areas are where plates jostle, collide with, or slide past each other, where relative plate movements may build up very large stresses, where major faults or breaks may already exist on which further movement may occur. On the other hand, intraplate earthquakes certainly occur and may be quite severe, as explored later in the chapter.

A map showing only deep-focus earthquakes would look somewhat different: the spreading ridges would have disappeared, while subduction zones would still show by their frequent deeper earthquakes. The explanation is that earthquakes occur in the lithosphere, where rocks are elastic, capable of storing energy as they deform and then rupturing or slipping suddenly. In the plastic asthenosphere, material simply flows under stress. Therefore, the deep-focus earthquakes are concentrated in subduction zones, where elastic lithosphere is pushed deep into the mantle.

It was, in fact, the distribution of earthquake foci at subduction boundaries that helped in the recognition of the subduction process. Look, for example, at the northern and western margins of the Pacific Ocean in figure 4.6. Adjacent to each seafloor trench is a region in which earthquake foci are progressively deeper with increasing distance from the trench. A similar distribution is seen along the west side of Central and South America. This pattern is called a *Benioff zone,* named for the scientist who first mapped extensively these dipping planes of earthquake foci that we now realize reveal subducting plates.

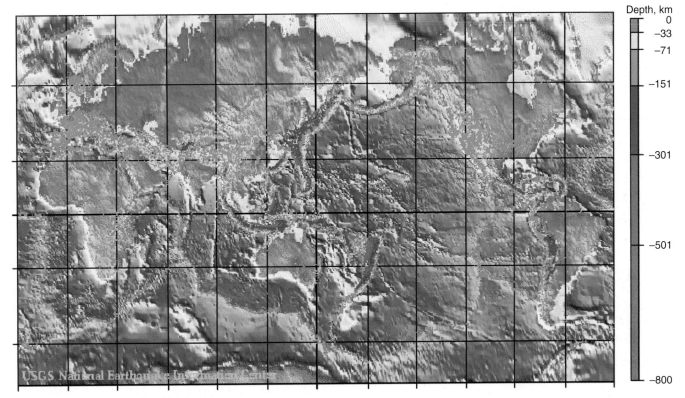

Figure 4.6

World seismicity, 1979–1995. Color of symbol indicates depth of focus. *Image courtesy of U.S. Geological Survey.*

Seismic Waves and Earthquake Severity

In the broadest sense, a *wave* is a disturbance that travels through a medium—air, water, rock, or whatever. Drop a pebble into a still pond, and the water is disturbed; we see the effects at the surface as ripples traveling away from that spot, but water below the surface is affected too. When sound waves travel through air, the air is alternately compressed and expanded, though we cannot see this happening. A visible analog can be created with a Slinky toy: Stretch the coil out on a smooth surface, hold one end in place, push the other end in quickly and then hold it still also. A pulse of compressed coil travels along the length of the Slinky. Waves can involve other types of motions, too. Take the Slinky and stretch it as before, but now twitch one end sideways, perpendicular to its length. The wave that travels along the Slinky this time will involve a shearing motion, loops of coil moving side-to-side relative to each other rather than moving closer together and farther apart. Rocks also can be affected by different types of waves.

Seismic Waves

When an earthquake occurs, it releases the stored-up energy in **seismic waves** that travel away from the focus. There are several types of seismic waves. **Body waves** (P waves and S waves) travel through the interior of the earth. **P waves** are compressional waves. As P waves travel through matter, the matter is alternately compressed and expanded. P waves travel through the earth, then, much as sound waves travel through air. **S waves** are shear waves, involving a side-to-side motion of molecules, as with the second of the Slinky demonstrations described above.

Seismic **surface waves** are somewhat similar to surface waves on water, discussed in chapter 7. That is, they cause rocks and soil to be displaced in such a way that the ground surface ripples or undulates. Surface waves also come in two types: Some cause vertical ground motions, like ripples on a pond, while others cause horizontal shearing motions, the ground surface rocking side-to-side. The surface waves are larger in amplitude—amount of ground displacement—than the body waves from the same earthquake. Therefore, most of the shaking and resultant structural damage from earthquakes is caused by the surface waves.

Figure 4.7 compares the actions of different types of seismic waves on rock.

Locating the Epicenter

Earthquake epicenters can be located using seismic body waves. Both types of body waves cause ground motions that are

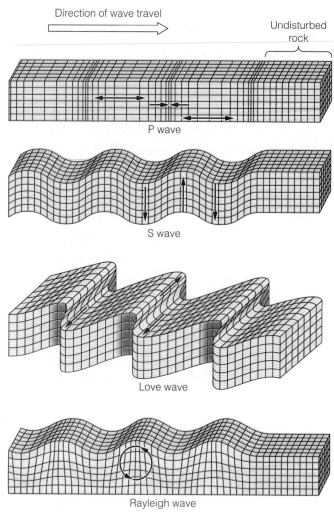

Direction of wave travel

Undisturbed rock

P wave

S wave

Love wave

Rayleigh wave

Figure 4.7

P waves and S waves are types of seismic body waves, one compressional, one shear. Love and Rayleigh waves are surface waves that differ in the type of ground motion they cause: Love waves involve a shearing motion on the surface, while Rayleigh waves resemble ripples on a pond. In all cases, waves are traveling left to right.

detectable using a **seismograph** (figure 4.8). P waves travel faster through rocks than do S waves. Therefore, at points some distance from the scene of an earthquake, the first P waves arrive somewhat before the first S waves; indeed, the two types of body waves are actually known as *primary* and *secondary* waves, reflecting these arrival-time differences.

The difference in arrival times of the first P and S waves is a function of distance to the earthquake's epicenter. The effect can be illustrated by considering a pedestrian and a bicyclist traveling the same route, starting at the same time. Assuming that the cyclist can travel faster, he or she will arrive at the destination first. The longer the route to be traveled, the greater the difference in time between the arrival of the cyclist and the later arrival of the pedestrian. Likewise, the farther the receiving seismograph is from the earthquake epicenter, the greater the time lag between the first arrivals of P waves and S waves. This is illustrated graphically in figure 4.9A. Once several recording stations have determined their distances from the epicenter in this way, the epicenter can be located on a map (figure 4.9B).

In practice, complicating factors such as inhomogeneities in the crust make epicenter location somewhat more difficult. Results from more than three stations are usually required, and computers assist in reconciling all the data. The general principle, however, is as described in the figure.

Magnitude and Intensity

All of the seismic waves represent energy release and transmission; they cause the ground shaking that people associate with earthquakes. The amount of ground motion is related to the **magnitude** of the earthquake. Historically, earthquake magnitude has most commonly been reported in this country using the *Richter magnitude scale,* named after geophysicist Charles F. Richter, who developed it.

A Richter magnitude number is assigned to an earthquake on the basis of the amount of ground displacement or shaking

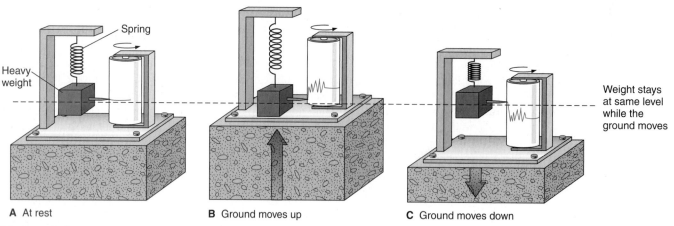

Spring

Heavy weight

Weight stays at same level while the ground moves

A At rest   **B** Ground moves up   **C** Ground moves down

Figure 4.8

This simple seismograph measures vertical ground motion. As the ground shifts up and down, the weight holds the attached pen steady, and the *seismogram* showing that motion is traced by the pen on the paper on the rotating cylinder, which is moving with the ground.

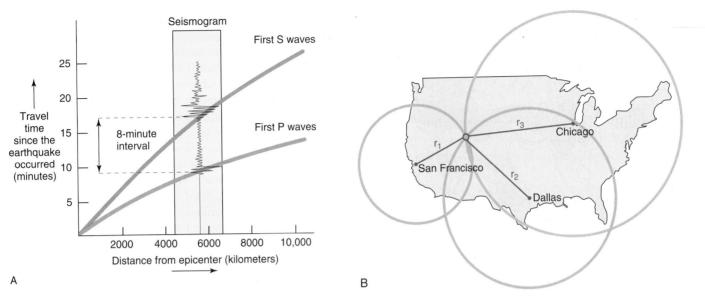

A

B

Figure 4.9

Use of seismic waves in locating earthquakes. (A) Difference in times of first arrivals of P waves and S waves is a function of the distance from the earthquake focus. This seismogram has a difference in arrival times of 8 minutes; the earthquake occurred about 5500 km away. (B) Triangulation using data from several seismograph stations allows location of an earthquake's epicenter: If the epicenter is 1350 km from San Francisco, it is on a circle of that radius around the city. If it is also found to be 1900 km from Dallas, it must fall at one of only two points. If it is further determined to be 2400 km from Chicago, its location is uniquely determined: northeastern Utah, near Salt Lake City.

that it produces near the epicenter. The amount of ground motion is measured by a seismograph, and adjusted for the particular type of instrument and the distance of the station from the earthquake epicenter (because ground motion naturally tends to decrease with increasing distance from the site of the earthquake) so that different measuring stations in different places will arrive at approximately the same estimate of the ground displacement as it would have been measured close to the epicenter. The Richter scale is logarithmic, which means that an earthquake of magnitude 4 causes ten times as much ground movement as one of magnitude 3, one hundred times as much as one of magnitude 2, and so on. The amount of energy re-

leased rises even faster with increased magnitude, by a factor of about 30 for each unit of magnitude: An earthquake of magnitude 4 releases approximately thirty times as much energy as one of magnitude 3, and nine hundred times as much as one of magnitude 2. Although we only hear of the very severe, damaging earthquakes, there are, in fact, hundreds of thousands of earthquakes of all sizes each year. Table 4.2 summarizes the frequency and energy release of earthquakes in different Richter magnitude ranges.

As seismologists studied more earthquakes in more places, it became apparent that the Richter magnitude scale had some limitations. The scale was developed in California, where

Table 4.2	Frequency of Earthquakes of Various Magnitudes		
Descriptor	**Magnitude**	**Number per Year**	**Approximate Energy Released per Earthquake (Joules)**
great	8 and over	1	over 6.3×10^{16}
major	7–7.9	15	$2–45 \times 10^{15}$
strong	6–6.9	134	$6–140 \times 10^{13}$
moderate	5–5.9	1320	$2–45 \times 10^{12}$
light	4–4.9	13,000	$6–140 \times 10^{10}$
minor	3–3.9	130,000	$2–45 \times 10^{9}$
very minor	2–2.9	1,300,000 (estimated)	$6–140 \times 10^{7}$

Source: Frequency data and descriptors from National Earthquake Information Center, based on data since 1900.

Because energy release increases by a factor of 30 per unit increase in magnitude, most of the energy released by earthquakes each year is released not by the hundreds of thousands of small tremors, but by the handful of earthquakes of magnitude 7 or larger. For reference, exploding 1 ton of TNT releases about 4.2×10^{9} Joules of energy.

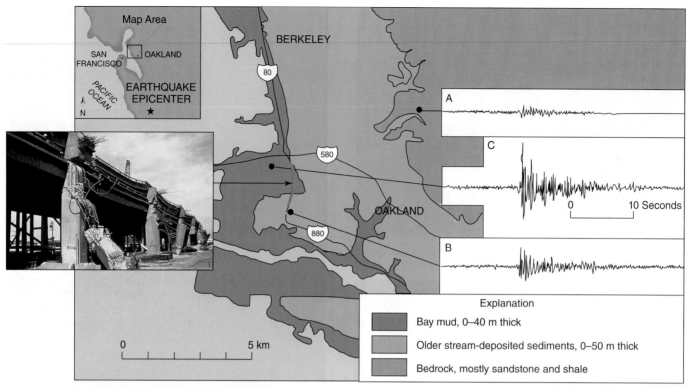

Figure 4.15

Seismograms show that ground shaking from the Loma Prieta earthquake, only moderate in bedrock (trace A), was amplified in younger river sediments (B) and especially in more recent, poorly consolidated mud (C). Inset photo is collapsed double-decker section of I-880 indicated by red line on map.

Source: R. A. Page et al., Goals, Opportunities, and Priorities for the USGS Earthquake Hazards Reduction Program, *U.S. Geological Survey Circular 1079. Inset: Photograph by D. Keefer, from U.S. Geological Survey Open-File Report 89-687.*

shaking is commonly amplified in unconsolidated materials (figure 4.15). The fact that underlying geology influences surface damage is demonstrated by the observation that intensity is not directly correlated with proximity to the epicenter (figures 4.16, 4.10)

The characteristics of the earthquakes in a particular region also must be taken into account. For example, severe earthquakes are generally followed by many **aftershocks,** earthquakes that are weaker than the principal tremor. The main shock usually causes the most damage, but when aftershocks are many and are nearly as strong as the main shock, they may also cause serious destruction, particularly to structures already weakened by the main shock.

The duration of an earthquake also affects how well a building survives it. In reinforced concrete, ground shaking leads to the formation of hairline cracks, which then widen and develop further as long as the shaking continues. A concrete building that can withstand a one-minute main shock might collapse in an earthquake in which the main shock lasts three minutes. Many of the California building codes, used as models around the world, are designed for a 25-second main shock, but earthquake main shocks can last ten times that long.

Even the best building codes are typically applied only to new construction. Adobe is no longer used for construction in Chile, but older adobe buildings remain at risk; recall figure 4.12A. When a major city is located near a fault zone, thousands of vulnerable older buildings may already have been built in high-risk areas. The costs to redesign, rebuild, or even modify all of these buildings would be staggering. Most legislative bodies are reluctant to require such efforts; indeed, many do nothing even about municipal buildings built in fault zones.

There is also the matter of public understanding of what "earthquake-resistant" construction means. It is common in areas with building codes or design guidelines that address earthquake resistance in detail to identify levels of expected structural soundness. For most residential and commercial buildings, "substantial life safety" may be the objective: the building may be extensively damaged, but casualties should be few (see figure 4.17). The structure, in fact, may have to be razed and rebuilt. Designing for "damage control" means that in addition to protecting those within, the structure should be repairable after the earthquake. This will add to the up-front costs, though it should save both repair costs and "down time" after a major quake.

PERCEIVED SHAKING	Not felt	Weak	Light	Moderate	Strong	Very strong	Severe	Violent	Extreme
POTENTIAL DAMAGE	none	none	none	Very light	Light	Moderate	Moderate/Heavy	Heavy	Very Heavy
PEAK ACC.(%g)	<.17	.17-1.4	1.4-3.9	3.9-9.2	9.2-18	18-34	34-65	65-124	>124
INSTRUMENTAL INTENSITY	I	II-III	IV	V	VI	VII	VIII	IX	X+

Figure 4.16

Distribution of intensities for the 2011 magnitude-5.8 Virginia earthquake (epicenter marked by star). These intensities are determined instrumentally from the maximum measured ground acceleration, reported as a percentage of *g*, earth's gravitational acceleration constant (the acceleration experienced by an object dropped near the surface).

Image courtesy U.S. Geological Survey Earthquake Hazards Program.

Figure 4.17

The concept of "substantial life safety"—and its limitations—are illustrated by this relatively modern building in the 1995 Kobe earthquake. Although the first story did not fare well, the majority of the structure remained intact, meeting the "substantial life safety" goal. Cases in which this goal clearly was not met are illustrated in such examples as figure 4.12B and D and the inset photo in figure 4.15.

Photograph by Dr. Roger Hutchison, courtesy NOAA /National Geophysical Data Center.

The highest (and most expensive) standard is design for "continued operation," meaning little or no structural damage at all; this would be important for hospitals, fire stations, and other buildings housing essential services, and also for facilities that might present hazards if the buildings failed (nuclear power plants, manufacturing facilities using or making highly toxic chemicals). This is the sort of standard to which the Trans-Alaska Pipeline was (fortunately) designed: it was built with a possible magnitude-8 earthquake on the Denali fault in mind, as shown in figure 4.11, so it came through the magnitude-7.9 earthquake in 2002, with its 14 feet of horizontal fault displacement, without a break. Thus, anyone living or buying property in an area where earthquakes are a significant concern would be well-advised to pin down just what level of earthquake resistance is promised in a building's design.

Additional factors can leave some populations at particular risk. Haiti is a poor country, where other pressing problems mean little attention is paid to developing building codes for earthquake resistance, few contractors are qualified to build structures appropriately, and most people cannot afford even minimal added building costs for enhanced earthquake

safety. Further, because hurricanes are a regular threat, concrete roofs have been viewed as superior to lighter sheet-metal roofs that are more easily blown off by high winds. Unfortunately, in the 2010 earthquake, buildings collapsing under those heavier concrete roofs resulted in unusually high numbers of casualties for a quake that size; see the opening photograph for chapter 3.

Ground Failure

Landslides (figure 4.18) can be a serious secondary earthquake hazard in hilly areas. As will be seen in chapter 8, earthquakes are one of the major events that trigger slides on unstable slopes. The best solution is not to build in such areas. Even if a whole region is hilly, detailed engineering studies of rock and soil properties and slope stability may make it possible to avoid the most dangerous sites. Visible evidence of past landslides is another indication of especially dangerous areas.

Ground shaking may cause a further problem in areas where the ground is very wet—in filled land near the coast or in places with a high water table. This problem is **liquefaction.** When wet soil is shaken by an earthquake, the soil particles may be jarred apart, allowing water to seep in between them, greatly reducing the friction between soil particles that gives the soil strength, and causing the ground to become somewhat like quicksand. When this happens, buildings can just topple over or partially sink into the liquefied soil; the soil has no strength to support them. The effects of liquefaction were dramatically illustrated in Niigata, Japan, in 1964. One multistory apartment building tipped over to settle at an angle

A

B

Figure 4.18

(A) Landslide from the magnitude-6.5 earthquake in Seattle, Washington, in 1965 made Union Pacific Railway tracks unuseable.
(B) Muzaffarabad highway blocked by landslides in the 2005 Pakistan earthquake.

(A) Photograph courtesy University of California at Berkeley/NOAA. (B) Photograph by John Beba, Geological Survey of India; courtesy NOAA.

of 30 degrees to the ground while the structure remained intact! (See figure 4.19A; again, "substantial life safety" was achieved, though the building was no longer habitable.) Liquefaction was likewise a major cause of damage from the Loma Prieta, Kobe, and Christchurch earthquakes (figure 4.19B). Telltale signs of liquefaction include "sand boils," formed as liquefied soil bubbles to the surface during the quake (figure 4.19C). In some areas prone to liquefaction, improved underground drainage systems may be installed to try to keep the soil drier, but little else can be done about this hazard, beyond avoiding the areas at risk. Not all areas with wet soils are subject to liquefaction; the nature of the soil or fill plays a large role in the extent of the danger.

Tsunamis and Coastal Effects

Coastal areas, especially around the Pacific Ocean basin where so many large earthquakes occur, may also be vulnerable to **tsunamis.** These are seismic sea waves, sometimes improperly called "tidal waves," although they have nothing to do with tides. The name derives from the Japanese for "harbor wave," which is descriptive of their behavior. When an undersea or near-shore earthquake occurs, sudden movement of the sea floor may set up waves traveling away from that spot, like ripples in a pond caused by a dropped pebble. Contrary to modern movie fiction, tsunamis are not seen as huge breakers in the open ocean that topple ocean liners in one sweep. In the open sea, tsunamis are only unusually broad swells on the water surface, but they travel extremely rapidly—speeds up to 1000 km/hr (about 600 mph) are not

uncommon. Because they travel so fast, as tsunamis approach land, the water tends to pile up against the shore. The tsunami may come ashore as a very high, very fast-moving wall of water that acts like a high tide run amok (perhaps the origin of the misnomer "tidal wave"), or it may develop into large breaking waves, just as ordinary ocean waves become breakers as the undulating waters "touch bottom" near shore. Tsunamis, however, can easily be over 15 meters high in the case of larger earthquakes. Several such waves or water surges may wash over the coast in succession; between waves, the water may be pulled swiftly seaward, emptying a harbor or bay, and perhaps pulling unwary onlookers along. Tsunamis can also travel long distances in the open ocean. Tsunamis set off on one side of the Pacific may still cause noticeable effects on the other side of the ocean.

Given the speeds at which tsunamis travel, little can be done to warn those near the earthquake epicenter, but people living some distance away can be warned in time to evacuate, saving lives, if not property. In 1948, two years after a devastating tsunami hit Hawaii, the U.S. Coast and Geodetic Survey established the Pacific Tsunami Early Warning System, based in Hawaii. Whenever a major earthquake occurs in the Pacific region, sea-level data are collected from a series of monitoring stations around the Pacific. If a tsunami is detected, data on its source, speed, and estimated time of arrival can be relayed to areas in danger, and people can be evacuated as necessary. Unfortunately, individuals' responses to such warnings may be variable: Some ignore the warnings, and some even go closer to shore to watch the waves, often with tragic consequences.

A

C

B

Figure 4.19

(A) Effects of soil liquefaction during an earthquake in Niigata, Japan, 1964. The buildings, which were designed to be earthquake-resistant, simply tipped over intact. (B) Liquefaction blew holes in roads during the 2011 earthquake in Christchurch, New Zealand; this van became trapped in one. (C) Sediment-laden water bubbled to the surface near El Centro, California, as a result of a 1979 earthquake in the Imperial Valley. Object near center of photo is pocketknife for scale.

(A) Photograph courtesy of National Geophysical Data Center. (B) Photograph by Steve Taylor (Ray White), courtesy NOAA/National Geophysical Data Center. (C) Photograph by G. Reagor, U.S. Geological Survey.

There was no tsunami warning system in the Indian Ocean prior to the devastation of the tsunami from the 2004 Sumatran earthquake (see Case Study 4.1). Even so, seismologists outside the area, noting the size of the quake, alerted local governments to the possibility that such an enormous earthquake might produce a sizeable tsunami. The word did not reach all affected coastal areas in time. In other cases, local officials chose not to issue public warnings, in part for fear of frightening tourists. Such behavior likely contributed to the death toll.

Five years after the Sumatran disaster, a tsunami warning system had been established for the Indian Ocean. A similar warning system for the Caribbean region is in the final stages of development as this is written, with the Pacific center providing interim coverage. Better warnings, together with public education, should minimize casualties from future tsunamis, especially with the aid of new monitoring techniques. In addition to surface buoys long used to detect the sea-surface undulations that might be due to a tsunami, the warning systems now also use deep buoys that are sensitive to deep-water pressure changes that could indicate a tsunami passing through. All the data are then relayed by satellite to stations where computers can quickly put the pieces together for prompt detection of the existence, location, and velocity of a tsunami so that appropriate warnings can be issued to areas at risk.

Even in the absence of tsunamis, there is the possibility of coastal flooding from sudden subsidence as plates shift during an earthquake. Areas that were formerly dry land may be permanently submerged and become uninhabitable

Megathrusts Make Mega-Disasters

The tectonics of southern and southeast Asia are complex. East of the India and Australia plates lies the Sunda Trench, marking a long subduction zone. The plate-boundary fault has been called a "megathrust"—a very, very large thrust fault—which is typical of subduction-zone boundaries.

For centuries, the fault at the contact between the small Burma Plate and the India Plate had been locked. The edge of the Burma Plate had been warped downward, stuck to the subducting India Plate. The fault let go on 26 December 2004, in an earthquake with moment magnitude of 9.1, making it the third-largest earthquake in the world since 1900. When the fault ruptured, the freed leading edge of the Burma Plate snapped upward by as much as 5 meters. The resulting abrupt shove on the water column above set off a deadly tsunami. It reached the Sumatran shore in minutes, but took hours to reach India and, still later, Africa.

How high onshore the tsunami surged varied not only with distance from the epicenter but with coastal geometry. The highest runup was observed on gently sloping shorelines and where water was funneled into a narrowing bay. Detailed studies after the earthquake showed that in some places the tsunami waves reached heights of 31 meters (nearly 100 feet). Destruction was profound and widespread (figure 1). An irony of tsunami behavior is that fishermen well out at sea were unaffected by, and even unaware of, the tsunami until they returned to port.

Most of the estimated 283,000 deaths from this earthquake were in fact due to the tsunami, not building collapse or other consequences of ground shaking. Thousands of people never found or accounted for were presumed to have been swept out to sea and drowned. Tsunami deaths occurred as far away as Africa.

And the Sunda Trench had more in store. When one section of a fault zone slips in a major earthquake, regional stresses are shifted, which can bring other sections closer to rupture. On 28 March 2005, a magnitude 8.6 earthquake (itself the seventh-largest worldwide since 1900) ruptured the section of the fault zone just south of the 2004 break (figure 2). Though no major tsunami resulted this time, more than 1300 people died. Many more, already traumatized by the 2004 disaster, were badly frightened again. On 12 September 2007, two earthquakes, of magnitude 8.4 and 7.9, ruptured parts of the subduction zone along the 1833 slip. Half a dozen earthquakes over magnitude 7, and many smaller ones, occurred along the trench over the next five years, and the seismicity continues.

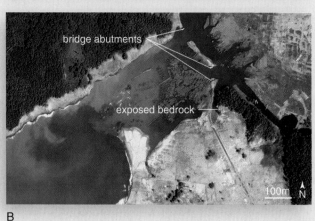

A

B

Figure 1

Gleebruk, Indonesia before (A) and after (B) the tsunami. Photographs were taken 12 April 2004 and 2 January 2005, respectively.

Photographs © Digital Globe/Getty Images.

Figure 2

Colored areas show ruptures associated with earthquakes in 1833, 1861, and 2004 along the Sunda Trench. Stress shifts following the December 2004 earthquake (epicenter at red star) may well have contributed to the March 2005 earthquake just to the south (yellow star). Aftershocks from the latter quake (yellow dots) suggest rupture of both the gap between the 1861 and 2004 rupture zones, and a large part of the former.

Map courtesy U.S. Geological Survey Earthquake Hazards Program.

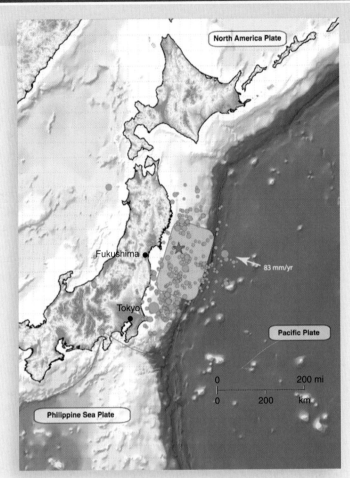

Figure 3

Location and size of the section of thrust fault plane that slipped in the magnitude-9 Japan earthquake; red star marks the epicenter.

Map after U.S. Geological Survey National Earthquake Information Center.

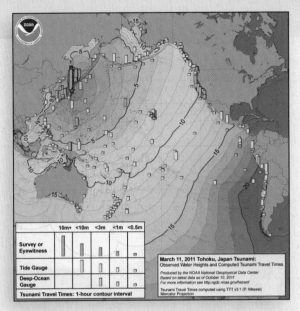

Figure 4

Travel times and heights for the tsunami produced by the 11 March 2011 Japan earthquake.

Map by NOAA/National Geophysical Data Center.

Figure 5

At Aichi, southwest of Tokyo, the force of the tsunami swept this large ferry and other boats inland, dumping them amid the rubble of houses. Note the remaining houses and cars, for scale.

Photograph by Lance Cpl. Garry Welch, U. S. Marine Corps, courtesy NOAA/NGDC.

Northern Japan also sits along a subduction zone. On 11 March 2011, a large section of that megathrust fault slipped by an estimated 30–40 meters, producing a magnitude-9.0 earthquake (figure 3). The quake caused severe ground shaking, some landslides, local liquefaction, fires—and a massive tsunami. The maximum runup height of that tsunami was measured at nearly 38 meters (125 feet) in Japan, and it was eventually detected all around the Pacific basin (figure 4), still several meters high when it reached South America some 20 hours after the quake.

Predictably, damage was extensive (figure 5 and chapter-opening photograph), but some was more serious than might have been expected. Perhaps the worst surprise occurred at Fukushima (see location in figure 3). The nuclear-power station there was planned with earthquake hazards in mind. Scientists studying the region's earthquake history—including tsunami deposits over a thousand years old—had projected the largest likely earthquake and tsunami. The nuclear plants at Fukushima were accordingly designed to withstand an earthquake of magnitude 8.2, and the tsunami wall protecting them was built 5.7 meters high. But the 2011 earthquake was far larger, and estimates of the tsunami's height when it hit Fukushima ranged from 8 to 14 meters. The six reactors lost electrical power to circulate vital cooling water; several suffered core meltdown and/or hydrogen explosions; considerable radiation was released, and hundreds of thousands of people were evacuated from areas most affected by that radiation.

Nuclear power is explored further in chapter 15. Here we can note that the Fukushima disaster has prompted a closer look at reactor design and safety around the world.

(figure 4.20). Conversely, uplift of sea floor may make docks and other coastal structures useless, though this rarely results in casualties.

Fire

A secondary hazard of earthquakes in cities is *fire,* which may be more devastating than ground movement (figure 4.21). In the 1906 San Francisco earthquake, 70% of the damage was due to fire, not simple building failure. As it was, the flames were confined to a 10-square-kilometer area only by dynamiting rows of buildings around the burning section. Fires occur because fuel lines and tanks and power lines are broken, touching off flames and fueling them. At the same time, water lines also are broken, leaving no way to fight the fires effectively, and streets fill with rubble, blocking fire-fighting equipment. In the 1995 Kobe earthquake, broken pipelines left firefighters with only the water in their trucks' tanks to battle more than 150 fires. Putting numerous valves in all water and fuel pipeline systems helps to combat these problems because breaks in pipes can then be isolated before too much pressure or liquid is lost.

Earthquake Prediction and Forecasting

Millions of people already live near major fault zones. For that reason, prediction of major earthquakes could result in many saved lives. Some progress has been made in this direction, but as the complexity of seismic activity is more fully recognized, initial optimism has been diminished considerably.

Seismic Gaps

Maps of the locations of earthquake epicenters along major faults show that there are stretches with little or no seismic activity, while small earthquakes continue along other sections of the same fault zone. Such quiescent, or dormant, sections of otherwise-active fault zones are called **seismic gaps.** They apparently represent "locked" sections of faults along which friction is preventing slip. These areas may be sites of future serious earthquakes. On either side of a locked section, stresses are being released by earthquakes. In the seismically quiet locked sections, friction is apparently sufficient to prevent the fault from slipping, so the stresses are simply building up. The concern, of course, is that the accumulated energy will become so great that, when that locked section of fault finally does slip again, a very large earthquake will result.

Recognition of these seismic gaps makes it possible to identify areas in which large earthquakes may be expected in the future. The subduction zone bordering the Banda Plate had been rather quiet seismically for many decades before the December 2004 quake; areas farther south along the Sunda Trench that are still locked may be the next to let go catastrophically. The 1989 Loma Prieta, California, earthquake occurred in what had been a seismic gap along the San Andreas fault. The

Figure 4.20

Flooding in Portage, Alaska, due to tectonic subsidence during 1964 earthquake.

Photograph by G. Plafker, courtesy USGS Photo Library, Denver, CO.

subduction zone beneath the Pacific Northwest is currently locked, having had its last major quake in 1700. Earthquake-cycle theory, discussed later in this chapter, may be a tool for anticipating major earthquakes that will fill such gaps abruptly.

Earthquake Precursors and Prediction

In its early stages, earthquake prediction was based particularly on the study of earthquake **precursor phenomena,** things that happen or rock properties that change prior to an earthquake. Many different possibilities have been examined. For example, the ground surface may be uplifted and tilted prior to an earthquake. Seismic-wave velocities in rocks near the fault may change before an earthquake; so may electrical resistivity (the resistance of rocks to electric current flowing through them). Changes have been observed in the levels of water in wells and/or in the content of radon (a radioactive gas that occurs naturally in rocks as a result of decay of trace amounts of uranium within them). Sensitive instruments can measure elastic strain accumulating across a fault.

The hope, with the study of precursor phenomena, has been that one could identify patterns of precursory changes that could be used with confidence to issue earthquake predictions precise enough as to time to allow precautionary evacuations or other preparations. Unfortunately, the precursors have not proven to be very reliable. Not only does the length of time over which precursory changes are seen before an earthquake vary, from minutes to months, but the pattern of precursors—which parameters show changes, and of what sorts—also varies. Many earthquakes, including large ones, seem quite unheralded by recognizable precursors at all. Loma Prieta was one of these; so

A

B

C

Figure 4.21

(A) Classic panorama of San Francisco in flames, five hours after the 1906 earthquake. (B) Five hours after the 1995 Kobe earthquake, the scene is similar, with many fires burning out of control. (C) In 2011, when the tsunami hit Otsuchi, Japan, it swept away the gas station, starting a fire that destroyed the entire town.

(B) Photograph by Dr. Roger Hutchison; (C) photograph by the Japanese Red Cross; all courtesy NOAA/National Geophysical Data Center.

was Northridge, and so was Kobe. After decades of precursor studies, no consistently useful precursors have been recognized. Even if more-reliable precursors are identified, too, as a practical matter it is always going to be easier to monitor shallow faults on land, like the San Andreas, than faults like the submarine Sumatran thrust fault or those ringing the Pacific Ocean.

Current Status of Earthquake Prediction

Only four nations—Japan, Russia, the People's Republic of China, and the United States—have had government-sponsored earthquake prediction programs. Such programs typically involve intensified monitoring of active fault zones to expand the observational data base, coupled with laboratory experiments designed to increase understanding of precursor phenomena and the behavior of rocks under stress. Even with these active research programs, scientists cannot monitor every area at once, and, as already noted, earthquake precursors are inconsistent at best, and at worst, often absent altogether.

Since 1976, the director of the U.S. Geological Survey has had the authority to issue warnings of impending earthquakes

and other potentially hazardous geologic events (volcanic eruptions, landslides, and so forth). An Earthquake Prediction Panel reviews scientific evidence that might indicate an earthquake threat and makes recommendations to the director regarding the issuance of appropriate public statements. These statements could range in detail and immediacy from a general notice to residents in a fault zone of the existence and nature of earthquake hazards there, to a specific warning of the anticipated time, location, and severity of an imminent earthquake. In early 1985, the panel made its first endorsement of an earthquake prediction; the results are described in Case Study 4.2.

In the People's Republic of China, tens of thousands of amateur observers and scientists work on earthquake prediction. In February 1975, after months of smaller earthquakes, radon anomalies, and increases in ground tilt followed by a rapid increase in both tilt and microearthquake frequency, the scientists predicted an imminent earthquake near Haicheng in northeastern China. The government ordered several million people out of their homes into the open. Nine and one-half hours later, a major earthquake struck, and many lives were saved because people were not crushed by collapsing buildings. The next year, they concluded that a major earthquake could be expected near Tangshan, but could only say that the event was likely to occur sometime during the following two months. When the earthquake—magnitude over 8.0 with aftershocks up to magnitude 7.9—did occur, there had been no sudden change in precursor phenomena to permit a warning, and hundreds of thousands of people died.

There had been some indication of increasing strain in the region of the Japan trench subduction zone prior to the 2011 earthquake, but undersea monitoring was minimal, and surface GPS measurements were of limited use in evaluating developments at depth; so, no warning could be issued.

In the United States, earthquake predictions have not yet become reliable enough to prompt such actions as large-scale evacuations. In the near term, it seems that the more feasible approach is earthquake *forecasting,* identifying levels of earthquake probability in fault zones within relatively broad time windows, as described below. This at least allows for long-term preparations, such as structural improvements. Ultimately, short-term, precise predictions that can save more lives are still likely to require better recognition and understanding of precursory changes, though recent studies have suggested that very short-term "early warnings" may have promise.

The Earthquake Cycle and Forecasting

Studies of the dates of large historic and prehistoric earthquakes along major fault zones have suggested that they may be broadly periodic, occurring at more-or-less regular intervals (figure 4.22). This is interpreted in terms of an **earthquake cycle** that would occur along a (non-creeping) fault segment: a period of stress buildup, sudden fault rupture in a major earthquake, followed by a brief interval of aftershocks reflecting minor lithospheric

adjustments, then another extended period of stress buildup (figure 4.23). The fact that major faults tend to break in segments is well documented: The San Andreas is one example; the thrust fault along the Sunda Trench is another; so is the Anatolian fault zone in Turkey. The rough periodicity can be understood in terms of two considerations. First, assuming that the stress buildup is primarily associated with the slow, ponderous, inexorable movements of lithospheric plates, which at least over decades or centuries will move at fairly constant rates, one might reasonably expect an approximately constant rate of buildup of stress (or accumulated elastic strain). Second, the rocks along a given fault zone will have particular physical properties, which allow them to accumulate a certain amount of strain energy before failure, or fault rupture, and that amount would be approximately constant from earthquake to earthquake. Those two factors together suggest that periodicity is a reasonable expectation. Once the pattern for a particular fault zone is established, one may use that pattern, together with measurements of strain accumulation in rocks along fault zones, to project the time window during which the next major earthquake is to be expected along that fault zone and to estimate the likelihood of the earthquake's occurrence in any particular time

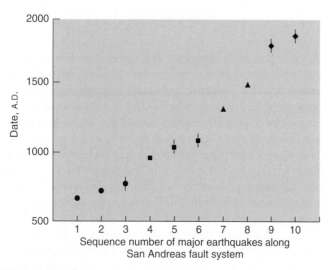

Figure 4.22

Periodicity of earthquakes assists in prediction/forecasting efforts. Long-term records of major earthquakes on the San Andreas show both broad periodicity and some tendency toward clustering of 2 to 3 earthquakes in each active period. Prehistoric dates are based on carbon-14 dating of faulted peat deposits.

Source: Data from U.S. Geological Survey Circular 1079.

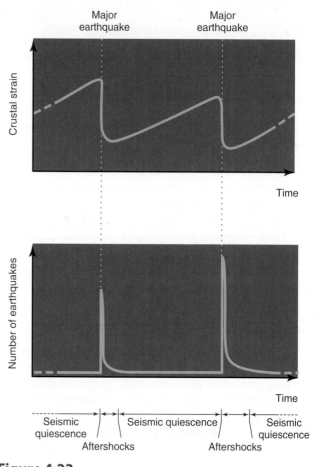

Figure 4.23

The "earthquake cycle" concept.

Source: After U.S. Geological Survey Circular 1079, modified after C. DeMets (pers. comm. 2006).

period. Also, if a given fault zone or segment can store only a certain amount of accumulated energy before failure, the maximum size of earthquake to be expected can be estimated.

This is, of course, a simplification of what can be a very complex reality. For example, though we speak of "The San Andreas fault," it is not a single simple, continuous slice through the lithosphere, nor are the rocks it divides either identical or homogeneous on either side. Both stress buildup and displacement are distributed across a number of faults and small blocks of lithosphere. Earthquake forecasting is correspondingly more difficult. Nevertheless, the U.S. Geological Survey's Working Group on California Earthquake Probabilities publishes forecasts for different segments of the San Andreas (figure 4.24). Note that in this example, the time window in question is a 30-year period. Such projections, longer-term analogues of the meteorologist's forecasting of probabilities of precipitation in coming days, are revised as large and small earthquakes occur and more data on slip in creeping sections are collected. For example, after the 1989 Loma Prieta earthquake, which substantially shifted stresses in the region, the combined probability of a quake of magnitude 7 or greater either on the peninsular segment of the San Andreas or on the nearby Hayward fault (locked since 1868) was increased from 50 to 67%. Forecasts are also being refined through recent research on fault interactions and the ways in which one earthquake can increase or decrease the likelihood of slip on another segment, or another fault entirely.

Earthquake Early Warnings?

When hurricanes threaten, public officials agonize over issues such as when or whether to order evacuation of threatened areas; how, logistically, to make it happen; and how to secure property in evacuated areas. Similar issues will arise if and when scientific earthquake predictions become feasible. In the meantime, a new idea has surfaced for saving lives and property damage: Earthquake early warnings issued a few seconds, or perhaps tens of seconds, before an earthquake strikes.

What good can that do? Potentially, quite a lot: Trains can be slowed and stopped, traffic lights adjusted to get people off vulnerable bridges, elevators stopped at the nearest floor and their doors opened to let passengers out. Automated emergency systems can shut off valves in fuel or chemical pipelines and initiate shutdown of nuclear power plants. People in homes or offices can quickly take cover under sturdy desks and tables. Those in coastal areas at risk of tsunamis can seek higher ground. And so on.

The idea is based on the relative travel times of seismic waves. The damaging surface waves travel more slowly than the P and S waves used to locate earthquakes. Therefore, earthquakes can be located seconds after they occur, and—at least for places at some distance from the epicenter—warnings transmitted to areas at risk before the surface waves hit. However, to decide when these warnings should be issued and automated-response systems activated, one wants to know the earthquake's magnitude as well; scrambling to react to every one of the thousands of earthquakes that occur each year would create chaos. And it is currently more difficult to get a very accurate measure of magnitude in the same few seconds that it takes to determine an earthquake's location.

Japan, at least, has taken this concept seriously. The Japan Meteorological Agency (which also deals with earthquakes and tsunamis) has already developed the capacity to issue these alerts, using a network of over 1000 seismographs, and public alerts have been issued for larger quakes since 2007. The effectiveness of the system is still limited by the fact that many facilities have not been retrofitted for automatic response, largely because of cost. Still, the system offers great promise for reducing casualties in a very earthquake-prone part of the world, and did so in 2011. The first P waves hit the network about 23 seconds after the quake occurred, and public warnings were issued 8 seconds later. This gave Tokyo residents about a minute before the surface waves hit, and the slower-moving tsunami waves reached the coast more than 20 minutes later. The system is credited with saving many lives.

Public Response to Earthquake Hazards

A step toward making any earthquake-hazard warning system work would be increasing public awareness of earthquakes as a hazard. In the People's Republic of China, vigorous public-education programs (and several major modern earthquakes)

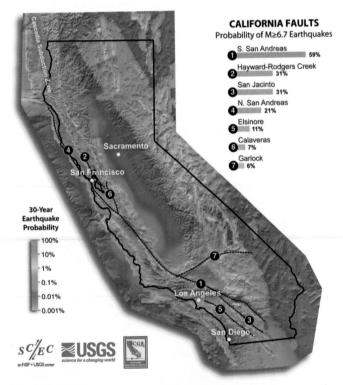

Figure 4.24

Earthquake forecast map issued by the 2007 Working Group on California Earthquake Probabilities, indicating the 30-year probability of a magnitude 6.7 or greater earthquake. Over the whole state, the probability of such a large earthquake by 2038 was estimated at 99.7%; the probability of at least one quake of magnitude 7.5 or greater was set at 46%.

Image courtesy: Southern California Earthquake Center/USGS/CGS.

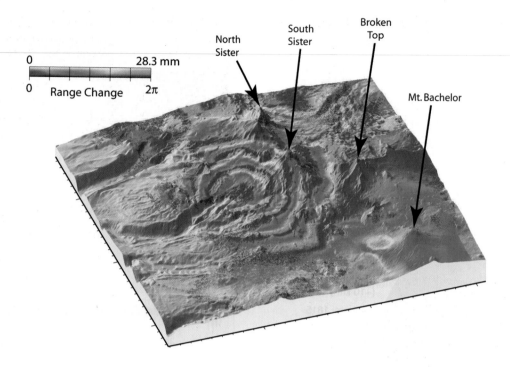

North Sister

South Sister

Broken Top

Mt. Bachelor

0   28.3 mm

0   Range Change   2π

Figure 5.28

"Interferogram" showing uplift, 1997–2001, in Three Sisters volcanic area in the Oregon Cascades. Uplift is attributed to magma emplacement several miles deep in the crust. Each full color cycle represents close to 1" of uplift; total uplift shown is nearly 6" (14 cm). This method is now being applied at Yellowstone and elsewhere.

Interferogram by C. Wicks, USGS Volcano Hazards Program.

Bulging, tilt, or uplift of the volcano's surface is also a warning sign. It often indicates the presence of a rising magma mass, the buildup of gas pressure, or both. On Kilauea, eruptions are preceded by an inflating of the volcano as magma rises up from the asthenosphere to fill the shallow magma chamber. Unfortunately, it is not possible to specify exactly when the swollen volcano will crack to release its contents. That varies from eruption to eruption with the pressures and stresses involved and the strength of the overlying rocks. This limitation is not unique to Kilauea. Uplift, tilt, and seismic activity may indicate that an eruption is approaching, but geologists do not yet have the ability to predict its exact timing. However, their monitoring tools are becoming more sophisticated. Historically, tilt was monitored via individual meters placed at a limited number of points on the volcano. Now, satellite radar interferometry allows a richer look at the regional picture (figure 5.28).

Changes in the mix of gases coming out of a volcano may give clues to impending eruptions. The SO_2 content of the escaping gas shows promise as a precursor, perhaps because more SO_2 can escape as magma nears the surface. Surveys of ground surface temperatures may reveal especially warm areas where magma is particularly close to the surface and about to break through.

Certainly, much more work is needed before the exact timing and nature of major volcanic eruptions can be anticipated consistently. The recognition of impending eruptions of Mount St. Helens, Pinatubo, and Soufrière Hills was obviously successful in saving many lives through evacuations and restrictions on access to the danger zones. Those danger zones, in turn, were effectively identified by a combination of mapping of debris from past eruptions and monitoring the geometry of the current bulging. Mount Pinatubo displayed increasing evidence of forthcoming activity for two months before its major eruption of 15 June 1991—local earthquakes, increasing in number, and becoming concentrated below a bulge on the volcano's summit a week before that eruption; preliminary ash emissions; a sudden drop in gas output that suggested a plugging of the volcanic "plumbing" that would tend to increase internal gas pressure and explosive potential. Those warnings, combined with aggressive and comprehensive public warning procedures and, eventually, the evacuation of over 80,000 people, greatly reduced the loss of life. In 1994, geoscientists monitoring Rabaul volcano in Papua New Guinea took a magnitude-5 earthquake followed by increased seismic activity generally as the signal to evacuate a town of 30,000 people. The evacuation, rehearsed before, was orderly—and hours later, a five-day explosive eruption began. The town was covered by a meter or more of wet volcanic ash, and a quarter of the buildings were destroyed, but no one died.

These successes clearly demonstrate the potential for reduction in fatalities when precursory changes occur and local citizens are educated to the dangers. On the other hand, the exact moment of the 1980 Mount St. Helens explosion was not known until seconds beforehand, and it is not currently possible to tell whether or just when a pyroclastic flow might emerge from a volcano like Soufrière Hills. Nor can volcanologists readily predict the volume of lava or pyroclastics to expect from an eruption or the length of the eruptive phase, although they can anticipate the likelihood of an explosive eruption or other eruption of a particular style based on tectonic setting and past behavior.

Recently, scientists have discovered another possible link between seismicity and volcanoes. In May 2006, a magnitude-6.4 earthquake occurred in Java, Indonesia. Several days later, the active Javanese volcanoes Merapi and Semeru showed increased output of heat and lava that persisted for over a week. It has been suggested that the earthquake contributed to the rise of additional magma into the volcanoes, enhancing their eruptions. Research has also suggested that a magnitude-5.2 earthquake near Chaitén may have accounted for its abrupt reawakening, by opening a crustal fissure through which magma could rise 5 km (3 miles) to the surface in a matter of hours. Erupting volcanoes, and perhaps dormant ones also, should therefore be monitored especially carefully following a strong earthquake nearby.

Evacuation as Response to Eruption Predictions

When data indicate an impending eruption that might threaten populated areas, the safest course is evacuation until the activity subsides. However, a given volcano may remain more or less dangerous for a long time. An active phase, consisting of a series of intermittent eruptions, may continue for months, years, or even decades. In these instances, either property must be abandoned for prolonged periods, or inhabitants must be prepared to move out, not once but many times. The Soufrière Hills volcano on Montserrat began its latest eruptive phase in July 1995. By the end of 2000, dozens of eruptive events—ash venting, pyroclastic flows, ejection of rocks, building or collapse of a dome in the summit—had occurred. Eruptive activity is continuing, and more than half the island remains off-limits to inhabitants.

Accurate prediction and assessment of the hazard is particularly difficult with volcanoes reawakening after a long dormancy because historical records for comparison with current data are sketchy or nonexistent. Such volcanoes may not even be monitored, because no significant threat is anticipated, or perhaps because the country in question has no resources to spare for volcano monitoring. And while some volcanoes, like Mount St. Helens or Mount Pinatubo, may produce weeks or months of warning signals as they emerge from dormancy to full-blown eruption, allowing time for scientific analysis, public education, and reasoned response, not all volcanoes' awakenings are so leisurely, as we have noted.

Given the uncertain nature of eruption prediction at present, too, some precautionary evacuations will continue to be shown, in retrospect, to have been unnecessary, or unnecessarily early. There are other volcanoes in the Caribbean similar in character to Mont Pelée and Soufrière Hills. In 1976, La Soufrière on Guadeloupe began its most recent eruption. Some 70,000 people were evacuated and remained displaced for months, but only a few small explosions occurred. Government officials and volcanologists were blamed for disruption of people's lives, needless as it turned out to be.

On the other hand, it was fortunate that officials took the 1980 threat of Mount St. Helens seriously after its first few signs of reawakening. The immediate area near the volcano was cleared of all but a few essential scientists and law enforcement personnel. Access to hundreds of square kilometers of terrain around the volcano was severely restricted for both residents and workers (mainly loggers). Others with no particular business in the area were banned altogether. Many people grumbled at being forced from their homes. A few flatly refused to go. Numerous tourists and reporters were frustrated in their efforts to get a close look at the action. When the major explosive eruption came, there were casualties, but far fewer than there might have been. On a normal spring weekend, 2000 or more people would have been on the mountain, with more living or camping nearby. Considering the influx of the curious once eruptions began and the numbers of people turned away at roadblocks leading into the restricted zone, it is possible that, with free access to the area, tens of thousands could have died. Many lives were likewise saved by timely evacuations near Mount Pinatubo in 1991, made possible by careful monitoring and supported by intensive educational efforts by the government.

More on Volcanic Hazards in the United States

Cascade Range

Mount St. Helens is only one of a set of volcanoes that threaten the western United States and southwestern Canada. There is subduction beneath the Pacific Northwest, which is the underlying cause of continuing volcanism there (figure 5.29). Several more of the Cascade Range volcanoes have shown signs of reawakening. Lassen Peak was last active between 1914 and 1921, not so very long ago geologically. Its products are very similar to those of Mount St. Helens; violent eruptions are possible. Mount Baker (last eruption, 1870), Mount Hood (1865), and Mount Shasta (active sometime within the last 200 years)

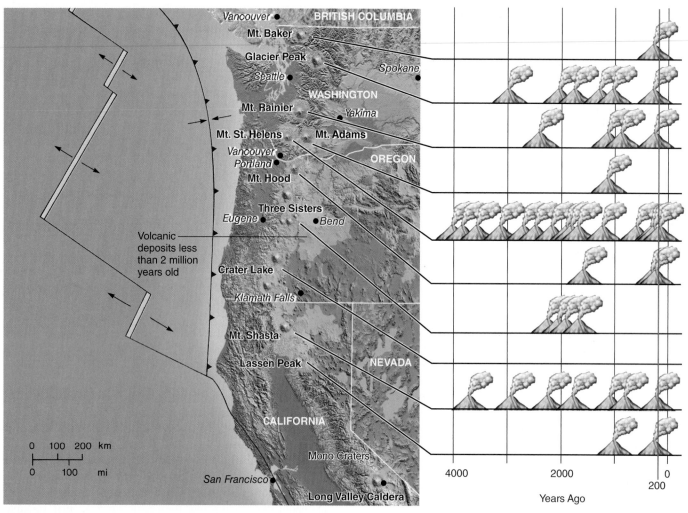

Figure 5.29

The Cascade Range volcanoes and their spatial relationship to the subduction zone and to major cities (plate-boundary symbols as in figure 3.15). Tan shaded area is covered by young volcanic deposits (less than 2 million years old).

Eruptive histories from USGS Open-File Report 94–585.

have shown seismic activity, and steam is escaping from these and from Mount Rainier, which last erupted in 1882. In fact, at least nine of the Cascade peaks are presently showing thermal activity of some kind (steam emission or hot springs). The eruption of Mount St. Helens has in one sense been useful: It has focused attention on this threat. Scientists are watching many of the Cascade Range volcanoes very closely now. There is particular concern about possible mudflow dangers from Mount Rainier (figure 5.30). This snow-capped stratovolcano has generated voluminous lahars in the geologically recent past; many towns are now potentially at risk.

The Aleutians

South-central Alaska and the Aleutian island chain sit above a subduction zone. This makes the region vulnerable not only to earthquakes, as noted in chapter 4, but also to volcanic

eruptions (figure 5.31). The volcanoes are typically andesitic with composite cones, and pyroclastic eruptions are common. There are more than 50 active volcanoes, and another 90 that are geologically young. Thirty of the volcanoes have erupted since 1900; collectively, these volcanoes average one eruption a year. Moreover, though Alaska is a region of relatively low population density, these volcanoes present a significant special hazard.

As with Eyjafjallajökull, the concern is the threat that volcanic ash poses to aviation. This is especially important with respect to the Alaskan volcanoes as Anchorage is a key stopover on long-distance cargo and passenger flights in the region, and many transcontinental flights follow great-circle routes over the area. It has been estimated that 80,000 large aircraft a year, and 30,000 passengers a day, pass through the skies above, or downwind of, these volcanoes. Since the hazard was first realized in the mid-1980s, a warning system has been

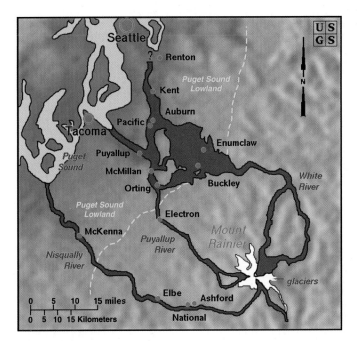

Figure 5.30

Over the last 5000 to 6000 years, huge mudflows have poured down stream valleys and into low-lying areas, tens of miles from Mount Rainier's summit. Even Tacoma or Seattle might be threatened by renewed activity on a similar scale.

Source: Data from T. W. Sisson, USGS Open-File Report 95–642.

put in place by the National Oceanic and Atmospheric Administration. However, much monitoring of the ash clouds is done by satellite, and ash and ordinary clouds cannot always be distinguished in the images (especially at night!). Volcanologists now coordinate with the National Weather Service to issue forecast maps of airborne ash plumes based on volcanic activity and wind velocity at different altitudes.

Case Study 5.2 focuses on Redoubt, cause of the most frightening of these airline ash incidents, and a great deal of additional damage.

Long Valley and Yellowstone Calderas

Finally, there are areas that do not now constitute an obvious hazard but that are causing some geologists concern because of increases in seismicity or thermal activity, and because of their past histories. One of these is the Mammoth Lakes area of California. In 1980, the area suddenly began experiencing earthquakes. Within one forty-eight-hour period in May 1980, four earthquakes of magnitude 6 rattled the region, interspersed with hundreds of lesser shocks. Mammoth Lakes lies within the Long Valley Caldera, a 13-kilometer-long oval depression formed during violent pyroclastic eruptions 700,000 years ago. Smaller eruptions occurred around the area as recently as 50,000 years ago.

A **caldera** is an enlarged volcanic crater, which may be formed either by an explosion enlarging an existing crater or by collapse of a volcano after a magma chamber within has emptied. The summit calderas of Mauna Loa and Kilauea formed predominantly by collapse as the magma chambers have become depleted, and enlarged as the rocks on the rim have weathered and crumbled. The caldera now occupied by Crater Lake would have formed by a combination of collapse and explosion (figure 5.32). In the case of a massive

Figure 5.31

The Alaska Volcano Observatory faces a large monitoring challenge, with 51 active Alaskan/Aleutian volcanoes.

Image by C. J. Nye, courtesy Alaska Volcano Observatory.

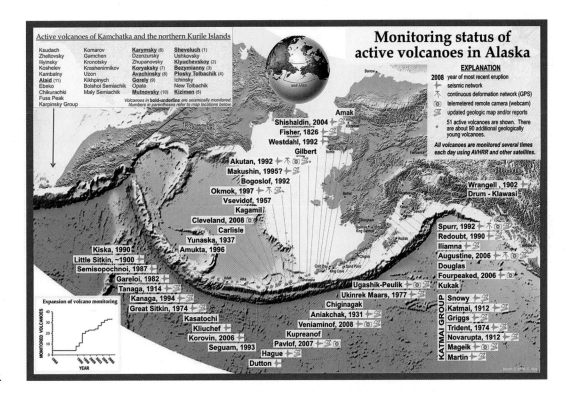

Redoubt Volcano, Alaska

Redoubt's location (figure 1) puts 60% of Alaska's population, and its major business centers, within reach of the volcano's eruptions. Nearby Augustine had erupted in 1986, so increasing seismicity at Redoubt in late 1989 drew prompt attention, though it was not obvious just when or how it might actually erupt. The first event, on 14 December, was an explosive one, spewing ash and rock downwind. Within a day thereafter, fresh glass shards, indicating fresh magma feeding the eruption, appeared in the ash. Redoubt would shortly wreak its costliest and perhaps most frightening havoc of this eruptive phase.

On 15 December, a KLM Boeing 747 flew into a cloud of ash from Redoubt, and all four engines failed. The plane plunged over 13,000 feet before the pilot somehow managed to restart the engines. Despite a windshield frosted and pitted by the abrasive ash, he managed to land the plane, and its 231 passengers, safely in Anchorage. However, the aircraft had suffered $80 million in damage.

Soon thereafter, a lava dome began to build in Redoubt's summit crater. Near-surface seismicity was mostly replaced by deeper events through the balance of December (figure 2). Then the regular tremors that volcanologists had come to associate with impending eruptions resumed on 29 December. Suspecting that the new lava dome had plugged the vent and that pressure was therefore building in the volcano, on 1–2 January 1990, the Alaska Volcano Observatory issued warnings of "moderate to strong explosive activity" expected

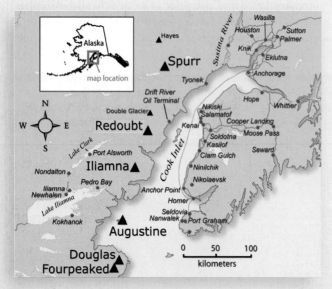

Figure 1

Alaska's population is concentrated around Cook Inlet; Redoubt, Augustine, and Spurr have all erupted since 1990.

J. Schaeffer and C. Cameron, courtesy Alaska Volcano Observatory/Alaska Division of Geologic and Geophysical Surveys.

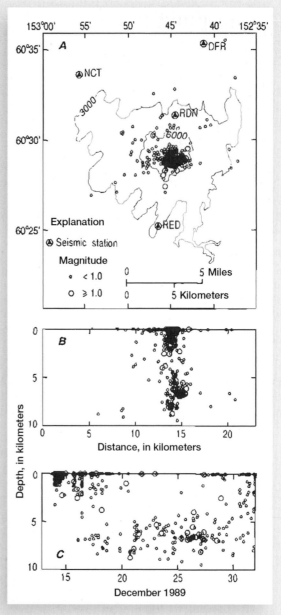

Figure 2

Seismicity at Redoubt, 13 December through 31 December 1989. (A) In map view, seismic activity is strongly concentrated below the summit; depth (B) is variable. (C) Depth of seismic events varies over time; mainly shallow during the first explosions, then deeper, with shallow events resuming in late December.

After USGS Circular 1061.

A

Figure 4

USGS geologists study pyroclastic debris from Redoubt.

Photograph by K. Bull, Alaska Volcano Observatory, U.S. Geological Survey.

and assorted damage due to the multiple pyroclastic eruptions, total cost was over $160 million.

In early 2009, Redoubt became active again for several months, with multiple pyroclastic eruptions. This time, new tools were available to help track the ash and keep aircraft safe (figure 6). As this is being written, the volcano is quiet, but monitoring will certainly continue, to ensure that any further eruption will not be a surprise.

B

Figure 3

(A) Mudflow deposits in the Drift River valley; Redoubt at rear of image. (B) Some lahars completely flooded structures in the valley.

Photographs by C. A. Gardner, Alaska Volcano Observatory, U.S. Geological Survey.

shortly. The Cook Inlet Pipeline Company evacuated the Drift River oil-tanker terminal. Strong explosions beginning on 2 January demolished the young lava dome in the crater and sent pyroclastic flows across Redoubt's glaciers. Lahars swept down the Drift River valley toward the terminal (figure 3). Larger blocks of hot rock crashed nearer the summit (figure 4). After the remains of the dome were blown away on 9 January, eruptions proceeded less violently, though ash and pyroclastic flows continued intermittently, interspersed with dome-building, for several more months (figure 5). Altogether, this eruption was the second costliest volcanic eruption in U.S. history, behind only the 1980 eruption of Mount St. Helens. Between aircraft damage, disruption of oil drilling in Cook Inlet and work at the pipeline terminal,

Figure 5

On 21 April 1990, ash rises from Redoubt. Note the classic stratovolcano shape.

Photograph by R. Clucas, Alaska Volcano Observatory, U.S. Geological Survey.

(Continued)

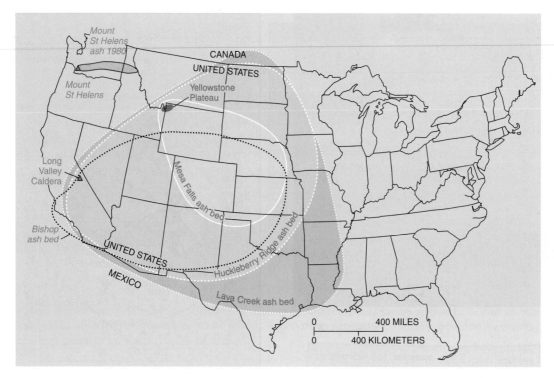

Figure 5.35

Major explosive eruptions of Yellowstone caldera produced the Mesa Falls, Huckleberry Ridge, and Lava Creek ash falls. The Lava Creek event was shown at VEI 8 in figure 5.26. Dashed black outline is Bishop ash fall from the Long Valley Caldera in eastern California.

After USGS Fact Sheet 2005–3024.

this is confirmed by the volume of pyroclastics ejected in three huge explosive eruptions over the last 2 million years or so (figure 5.35). Eruptions of this scale, with a VEI of 8, lead to the term "supervolcano," which has caught public attention.

Whether eruption on that scale is likely in the foreseeable future, no one knows. Scientists continue to monitor uplift and changes in seismic and thermal activity at Yellowstone. The research goes on.

Summary

Most volcanic activity is concentrated near plate boundaries. Volcanoes differ in eruptive style and in the kinds of dangers they present. Those along spreading ridges and at oceanic hot spots tend to be more placid, usually erupting a fluid, basaltic lava. Subduction-zone volcanoes are supplied by more-viscous, silica-rich, gas-charged andesitic or rhyolitic magma, so, in addition to lava, they may emit large quantities of pyroclastics and pyroclastic flows, and they may also erupt explosively. The type of volcanic structure built is directly related to the composition/physical properties of the magma; the fluid basaltic lavas build shield volcanoes, while the more-felsic magmas build smaller, steeper stratovolcanoes of lava and pyroclastics. At present, volcanologists can detect the early signs that a volcano may erupt in the near future (such as bulging, increased seismicity, or increased thermal activity), but they cannot predict the exact timing or type of eruption. Individual volcanoes, however, show characteristic eruptive styles (as a function of magma type) and patterns of activity. Therefore, knowledge of a volcano's eruptive history allows anticipation of the general nature of eruptions and of the likelihood of renewed activity in the near future.

Key Terms and Concepts

active volcano 108
andesite 92
basalt 92
caldera 113
cinder cone 96

composite volcano 96
dormant volcano 108
extinct volcano 108
fissure eruption 93
lahar 101

lava dome 96
phreatic eruption 105
pyroclastic flow 104
pyroclastics 96
rhyolite 92

shield volcano 95
stratovolcano 96
Volcanic Explosivity Index 109

Exercises

Questions for Review

1. Define a fissure eruption, and give an example.

2. The Hawaiian Islands are all shield volcanoes. What are shield volcanoes, and why are they not especially hazardous to life?

3. The eruptive style of Mount St. Helens is quite different from that of Kilauea in Hawaii. Why?

4. What are pyroclastics? Identify a kind of volcanic structure that pyroclastics may build.

5. Describe two strategies for protecting an inhabited area from an advancing lava flow.

6. Describe a way in which a lahar may develop, and a way to avoid its most likely path.

7. What is a nuée ardente, and why is a volcano known for producing these hot pyroclastic flows a special threat during periods of activity?

8. Explain the nature of a phreatic eruption, and give an example.

9. How may volcanic eruptions influence global climate?

10. Why is volcanic ash a particular hazard to aviation? Describe one strategy for reducing the risk.

11. Discuss the distinctions among active, dormant, and extinct volcanoes, and comment on the limitations of this classification scheme.

12. What is the Volcanic Explosivity Index?

13. Describe two precursor phenomena that may precede volcanic eruptions.

14. What is the underlying cause of present and potential future volcanic activity in the Cascade Range of the western United States?

15. What is the origin of volcanic activity at Yellowstone, and why is it sometimes described as a "supervolcano"?

Exploring Further

(For investigating volcanic activity, two excellent websites are the U.S. Geological Survey's Volcano Hazards Program site, http://volcanoes.usgs.gov/, and the Global Volcanism program site, http://www.volcano.si.edu/. The USGS site includes links to all of the U.S. regional volcano observatories.)

1. As this is written, Kilauea continues an active phase that has lasted close to three decades. Check out the current activity report of the Hawaii Volcanoes Observatory.

2. Go to the USGS site mentioned above. How many volcanoes are currently featured as being watched? Which one(s) have the highest Aviation Watch color codes, and why?

3. The Global Volcanism Program site highlights active volcanoes. How many volcanoes are currently on that list? Choose a volcano from the list with which you are *not* familiar, and investigate more closely: What type of volcano is it, and what type of activity is currently in progress? What is its tectonic setting, and how does that relate to its eruptive style?

4. Before the explosion of Mount St. Helens in May 1980, scientists made predictions about the probable extent of and kinds of damage to be expected from a violent eruption. Investigate those predictions, and compare them with the actual effects of the blast.

5. From figure 5.35, estimate the area covered by the Lava Creek ash bed. If the volume of ash was about 1000 km^3, as indicated in figure 5.26, calculate the approximate average thickness of ash deposited. Now go to vulcan.wr.usgs.gov/Volcanoes/Cascades/Hazards/tephra_plot_distance.html, see how ash thickness changed with distance from vent for several other explosive eruptions, and speculate on the thickness of the Lava Creek ash near its source.

6. Consider a region of dry upper mantle at the pressure and temperature of the starting point in figure 5.3. As a warm plume of this mantle rises, the pressure on it is reduced. If the density of the upper mantle is about 3.3 g/cm^3, and 1 atmosphere of pressure approximately equals 1 kg/cm^2, how far must the plume rise before melting begins, assuming no water is added?

Streams and Flooding

CHAPTER 6

W ater is the single most important agent sculpturing the earth's surface. Mountains may be raised by the action of plate tectonics and volcanism, but they are shaped primarily by water. Streams carve valleys, level plains, and move tremendous amounts of sediment from place to place.

They have also long played a role in human affairs. Throughout human history, streams have served as a vital source of fresh water, and often of fish for food. Before there were airplanes, trucks, or trains, there were boats to carry people and goods over water. Before the widespread use of fossil fuels, flowing water pushing paddlewheels powered mills and factories. For these and other reasons, many towns sprang up and grew along streams. This, in turn, has put people in the path of floods when the streams overflow.

Floods are probably the most widely experienced catastrophic geologic hazards (figure 6.1). On the average, in the United States alone, floods annually take 140 lives and cause well over $6 billion in property damage. The 1993 flooding in the Mississippi River basin took 48 lives and caused an estimated $15–20 billion in damages.

Most floods do not make national headlines, but they may be no less devastating to those affected by them. Some floods are the result of unusual events, such as the collapse of a dam, but the vast majority are a perfectly normal, and to some extent predictable, part of the natural functioning of streams. Before discussing flood hazards, therefore, we will examine how water moves through the hydrologic cycle, and we will also look at the basic characteristics and behavior of streams.

Record flooding in the upper Mississippi River basin in spring 2011 led to deliberate breaching of levees by the U.S. Army Corps of Engineers. The city of Cairo, Illinois, sits at the confluence of the Mississippi (flowing in from the northwest, and obviously flooding here) and the Ohio River (joining it from the northeast). Arrow indicates the Bird's Point levee, breached on 2 May 2011, allowing a flow of over 500,000 cu. ft./sec into the New Madrid floodway (broad brown area west of the river below the levee), which was created for just such an emergency after flooding in 1927 but had not been used since 1937.

NASA image created by Jesse Allen and Robert Simmon, using Landsat data provided by the U. S. Geological Survey.

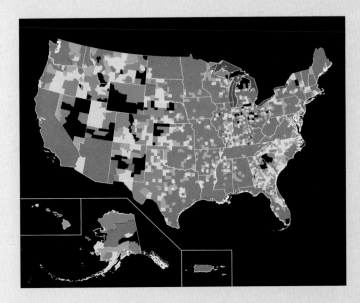

Figure 6.1
Presidential disaster declarations related to flooding, by county, from 1 June 1965 through 1 June 2003: green, one; yellow, two; orange, three; red, four or more. Some flood disasters are related to coastal storms, but many are due to stream flooding.

Source: U.S. Geological Survey

The Hydrologic Cycle

The **hydrosphere** includes all the water at and near the surface of the earth. Most of it is believed to have been outgassed from the earth's interior early in its history, when the earth's temperature was higher; icy planetesimals added late to the accreting earth's surface may have contributed some, too. Now, with only occasional minor additions from volcanoes bringing up "new" water from the mantle, and small amounts of water returned to the mantle with subducted lithosphere, the quantity of water in the hydrosphere remains essentially constant.

All of the water in the hydrosphere is caught up in the **hydrologic cycle,** illustrated in figure 6.2. The largest single reservoir in the hydrologic cycle, by far, consists of the world's oceans, which contain over 97% of the water in the hydrosphere; lakes and streams together contain only 0.016% of the water. The main processes of the hydrologic cycle involve evaporation into and precipitation out of the atmosphere. Precipitation onto land can reevaporate (directly from the ground surface or indirectly through plants by evapotranspiration), infiltrate into the ground, or run off over the ground surface. Surface runoff may occur in streams or by unchanneled overland flow. Water that percolates into the ground may also flow (see chapter 11) and commonly returns, in time, to the oceans. The oceans are the principal source of evaporated water because of their vast areas of exposed water surface.

The total amount of water moving through the hydrologic cycle is large, more than 100 million billion gallons per year. A portion of this water is temporarily diverted for human use, but it ultimately makes its way back into the natural global water cycle by a variety of routes, including release of municipal sewage, evaporation from irrigated fields, or discharge of industrial wastewater into streams. Water in the hydrosphere may spend extended periods of time in storage in one or another of the water reservoirs shown in figure 6.2—a few months in a river, a few thousand years in the oceans, centuries to millennia in glaciers or as ground water—but from the longer perspective of geologic history, it is still regarded as moving continually through the hydrologic cycle. In the process, it may participate in other geologic cycles and processes: for example, streams eroding rock and moving sediment, and subsurface waters dissolving rock and transporting dissolved chemicals, are also contributing to the rock cycle. We revisit the hydrologic cycle in the context of water resources in chapter 11.

Streams and Their Features

Streams—General

A **stream** is any body of flowing water confined within a channel, regardless of size, although people tend, informally, to use the term *river* for a relatively large stream. It flows downhill through local topographic lows, carrying away water over the earth's surface. The region from which a stream draws water is its **drainage basin** (figure 6.3), or *watershed.* A *divide* separates drainage basins. The size of a stream at any point is related in part to the size (area) of the drainage basin upstream from that point, which determines how much water from falling snow or rain can be collected into the stream. Stream size is also influenced by several other factors, including climate (amount of precipitation and evaporation), vegetation or lack of it, and the underlying geology, all of which are discussed in more detail later in the chapter. That is, a stream carves its channel (in the absence of human intervention), and the channel carved is broadly proportional to the volume of water that must typically be accommodated. Long-term, sustained changes in precipitation, land use, or other factors that change the volume of water customarily flowing in the stream will be reflected in corresponding changes in channel geometry.

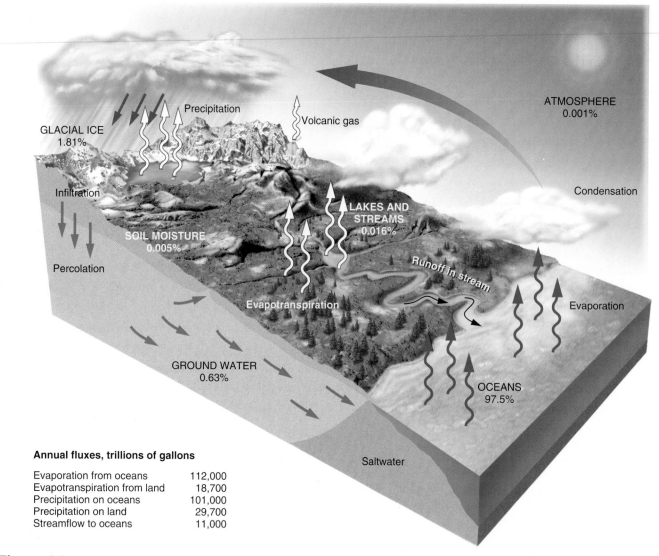

Annual fluxes, trillions of gallons

Evaporation from oceans	112,000
Evapotranspiration from land	18,700
Precipitation on oceans	101,000
Precipitation on land	29,700
Streamflow to oceans	11,000

Figure 6.2

Principal processes and reservoirs of the hydrologic cycle (reservoirs in capital letters). Straight lines show movement of liquid water; wavy lines show water vapor. (Limited space prevents showing all processes everywhere they occur.)

The size of a stream may be described by its **discharge,** the volume of water flowing past a given point (or, more precisely, through a given cross section) in a specified length of time. Discharge is the product of channel cross section (area) times average stream velocity (figure 6.4). Historically, the conventional unit used to express discharge is cubic feet per second; the analogous metric unit is cubic meters per second. Discharge may range from less than one cubic foot per second on a small creek to millions of cubic feet per second in a major river. For a given stream, discharge will vary with season and weather, and may be influenced by human activities as well.

Sediment Transport

Water is a powerful agent for transporting material. Streams can move material in several ways (figure 6.5). Heavier debris may be rolled, dragged, or pushed along the bottom of the stream

bed as its *traction load.* Material of intermediate size may be carried in short hops along the stream bed by a process called *saltation.* All this material is collectively described as the *bed load* of the stream. The *suspended load* consists of material that is light or fine enough to be moved along suspended in the stream, supported by the flowing water. Suspended sediment clouds a stream and gives the water a muddy appearance. Finally, some substances may be completely dissolved in the water, to make up the stream's *dissolved load.*

The total quantity of material that a stream transports by all these methods is called, simply, its **load.** Stream **capacity** is a measure of the total load of material a stream can move. Capacity is closely related to discharge: the faster the water flows, and the more water is present, the more material can be moved. How much of a load is actually transported also depends on the availability of sediments or soluble material: a stream flowing over solid bedrock will not be able to dislodge

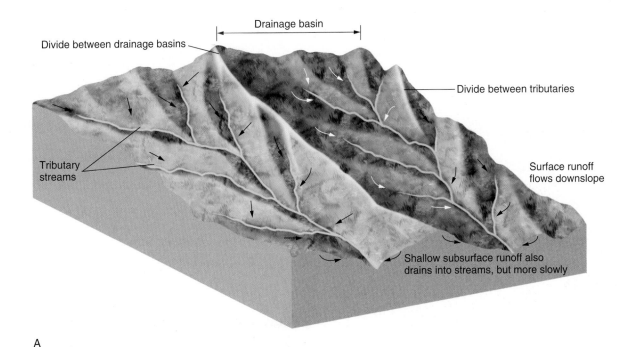

A

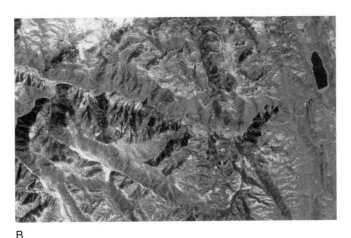

B

Figure 6.3

Streams and their drainage basins. (A) A topographic high creates a divide between the drainage basins of adjacent streams. (B) Mountain ridges separate drainage basins in the Colorado Rockies. The largest stream valley in this image is Big Thompson Canyon, site of serious flooding decades ago.

(B) NASA image by Jesse Allen and Robert Simmon, using EO-1 ALI data provided by the NASA EO-1 team.

much material, while a similar stream flowing through sand or soil may move considerable material. Vegetation can also influence sediment transport, by preventing the sediment from reaching the stream at all, or by blocking its movement within the stream channel.

Velocity, Gradient, and Base Level

Stream velocity is related partly to discharge and partly to the steepness (pitch or angle) of the slope down which the stream flows. The steepness of the stream channel is called its **gradient.** It is calculated as the difference in elevation between two points along a stream, divided by the horizontal distance between them along the stream channel (the water's flow path). All else being equal, the higher the gradient, the steeper the channel (by definition), and the faster the stream flows. Gradient and velocity commonly vary along the length of a stream (figure 6.6). Nearer its source, where the first perceptible trickle

of water signals the stream's existence, the gradient is usually steeper, and it tends to decrease downstream. Velocity may or may not decrease correspondingly: The effect of decreasing gradient may be counteracted by other factors, including increased water volume as additional tributaries enter the stream, which adds to the mass of water being pulled downstream by gravity. Velocity is also influenced by friction between water and stream bed, and changes in the channel's width and depth. Overall, discharge does tend to increase downstream, at least in moist climates, primarily because of water added by tributaries along the way.

By the time the stream reaches its end, or mouth, which is usually where it flows into another body of water, the gradient is typically quite low. Near its mouth, the stream is approaching its **base level,** which is the lowest elevation to which the stream can erode downward. For most streams, base level is the water (surface) level of the body of water into which they flow. For streams flowing into the ocean, for instance, base level is sea level. The closer a stream is to its base level, the lower the stream's gradient. As noted above, it may flow more slowly in consequence, although the slowing effect of decreased gradient may be compensated by an increase in discharge due to widening of the channel and a corresponding increase in channel cross section. The downward pull of gravity not only pulls water down the channel, but

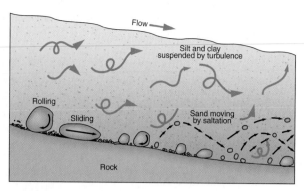

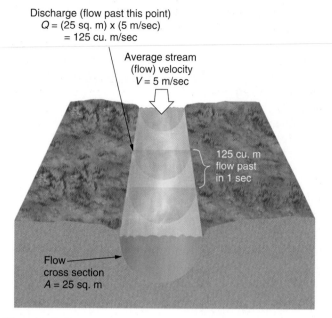

Discharge (flow past this point)
Q = (25 sq. m) x (5 m/sec)
  = 125 cu. m/sec

Average stream
(flow) velocity
V = 5 m/sec

125 cu. m
flow past
in 1 sec

Flow
cross section
A = 25 sq. m

Figure 6.4

Discharge (Q) equals channel cross section (area) times average velocity: $Q = A \times V$. In a real stream, both channel shape and flow velocity vary, so reported discharge is necessarily an average. In any cross section, velocity near the center of the channel will be greater than along bottom and sides of the channel. Discharge will also vary seasonally as conditions become wetter or dryer, and in response to single events such as rainstorms.

Flow

Silt and clay
suspended by turbulence

Rolling

Sliding

Sand moving
by saltation

Rock

Figure 6.5

Processes of sediment transport in a stream. How material can be moved depends on its size and weight, and the flow velocity.

of the stream's elevation from source to mouth) assumes a characteristic concave-upward shape, as shown in figure 6.6.

Velocity and Sediment Sorting and Deposition

Variations in a stream's velocity along its length are reflected in the sizes of sediments deposited at different points. The more rapidly a stream flows, the larger and denser are the particles that can be moved. The sediments found motionless in a stream bed at any point are those too big or heavy for that stream to move at that point. Where the stream flows most quickly, it carries gravel and even boulders along with the finer sediments. As the stream slows down, it starts leaving behind the heaviest, largest particles—the boulders and gravel—and continues to move the lighter, finer materials along. If stream velocity continues to decrease, successively smaller particles

also causes a stream to cut down vertically toward its base level. Counteracting this erosion is the influx of fresh sediment into the stream from the drainage basin. Over time, natural streams tend toward a balance, or equilibrium, between erosion and deposition of sediments. The **longitudinal profile** of such a stream (a sketch

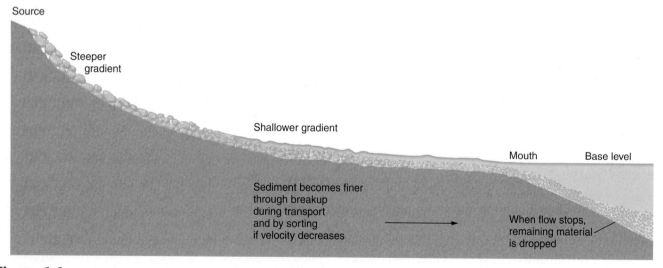

Source

Steeper
gradient

Shallower gradient

Mouth   Base level

Sediment becomes finer
through breakup
during transport
and by sorting
if velocity decreases

When flow stops,
remaining material
is dropped

Figure 6.6

Typical longitudinal profile of a stream in a temperate climate. Note that the gradient decreases from source to mouth. Particle sizes may decrease with breakup downstream. Total discharge, capacity, and velocity may all increase downstream despite the decrease in gradient (though commonly, velocity decreases with decreasing gradient). Conversely, in a dry climate, water loss by evaporation and infiltration into the ground may *decrease* discharge downstream.

are dropped: the sand-sized particles next, then the clay-sized ones. In a very slowly flowing stream, only the finest sediments and dissolved materials are still being carried. The relationship between the velocity of water flow and the size of particles moved accounts for one characteristic of stream-deposited sediments: They are commonly **well sorted** by size or density, with materials deposited at a given point tending to be similar in size or weight. If velocity decreases along the length of the stream profile, then so will the coarseness of sediments deposited, as also shown in figure 6.6.

If a stream flows into a body of standing water, like a lake or ocean, the stream's flow velocity drops to zero, and all the remaining suspended sediment is dropped. If the stream is still carrying a substantial load as it reaches its mouth, and it then flows into still waters, a large, fan-shaped pile of sediment, a **delta,** may be built up (figure 6.7). A similarly shaped feature,

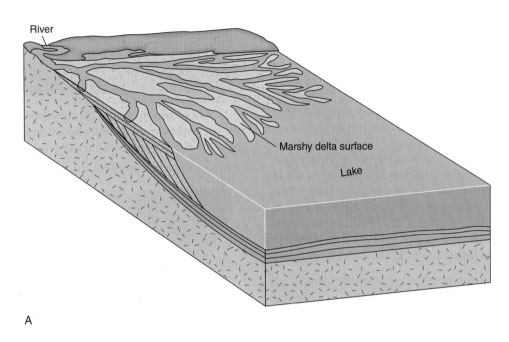

A

B

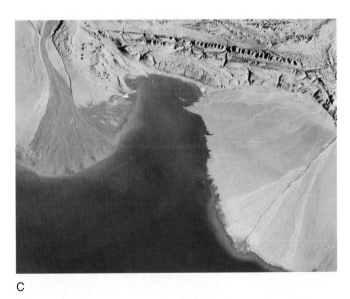

C

Figure 6.7

(A) A river like the Mississippi builds a delta by depositing its sediment load when it flows into a still lake or ocean; the stream may split up into many small channels as it flows across the delta to open water. (B) The Mississippi River delta (satellite photo). The river carries a million tons of sediment each day. Here, the suspended sediment shows as a milky, clouded area in the Gulf of Mexico. The sediment deposition gradually built Louisiana outward into the gulf. Now, upstream dams have cut off sediment influx, and the delta is eroding in many places. (C) Deltas formed by streams flowing into Lake Ayakum on the Tibetan Plateau.

(B) Photo courtesy NASA; (C) NASA image courtesy Image Science and Analysis Laboratory, Johnson Space Center.

A

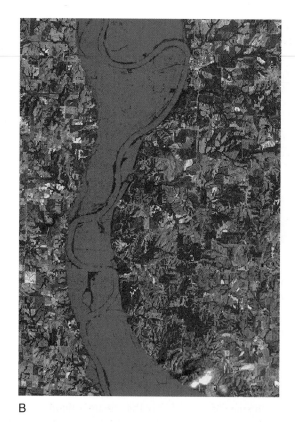

B

prehistoric channel

Mississippi River

oxbow lake

2km N

C

Figure 6.12

(A) Floodplain of the Missouri River near Glasgow, Missouri is bounded by steep bluffs. (B) During the 1993 flooding, the floodplain lived up to its name. The channel is barely visible, outlined by vegetation growing on levees along the banks.
(C) Meandering of the lower Mississippi River has produced many oxbows. Note point bars on the inside banks of several meanders.

(A) and (B) Images by Scientific Assessment and Strategy Team, courtesy USGS. (C) Image by Robert Simmon, courtesy NASA.

accumulations early in the winter, followed by thawing and persistent rains that both added yet more water and accelerated the melting of the snow.

The rate of surface runoff is influenced by the extent of infiltration, which, in turn, is controlled by the soil type and how much soil is exposed. Soils, and rocks, vary in *porosity* and *permeability,* properties that are explored further in chapter 11. A very porous and permeable soil allows a great deal of water to sink in relatively fast. If the soil is less permeable (meaning that water does not readily sink into or flow through it) or is covered by artificial structures, the proportion of water that runs off over the surface increases. Once even permeable soil is satu-

rated with water, any additional moisture is necessarily forced to become part of the surface runoff.

Topography also influences the extent or rate of surface runoff: The steeper the terrain, the more readily water runs off over the surface and the less it tends to sink into the soil. Water that infiltrates the soil, like surface runoff, tends to flow down-gradient, and may, in time, also reach the stream. However, the subsurface runoff water, flowing through soil or rock, generally moves much more slowly than the surface runoff. The more gradually the water reaches the stream, the better the chances that the stream discharge will be adequate to carry the water away without flooding. Therefore, the relative amounts of surface runoff and subsurface (groundwater) flow, which are strongly influenced by the near-surface geology of the drainage basin, are fundamental factors affecting the severity of stream flooding.

Vegetation may reduce flood hazards in several ways. The plants may simply provide a physical barrier to surface runoff, decreasing its velocity and thus slowing the rate at which water reaches a stream. Plant roots working into the soil loosen it, which tends to maintain or increase the soil's permeability, and hence infiltration, thus reducing the proportion of surface run-off. Plants also absorb water, using some of it to grow and releasing some slowly by evapotranspiration from foliage. All of these factors reduce the volume of water introduced directly into a stream system.

All other factors being equal, local flood hazard may vary seasonally or as a result of meteorologic fluctuations. Just as soil already saturated from previous storms cannot readily absorb more water, so the solidly frozen ground of cold regions prevents infiltration; a midwinter rainstorm in such a region may produce flooding with a quantity of rain that could be readily absorbed by the soil in summer. The extent and vigor of growth of vegetation varies seasonally also, as does atmospheric humidity and thus evapotranspiration.

Flood Characteristics

During a flood, the water level of a stream is higher than usual, and its velocity and discharge also increase as the greater mass of water is pulled downstream by gravity. The higher volume (mass) and velocity together produce the increased force that gives floodwaters their destructive power (figure 6.13). The elevation of the water surface at any point is termed the **stage** of the stream. A stream is at *flood stage* when stream stage exceeds bank height. The magnitude of a flood can be described by either the maximum discharge or maximum stage reached. The stream is said to **crest** when the maximum stage is reached. This may occur within minutes of the influx of water, as with flooding just below a failed dam. However, in places far downstream from the water input or where surface runoff has been slowed, the stream may not crest for several days after the flood episode begins. In other words, just because the rain has stopped does not mean that the worst is over, as was made abundantly evident in the 1993 Midwestern flooding, described in Case Study 6.2. The flood crested in late June in Minnesota; 9 July in the Quad Cities; 1 August in St. Louis; and 8 August in Cape Girardeau, Missouri, near the southern end of Illinois.

Flooding may afflict only a few kilometers along a small stream, or a region the size of the Mississippi River drainage basin, depending on how widely distributed the excess water is. Floods that affect only small, localized areas (or streams draining small basins) are sometimes called **upstream floods.** These are most often caused by sudden, locally intense rainstorms and by events like dam failure. Even if the total amount of water involved is moderate, the rapidity with which it enters the stream can cause it temporarily to exceed the stream channel capacity. The resultant flood is typically brief, though it can also be severe. Because of the localized nature of the excess water during an upstream flood, there is unfilled channel capacity farther downstream to accommodate the extra water. Thus, the water can be rapidly carried down from the upstream area, quickly reducing the stream's stage.

Flash floods are a variety of upstream flood, characterized by especially rapid rise of stream stage. They can occur anywhere that surface runoff is rapid, large in volume, and funneled into a relatively restricted area. Canyons are common natural sites of flash floods (figure 6.14). Urban areas may experience flash floods in a highway underpass or just a dip in the road that becomes a temporary stream as runoff washes over

Figure 6.13

House undercut during Big Thompson Canyon flood of 1976. Behind it is a landslide also caused by undercutting of the hillside. Note how low the normal flow is in August, when this photo was taken.

Photograph by R. R. Shroba, U.S. Geological Survey Photographic Library.

Figure 6.14

This canyon in Zion National Park is an example of a setting in which dangerous flash floods are possible, and steep canyon walls offer no route of quick retreat to higher ground. Damage from Big Thompson Canyon flash flooding was so severe in part because it too is steep-sided; recall figures 6.3B and 6.13.

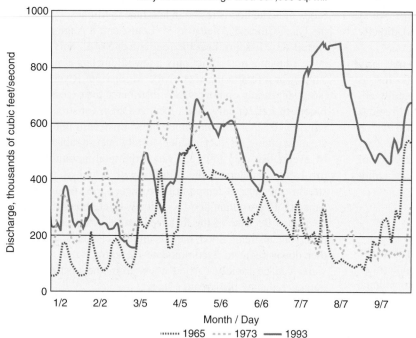

Mississippi River at St. Louis, Missouri
daily flows. Drainage area 697,000 sq. mi.

Month / Day

······· 1965   - - - 1973   —— 1993

Figure 6.15

The summer 1993 flooding in the upper Mississippi River basin was not only unusual in its timing (compare with 1965 and 1973 patterns) but, in many places, broke flood-stage and/or discharge records set in 1973 for *any* time of year. In a typical year, snowmelt in the north plus spring rains lead to a rise in discharge in late spring. Summer is generally dry. Heavy 1993 summer rains were exceptional and produced exceptional flooding.

Hydrograph data courtesy U.S. Geological Survey

it. Flash floods can even occur in deserts after a cloudburst, though in this case, the water level is likely to subside quickly as water sinks into parched ground rather than draining away in a stream.

Floods that affect large stream systems and large drainage basins are called **downstream floods.** These floods more often result from prolonged heavy rains over a broad area or from extensive regional snowmelt. Downstream floods usually last longer than upstream floods because the whole stream system is choked with excess water. The summer 1993 flooding in the Mississippi River basin is an excellent example of downstream flooding, brought on by many days of above-average rainfall over a broad area, with parts of the stream system staying above flood stage for weeks.

Stream Hydrographs

Fluctuations in stream stage or discharge over time can be plotted on a **hydrograph** (figure 6.15). Hydrographs spanning long periods of time are very useful in constructing a picture of the "normal" behavior of a stream and of that stream's response to flood-causing events. Logically enough, a flood shows up as a peak on the hydrograph. The height and width of that peak and its position in time relative to the water-input event(s) depend, in part, on where the measurements are taken relative to where the excess water is entering the system. Upstream, where the drainage basin is smaller and the water need not travel so far to reach the stream, the peak is likely to be sharper—with a higher crest and more rapid rise and fall of the water level—and to occur sooner after the influx of water. By the time that water pulse has moved downstream to a lower point in the drainage basin, it will

have dispersed somewhat, so that the arrival of the water spans a longer time. The peak will be spread out, so that the hydrograph shows both a later and a broader, gentler peak (see figure 6.16A). A short event like a severe cloudburst will tend to produce a sharper peak than a more prolonged event like several days of steady rain or snowmelt, even if the same amount of water is involved. Sometimes, the flood-causing event also causes permanent changes in the drainage system (figure 6.16B).

Hydrographs can be plotted with either stage or discharge on the vertical axis. Which parameter is used may depend on the purpose of the hydrograph. For example, stage may be the more relevant to knowing how much of a given floodplain area will be affected by the floodwaters. As a practical matter, too, measurement of stream stage is much simpler, so accurate stage data may be more readily available.

Flood-Frequency Curves

Another way of looking at flooding is in terms of the frequency of flood events of differing severity. Long-term records make it possible to construct a curve showing discharge as a function of recurrence interval for a particular stream or section of one. The sample **flood-frequency curve** shown in figure 6.17 indicates that for this stream, on average, an event producing a discharge of 675 cubic feet/second occurs once every ten years, an event with discharge of 350 cubic feet/second occurs once every two years, and so on. A flood event can then be described by its **recurrence interval:** how frequently a flood of that severity occurs, *on average,* for that stream.

Alternatively (and preferably), one can refer to the *probability* that a flood of given size will occur in any one year; this

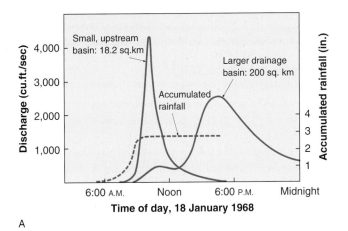

A

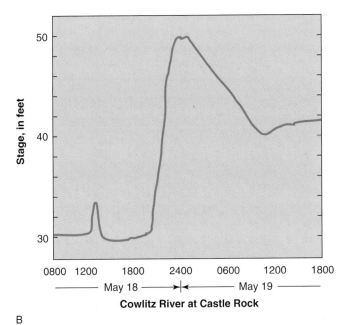

B

Figure 6.16

(A) Flood hydrographs for two different points along Calaveras Creek near Elmendorf, Texas. The flood has been caused by heavy rainfall in the upper reaches of the drainage basin. Here, flooding quickly follows the rain. The larger stream lower in the drainage basin responds more sluggishly to the input. (B) Mudflow into the Cowlitz River from the 18 May 1980 eruption of Mount St. Helens caused a sharp rise in stream stage (height), and flooding; infilling of the channel by sediment resulted in a long-term increase in stream stage—higher water level—for given discharge and net water depth.

(A) Source Water Resources Division, U.S. Geological Survey; (B) after U.S. Geological Survey.

probability is the inverse of the recurrence interval. For the stream of figure 6.17, a flood with a discharge of 675 cubic feet/second is called a "ten-year flood," meaning that a flood of that size occurs about once every ten years, or has a 10% (1/10) probability of occurrence in any year; a discharge of 900 cubic feet/second is called a "forty-year flood" and has a 2.5% (1/40) probability of occurrence in any year, and so on.

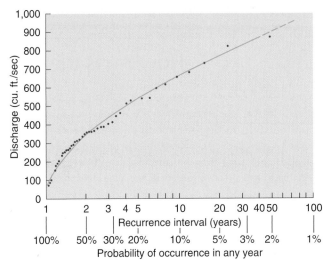

Figure 6.17

Flood-frequency curve for Eagle River at Red Cliff, Colorado. Records span 46 years, so estimates of the sizes of moderately frequent floods are likely to be fairly accurate, though the rare larger floods are hard to project.

USGS Open-File Report 79–1060

Flood-frequency (or flood-probability) curves can be extremely useful in assessing regional flood hazards. If the severity of the one-hundred- or two-hundred-year flood can be estimated, then, even if such a rare and serious event has not occurred within memory, scientists and planners can, with the aid of topographic maps, project how much of the region would be flooded in such events. They can speak, for example, of the "one-hundred-year floodplain" (the area that would be water-covered in a one-hundred-year flood). Such information is useful in preparing flood-hazard maps, an exercise that is also aided by the use of aerial or satellite photography. Such maps, in turn, are helpful in siting new construction projects to minimize the risk of flood damage or in making property owners in threatened areas aware of dangers, just as can be done for earthquake and volcano hazards.

Unfortunately, the best-constrained part of the curve is, by definition, that for the lower-discharge, more-frequent, higher-probability, less-serious floods. Often, considerable guesswork is involved in projecting the shape of the curve at the high-discharge end. Reliable records of stream stages, discharges, and the extent of past floods typically extend back only a few decades. Many areas, then, may never have recorded a fifty- or one-hundred-year flood. Moreover, when the odd severe flood does occur, how does one know whether it was a sixty-year flood, a one-hundred-year flood, or whatever? The high-discharge events are rare, and even when they happen, their recurrence interval can only be estimated. Considerable uncertainty can exist, then, about just how bad a particular stream's one-hundred- or two-hundred-year flood might be. This problem is explored further in Case Study 6.1, "How Big Is the One-Hundred-Year Flood?"

It is important to remember, too, that the recurrence intervals or probabilities assigned to floods are *averages*.

How Big Is the One-Hundred-Year Flood?

A common way to estimate the recurrence interval of a flood of given size is as follows: Suppose the records of maximum discharge (or maximum stage) reached by a particular stream each year have been kept for N years. Each of these yearly maxima can be given a rank M, ranging from 1 to n, 1 being the largest, n the smallest. Then the recurrence interval R of a given annual maximum is defined as

$$R = \frac{(N + 1)}{M}$$

This approach assumes that the N years of record are representative of "typical" stream behavior over such a period; the larger N, the more likely this is to be the case.

For example, table 1 shows the maximum one-day mean discharges of the Big Thompson River, as measured near Estes Park, Colorado, for twenty-five consecutive years, 1951–1975. If these values are ranked 1 to 25, the 1971 maximum of 1030 cubic feet/second is the seventh largest and, therefore, has an estimated recurrence interval of

$$\frac{(25 + 1)}{7} = 3.71 \text{ years.}$$

or a 27% probability of occurrence in any year.

Suppose, however, that only ten years of records are available, for 1966–1975. The 1971 maximum discharge happens to be the largest in that period of record. On the basis of the shorter record, its

Table 1	Calculated Recurrence Intervals for Discharges of Big Thompson River at Estes Park, Colorado				
		FOR TWENTY-FIVE-YEAR RECORD		FOR TEN-YEAR RECORD	
Year	Maximum Mean One-Day Discharge (cu. ft./sec)	M (rank)	R (years)	M (rank)	R (years)
1951	1220	4	6.50	3	3.67
1952	1310	3	8.67	2	5.50
1953	1150	5	5.20	4	2.75
1954	346	25	1.04	10	1.10
1955	470	23	1.13	9	1.22
1956	830	13	2.00	6	1.83
1957	1440	2	13.00	1	11.00
1958	1040	6	4.33	5	2.20
1959	816	14	1.86	7	1.57
1960	769	17	1.53	8	1.38
1961	836	12	2.17		
1962	709	19	1.37		
1963	692	21	1.23		
1964	481	22	1.18		
1965	1520	1	26.00		
1966	368	24	1.08	10	1.10
1967	698	20	1.30	9	1.22
1968	764	18	1.44	8	1.38
1969	878	10	2.60	4	2.75
1970	950	9	2.89	3	3.67
1971	1030	7	3.71	1	11.00
1972	857	11	2.36	5	2.20
1973	1020	8	3.25	2	5.50
1974	796	15	1.73	6	1.83
1975	793	16	1.62	7	1.57

Source: Data from U.S. Geological Survey Open-File Report 79–681.

estimated recurrence interval is (10 + 1)/1 = 11 years, corresponding to a 9% probability of occurrence in any year. Alternatively, if we look at only the first ten years of record, 1951–1960, the recurrence interval for the 1958 maximum discharge of 1040 cubic feet/second (a maximum discharge of nearly the same size as that in 1971) can be estimated at 2.2 years, meaning that it would have a 45% probability of occurrence in any one year.

Which estimate is "right"? Perhaps none of them, but their differences illustrate the need for long-term records to smooth out short-term anomalies in streamflow patterns, which are not uncommon, given that climatic conditions can fluctuate significantly over a period of a decade or two.

The point is further illustrated in figure 1. It is rare to have one hundred years or more of records for a given stream, so the magnitude of fifty-year, one-hundred-year, or larger floods is commonly estimated from a flood-frequency curve. Curves *A* and *B* in figure 1 are based, respectively, on the first and last ten years of data from the last two columns of table 1, the data points represented by X symbols for *A* and open circles for *B*. These two data sets give estimates for the size of the one-hundred-year flood that differ by more than 50%. These results, in

turn, both differ somewhat from an estimate based on the full 25 years of data (curve *C*, solid circles). This graphically illustrates how important long-term records can be in the projection of recurrence intervals of larger flood events. The flood-frequency curve in figure 6.17, based on nearly fifty years of record, inspires more confidence—if recent land-use changes have been minimal, at least.

In 1976, the stream on which this case study is based, the Big Thompson River, experienced catastrophic flash flooding, one consequence of which was shown in figure 6.13. A large thunderstorm dumped about a foot of rain on parts of the drainage basin, and the resultant flood took nearly 150 lives and caused $35 million in property damage. Research later showed that in some places, the flooding exceeded anything reached in the 10,000 years since the glaciers covered these valleys; yet at Estes Park, the gage used for the data in this case study, discharge peaked at 457 cu. ft. /sec, less than a two-year flood. This illustrates how localized and variable in severity such upstream floods can be.

For a further look at flood-frequency projection, see the online poster "100-Year Flood—It's All About Chance" from the U.S. Geological Survey at pubs.usgs.gov/gip/106/.

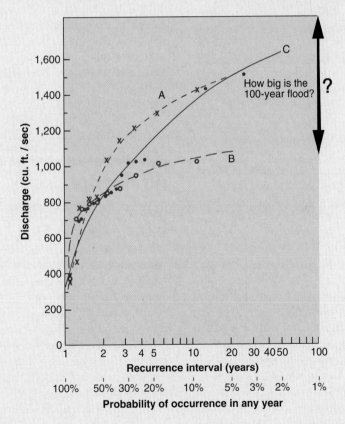

Figure 1

Flood-frequency curves for a given stream can look very different, depending on the length of record available and the particular period studied.

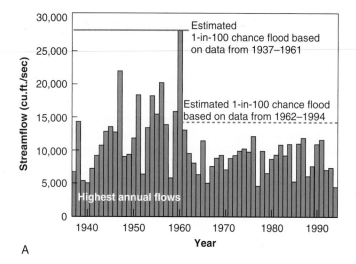

A

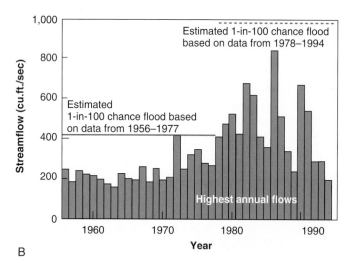

B

Figure 6.18

The size of the "1-in-100 chance flood" (the "hundred-year flood") can change dramatically as a result of human activities. (A) For the Green River at Auburn, Washington, the projected size of this flood has been halved by the Howard Hanson Dam. (B) Rapid urban development in the Mercer Creek Basin since 1977 has more than doubled the discharge of the projected hundred-year flood.

Modified after USGS Fact Sheet 229–96

The recurrence-interval terminology may be misleading in this regard, especially to the general public. Over many centuries, a fifty-year flood should occur an average of once every fifty years; or, in other words, there is a 2% chance that that discharge will be exceeded in any one year. However, that does not mean that two fifty-year floods could not occur in successive years, or even in the same year. The probability of two fifty-year flood events in one year is very low (2% × 2%, or .04%), but it is possible. Parts of the Chicago area, in 1986–1987, experienced two "one-hundred-year" floods within a ten-month period. Therefore, it is foolish to assume that, just because a severe flood has recently occurred in an area, it is in any sense "safe" for a while. Statistically, another such severe flood probably will not happen for

some time, but there is no guarantee of that. Misleading though it may be, the recurrence-interval terminology is still in common use by many agencies involved with flood-hazard mapping and flood-control efforts, such as the U.S. Army Corps of Engineers.

Another complication is that streams in heavily populated areas are affected by human activities, as we shall see further in the next section. The way a stream responded to 10 centimeters of rain from a thunderstorm a hundred years ago may be quite different from the way it responds today. The flood-frequency curves are therefore changing with time. A flood-control dam that can moderate spikes in discharge may reduce the magnitude of the hundred-year flood. Except for measures specifically designed for flood control, however, most human activities have tended to aggravate flood hazards and to decrease the recurrence intervals of high-discharge events. In other words, what may have been a one-hundred-year flood two centuries ago might be a fifty-year flood, or a more frequent one, today. Figure 6.18 illustrates both types of changes.

The challenge of determining flood recurrence intervals from historical data is not unlike the forecasting of earthquakes by applying the earthquake-cycle model to historic and prehistoric seismicity along a fault zone. The results inevitably have associated uncertainties, and the more complex the particular geologic setting, the more land use and climate have changed over time, and the sparser the data, the larger the uncertainties will be.

Economics plays a role, too. A 1998 report to Congress by the U.S. Geological Survey showed that over the previous 25 years, the number of active stream-gaging stations declined: Stations abandoned exceeded stations started or reactivated. Ongoing budget constraints continue to cause a net decrease in active gaging stations, from a peak of 7800 in 1968 to about 3100 in 2011. The ability to accumulate or extend long-term streamflow records in many places is thus being lost.

Consequences of Development in Floodplains

Why would anyone live in a floodplain? One reason might be ignorance of the extent of the flood hazard. Someone living in a stream's one-hundred- or two-hundred-year floodplain may be unaware that the stream could rise that high, if historical flooding has all been less severe. In mountainous areas, floodplains may be the only flat or nearly flat land on which to build, and construction is generally far easier and cheaper on nearly level land than on steep slopes. Around a major river like the Mississippi, the one-hundred- or two-hundred-year floodplain may include a major portion of the land for miles around, as became painfully evident in 1993, and it may be impractical to leave that much real estate entirely vacant. Farmers have settled in floodplains since ancient times because flooding streams deposit fine sediment over the lands flooded, replenishing nutrients in the soil and thus making the soil especially fertile. Where rivers are used for transportation, cities may have been built deliberately as close to the water as possible. And, of course,

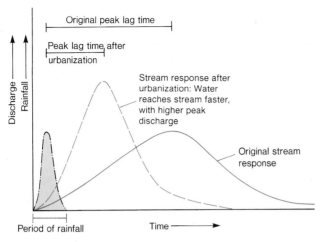

Figure 6.19

Hydrograph reflecting modification of stream response to precipitation following urbanization: Peak discharge increases, lag time to peak decreases.

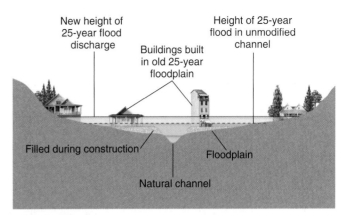

Figure 6.20

How floodplain development increases flood stage for a given discharge.

many streams are very scenic features to live near. Obviously, the more people settle and build in floodplains, the more damage flooding will do. What people often fail to realize is that floodplain development can actually increase the likelihood or severity of flooding.

As mentioned earlier, two factors affecting flood severity are the proportion and rate of surface runoff. The materials extensively used to cover the ground when cities are built, such as asphalt and concrete, are relatively impermeable and greatly reduce infiltration (and groundwater replenishment, as will be seen in chapter 11). Therefore, when considerable area is covered by these materials, surface runoff tends to be much more concentrated and rapid than before, increasing the risk of flooding. This is illustrated by figure 6.19. If *peak lag time* is defined as the time lag between a precipitation event and the peak flood discharge (or stage), then, typically, that lag time decreases with increased urbanization, where the latter involves covering land with impermeable materials and/or installing storm sewers. The size of the peak discharge (stage) also increases, as was evident in figure 6.18B.

Building in a floodplain also can increase flood heights (see figure 6.20). The structures occupy volume that water formerly could fill when the stream flooded, and a given discharge then corresponds to a higher stage (water level). Floods that occur are more serious. Filling in floodplain land for construction similarly decreases the volume available to stream water and further aggravates the situation.

Measures taken to drain water from low areas can likewise aggravate flooding along a stream. In cities, storm sewers are installed to keep water from flooding streets during heavy rains, and, often, the storm water is channeled straight into a nearby stream. This works fine if the total flow is moderate enough, but by decreasing the time normally taken by the water to reach the stream channel, such measures increase the probability of stream flooding. The same is true of the use of tile

drainage systems in farmland. Water that previously stood in low spots in the field and only slowly reached the stream after infiltration instead flows swiftly and directly into the stream, increasing the flood hazard for those along the banks.

Both farming and urbanization also disturb the land by removing natural vegetation, thus leaving the soil exposed. The consequences are twofold. As noted earlier, vegetation can decrease flood hazards somewhat by providing a physical barrier to surface runoff, by soaking up some of the water, and through plants' root action, which keeps the soil looser and more permeable. Vegetation also can be critical to preventing soil erosion. When vegetation is removed and erosion increased, much more soil can be washed into streams. There, it can fill in, or "silt up," the channel, decreasing the channel's volume and thus reducing the stream's capacity to carry water away quickly.

It is not obvious whether land-use or climatic changes, or both, are responsible, but it is interesting to note that long-term records for the Red River near Grand Forks, North Dakota, show an increase in peak discharge over time leading up to the catastrophic 1997 floods (figure 6.21).

Strategies for Reducing Flood Hazards

Restrictive Zoning and "Floodproofing"

Short of avoiding floodplains altogether, various approaches can reduce the risk of flood damage. Many of these approaches—restrictive zoning, special engineering practices— are similar to strategies applicable to reducing damage from seismic and other geologic hazards. And commonly, existing structures present particular challenges, as they are typically exempt from newer regulations.

A first step is to identify as accurately as possible the area at risk. Careful mapping coupled with accurate stream discharge data allows identification of those areas threatened by floods of different recurrence intervals. Land that could be

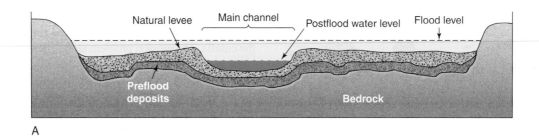

A

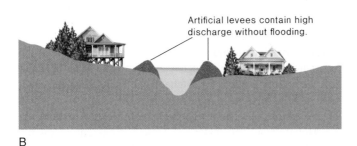

B

Figure 6.23

(A) When streams flood, waters slow quickly as they flow onto the floodplain, and therefore tend to deposit sediment, especially close to the channel where velocity first drops. (B) Artificial levees are designed to protect floodplain land from flooding by raising the height of the stream bank.

increasing flood risks downstream. It artificially raises the stage of the stream for a given discharge, which can increase the risks upstream, too, as the increase in stage is propagated above the confined section. This is why the Bird's Point levee was deliberately breached in 2011 to send water down the New Madrid floodway (chapter-opening image). Another large problem is that levees may make people feel so safe about living in the floodplain that, in the absence of restrictive zoning, development will be far more extensive than if the stream were allowed to flood naturally from time to time. If levees are overtopped by an unanticipated, severe flood, or if they simply fail, far more

lives and property may be lost as a result (figure 6.24A). Also, if levees are overtopped, water is then trapped outside the stream channel, where it may stand for some time after the flood subsides until infiltration and percolation return it to the stream (figure 6.24B).

Levees alter sedimentation patterns, too. During flooding, sediment is deposited in the floodplain outside the channel. If the stream and its load are confined by levees, increased sedimentation may occur in the channel. This will raise stream stage for a given discharge—the channel bottom is raised, so stage rises also—and either the levees must be continually

A

B

Figure 6.24

The negative sides of levees as flood-control devices. (A) If they are breached, as here in the 1993 Mississippi River floods, those living behind the levees may suddenly and unexpectedly find themselves inundated. (B) Breached or overtopped levees dam water behind them. The stream stage may drop rapidly after the flood crests, but the water levels behind the levees remain high. Sandbagging was ineffective in preventing flooding by the Kishwaukee River, DeKalb, Illinois, in July 1983. By the time this picture was taken from the top of the levee, the water level in the river (right) had dropped several meters, but water trapped behind the levee (left) remained high, prolonging the damage to property behind the levee.

(A) Photograph courtesy C. S. Melcher, U.S. Geological Survey

raised to compensate, or the channel must be dredged, an additional expense. Jackson Square in New Orleans stands at the elevation of the wharf in Civil War days; behind the levee beside the square, the Mississippi flows 10–15 feet *higher,* largely due to sediment accumulation in the channel. This is part of why the city was so vulnerable during Hurricane Katrina.

Flood-Control Dams and Reservoirs

Yet another approach to moderating streamflow to prevent or minimize flooding is through the construction of flood-control dams at one or more points along the stream. Excess water is held behind a dam in the reservoir formed upstream and may then be released at a controlled rate that does not overwhelm the capacity of the channel beyond. Additional benefits of constructing flood-control dams and their associated reservoirs (artificial lakes) may include availability of the water for irrigation, generation of hydroelectric power at the dam sites, and development of recreational facilities for swimming, boating, and fishing at the reservoir.

The practice also has its drawbacks. Navigation on the river—both by people and by aquatic animals—may be restricted by the presence of the dams. Also, the creation of a reservoir necessarily floods much of the stream valley behind the dam and may destroy wildlife habitats or displace people and their works. (China's massive Three Gorges Dam project, discussed more fully in chapter 20, has displaced over a million people and flooded numerous towns.) Fish migration can be severely disrupted. As this is being written, the issue of whether to breach (dismantle) a number of hydropower dams in the Pacific Northwest in efforts to restore populations of endangered salmon is being hotly debated as the Army Corps of Engineers weighs the alternatives; the final disposition in many cases may lie with Congress. Already, more dams are being removed than are being built each year in the United States, as their negative effects are more fully understood, and the numbers are rising, from 10–20 dams removed each year in the mid-1990s to 60 removed in 2010.

If the stream normally carries a high sediment load, further complications arise. The reservoir represents a new base level for the stream above the dam (figure 6.25). When the stream flows into that reservoir, its velocity drops to zero, and it dumps its load of sediment. Silting-up of the reservoir, in turn, decreases its volume, so it becomes less effective as a flood-control device. Some reservoirs have filled completely in a matter of decades, becoming useless. Others have been kept clear only by repeated dredging, which can be expensive and presents the problem of where to dump the dredged sediment. At the same time, the water released below the dam is free of sediment and thus may cause more active erosion of the channel there. Or, if a large fraction of water in the stream system is impounded by the dam, the greatly reduced water volume downstream may change the nature of vegetation and of habitats there. Recognition of some negative long-term impacts of the Glen Canyon Dam system led, in 1996, to the first of three experiments in controlled flooding of the Grand Canyon

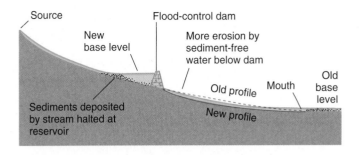

Figure 6.25

(A) Effects of dam construction on a stream: change in base level, sediment deposition in reservoir, reshaping of the stream channel. (B) Two of the four large dams and reservoirs that protect Santa Fe from flash floods of the San Gabriel River as it plunges down out of the mountains. Note sediment clouding reservoir water.

(B) Image courtesy NASA/GSFC/METI/ERSDAC/JAROS and the U.S./Japan ASTER Science Team.

to restore streambank habitat and shift sediment in the canyon, scouring silted-up sections of channel and depositing sediment elsewhere at higher elevations to restore beach habitats for wildlife. The long-term consequences of these actions are still being evaluated.

Some large reservoirs, such as Lake Mead behind Hoover Dam, have even been found to cause earthquakes. The reservoir water represents an added load on the rocks, increasing the stresses on them, while infiltration of water into the ground under the reservoir increases pore pressure sufficiently to reactivate old faults. Since Hoover Dam was built, at least 10,000 small earthquakes have occurred in the vicinity. Filling of the

Life on the Mississippi: The Ups and Downs of Levees

The Mississippi River is a long-term case study in levee-building for flood control. It is the highest-discharge stream in the United States and the third largest in the world (figure 1). The first levees were built in 1717, after New Orleans was flooded. Further construction of a patchwork of local levees continued intermittently over the next century. Following a series of floods in the mid-1800s, there was great public outcry for greater flood-control measures. In 1879, the Mississippi River Commission was formed to oversee and coordinate flood-control efforts over the whole lower Mississippi area. The commission urged the building of more levees. In 1882, when over $10 million (a huge sum for the time) had already been spent on levees, severe floods caused several hundred breaks in the system and prompted renewed efforts. By the turn of the century, planners were confident that the levees were high enough and strong enough to withstand any flood. Flooding in 1912–1913, accompanied by twenty

failures of the levees, proved them wrong. By 1926, there were over 2900 kilometers (1800 miles) of levees along the Mississippi, standing an average of 6 meters high; nearly half a billion cubic meters of earth had been used to construct them. Funds totalling more than twenty times the original cost estimates had been spent.

In 1927, the worst flooding recorded to that date occurred along the Mississippi. The levees were breached in 225 places; 183 people died; more than 75,000 people were forced from their homes; and there was an estimated $500 million worth of damage. This was regarded as a crisis of national importance, and thereafter, the federal government took over primary management of the flood-control measures. In the upper parts of the drainage basin, the government undertook more varied efforts—for example, five major flood-control dams were built along the Missouri River in the 1940s in response to flooding there—but the immense volume of the lower Mississippi

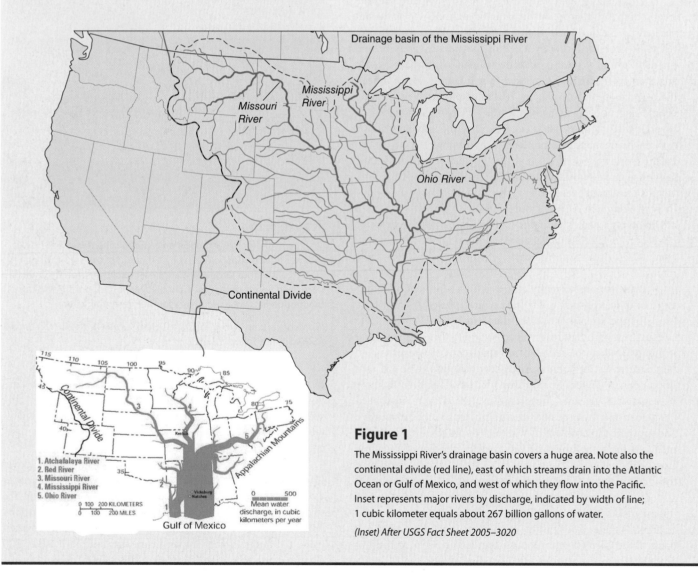

Figure 1

The Mississippi River's drainage basin covers a huge area. Note also the continental divide (red line), east of which streams drain into the Atlantic Ocean or Gulf of Mexico, and west of which they flow into the Pacific. Inset represents major rivers by discharge, indicated by width of line; 1 cubic kilometer equals about 267 billion gallons of water.

(Inset) After USGS Fact Sheet 2005–3020

A

B

Figure 2

(A) Flooding in Davenport, Iowa, Summer 1993. (B) On a regional scale, the area around the Mississippi became a giant lake. Ironically, in some places only the vegetation growing atop ineffective levees marked the margins of the stream channel, as here.

(A) © Doug Sherman/Geofile; (B) Photograph courtesy of C. S. Melcher, U.S. Geological Survey.

cannot be contained in a few reservoirs. So, the building up and reinforcing of levees continued.

A mild and wet fall and winter followed by prolonged rain in March and April contributed to record-setting flooding in much of the Mississippi's drainage basin in the spring of 1973. The river remained above flood stage for more than three months in some places. Over 12 million acres of land were flooded, 50,000 people were evacuated, and over $400 million in damage was done, not counting secondary effects, such as loss of wildlife and the estimated 240 million tons of sediment that were washed down the Mississippi, much of it fertile soil irrecoverably eroded from farmland.

Still worse flooding was yet to come. A sustained period of heavy rains and thunderstorms in the summer of 1993 led to flooding that caused 50 deaths, over $10 billion in property and crop damage, and, locally, record-breaking flood stages and/or levee breaches at various places in the Mississippi basin. Some places remained swamped for weeks before the water finally receded (figure 2; recall also figure 6.12). In spring 2011, flooding was so severe that spillways unused for decades were pressed into service (figure 3) and levees were deliberately breached, as at Bird's Point, to try to bring stream stages down. After nearly 300 years of ever more extensive (and expensive) flood-control efforts in the Mississippi basin, damaging floods are still happening—and we may now be facing possible increased risks from long-term climate change.

Inevitably, the question of what level of flood control is adequate or desirable arises. We could build still higher levees, bigger dams, and larger reservoirs and spillways, and (barring dam or levee failure) be safe even from a projected five-hundred-year flood or worse, but the costs would be astronomical and could exceed the property damage expected from such events. By definition, the

probability of a five-hundred-year or larger flood is extremely low. Precautions against high-frequency flood events seem to make sense. But at what point, for what probability of flood, does it make more economic or practical sense simply to accept the real but small risk of the rare, disastrous, high-discharge flood events?

The same questions are now being asked about New Orleans in the wake of Hurricane Katrina, discussed in chapter 7.

Figure 3

In mid-May 2011, floodgates were opened to divert over 100,000 cu. ft./sec of water from the Mississippi to the Atchafalaya drainage via the Morganza Spillway, relieving pressure on levees around Baton Rouge and New Orleans. Water churning into spillway causes white area in foreground. This spillway was used only once before, in 1973.

Photograph courtesy Team New Orleans, U.S. Army Corps of Engineers.

reservoir behind the Konya Dam in India began in 1963, and earthquakes were detected within months thereafter. In December 1967, an earthquake of magnitude 6.4 resulted in 177 deaths and considerable damage there. At least ninety other cases of reservoir-induced seismicity are known. Earthquakes caused in this way are usually of low magnitude, but their foci, naturally, are close to the dam. This raises concerns about the possibility of catastrophic dam failure caused, in effect, by the dam itself.

If levees increase residents' sense of security, the presence of a flood-control dam may do so to an even greater extent. People may feel that the flood hazard has been eliminated and thus neglect even basic floodplain-zoning and other practical precautions. But dams may fail; or, the dam may hold, but the reservoir may fill abruptly with a sudden landslide (as in the Vaiont disaster described in chapter 8) and flooding may occur anyway from the suddenly displaced water. Floods may result from intense rainfall or runoff *below* the dam. Also, a dam/reservoir complex may not necessarily be managed so as to maximize its flood-control capabilities. For obvious reasons, dams and reservoirs often serve multiple uses (e.g., flood control plus hydroelectric power generation plus water supply). But for maximally effective flood control, one wants maximum water-storage *potential,* while for other uses, it is generally desirable to maximize actual water *storage*—leaving little extra volume for flood control. The several uses of the dam/reservoir system may thus tend to compete with, rather than to complement, each other.

Flood Warnings?

Many areas of the United States have experienced serious floods (figure 6.26), and their causes are varied (table 6.1). Depending on the cause, different kinds of precautions are possible. The areas that would be affected by flooding associated with the 1980 explosion of Mount St. Helens could be anticipated from topography and an understanding of lahars, but the actual eruption was too sudden to allow last-minute evacuation. The coastal flooding from a storm surge—discussed further in chapter 7—typically allows time for evacuation as the storm approaches, but damage to coastal property is inevitable. Dam failures are sudden; rising water from spring snowmelt, much slower. As long as people inhabit areas prone to flooding, property will be destroyed when the floods

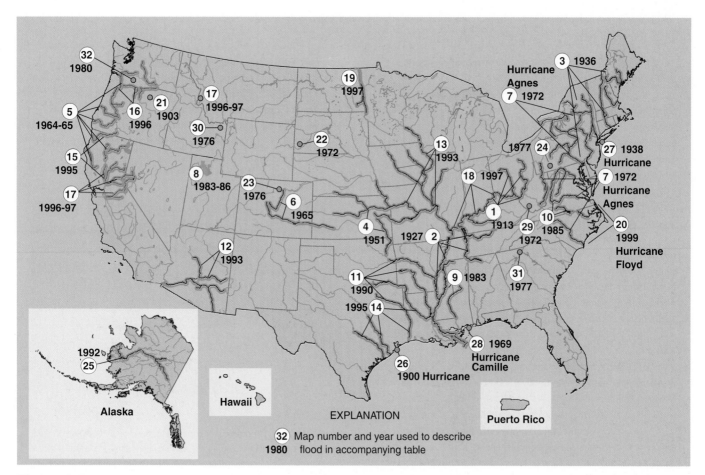

Figure 6.26

Significant U.S. floods of the twentieth century. Map number key is in table 6.1. Hurricane Katrina, in 2005, affected much the same area as Camille; in 2008, Hurricane Ike descended on Galveston, like the infamous hurricane of 1900.

Modified after USGS Fact Sheet 024–00

Flood Type	Map No.	Date	Area or Stream with Flooding	Reported Deaths	Approximate Cost (uninflated)*	Comments
Regional flood	1	Mar.–Apr. 1913	Ohio, statewide	467	$143M	Excessive regional rain.
	2	Apr.–May 1927	Mississippi River from Missouri to Louisiana	unknown	$230M	Record discharge downstream from Cairo, Illinois.
	3	Mar. 1936	New England	150+	$300M	Excessive rainfall on snow.
	4	July 1951	Kansas and Neosho River Basins in Kansas	15	$800M	Excessive regional rain.
	5	Dec. 1964–Jan. 1965	Pacific Northwest	47	$430M	Excessive rainfall on snow.
	6	June 1965	South Platte and Arkansas Rivers in Colorado	24	$570M	14 inches of rain in a few hours in eastern Colorado.
	7	June 1972	Northeastern United States	117	$3.2B	Extratropical remnants of Hurricane Agnes.
	8	Apr.–June 1983 June 1983–1986	Shoreline of Great Salt Lake, Utah	unknown	$621M	In June 1986, the Great Salt Lake reached its highest elevation and caused $268M more in property damage.
	9	May 1983	Central and northeast Mississippi	1	$500M	Excessive regional rain.
	10	Nov. 1985	Shenandoah, James, and Roanoke Rivers in Virginia and West Virginia	69	$1.25B	Excessive regional rain.
	11	Apr. 1990	Trinity, Arkansas, and Red Rivers in Texas, Arkansas, and Oklahoma	17	$1B	Recurring intense thunderstorms.
	12	Jan. 1993	Gila, Salt, and Santa Cruz Rivers in Arizona	unknown	$400M	Persistent winter precipitation.
	13	May–Sept. 1993	Mississippi River Basin in central United States	48	$20B	Long period of excessive rainfall.
	14	May 1995	South-central United States	32	$5–6B	Rain from recurring thunderstorms.
	15	Jan.–Mar. 1995	California	27	$3B	Frequent winter storms.
	16	Feb. 1996	Pacific Northwest and western Montana	9	$1B	Torrential rains and snowmelt.
	17	Dec. 1996–Jan. 1997	Pacific Northwest and Montana	36	$2–3B	Torrential rains and snowmelt.
	18	Mar. 1997	Ohio River and tributaries	50+	$500M	Slow moving frontal system.
	19	Apr.–May 1997	Red River of the North in North Dakota and Minnesota	8	$2B	Very rapid snowmelt.
	20	Sept. 1999	Eastern North Carolina	42	$6B	Slow-moving Hurricane Floyd.
Flash flood	21	June 14, 1903	Willow Creek in Oregon	225	unknown	City of Heppner, Oregon, destroyed.
	22	June 9–10, 1972	Rapid City, South Dakota	237	$160M	15 inches of rain in 5 hours.
	23	July 31, 1976	Big Thompson and Cache la Poudre Rivers in Colorado	144	$39M	Flash flood in canyon after excessive rainfall.
	24	July 19–20, 1977	Conemaugh River in Pennsylvania	78	$300M	12 inches of rain in 6–8 hours.
Ice-jam flood	25	May 1992	Yukon River in Alaska	0	unknown	100-year flood on Yukon River.
Storm-surge flood	26	Sept. 1900	Galveston, Texas	6000+	unknown	Hurricane.
	27	Sept. 1938	Northeast United States	494	$306M	Hurricane.
	28	Aug. 1969	Gulf Coast, Mississippi and Louisiana	259	$1.4B	Hurricane Camille.
Dam-failure flood	29	Feb. 2, 1972	Buffalo Creek in West Virginia	125	$60M	Dam failure after excessive rainfall.
	30	June 5, 1976	Teton River in Idaho	11	$400M	Earthen dam breached.
	31	Nov. 8, 1977	Toccoa Creek in Georgia	39	$2.8M	Dam failure after excessive rainfall.
Mudflow flood	32	May 18, 1980	Toutle and lower Cowlitz Rivers in Washington	60	unknown	Result of eruption of Mt. St. Helens.

Table 6.1 Significant Floods of the Twentieth Century

After U.S. Geological Survey Fact Sheet 024–00.

*[M, million; B, billion]

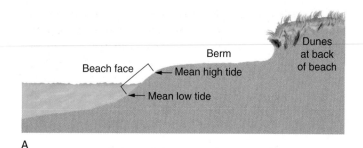

A

B

C

Figure 7.2

A beach. (A) Schematic beach profile. The beach may be backed by a cliff rather than by sand dunes. (B) The various components of the profile are clearly developed on this California beach. (C) Where the sediment is coarser, as on this cobbled Alaskan shore, the slope of the beach face may be much steeper.

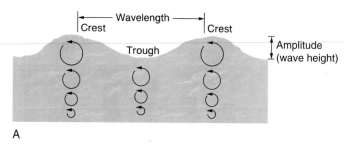

A

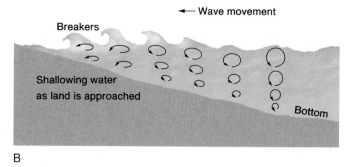

B

C

Figure 7.3

(A) Waves are undulations in the water surface, beneath which water moves in circular orbits. *Wave base* is the depth below which there is no orbital motion. (B) Breakers develop as waves approach shore and orbits are disrupted. (C) Evenly spaced natural breakers develop on a gently sloping shore off the California coast, as wind surfers take advantage of the wind that makes the waves.

and overall rotation of the earth-moon system produces a complementary bulge of water on the opposite side of the earth. (The solid earth actually deforms in response to these forces, too, but the oceans' response is much larger.) As the rotating earth spins through these bulges of water each day, overall water level at a given point on the surface rises and falls twice a day. This is the phenomenon recognized as tides. At high tide, waves reach higher up the beach face; more of the beach face is exposed at low tide. Tidal extremes are greatest when sun, moon, and earth are all aligned, and the sun and moon are thus pulling together, at times of full and new moons. The resultant tides are *spring tides* (figure 7.5A). (They have nothing to do with the spring

season of the year.) When the sun and moon are pulling at right angles to each other, the difference between high and low tides is minimized. These are *neap tides* (figure 7.5B). The magnitude of water-level fluctuations in a particular location is also controlled, in part, by the local underwater topography, and it may be modified by other features such as bays and inlets. Storms can create further variations in water height and in intensity of erosion, as discussed later in the chapter.

A

B

C

Figure 7.4

(A) Undercutting by wave action causes blocks of sandstone higher on this cliff to collapse into the water. Pictured Rocks National Lakeshore. (B) Sea arch formed by wave action along the California coast. Erosion is most intense at the waterline, undercutting the rock above. (C) Continued erosion causes the collapse of an arch, leaving monument-like *sea stacks*.

Over longer periods, water levels may shift systematically up or down as a result of tectonic processes or from changes related to glaciation, or human activities such as extraction of ground water or oil that can cause surface subsidence. These relative changes in land and water height may produce distinctive coastal features and may also aggravate coastal problems, as will be seen later in the chapter.

Sediment Transport and Deposition

Just as water flowing in streams moves sediment, so do waves and currents. The faster the currents and more energetic the waves, the larger and heavier the sediment particles that can be moved. As waves approach shore and the water orbits are disrupted, the water tumbles forward, up the beach face.

When waves approach a beach at an angle, as the water washes up onto and down off the beach, there is also net movement of water laterally along the shoreline, creating a **longshore current** (figure 7.6). Likewise, any sand caught up by and moved along with the flowing water is not carried straight up and down the beach. As the waves wash up the beach face at an angle to the shoreline, the sand is pushed along the beach as well. The net result is **littoral drift,** gradual sand movement down the beach in the same general direction as the motion of the longshore current. Currents tend to move consistently in certain preferred directions on any given beach, which means that, over time, they continually transport sand from one end of the beach to the other. On many beaches where this occurs, the continued existence of the beach is only assured by a fresh supply of sediment produced locally by wave erosion, delivered by streams, or supplied by dunes behind the beach.

The more energetic the waves and currents, the more and farther the sediment is moved. Higher water levels during storms (discussed below) also help the surf reach farther inland, perhaps all the way across the berm to dunes at the back of the beach. But a ridge of sand or gravel may survive the storm's fury, and may even be built higher as storm winds sweep sediment onto it. This is one way in which a *beach ridge* forms at the back of a beach. Where sediment supply is plentiful, as near the mouth of a sediment-laden river, or where the collapse of sandy cliffs adds sand to the beach, the beach may build outward into the water, new beach ridges forming seaward of the old, producing a series of roughly parallel ridges that may have marshy areas between them.

Storms and Coastal Dynamics

Unconsolidated materials, such as beach sand, are especially readily moved and rapidly eroded during storms. The low air pressure associated with a storm causes a bulge in the water surface. This, coupled with strong onshore winds piling water against the land, can result in unusually high water levels during the storm, a storm **surge.** The temporarily elevated water level associated with the surge, together with unusually energetic wave action and greater wave height (figure 7.7), combine to attack the coast with exceptional force. The gently sloping expanse of beach (berm) along the outer shore above the usual high-tide

A

B

C

Figure 7.16

The Cape Hatteras Lighthouse is a historic landmark. (A) It had been threatened for many years (note attempted shoreline stabilization structure in this 1996 photo). (B) Sandbagging to protect the dunes had also been tried. (C) Finally the lighthouse was moved, at a cost of $11.8 million.

(A) and (C) Photographs courtesy USGS Center for Coastal Geology; (B) Photograph by R. Dolan, U.S. Geological Survey

Figure 7.17

When groins or jetties perpendicular to the shoreline disrupt longshore currents, deposition occurs up-current, erosion below. Ocean City, New Jersey.

Photograph courtesy of S. Jeffress Williams

Figure 7.21

As waves approach this rugged section of the California coast, projecting rocks are subject to highest-energy wave action; sand accumulates only in recessed bays.

city after the deadly 1900 hurricane. It proved its value in 2008 when Hurricane Ike hit. However, one unintended result has been the total loss of the beach in front of the seawall during times of high seas.

Especially Difficult Coastal Environments

Many coastal environments are unstable, but some, including barrier islands and estuaries, are particularly vulnerable either to natural forces or to human interference.

Barrier Islands

Barrier islands are long, low, narrow islands paralleling a coastline somewhat offshore from it (figure 7.23). Exactly how or why they form is not known. Some models suggest that they have formed through the action of longshore currents on delta sands deposited at the mouths of streams. Their formation may also require changes in relative sea level. However they have formed, they provide important protection for the water and shore inland from them because they constitute the first line of defense against the fury of high surf from storms at sea. The barrier islands themselves are extremely vulnerable, partly as a result of their low relief. Water several meters deep may wash right over these low-lying islands during unusually high storm tides, such as occur during hurricanes

A

B

C

Figure 7.22

Some shore-protection structures. (A) Riprap near El Granada, California. Erosion of the sandy cliff will continue unabated on either side of riprap. (B) More substantial (and expensive) seawall protects estate at Newport, Rhode Island. (C) Patchwork of individual seawalls at base of cliff at Bodega Bay, California, shows varying degrees of success: Properties at left have already slumped downslope, while those at right are hanging on the edge of the cliff—for now. In the absence of protection structures, cliff-retreat rates here are of the order of a meter a year, sure sign of an unstable coastline.

(A) Photograph courtesy USGS Photo Library, Denver, CO.

A

May 21, 2009

Texas City Dike

to Houston

Bolivar
Peninsula

Intracoastal
Waterway

Galveston

Subsided
Wetlands

N

B

November 5, 2012

USGS

Figure 7.24

Photographs of Mantoloking, NJ, before and shortly after Hurricane Sandy. Note extreme erosion, landward shift of sand, loss of houses, and destruction of bridge. Yellow arrows point to same structure.

Photographs courtesy USGS Coastal and Marine Geology Program

Figure 7.23

(A) Barrier islands: North Carolina's Outer Banks. (B) One reason that Galveston is so vulnerable is that Galveston Island is a barrier island. The seawall runs along the southern edge of the island.

(A) Photograph by R. Dolan, USGS Photo Library, Denver, CO. (B) Image courtesy of Earth Science and Image Analysis Laboratory, NASA Johnson Space Center.

(figure 7.24). Strong storms may even slice right through narrow barrier islands, separating people from bridges to the mainland, as in Hurricane Isabel in 2003, Katrina in 2005, Irene in 2011, and Sandy in 2012.

Because they are usually subject to higher-energy waters on their seaward sides than on their landward sides, most barrier islands are retreating landward with time. Typical rates of retreat on the Atlantic coast of the United States are 2 meters (6 feet) per year, but rates in excess of 20 meters per year have been noted. Clearly, such settings represent particularly unstable locations in which to build, yet the aesthetic appeal of long beaches has led to extensive development on privately owned stretches of barrier islands. About 1.4 million acres of barrier islands exist in the United States, and approximately 20% of this total area has been developed. Thousands of structures, including homes, businesses, roads, and bridges, are at risk.

On barrier islands, shoreline-stabilization efforts—building groins and breakwaters and replenishing sand—tend to be especially expensive and, frequently, futile. At best, the benefits are temporary. Construction of artificial stabilization structures may easily cost tens of millions of dollars and, at the same time, destroy the natural character of the shoreline and even the beach, which was the whole attraction for developers in the first place. At least as costly an alternative is to keep moving buildings or rebuilding roads ever farther landward as the beach before them erodes (recall the Cape Hatteras lighthouse). Expense aside, this is clearly a "solution" for the short term only, and in many cases, there is no place left to retreat anyway (consider figure 7.24). Other human activities exacerbate the plight of barrier islands: dams on coastal rivers in Texas have trapped much sediment, starving fragile barrier islands there of vital sand supply. Considering the inexorable rise of global sea level and that more barrier-island land is being submerged more frequently, the problems will only get worse. See also Case Study 7.

Estuaries

An **estuary** is a body of water along a coastline, open to the sea, in which the tide rises and falls and in which fresh and salt water meet and mix to create brackish water. San Francisco Bay, Chesapeake Bay, Puget Sound, and Long Island Sound are examples. Some estuaries form at the lower ends of stream valleys, especially in drowned valleys. Others may be tidal basins in which the water is more salty than not.

The salinity reflects the balance between freshwater input, usually river flow, and salt water. It may vary with seasonal fluctuations in streamflow, and within the basin, with proximity to the freshwater sources(s). Over time, the complex communities of organisms in estuaries have adjusted to the salinity of the particular water in which they live. Any modifications that permanently alter this balance, and thus the salinity, can have a catastrophic impact on the organisms. In many places, demand for the fresh water that would otherwise flow into estuaries is diminishing that flow. The salinity of San Francisco Bay has been rising markedly—in some places, tenfold—as fresh water from rivers flowing into the bay is consumed en route. Also, water circulation in estuaries is often very limited. This makes them especially vulnerable to pollution; because they are not freely flushed out by vigorous water flow, pollutants can accumulate. Many of the world's largest coastal cities—San Francisco and New York, for example—are located beside estuaries that may receive polluted effluent water.

Unfortunately, many vital wetland areas are estuaries under pressure from environmental changes and human activities. For example, where land is at a premium, estuaries are frequently called on to supply more. They may be isolated from the sea by dikes and pumped dry or, where the land is lower, wholly or partially filled in. Naturally, this further restricts water flow and also generally changes the water's chemistry. In addition, development may be accompanied by pollution; recall also the chapter-opening photograph. All of

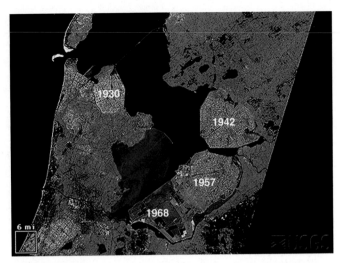

Figure 7.25

The Zuider Zee, a partially filled estuary in the Netherlands (1973 Landsat satellite photo). Patches of reclaimed land (polders), with date on which each was drained. Darker blue area at center, Markerwaard, was diked but not drained, for use as a freshwater reserve.

Image courtesy U.S. Geological Survey, EROS Data Center

this greatly stresses the organisms of the remaining estuary. Depending on the local geology, additional problems may arise. It was already noted in chapter 4 that buildings erected on filled land rather than on bedrock suffer much greater damage from ground shaking during earthquakes. This was abundantly demonstrated in San Francisco in 1906 and again in 1989. Yet the pressure to create dry land for development where no land existed before is often great.

One of the most ambitious projects involving land reclamation from the sea in an estuary is that of the Zuider Zee in the Netherlands (figure 7.25). The estuary itself did not form until a sandbar was breached in a storm during the thirteenth century. Two centuries later, with a growing population wanting more farmland, the reclamation was begun. Initially, the North Sea's access to the estuary was dammed up, and the flow of fresh water into the Zuider Zee from rivers gradually changed the water in it from brackish to relatively fresh. (This was necessary if the reclaimed land was to be used for farming; most crops grow poorly, if at all, in salty soil.) Portions were filled in to create dry land, while other areas were isolated by dikes and then pumped dry. More than half a million acres of new land have been created here.

Certainly, this has been a costly and prolonged effort. The marine organisms that had lived in the Zuider Zee estuary were necessarily destroyed when it was converted to a freshwater system. Continual vigilance over and maintenance of the dikes is required because if the farmlands were flooded with salt water, they would be unsuitable for growing crops for years.

Hurricanes and Coastal Vulnerability

Even in a year notable for the number of major hurricanes that occurred, Katrina stood out. Though it was only category 4 at landfall and weakened quickly thereafter, it was unusually large, with hurricane-force winds extending 200 km (125 miles) outward from its center. Hurricane Camille, a category-5 storm in 1969, was less than half Katrina's size.

The height of storm surge depends not only on the intensity (category) of a hurricane, but on its size, speed, air pressure, and other factors, including local coastal topography. A big, slow-moving storm like Katrina can build up a tremendous surge, locally reported at up to 10 meters (over 30 feet)! Barrier islands, predictably, were hard hit. On the mainland, things weren't necessarily any better (recall figure 7.10).

And then there was New Orleans. Much of the city actually lies below sea level, by as much as 5 meters, the result of decades of slow sinking. It has been protected by levees, not just along the Mississippi, but along the shore of Lake Pontchartrain, which is really a large estuary. The levees around Lake Pontchartrain and its canals had been built with a category 3 hurricane in mind. Unfortunately, Katrina was worse. Once some poorly constructed levees were overtopped, too, erosion behind them caused them to fail altogether. The result was inundation of a huge area of the city (figure 1) and standing water for weeks thereafter, as levees were patched and the water then pumped out.

The remaining question for many is: Now what? Some have advocated abandoning the lowest, most flood-prone sections of New Orleans. Some suggest adding fill to raise rebuilt structures in such areas. Others advocate higher, stronger levees, and/or systems of gates to stem the influx of storm surge into Lake Pontchartrain. The costs of such enhanced flood-control measures are in the tens of

billions of dollars. Some floodwalls have been rebuilt higher than before, but the city is as vulnerable as the weakest of the protection structures. Debate continues over how much protection is enough, and when stronger measures are no longer cost-effective. Meanwhile, New Orleans continues to sink, at rates recently measured at up to 28 mm (over 1 inch) per year. Likely causes include subsidence related to groundwater extraction, draining of wetlands, loss of sediments from the Mississippi, gradual displacement along a local fault, and settling of sediments; the latter may be exacerbated by the weight of additional fill or levees. Global sea-level rise will only worsen the problems.

Three years later, Hurricane Ike—another large, slow-moving storm—slammed into Galveston. While the seawall protected the city from waves and surge, anything in front of it was lost. And on the western end of Galveston Island, where there is no seawall and the land is sinking (largely due to groundwater withdrawal by the petrochemical industry), water damage was far worse, as is typical on barrier islands (figure 2). Here too, the risks will only increase with a rising sea level; but adding or strengthening shoreline-protection structures will be costly.

Figure 1

Flooding in New Orleans from Katrina: Flooded areas in the city appear as dark blue or purple, between Lake Pontchartrain at the top and the Mississippi River winding through the city across center of image.

Image courtesy Lawrence Ong, EO-1 Mission Science Office, NASA GSFC

Figure 2

Without a seawall, Crystal Beach, west of Galveston, took the full brunt of the storm surge in addition to the wind. Houses and vegetation were swept away, and erosion and overwash have also pushed the shoreline significantly landward. Yellow arrows point to same structure.

Photographs courtesy U.S. Geological Survey Coastal and Marine Geology Program

A

May 6, 2008

August 31, 2011

B

Figure 3

Outer Banks near Rodanthe, NC. Yellow arrows point to same structure. (A) Three days after Irene's landfall. U.S. Highway 12 is completely severed; note futile sandbagging in foreground. (B) Before and after closeup views. After Irene, the large building on the beach is gone, along with most of the beach. Water is still murky with suspended sediment churned up by the storm. Interestingly, a new building had been built in this vulnerable area since 2008: look closely just right of building indicated by arrow in both pictures.

Photographs courtesy U.S. Geological Survey Coastal and Marine Geology Program.

The year 2011 brought Hurricane Irene. After striking Puerto Rico, it moved north to landfall as a Category 1 storm at North Carolina's Outer Banks on 27 August, wreaking typical havoc on these barrier islands (figure 3). The storm continued up the eastern seaboard to Atlantic Canada. Though it had weakened to a tropical storm when it reached New England, Irene still brought strong winds and heavy rain—locally over ten inches—which produced severe flooding, and sediment runoff, much of it into the Long Island Sound estuary (figure 4). Many of the same areas of the northeastern U.S. were hit by Hurricane Sandy in 2012. It too was only a Category I storm at U.S. land fall, near Atlantic City, NJ. However, the hurricane merged with two other storm systems, to become the largest Atlantic storm on record, over 1000 miles across, and its record-setting storm surges and heavy rain produced massive damage, such as was seen in figure 7.24.

Hurricanes are not the only storms that threaten coastal areas, and some countries, such as the Netherlands, are essentially *all* low-lying land at risk (recall figure 7.25). Motivated by disastrous flooding in 1953 and the increasing threat due to global sea-level rise, that nation is embarking on a decades-long, multibillion-dollar plan of coastal fortification in an effort to "climate-proof" its land, and protect its people and their economy. How successful they will be, and whether other nations will undertake coastal protection on that scale, remains to be seen.

Figure 4

The Connecticut River provides about 70% of the freshwater input to Long Island Sound. Here, nearly a week after Irene struck, it also provides a load of sediment, clouding the water of the estuary.

NASA Earth Observatory image by Robert Simmon using Landsat 5 data from U.S. Geological Survey Global Visualization Viewer.

In the Netherlands, where land is at a premium, the benefits of filling in the Zuider Zee may outweigh the costs, environmental and economic. Destruction of *all* estuaries would, however, result in the elimination of many life forms uniquely adapted to their brackish-water environment. Even organisms that do not live their whole lives in estuaries may use the estuary's quiet waters as breeding grounds, and so these populations also would suffer from habitat destruction. Where other land exists for development, it may be argued that the estuaries are too delicate and valuable a system to tamper with.

Costs of Construction—and Reconstruction—in High-Energy Environments

Historically, the most concentrated damage has resulted from major storms, and the number of people and value of property at risk only increases with the shift in population toward the coast, illustrated in figure 7.1. (Recall also table 7.2.) U.S. federal disaster relief funds, provided in barrier-island and other coastal areas, can easily run into hundreds of millions of dollars after each such storm. Increasingly, the question is asked, does it make sense to continue subsidizing and maintaining development in such very risky areas? More and more often, the answer being given is "No!" From 1978 to 1982, $43 million in federal flood insurance (see chapter 19) was paid in damage claims to barrier-island residents, which far exceeded the premiums they had paid. So, in 1982, Congress decided to remove federal development subsidies from about 650 miles of beachfront property and to eliminate in the following year the availability of federal flood insurance for these areas. It simply did not make good economic sense to encourage unwise development through such subsidies and protection. Still, more than a million flood-insurance policies representing over $1 billion in coverage were still outstanding half a dozen years later in vulnerable coastal areas.

A continuing (and growing) concern is one of fairness in the distribution of costs of coastal protection. While some stabilization and beach-replenishment projects are locally funded, many are undertaken by the U.S. Army Corps of Engineers. Thus, millions of federal taxpayer dollars may be spent—over and over—to protect a handful of private properties of individuals choosing to locate in high-risk areas. A national study done in March 2000 reported that beach renourishment must typically be redone every 3 to 7 years. For example, from 1962–1995, the beach at Cape May, New Jersey, had to be renourished ten times, at a total cost of $25 million; from 1952–1995, the Ocean City, New Jersey beach was supplied with fresh sand 22 times, costing over $83 million altogether. Reported costs of beach maintenance and renourishment in developed areas along the Florida and Carolina coasts ranged up to $17.5 million per mile. When Hurricane Katrina hit, about $150 million in beach-replenishment projects were already underway in Florida; Katrina undid some of that work and created a need for more. Many beach restorations begun after Hurricane Irene were still incomplete when Hurricane

Sandy came through to make the damage worse than before. Various beach-replenishment projects have been undertaken at Hawaii's Waikiki Beach, since at least the 1930s. The most recent, begun in 2012, will pump offshore sand to a 1/3-mile section of Waikiki now eroding at 1–2 feet per year, at an estimated cost of $2.5 million. Clearly, the economic importance of tourism there means that such investments will continue in future.

Most barrier-island areas are owned by federal, state, or local governments. Even where they are undeveloped, much money has been spent maintaining access roads and various other structures, often including beach-protection structures. It is becoming increasingly obvious that the most sensible thing to do with many such structures is to abandon the costly and ultimately doomed efforts to protect or maintain them and simply to leave these areas in their dynamic, natural, rapidly changing state, most often as undeveloped recreation areas. Certainly there are many safer, more stable, and more practical areas in which to build. However, considerable political pressure may be brought to bear in favor of continuing, and even expanding, shoreline-protection efforts.

Recognition of Coastal Hazards

Conceding the attraction of beaches and coastlines and the likelihood that people will want to continue living along them, can at least the most unstable or threatened areas be identified so that the problems can be minimized? That is often possible, but it depends both on observations of present conditions and on some knowledge of the area's history (figure 7.26).

The best setting for building near a beach or on an island, for instance, is at a relatively high elevation (5 meters or more above normal high tide, to be above the reach of most storm tides) and in a spot with many high dunes between the proposed building site and the water, to supply added protection. Thick vegetation, if present, will help to stabilize the beach sand. Also, information about what has happened in major storms in the past is very useful. Was the site flooded? Did the overwash cover the whole island? Were protective dunes destroyed? A key factor in determining a "safe" elevation, too, is overall water level. On a lake, not only the short-term range of storm tide heights but also the long-term range in lake levels must be considered. On a seacoast, it would be important to know if that particular stretch of coastline was emergent or submergent over the long term. Around the Pacific Ocean basin, possible danger from tsunamis should not be overlooked.

On either beach or cliff sites, one very important factor is the rate of coastline erosion. Information might be obtained from people who have lived in the area for some time. Better and more reliable guides are old aerial photographs, if available from the U.S. or state geological survey, county planning office, or other sources, or detailed maps made some years earlier that will show how the coastline looked in past times. Comparison with the present configuration and knowledge of when the photos were taken or maps made allows estimation of the rate of erosion. It also should be kept in mind that the shoreline retreat may be more rapid in the future than it has been in the past as a consequence of a faster-rising sea level, as discussed earlier.

A

B

Figure 7.26

(A) Sometimes coastal hazards are obvious, if you look. Here, the houses at the cliff base provide unintentional—and vulnerable—protection for the cliff-edge structures above. Northern Monterey Bay, California. (B) This cliff at Moss Beach in San Mateo County, California, retreated more than 50 meters in a century. Past cliff positions are known from old maps and photographs; arrows indicate position of cliff at corresponding dates.

(A) Photograph by Cheryl Hapke, U.S. Geological Survey. (B) Photograph by K. R. LaJoie, USGS Photo Library, Denver, CO.

On cliff sites, too, there is landslide potential to consider, and in a seismically active area, the dangers are greatly magnified.

Storms, of course, commonly accelerate change. Aerial photographs and newer tools such as scanning airborne laser altimetry allow monitoring of changes and suggest the possible magnitude of future storm damage. The U.S. Geological Survey and other agencies maintain archives of images that can be useful here.

It is advisable to find out what shoreline modifications are in place or are planned, not only close to the site of interest but elsewhere along the coast. These can have impacts far away from where they are actually built. Sometimes, aerial or even satellite photographs make it possible to examine the patterns of sediment distribution and movement along the coast, which should help in the assessment of the likely impact of any shoreline modifications. The fact that such structures exist or are being contemplated is itself a warning sign! A history of repairs to or rebuilding of structures not only suggests a very dynamic coastline, but also the possibility that protection efforts might have to be abandoned in the future for economic reasons.

Summary

Many coastal areas are rapidly changing. Accelerated by rising sea levels worldwide, erosion is causing shorelines to retreat landward in most areas, often at rates of more than a meter a year. Sandy cliffs and barrier islands are especially vulnerable to erosion. Storms, with their associated surges and higher waves, cause especially rapid coastal change. Efforts to stabilize beaches are generally expensive, often ineffective over the long (or even short) term, and frequently change the character of the shoreline. They may also cause unforeseen negative consequences to the coastal zone and its organisms. Demand for development has led not only to construction on unstable coastal lands, but also to the reclamation of estuaries to create more land, to the detriment of plant and animal populations.

Key Terms and Concepts

active margin 147
barrier islands 159
beach 147
beach face 147
drowned valley 153
estuary 161
littoral drift 149
longshore current 149
milling 147
passive margin 147
surge 149
wave-cut platform 153
wave refraction 158

Exercises

Questions for Review

1. High storm tides may cause landward recession of dunes. Explain this phenomenon, using a sketch if you wish.

2. Evaluate the use of riprap and seawalls as cliff protection structures.

3. Explain longshore currents and how they cause littoral drift.

4. Sketch a shoreline on which a jetty has been placed to restrict littoral drift; indicate where sand erosion and deposition will subsequently occur and how this will reshape the shoreline.

5. What is storm surge, and how does it exacerbate coastal erosion?

6. Discuss the pros and cons of sand replenishment as a strategy for stabilizing an eroding beach.

7. Describe three ways in which the relative elevation of land and sea may be altered. What is the present trend in global sea level?

8. Briefly explain the formation of (a) wave-cut platforms and (b) drowned valleys.

9. What are barrier islands, and why have they proven to be particularly unstable environments for construction?

10. What is an estuary? Why do estuaries constitute such distinctive coastal environments?

11. Briefly describe at least two ways in which the dynamics of a coastline over a period of years can be investigated.

Exploring Further

1. Choose any major coastal city, find out how far above sea level it lies, and determine what proportion of it would be inundated by a relative rise in sea level of (a) 1 meter and (b) 5 meters. Consider what defensive actions might be taken to protect threatened structures in each case. (It may be instructive to investigate what has happened to Venice, Italy, under similar circumstances.)

2. The fictional film *The Day After Tomorrow* depicts a supposed storm surge that immerses the Statue of Liberty up to her neck. In round numbers, that indicates a surge of about 250 feet. Use the data from table 7.1 to make a graph of surge height versus wind velocity, and extrapolate to estimate the wind speed required to create a surge that large. Compare your estimate with the highest measured natural wind speed, 231 mph, and consider the plausibility of the movie's depiction.

3. Consider the vulnerability of barrier-island real estate, and think about what kinds of protections and controls you might advocate: How much should a prospective purchaser of the property be told about beach stability? Should local property owners pay for shoreline-protection structures, or should state or federal taxes be used? Should property owners be prevented from developing the land, for their own protection? What would you consider the most important factors in your decisions? What are the main unknowns?

4. One feature of the Netherlands long-term plan mentioned in Case Study 7 is that it involves a conscious decision that the level of protection afforded each region will be proportional to its economic value. (See the paper by D. Wolman in the online suggested readings.) Consider and comment on the implications of this policy.

Mass Movements

CHAPTER 8

While the internal heat of the earth drives mountain-building processes, just as inevitably, the force of gravity acts to tear the mountains down. Gravity is the great leveler. It tugs constantly downward on every mass of material everywhere on earth, causing a variety of phenomena collectively called **mass wasting,** or **mass movements,** whereby geological materials are moved downward, commonly downslope, from one place to another. The movement can be slow, subtle, almost undetectable on a day-to-day basis but cumulatively large over days or years. Or the movement can be sudden, swift, and devastating, as in a rockslide or avalanche. Mass movements need not even involve slipping on a slope: vertical movements include surface subsidence, as from extraction of oil or ground water, or collapse into sinkholes as described further in chapter 11.

Landslide is a general term for the results of rapid mass movements. Overall, landslide damage is greater than one might imagine from the usually brief coverage of landslides in the news media. In the United States alone, landslides and other mass movements cause over $3.5 billion in property damage every year, and 25 to 50 deaths. Many landslides occur quite independently of human activities. In some areas, active steps have been taken to control downslope movement or to limit its damage. On the other hand, certain human activities aggravate local landslide dangers, usually as a result of failure to take those hazards into account.

Large areas of this country—and not necessarily only mountainous regions—are potentially at risk from landslides (figure 8.1). It is important to realize, too, that given the area represented in that figure, it is necessarily a generalization, representing *regional* landslide hazard levels associated with the characteristic regional topography, soil type, etc. One can live in a high-risk spot within an area generally classified as low-risk on this map!

On 4 January 2010, a landslide buried the town of Attabad, Pakistan, killing 20 and blocking both the Hunza River and the Karakoram Highway, the only road linking Pakistan and China, thereby disrupting trade between the two countries. River water dammed by the slide (bottom center of image) accumulated in a lake that grew to 23 km (nearly 15 miles) long, drowning several villages and displacing 6000 people. More than two years later, 25,000 remain isolated by the slide and lake.

NASA Earth Observatory image by Jesse Allen using EO-1 ALI data provided by the NASA EO-1 team.

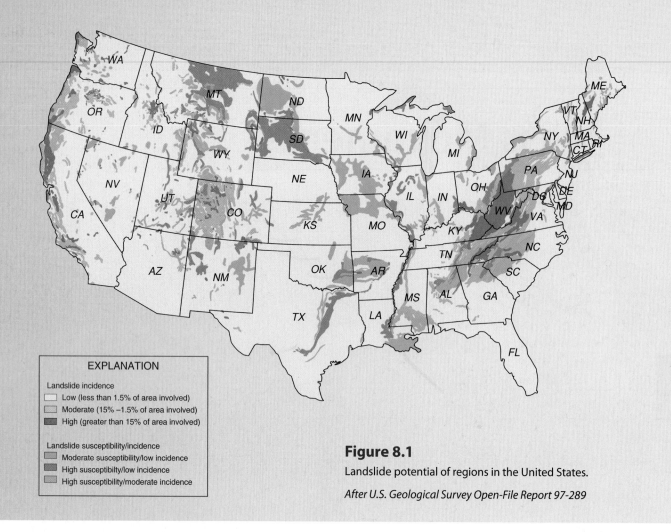

EXPLANATION

Landslide incidence
- ☐ Low (less than 1.5% of area involved)
- ☐ Moderate (15% –1.5% of area involved)
- ■ High (greater than 15% of area involved)

Landslide susceptibility/incidence
- ☐ Moderate susceptibility/low incidence
- ■ High susceptibilty/low incidence
- ☐ High susceptibility/moderate incidence

Figure 8.1

Landslide potential of regions in the United States.

After U.S. Geological Survey Open-File Report 97-289

Factors Influencing Slope Stability

Basically, mass movements occur whenever the downward pull caused by gravity overcomes the forces resisting it. As first described in chapter 3, a **shearing stress** is one tending to cause parts of an object to slide past each other across a plane, as the plates on opposite sides of a transform fault do. The downslope pull leading to mass movements is also a shearing stress, its size related to the mass of material involved and the slope angle. With a block of rock sitting on a slope, *friction* between the block and the underlying slope counteracts the shearing stress. A single body of rock can also be subjected to shearing stress; its **shear strength** is its ability to resist being torn apart along a plane by the shearing stress. When shearing stress exceeds frictional resistance or the shear strength of the material, as applicable, sliding occurs. Therefore, factors that increase shearing stress, decrease friction, or decrease shear strength tend to increase the likelihood of sliding, and vice versa. Application of these simple principles is key to reducing landslide hazards.

Effects of Slope and Materials

The mass of the material involved is one key factor in slide potential: The gravitational force pulling it downward, and thus the shearing stress, is directly proportional to that mass. Anything that increases the mass (for example, saturating soil with rain or snowmelt) increases the risk of a slide.

All else being equal, the steeper the slope, the greater the shearing stress (figure 8.2) and therefore the greater the likelihood of slope failure. For dry, unconsolidated material, the **angle of repose** is the maximum slope angle at which the material is stable (figure 8.3). This angle varies with the material. Smooth, rounded particles tend to support only very low-angle slopes (imagine trying to make a heap of marbles or ball bearings), while rough, sticky, or irregular particles can be piled more steeply without becoming unstable. Other properties being equal, coarse fragments can usually maintain a steeper slope angle than fine ones. The tendency of a given material to assume a constant characteristic slope can be seen in such diverse geologic forms as cinder cones (figure 8.4A), sand dunes (figure 8.4B), and beach faces (recall figure 7.2 B, C).

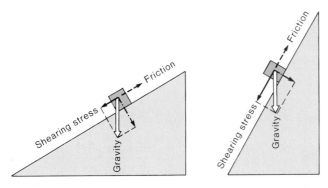

Figure 8.2

Effects of slope geometry on slide potential for a block of rock on a slope. The mass of the block and thus the total downward pull of gravity are the same in both cases, but the steeper the slope, the greater the shearing stress component. An increase in pore pressure can decrease frictional resistance to shearing stress.

Solid rock can be perfectly stable even at a vertical slope but may lose its strength if it is broken up by weathering or fracturing. Also, in layered sedimentary rocks, there may be weakness along bedding planes, where different rock units are imperfectly held together; some units may themselves be weak or even slippery (clay-rich layers, for example). Such planes of weakness are potential slide or failure planes.

Slopes may be steepened to unstable angles by natural erosion by water or ice. Erosion also can undercut rock or soil, removing the support beneath a mass of material and thus leaving it susceptible to falling or sliding. This is a common contributing factor to landslides in coastal areas and along stream valleys; recall figures 7.4A and 7.22C.

Over long periods of time, slow tectonic deformation also can alter the angles of slopes and bedding planes, making them steeper or shallower. This is most often a significant factor in young, active mountain ranges, such as the Alps or the coast ranges of California. It might have been a factor in the Pakistan landslide shown in the chapter-opening image, as that area falls within the Himalayan mountain system. In such a case, past evidence of landslides may be limited in the immediate area because the dangerously steep slopes have not existed long in a geologic sense.

Effects of Fluid

The role of fluid in mass movements is variable. Addition of some moisture to dry soils may increase adhesion, helping the particles to stick together (it takes damp sand to make a sand

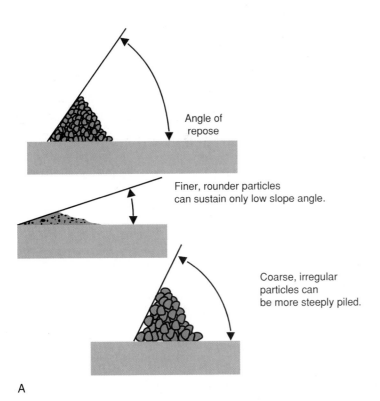

Angle of repose

Finer, rounder particles can sustain only low slope angle.

Coarse, irregular particles can be more steeply piled.

A

B

Figure 8.3

(A) Angle of repose indicates an unconsolidated material's resistance to sliding. Coarser and rougher particles can maintain a steeper stable slope angle. (Variation in angle of repose in natural materials is less than the range illustrated here.) (B) This California cliff holds its near-vertical slope only temporarily; eventually the weakly cohesive sediment collapses to assume its natural angle of repose, illustrated by the slope of the surface of the sediment pile at the base of the cliff.

(B) Photograph by J. T. McGill, USGS Photo Library, Denver, CO.

A

B

Figure 8.4

(A) Angle of repose of volcanic cinders is shown not only by the smoothly sloping sides of each individual cinder cone in the summit crater of Haleakala volcano in Hawaii, but also by the consistent slope and shape from cone to cone. (B) The uniform slope on the face of this sand dune is maintained by slumping to the sand's angle of repose. Oregon Dunes National Recreation Area.

castle). However, saturation of unconsolidated materials reduces the friction between particles that otherwise provides cohesion and strength, and the reduced friction can destabilize a slope. As was described in chapter 4, elevated pore-fluid pressure can trigger earthquakes by facilitating slip in faulted rocks under tectonic stress. It is also effective in promoting sliding in rocks under stress due to gravity, reducing the force holding the rock to the plane below (vector perpendicular to slope in figure 8.2). As already noted, the very mass of water in saturated soil adds extra weight, and thus extra downward pull.

Aside from its role in erosion, water can greatly increase the likelihood of mass movements in other ways. It can seep

Figure 8.5

Instability in expansive clay seriously damaged this road near Boulder, Colorado.

Photograph courtesy USGS Photo Library, Denver, CO.

Figure 8.6

Laguna Beach, CA landslide of 1 June 2005.

Photograph by Gerald Bawdeu, U.S. Geological Survey.

along bedding planes in layered rock, reducing friction and making sliding more likely. The expansion and contraction of water freezing and thawing in cracks in rocks or in soil can act as a wedge to drive chunks of material apart *(frost wedging)*. *Frost heaving,* the expansion of wet soil as it freezes and the ice expands, loosens and displaces the soil, which may then be more susceptible to sliding when it next thaws.

Some soils rich in clays absorb water readily; one type of clay, montmorillonite, may take up twenty times its weight in water and form a very weak gel. Such material fails easily under stress. Other clays expand when wet, contract when dry, and can destabilize a slope in the process (figure 8.5; see also chapter 20).

Sudden, rapid landslides commonly involve a triggering mechanism. Heavy rainfall or rapid melting of abundant snow can be a very effective trigger, quickly adding weight, decreasing friction, and increasing pore pressure (figure 8.6).

A

B

Figure 8.7

(A) Debris flows in Deming, Washington, following two days of heavy storms. Note buildings mired in the mud near the road. More than 1500 new landslides resulted from these storms. (B) Closeup of debris flow in Whatcom County, Washington.

Photographs courtesy Washington Division of Geology & Earth Resources, U.S. Geological Survey

As with flooding, the key is not simply how much water is involved, but how rapidly it is added and how close to saturated the system was already. The January 2009 landslides in the Pacific Northwest (figure 8.7) accompanied severe flooding, both problems caused by heavy rainfall onto—and contributing to rapid melting of—unusually heavy snow that had fallen earlier in the winter. Tropical regions subject to intense cloudbursts may also be subject to frequent slides in thick tropical soils—which may tend to be clay-rich—and in severely weathered rock.

Effects of Vegetation

Plant roots, especially those of trees and shrubs, can provide a strong interlocking network to hold unconsolidated materials together and prevent flow (figure 8.8A). In addition, vegetation takes up moisture from the upper layers of soil and can thus reduce the overall moisture content of the mass, increasing its shear strength. Moisture loss through the vegetation by transpiration also helps to dry out sodden soil more quickly. Commonly, then, vegetation tends to increase slope stability. However, the plants also add weight to the slope. If the added weight is large and the root network of limited extent, the vegetation may have a destabilizing effect instead. Also, plants that take up so much water that they dry out soil until it loses its adhesion can aggravate the slide hazard.

A relatively recently recognized hazard is the fact that when previously forested slopes are bared by fires or by logging, they may become much more prone to sliding than before. Previous illegal logging on slopes above the town of Guinsaugon in the Philippines was blamed for the February 2006 mudslide, triggered by heavy rains, that engulfed the town and killed an estimated 1800 of its 1857 residents. Wildfires in California frequently strip slopes of stabilizing vegetation, and deadly landslides may follow (figure 8.8B).

Earthquakes

Landslides are a common consequence of earthquakes in hilly terrain, as noted in chapter 4. Seismic waves passing through rock stress and fracture it. The added stress may be as much as half that already present due to gravity. Ground shaking also jars apart soil particles and rock masses, reducing the friction that holds them in place. In the 1964 Alaskan earthquake, the Turnagain Heights section of Anchorage was heavily damaged by landslides (recall figure 4.27C). California contains not only many fault zones but also many sea cliffs and hillsides prone to landslides during earthquakes.

One of the most lethal earthquake-induced landslides occurred in Peru in 1970. An earlier landslide had already

A

B

Figure 8.8

(A) Examination of this California roadcut shows surface soil held in place by grass roots, erosion and slope failure below. (B) October 2003 wildfires removed vegetation in Cable Canyon, California; rains sent this debris flow through a campground on 25 December, killing two people.

(B) Photograph by Sue Cannon, U.S. Geological Survey

Figure 8.9

The Nevados Huascarán debris avalanche, Peru, 1970.

Photograph courtesy USGS Photo Library, Denver, CO.

occurred below the steep, snowy slopes of Nevados Huascarán, the highest peak in the Peruvian Andes, in 1962, without the help of an earthquake. It had killed approximately 3500 people. In 1970, a magnitude-7.7 earthquake centered 130 kilometers to the west shook loose a much larger debris avalanche that buried most of the towns of Yungay and Ranrachira and more than 18,000 people (figure 8.9). Once sliding had begun, escape was impossible: Some of the debris was estimated to have moved at 1000 kilometers per hour (about 600 miles per hour). The steep mountains of Central and South America, sitting above subduction zones, continue to suffer earthquake-induced slides (figure 8.10).

Earthquakes in and near ocean basins may also trigger submarine landslides. Sediment from these may be churned up into a denser-than-water suspension that flows downslope to the sea floor, much as pyroclastic flows flow down from a volcano. These *turbidity currents* not only carry sediment to the deep-sea floor; they are forceful enough to break cables, and they could therefore disrupt transatlantic communications in the days when all transatlantic telephone calls were carried by undersea cables!

Quick Clays

A geologic factor that contributed to the 1964 Anchorage landslides and that continues to add to the landslide hazards in Alaska, California, parts of northern Europe, and elsewhere is a material known as "quick" or "sensitive" clay. True **quick clays** (figure 8.11) are most common in northern polar latitudes. The grinding and pulverizing action of massive glaciers can produce a **rock flour** of clay-sized particles, less than

Figure 8.10

Slide on Pan-American Highway east of Ropango, El Salvador, triggered by magnitude-7.6 earthquake in January 2001.

Photograph by E. L. Harp, U.S. Geological Survey

Figure 8.11

Example of quick clay. Turnagain Heights, Anchorage, Alaska.

0.02 millimeter (0.0008 inch) in diameter. When this extremely fine material is deposited in a marine environment, and the sediment is later uplifted above sea level by tectonic movements, it contains salty pore water. The sodium chloride in the pore water acts as a glue, holding the clay particles together. Fresh water subsequently infiltrating the clay washes out the salts, leaving a delicate, honeycomblike structure of particles. Seismic-wave vibrations break the structure apart, reducing the strength of the quick clay by as much as twenty to thirty times,

creating a finer-grained equivalent of quicksand that is highly prone to sliding. Failure of the Bootlegger Clay, a quick clay underlying Anchorage, Alaska, was responsible for the extent of damage from the 1964 earthquake. Nor is a large earthquake necessary to trigger failure; vibrations from passing vehicles can also do it. So-called **sensitive clays** are somewhat similar in behavior to quick clays but may be formed from different materials. Weathering of volcanic ash, for example, can produce a sensitive-clay sediment. Such deposits are not uncommon in the western United States.

Types of Mass Wasting

In the broadest sense, even subsidence of the ground surface is a form of mass wasting because it is gravity-driven. Subsidence includes slow downward movement as ground water or oil is extracted, and the rapid drop of sinkhole formation, described in chapter 11. Here we focus on downslope movements and displacement of distinct masses of rock and soil.

When downslope movement is quite slow, even particle by particle, the motion is described as **creep.** Soil creep, which is often triggered by frost heaving, occurs more commonly than rock creep. Though gradual, creep may nevertheless leave telltale signs of its occurrence so that areas of particular risk can be avoided. For example, building foundations may gradually weaken and fail as soil shifts; roads and railway lines may be disrupted. Often, soil creep causes serious property damage, though lives are rarely threatened. A related phenomenon, *solifluction,* describes slow movement of wet soil over impermeable material. It will be considered further in connection with permafrost, discussed in chapter 10.

Mass movements may be subdivided on the basis of the type of material moved and the characteristic type or form of movement. The material moved can range from unconsolidated, fairly fine material (for example, soil or snow) to large, solid masses of rock. *Landslide* is a nonspecific term for rapid mass movements in rock or soil. There are several types of landslides, and a particular event may involve more than one type of motion. The rate of motion is commonly related to the proportion of moisture: frequently, the wetter the material, the faster the movement. In general, the more rapid the movement, the greater the likelihood of casualties. A few examples may clarify the different types of mass movements, which are summarized in figure 8.12.

Falls

A **fall** is a free-falling action in which the moving material is not always in contact with the ground below. Falls are most often **rockfalls** (figure 8.13). They frequently occur on very steep slopes when rocks high on the slope, weakened and broken up by weathering, lose support as materials under them erode away. Falls are common along rocky coastlines where cliffs are undercut by wave action and also at roadcuts through

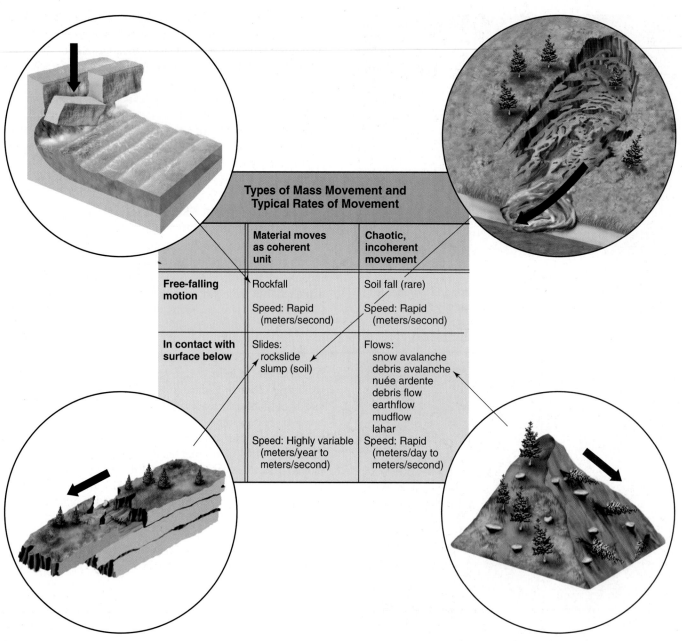

Types of Mass Movement and Typical Rates of Movement		
	Material moves as coherent unit	Chaotic, incoherent movement
Free-falling motion	Rockfall Speed: Rapid (meters/second)	Soil fall (rare) Speed: Rapid (meters/second)

In contact with surface below	Slides:
	  rockslide
	  slump (soil)
	
	Speed: Highly variable (meters/year to meters/second)
		Flows:
		  snow avalanche
		  debris avalanche
		  nuée ardente
		  debris flow
		  earthflow
		  mudflow
		  lahar
		Speed: Rapid (meters/day to meters/second)

Figure 8.12

Comparison of nature of various types of mass movements.

solid rock. The coarse rubble that accumulates at the foot of a slope prone to rockfalls is **talus** (figure 8.14). As the pounding water undercuts the resistant rock layer that has created Niagara Falls (figure 8.15), repeated rockfalls cause the falls to move slowly upstream, while talus accumulates at the base of the falls.

Slumps and Slides

In a **slide,** a fairly cohesive unit of rock or soil slips downward along a clearly defined surface or plane. Rockslides most often involve movement along a planar fracture or a bedding plane between successive layers of sedimentary rocks, or slippage where differences in original composition of the sedimentary layers results in a weakened layer or a surface with little cohesion (figure 8.16). Rockslides may be complex, with multiple rock blocks slipping along the same plane.

    Slumps can occur in rock or soil. In a soil slump, a rotational movement of the soil mass typically accompanies the downslope movement. A *scarp* may be formed at the top of the slide. Usually, the surface at the top of the slide is relatively undisturbed. The lower part of the slump block may end in a flow (figure 8.17). A rockslide that does not move very far may also be a slump.

Figure 8.13

Rockfall in Colorado National Monument. Note accumulation of talus at base of cliff, from past falls.

Flows and Avalanches

In a **flow,** the material moved is not coherent but moves in a more chaotic, disorganized fashion, with mixing of particles within the flowing mass, as a fluid flows. Flows of unconsolidated material are extremely common. They need not involve only soils. Snow avalanches are one kind of flow; nuées ardentes, or pyroclastic flows (see chapter 5), are another, associated only with volcanic activity. Where soil is the flowing material, these phenomena may be described as *earthflows* (fairly dry soil; figure 8.18) or *mudflows* (if saturated with water), which include the volcanic lahars. A flow involving a wide variety of materials—soil, rocks, trees, and so on—together in a single flow is a **debris avalanche,** or **debris flow,** the latter term more often used for water-saturated debris (figure 8.19; recall also figures 8.7 and 8.8B). Regardless of the nature of the materials moved, all flows have in common the chaotic, incoherent movement of the particles or objects in the flow.

Consequences of Mass Movements

Just as landslides can be a result of earthquakes, other unfortunate events, notably floods, can be produced by landslides. A stream in the process of cutting a valley may create unstable slopes. Subsequent landslides into the valley can dam up the stream flowing through it, creating a natural reservoir. The filling of the reservoir makes the area behind the earth dam uninhabitable, though it usually happens slowly enough that lives are not lost.

Heavy rains and snows in Utah in 1982–1983 caused landslides in unprecedented numbers. In one case, an old 3-million-cubic-meter slide near Thistle, Utah, was reactivated. It blocked Spanish Fork Canyon, creating a large lake and cutting off one transcontinental railway line and a major highway route into central Utah (figure 8.20). Shipments of coal and other supplies were disrupted. Debris flows north of Salt Lake City also forced hundreds of residents to evacuate and destroyed some homes, and water levels in reservoirs in the Wasatch Plateau were lowered to reduce dangers of landslides into the reservoirs or breaching of the dams.

A further danger is that the unplanned dam formed by a landslide will later fail. In 1925, an enormous rockslide a kilometer wide and over 3 kilometers (2 miles) long in the valley of the Gros Ventre River in Wyoming blocked that valley, and the resulting lake stretched to 9 kilometers long. After spring rains and snowmelt, the natural dam failed. Floodwaters swept down the valley below. Six people died, but the toll might have been far worse in a more populous area.

The 2010 Pakistan landslide of the chapter-opening image likewise created an unintended lake, and with it, flooding problems above and below the dam. The accumulated water began to flow over the slide a few months after the event, flooding towns downstream as well. Plans were made to cut a spillway through the slide to drain the lake, but more than two years after the slide, this has not occurred. Evidence of water seeping into the slide is raising fears that it may wash out catastrophically.

Impact of Human Activities

Given the factors that influence slope stability, it is possible to see many ways in which human activities increase the risk of mass movements. One way is to clear away stabilizing vegetation. As noted earlier, where clear-cutting logging operations have exposed sloping soil, landslides of soil and mud may occur far more frequently or may be more severe than before.

Many types of construction lead to oversteepening of slopes. Highway roadcuts, quarrying or open-pit mining operations, and construction of stepped home-building sites on hillsides are among the activities that can cause problems (figure 8.21). Where dipping layers of rock are present, removal of material at the bottom ends of the layers may leave large masses of rock unsupported, held in place only by friction between layers. Slopes cut in unconsolidated materials at angles higher than the angle of repose of those materials are by nature unstable, especially if there is no attempt to plant stabilizing vegetation (figure 8.22). In addition, putting a house above a naturally unstable or artificially steepened slope adds weight to the slope, thereby increasing the shear stress acting on the slope. Other activities connected with the

A

B

C

Figure 8.14

Rockfalls occur in many settings, creating talus from many different rock types: (A) the sandstones of Pictured Rocks National Lakeshore; (B) gneiss boulders that tumble down a slope crossed by a trail in the Colorado Rockies; (C) chunks of volcanic rock fallen from the columns of Devil's Tower. Note in (B) and (C) that talus, too, tends to assume a consistent slope angle.

presence of housing developments on hillsides can increase the risk of landslides in more subtle ways. Watering the lawn, using a septic tank for sewage disposal, and even putting in an in-ground swimming pool from which water can seep slowly out are all activities that increase the moisture content of the soil and render the slope more susceptible to slides. On the other hand, the planting of suitable vegetation can reduce the risk of slides.

Recent building-code restrictions have limited development in a few unstable hilly areas, and some measures have been taken to correct past practices that contributed to landslides. (Chapter 19 includes an illustration of the effectiveness of legislation to restrict building practices in Los Angeles County.) However, landslides do not necessarily cease just because ill-advised practices that caused or accelerated them have been stopped. Once activated or reactivated, slides may continue to slip for decades. The area of Portuguese Bend in the Palos Verdes Peninsula in Los Angeles County is a case in point.

In the 1950s, housing developments were begun in Portuguese Bend. In 1956, a 1-kilometer-square area began to slip, though the slope in the vicinity was less than 7 degrees, and within months, there had been 20 meters of movement. What activated this particular slide is unclear. Most of the homes used cesspools for sewage disposal, which added fluid to

Figure 8.15

Niagara Falls are moving slowly upstream as a consequence of rockfalls.

Photograph by Ken Winters, courtesy U.S. Army Corps of Engineers

A

Figure 8.16

Rockslides occur along steeply sloping fractures in this granite in the Colorado Rocky Mountains.

B

Figure 8.17

(A) Slump and flow at La Conchita, California, in spring 1995. Houses at the toe were destroyed by the debris flow, but no one was killed. (B) Ten years later, part of the 1995 slide remobilized after a period of heavy rain, flowing out over four blocks of the town, and ten people died. What is perhaps more disturbing is that recent studies suggest that the La Conchita slides are just a small part of a much larger complex slide; see scarp at upper left.

(A) Photograph by R. L. Schuster, (B) photograph by Jonathan Godt, both courtesy U.S. Geological Survey

the ground, potentially increasing fluid pressure. On the other hand, the county highway department, in building a road across what became the top of this slide, added a lot of fill, and thus weight. In any event, the resultant damage was extensive: cracks developed within the sliding mass; houses on and near

Figure 8.18

Recent earthflows in grassland, Gilroy, California.

Photograph © The McGraw-Hill Companies, Inc./Doug Sherman, photographer

Figure 8.19

Closeup of 1970 debris avalanche at Nevados Huascarán (figure 8.9).

Photograph by B. Bradley, University of Colorado, Courtesy NOAA/NGDC.

Figure 8.20

An old, 3-million-cubic-meter slide near Thistle, Utah, was reactivated in the wet spring of 1983. It blocked Spanish Fork Canyon and cut off highway and rail routes (note where they continue up the valley near top of photo). Repairing the damage cost over $200 million (in 1984 dollars), making this the single most expensive landslide to "fix" in U.S. history.

Photograph courtesy USGS Photo Library, Denver, CO.

Human activities can increase the hazard of landslides in still other ways. Irrigation and the use of septic tanks increase the flushing of water through soils and sediments. In areas underlain by quick or sensitive clays, these practices may hasten the washing-out of salty pore waters and the destabilization of the clays, which may fail after the soils are drained. Even cleanup after one slide may reduce stability and contribute to the next slide: Often, the cleanup involves removing the toe of a slump or flow—for example, where it crosses a road. This removes support for the land above and may oversteepen the local slope, too. Artificial reservoirs may cause not only earthquakes, but landslides, too. As the reservoirs fill, pore pressures in rocks along the sides of the reservoir increase, and the strength of the rocks to resist shearing stress can be correspondingly decreased. The case of the Vaiont reservoir disaster (Case Study 8) is a classic in this respect. Chapter 20 describes other examples of ways in which engineering activities may influence land stability.

A Compounding of Problems: The Venezuelan Coast

On the north coast of Venezuela, the terrain is mountainous, with steep slopes and narrow stream channels. A few people settled on the slopes, risking landslides; larger towns developed

the slide were damaged or destroyed; a road built across the base of the slide had to be rebuilt repeatedly. Over $10 million in property damage resulted from the slippage. Worse, the movement has continued for decades since, slow but unstoppable, at a rate of about 3 meters per year. Some portions of the slide have moved 70 meters.

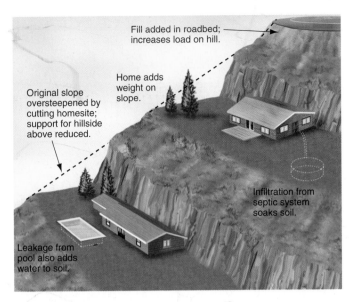

Figure 8.21

Effects of construction and human habitation on slope stability.

Figure 8.23

Quebrada San Julián (foreground) and Caraballeda (rear) were built, respectively, in a floodplain and on an alluvial plain; but the mountainous terrain leaves little choice.

Photograph by M. C. Larsen and H. T. Sierra, U.S. Geological Survey

Figure 8.22

Failure of steep slope along a roadcut now threatens the cabin above.

in the only flat areas available, beside the streams or in alluvial plains at the mouths of the valleys (figure 8.23).

The first two weeks of December 1999 were hardly dry, with 293 millimeters (nearly a foot) of rain, but the real problem was the deluge of another 911 millimeters (close to *3 feet*) of rain that fell December 14–16. Runoff from the steep slopes was swift, and flash floods quickly rose in the valleys. Worse yet, thousands of debris avalanches from the hillsides cascaded into the same valleys, mingling with floodwaters to make huge debris flows that swept out into the alluvial plains and through the towns (figure 8.24). An estimated 30,000 people died from the consequences of flooding and landslides combined.

Nor is the danger now past, for much rock and soil remains in those stream channels, to be remobilized by future runoff. And there are few safe sites in the region for those who choose to stay and rebuild.

Possible Preventive Measures

Avoiding the most landslide-prone areas altogether would greatly limit damages, of course, but as is true with fault zones, floodplains, and other hazardous settings, developments may already exist in areas at risk, and economic pressure for more development can be strong. In areas such as the Venezuelan coast, there may be little buildable land *not* at risk. Population density, too, coupled with lack of "safe" locations, may push development into unsafe areas.

Sometimes, local officials must resign themselves to the existence of a mass-movement problem and may take some steps to limit the resulting damage. In places where the structures to be protected are few or small, and the slide zone is narrow, it may be economically feasible to bridge structures and simply let the slides flow over them. This may be done to protect a railway line or road running along a valley from

A

B

Figure 8.24

Damage in Caraballeda: (A) Overview of Los Corales sector. (B) A close look at this apartment building shows that at their peak, debris flows flowed through the lowest three floors. Note sizes of some of the boulders moved; the largest were estimated to weigh up to 400 tons, confirming the force of the flow.

Photographs by M. C. Larsen and H. T. Sierra, U.S. Geological Survey

avalanches—either snow or debris—from a particularly steep slope (figure 8.25). This solution would be far too expensive on a large scale, however, and no use at all if the base on which the structure was built were sliding also.

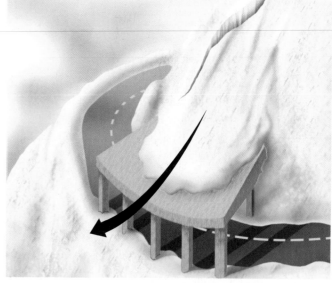

A

B

Figure 8.25

Avalanche-protection structures reduce the consequences of slope instability. (A) Shelter built over railroad track or road in snow avalanche area diverts snow flow. (B) Example of avalanche-protection structure in the Canadian Rockies.

Slope Stabilization

If a slope is too steep to be stable under the load it carries, any of the following steps will reduce slide potential: (1) reduce the slope angle, (2) place additional supporting material at the foot of the slope to prevent a slide or flow at the base of the slope, or (3) reduce the load (weight, shearing stress) on the slope by removing some of the rock or soil (or artificial structures) high on the slope (figure 8.26). These measures may be used in combination. Depending on just how unstable the slope, they may need to be executed cautiously. If earthmoving equipment is being used to remove soil at the top of a slope, for example, the added weight of the equipment and vibrations from it could possibly trigger a landslide.

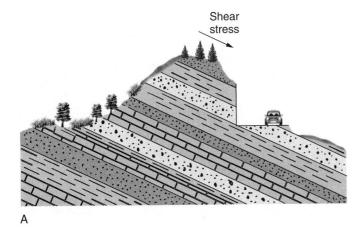

Shear stress

A

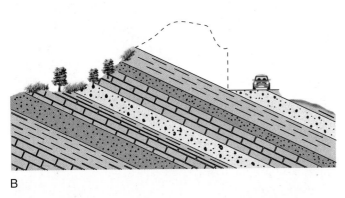

B

C

Figure 8.26

Slope stabilization by slope reduction and removal of unstable material along roadcut. (A) Before: Roadcut leaves steep, unsupported slope. If the shale layers (grey) are relatively impermeable, fluid may accumulate along bedding planes, promoting sliding. (B) After: Material removed to reduce slope angle and load. (C) This roadcut near Denver, CO has been "terraced," cut into steps to break up the long slope while reducing the overall slope angle.

To stabilize exposed near-surface soil, ground covers or other vegetation may be planted (preferably fast-growing materials with sturdy root systems). But sometimes plants are insufficient, and the other preventive measures already described impractical, in a particular situation. Then, retaining walls or other stabilization structures can be built against the slope to try to hold it in place (figure 8.27). Given the distribution of stresses acting on retaining walls, the greatest successes of this kind have generally been low, thick walls placed at the toe of a fairly coherent slide to stop its movement. High, thin walls have been less successful (figure 8.28).

Since water can play such a major role in mass movements, the other principal strategy for reducing landslide hazards is to decrease the water content or pore pressure of the rock or soil. This might be done by covering the surface completely with an impermeable material and diverting surface runoff above the slope. Alternatively, subsurface drainage might be undertaken. Systems of underground boreholes can be drilled to increase drainage, and pipelines installed to carry the water out of the slide area (figure 8.29). All such moisture-reducing techniques naturally have the greatest impact where rocks or soils are relatively permeable. Where the rock or soil is fine-grained and drains only slowly, hot air may be blown through boreholes to help dry out the ground. Such moisture reduction reduces pore pressure and increases frictional resistance to sliding.

Other slope-stabilization techniques that have been tried include the driving of vertical piles into the foot of a shallow slide to hold the sliding block in place. The procedure works only where the slide is comparatively solid (loose soils may simply flow between piles), with thin slides (so that piles can be driven deep into stable material below the sliding mass), and on low-angle slopes (otherwise, the shearing stresses may simply snap the piles). So far, this technique has not been very effective.

The use of rock bolts to stabilize rocky slopes and, occasionally, rockslides has had greater success (figure 8.30). Rock bolts have long been used in tunneling and mining to stabilize rock walls. It is sometimes also possible to anchor a rockslide with giant steel bolts driven into stable rocks below the slip plane. Again, this works best on thin slide blocks of very coherent rocks on low-angle slopes.

Procedures occasionally used on unconsolidated materials include hardening unstable soil by drying and baking it with heat (this procedure works well with clay-rich soils) or by treating with portland cement. By far the most common strategies, however, are modification of slope geometry and load, dewatering, or a combination of these techniques. The more ambitious engineering efforts are correspondingly expensive and usually reserved for large construction projects, not individual homesites.

Determination of the absolute limits of slope stability is an imprecise science. This was illustrated in Japan in 1969, when a test hillside being deliberately saturated with water unexpectedly slid prematurely, killing several of the researchers. The same point is made less dramatically every time a slope stabilization effort fails.

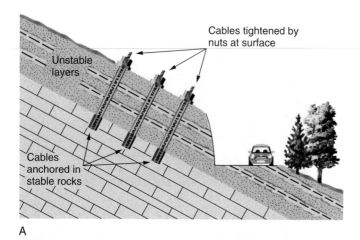

A

B

Figure 8.30

(A) Installation of rock bolts to stabilize a slope. The bolts are steel cables anchored in cement. Tightening nuts at the surface pulls unstable layers together and anchors them to the bedrock below. (B) Rock bolts used to stabilize a rock face above a tunnel in Alaska. See also figures 8.27 C, D.

A

B

Figure 8.31

(A) A close look at the granite domes of Yosemite shows curved slabs of rock at the surface; some above valleys are poised to fall. (B) In 1996, a 200-ton block crashed down, shattering on impact with a blast that leveled thousands of trees.

(B) Photograph by Gerry Wieczorek, U.S. Geological Survey

prolonged over a considerable period of time, curved tree trunks may result (figure 8.33A). (Trees growing in slanted fractures in rock, however, may develop curved trunks through similar growth tendencies in the absence of creep; this is a better indicator of movement in unconsolidated material.) Inanimate objects can reflect soil creep, too. Slanted utility poles and fences and the tilting-over of once-vertical gravestones or other monuments also indicate that the soil is moving (figure 8.33B). The ground surface itself may show cracks parallel to (across) the slope.

A prospective home buyer can look for additional signs that might indicate unstable land underneath (figure 8.34). Ground slippage may have caused cracks in driveways, garage floors, freestanding brick or concrete walls, or buildings;

cracks in walls or ceilings are especially suspicious in newer buildings that would not yet normally show the settling cracks common in old structures. Doors and windows that jam or do not close properly may reflect a warped frame due to differential movement in the soil and foundation. Sliding may have caused leaky swimming or decorative pools, or broken utility lines or pipes. If movement has already been sufficient to cause obvious structural damage, it is probable that the slope cannot be stabilized adequately, except perhaps at very great

A

B

Figure 8.32

Areas prone to landslides may be recognized by the failure of vegetation to establish itself on unstable slopes. (A) Rock Creek Valley, Montana; note road snaking across slide-scarred surfaces. (B) This road was cut into the hillside to dodge around a bay on the California coast, which probably contributed to the sliding now revealed by the fresh, raw slope above and the fresh asphalt on the road. Reducing the slope angle further would require moving huge amounts of material, and the slope is very large for trying slope stabilization.

A

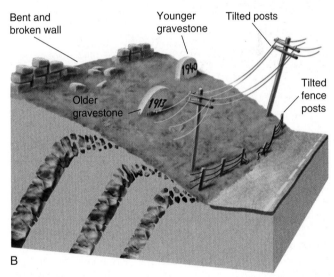

B

Figure 8.33

(A) Tilted and curved tree trunks are a consequence of creep, as seen here on a weathered slope in Bryce Canyon National Park. (B) Other signs of creep: tilted monuments, fences, utility poles.

expense. Possible sliding should be investigated particularly on a site with more than 15% slope (15 meters of rise over 100 meters of horizontal distance), on a site with much steeper slopes above or below it, or in any area where landslides are a recognized problem. (Given appropriate conditions, slides may occur on quite shallow slopes.)

Landslide Warnings?

In early 1982, severe rain-triggered landslides in the San Francisco Bay area killed 25 people and caused over $66 million in damages. In response to this event, the U.S. Geological Survey began to develop a landslide warning system. The basis of the warning system was the development of quantitative relationships among rainfall intensity (quantity of water per unit time), storm duration, and a variety of slope and soil characteristics relating to slope stability—slope angle, pore fluid pressure, shear strength, and so on. These relationships were formulated using statistical analyses of observational data on past landslides. For a given slope, it was possible to approximate threshold values of storm intensity and duration above which landsliding would become likely, given the recent precipitation history at the time (how saturated the ground was prior to the particular storm of concern).

The system, though incomplete, was first tested in February of 1986. Warnings were broadcast as special weather advisories on local radio and television stations. These warnings were rather specific as to time, more general as to area at risk (necessarily so, given the complexity of the geology and terrain and lack of highly detailed data on individual small areas). Some local government agencies recommended evacuations, and many were able to plan emergency responses to the landslides before they occurred. Many landslides did occur; total estimated landslide damage was $10 million, with one death. Of ten landslides for which the times of occurrence are known precisely, eight took place when forecast.

Similar studies in the Seattle area have yielded general relationships between rainfall and the probability of landslides (figure 8.35). These allow local residents to anticipate when

A

B

Figure 8.34

Evidence of slope instability in the Pacific Palisades of California includes broken walls (A) and, in extreme cases, slumped yards (B).

Photographs by J. T. McGill, USGS Photo Library, Denver, CO.

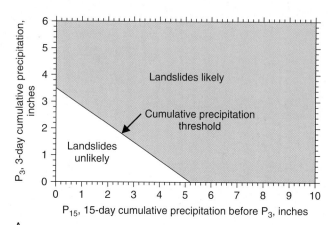

A

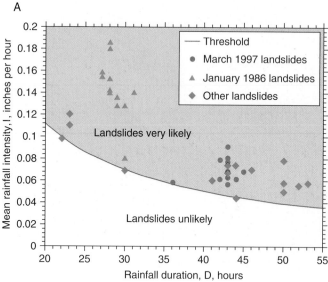

B

Figure 8.35

(A) The probability of landslides depends both on how wet the soil is from prior precipitation (represented by P_{15}) and on recent rainfall (P_3). (B) The more intense the rainfall, the shorter the duration required to trigger landslides. Graphs A and B are both specific to the Seattle area, but the same basic principles apply elsewhere as well.

After USGS Fact Sheet 2007-3005

landslides may occur, and thus to be alert for local landslide warnings. The National Oceanic and Atmospheric Administration (NOAA) has collaborated with the U.S. Geological Survey to develop a flash-flood and debris-flow early-warning system for areas of southern California that have been bared by wildfires. It, too, relies on rainfall intensity/duration relationships specific to the affected areas.

The models need refinement. The ultimate goal would be predictions that are precise both as to time and as to location. To achieve this, too, more extensive data on local geology and topography are needed. Certainly the public is not yet accustomed to landslide warnings, so response is uneven. But the successes of the 1986 and subsequent efforts have suggested that landslide prediction has considerable potential to reduce casualties and enhance the efficiency of agencies' responses to such events. The U.S. Geological Survey and other agencies have continued to expand research into the relationships between rain and landslides, and monitoring of potentially unstable slopes in key areas.

One such area was in Puget Sound. A wet winter in 1996–1997 produced numerous landslides in coastal Washington (figure 8.36). Thereafter, monitoring stations were established on particularly unstable slopes, to track pore pressure in relation to rainfall (figure 8.37). Data need to be collected for some time before the relationship between pore pressure and sliding becomes clear enough that specific landslide warnings for a particular area can be based on these data. Currently, more than half a dozen sites are subject to such detailed monitoring.

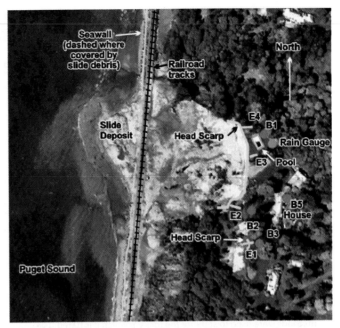

A

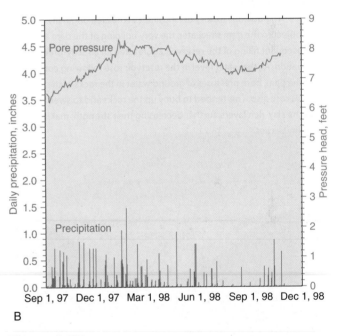

B

Figure 8.37

(A) Landslide-monitoring site at Woodway, Washington. Note proximity of house and pool to slide scarps. (B) Seasonal fluctuations in rainfall are reflected in seasonal fluctuations in pore pressure. Such observations over time help to distinguish dangerously elevated pore pressures likely to trigger sliding.

Courtesy U.S. Geological Survey

Figure 8.36

Head of landslide in Magnolia area of Seattle, Washington. One house has been destroyed and others are at risk. Plastic sheeting has been draped over top of scarp to deflect additional rainwater.

Photograph courtesy U.S. Geological Survey

Wind and Its Geologic Impacts

Air moves from place to place over the earth's surface mainly in response to differences in pressure, which are often related to differences in surface temperature. As noted in figure 9.16, solar radiation falls most intensely on the earth near the equator, and, consequently, solar heating of the atmosphere and surface is more intense there. The sun's rays are more dispersed near the poles. On a non-rotating earth of uniform surface, that surface would be heated more near the equator and less near the poles; correspondingly, the air over the equatorial regions would be warmer than the air over the poles. Because warm, less-dense (lower-pressure) air rises, warm near-surface equatorial air would rise, while cooler, denser polar air would move into the low-pressure region. The rising warm air would spread out, cool, and sink. Large circulating air cells would develop, cycling air from equator to poles and back. These would be atmospheric convection cells, analogous to the mantle convection cells associated with plate motions.

The actual picture is considerably more complicated. The earth rotates on its axis, which adds a net east-west component to air flow as viewed from the surface. Land and water are heated differentially by sunlight, with surface temperatures on the continents generally fluctuating much more than temperatures of adjacent oceans. Thus, the pattern of distribution of land and water influences the distribution of high- and low-pressure regions and further modifies air flow. In addition, the earth's surface is not flat; terrain irregularities introduce further complexities in air circulation. Friction between moving air masses and land surfaces can also alter wind direction and speed; tall, dense vegetation may reduce near-surface wind speeds by 30 to 40% over vegetated areas. A generalized view of actual global air circulation patterns is shown in figure 9.17. Different regions of the earth are characterized by different prevailing wind directions. Most of the United States is in a zone of westerlies, in which winds generally blow from west/southwest to east/northeast. Local weather conditions and the details of local geography produce regional deviations from this pattern on a day-to-day basis.

Air and water have much in common as agents shaping the land. Both can erode and deposit material; both move material more effectively the faster they flow; both can move particles by rolling them along, by saltation, or in suspension (see chapter 6). Because water is far denser than air, it is much more efficient at eroding rocks and moving sediments, and has the added abilities to dissolve geologic materials and to attack them physically by freezing and thawing. On average worldwide, wind erosion moves only a small percentage of the amount of material moved

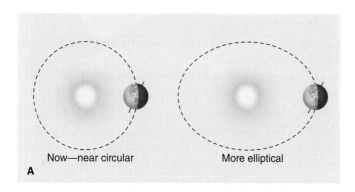

Figure 9.16

(A) Earth's orbit varies in shape from almost circular to more elliptical and back, in cycles of about 100,000 years. (B) The angle of tilt of Earth's axis relative to its orbital plane (the ecliptic), now 23.5°, varies between 22.1° and 24.5° in cycles of about 4000 years. In addition, the axis "wobbles," changing its orientation in space as a spinning top does when slowing down, here illustrated by its pointing toward different stars (North Star at present); this *precession* cycle is about 20,000 years long.

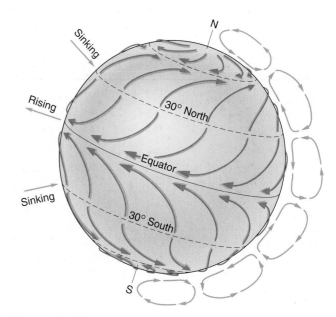

Figure 9.17

Principal present atmospheric circulation patterns. On a uniform, nonrotating earth, a single set of convection cells would circulate directly from equator to poles and back.

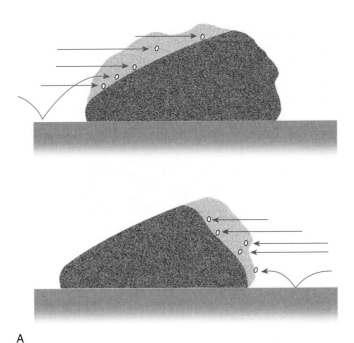

A

Figure 9.18

Wind abrasion on low-lying rocks results in the planing of rock surfaces. (A) If the wind is predominantly from one direction, rocks are planed or flattened on the upwind side. With a persistent shift in wind direction, additional facets are cut in the rock. (B) Example of a ventifact: rock polished and faceted by windblown sand at San Gorgonio Pass near Palm Springs, California.

Photograph © Doug Sherman/Geofile

B

by stream erosion. Historically, the cumulative significance of wind erosion is approximately comparable to that of glaciers. Nevertheless, like glaciers, winds can be very important in individual locations particularly subject to their effects.

Wind Erosion

Like water, wind erosion acts more effectively on sediment than on solid rock, and wind-related processes are especially significant where the sediment is exposed, not covered by structures or vegetation, in such areas as deserts, beaches, and unplanted (or incorrectly planted) farmland. In dry areas like deserts, wind may be the major or even the sole agent of sediment transport.

Wind erosion consists of either abrasion or deflation. *Wind abrasion* is the wearing-away of a solid object by the impact of particles carried by wind. It is a sort of natural sandblasting, analogous to milling by sand-laden waves. Where winds blow consistently from certain directions, exposed boulders may be planed off in the direction(s) from which they have been abraded, becoming **ventifacts** ("wind-made" rocks; figure 9.18). If wind velocity is too low to lift the largest transported particles very high above the ground, tall rocks may show undercutting close to ground level (figure 9.19). Abrasion can also cause serious property damage. Desert travelers caught in windstorms have

Figure 9.19

Granite boulder undercut by wind abrasion, Llano de Caldera, Atacama Province, Chile.

Photograph by K. Segerstrom, U.S. Geological Survey

been left with cars stripped of paint and windshields so pitted and frosted that they could no longer be seen through. Abrasion likewise scrapes paint from buildings and can erode construction materials such as wood or soft stone.

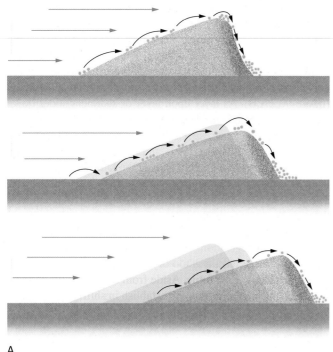

A

C

B

Figure 9.22

Dune migration and its consequences. (A) A schematic of dune migration. (B) With wind coming from the left across these sand ripples, fine sediment blown up the windward slopes slides down the slip faces. There it accumulates, sheltered from the wind, while coarser sediment remains exposed on the windward slopes. To move the coarser sediment will take stronger winds. (C) The huge dune named "Mt. Baldy" marches through full-grown trees at Indiana Dunes National Lakeshore.

Figure 9.23

Crossbeds in a sand dune. Here, several sets of beds meeting at oblique angles show the effects of shifting wind directions, changing sediment deposition patterns. Note also wind-produced ripples in foreground. Oregon Dunes National Recreation Area.

when dry and not heavily loaded, it may not make suitable foundation material. Loess is subject to hydrocompaction, a process by which it settles, cracks, and becomes denser and more consolidated when wetted, to the detriment of structures built on top of it. The very weight of a large structure can also cause settling and collapse.

Deserts and Desertification

Many of the features of wind erosion and deposition are most readily observed in **deserts**. Deserts can be defined in a variety of ways. A desert may be defined as a region with so little vegetation that only a limited population (human or animal) can be supported on that land. It need not be hot or even, technically, dry. Ice sheets are a kind of desert with plenty of water—if not

liquid—and temperatures too low for much life to survive. In more temperate climates, deserts are characterized by very little precipitation, commonly less than 10 centimeters (about 4 inches) a year, but they may be consistently hot, cold, or variable in temperature depending on the season or time of day. The distribution of the arid regions of the world (exclusive of polar deserts) is shown in figure 9.26.

Figure 9.24

Example of loess. Norton County, KS

Photograph by L. B. Buck, USGS Photo Library, Denver, CO.

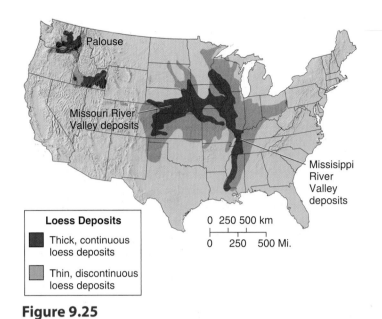

Figure 9.25

Much of the loess distribution in the central United States is close to principal stream valleys supplied by glacial meltwater. Loess to the southwest may have been derived from western deserts.

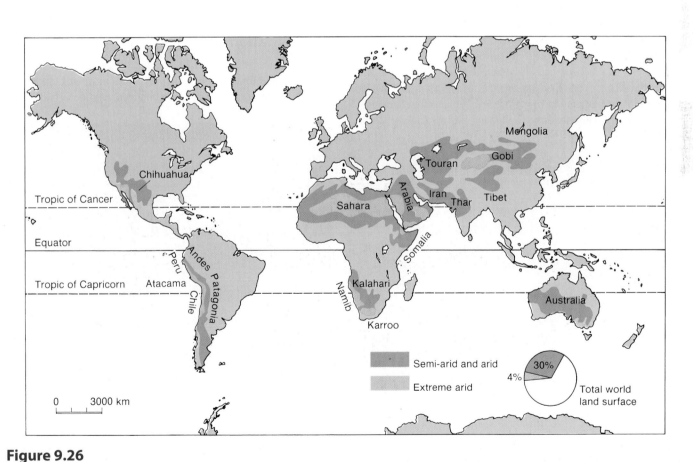

Figure 9.26

Distribution of the world's arid lands.

Data from A. Goudie and J. Wilkinson, The Warm Desert Environment. *Copyright © 1977 by Cambridge University Press*

Causes of Natural Deserts

A variety of factors contribute to the formation of a desert, so deserts come in several different types, depending on their origins.

One factor is moderately high surface temperatures. Most vegetation, under such conditions, requires abundant rainfall and/or slow evaporation of what precipitation does fall. The availability of precipitation is governed, in part, by the global air-circulation patterns shown in figure 9.17. Warm air holds more moisture than cold. Similarly, when the pressure on a mass of air is increased, the air can hold more moisture. Saturated warm air rises at the equator and spreads outward. Air pressure and temperature both decrease with increasing altitude, so as the air rises and cools, it must dump much of its moisture, producing the heavy downpours common in the tropics. When that air then circulates downward, at about 30 degrees north and south latitudes, it is warmed as it approaches the surface, and also subjected to increasing pressure from the deepening column of air above it. It can then hold considerably more water, so when it reaches the earth's surface, it causes rapid evaporation. Note in figure 9.26 that many of the world's major deserts fall in belts close to these zones of sinking air at 30 degrees north and south of the equator. These are the subtropical-latitude deserts.

Topography also plays a role in controlling the distribution of precipitation. A high mountain range along the path of principal air currents between the ocean and a desert area may be the cause of the latter's dryness. As moisture-laden air from over the ocean moves inland across the mountains, it is forced to higher altitudes, where the temperatures are colder and the air thinner (lower pressure). Under these conditions, much of the moisture originally in the air mass is forced out as precipitation, and the air is much drier when it moves farther inland and down out of the mountains. In effect, the mountains cast a **rain shadow** on the land beyond (figure 9.27).

Rain shadows cast by the Sierra Nevada of California and, to a lesser extent, by the southern Rockies contribute to the dryness and desert regions of the western United States.

Because the oceans are the major source of the moisture in the air, distance from the ocean (in the direction of air movement) can by itself be a factor contributing to the formation of a desert. The longer an air mass is in transit over dry land, the greater chance it has of losing some of its moisture through precipitation. This contributes to the development of deserts in continental interiors. On the other hand, even coastal areas can have deserts under special circumstances. If the land is hot and the adjacent ocean cooled by cold currents, the moist air coming off the ocean will be cool and carry less moisture than air over a warmer ocean. As that cooler air warms over the land and becomes capable of holding still more moisture, it causes rapid evaporation from the land rather than precipitation. This phenomenon is observed along portions of the western coasts of Africa and South America (figure 9.28). Polar deserts can also be attributed to the differences in moisture-holding capacity between warm and cold air: Air traveling from warmer latitudes to colder near-polar ones will tend to lose moisture by precipitation, so less remains to fall as snow near the poles, and the limited evaporation from cold high-latitude oceans contributes little additional moisture to enhance local precipitation. Thick polar ice caps, then, reflect effective preservation of what snow does fall, rather than heavy precipitation.

Desertification

Climatic zones shift over time. In addition, topography changes, global temperatures change, and plate motions move landmasses to different latitudes. Amid these changes, new deserts develop

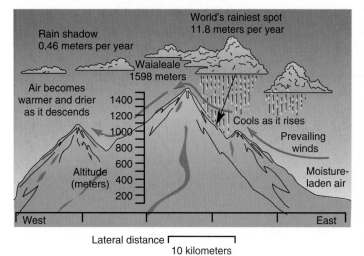

Figure 9.27

Rain shadows can occur even on an island in the ocean! Mt. Waialeale, on the Hawaiian island of Kauai, is described as the "world's rainiest spot," receiving over 450 inches of rain a year. But downwind from it, in its rain shadow, annual rainfall is just 18 inches. With air that is drier to start with—as in most parts of the world—the dryness in the rain shadow is correspondingly more extreme.

Figure 9.28

The Namib Desert lies along the west coast of Namibia (see figure 9.26). It receives an average of only 63 mm (about 2½ in.) of rain per year, plus some additional moisture as fog off the adjacent Atlantic Ocean.

Image courtesy NASA/GSFC/METI/ERSDAC/JAROS and the U.S./Japan ASTER Science Team.

in areas that previously had more extensive vegetative cover. The term **desertification**, however, is generally restricted to apply only to the relatively rapid development or expansion of deserts caused or accelerated by the impact of human activities.

The exact definition of the lands at risk is difficult. *Arid* lands are commonly defined as those with annual rainfall of less than 250 millimeters (about 10 inches); *semiarid* lands, 250 to 500 millimeters (10–20 inches). The extent to which vegetation will thrive in low-precipitation areas depends on such additional factors as temperature and local evaporation rates. Many of the arid lands border true desert regions. Desertification does not involve the advance of desert into nondesert regions as a result of forces originating within the desert. Rather, desertification is a patchy conversion of dry-but-habitable land to uninhabitable desert as a consequence of land-use practices (perhaps accelerated by such natural factors as drought).

The limited vegetation of the arid and semiarid lands is critical to preventing their becoming deserts. Plants shade the soil, and their roots help break it up. In the absence of the plants, under the baking sun typical of many drylands, the soil may crust over and harden, becoming less permeable. This decreases infiltration by what little rain does fall, and increases water loss by surface runoff. That, in turn, decreases reserves of soil moisture on which future plant growth may depend. Vegetation also shields the soil from erosion by wind. This is key to preserving soil fertility, for it is typically the topmost layer of soil that is richest in organic matter and most fertile. Thus, loss of vegetation leads to soil degradation that permanently diminishes the ability of the land to support future plant growth. Natural drought cycles may be short enough that the land will recover when the rains return. But the dryland environment is not resilient. If human activities increase the pressure on these marginal lands, the degradation may be irreversible, and desertification has begun.

The vegetation in dry lands is a precious resource to people living there. It may provide food for people or livestock, wood for shelter or energy. Just trying to support too large a population in such a region may result in stripping the vegetative cover to the point of initiating desertification. As it progresses, the land can support ever-fewer people, and the situation worsens.

On land used for farming, native vegetation is routinely cleared to make way for crops. The native vegetation would have adapted to the dry conditions; the crops often require more moisture or lack the ability to survive a natural local drought cycle. While the crops thrive, all may be well. If the crops fail, perhaps during a drought, the land is left bare of vegetation and vulnerable to the types of degradation described earlier. It becomes harder to grow future crops, and desertification progresses.

Similar results follow from the raising of numerous livestock on the dry lands. In drier periods, vegetation may be reduced or stunted. Yet it is precisely during those periods that livestock, needing the vegetation not only for food but also for the moisture it contains, put the greatest grazing pressure on the land. The soil may again be stripped bare, and the deterioration of desertification follows.

Desertification is cause for concern because it effectively reduces the amount of arable (cultivatable) land on which the world depends for food. The U.N. has estimated that one-third of the world's population lives in "drylands," and that over a billion people, half in Africa alone, live in areas prone to desertification. Overall, up to 70% of the drylands are undergoing significant degradation (figure 9.29). Some projections suggest that, early in this century, one-third of the world's once-arable land will be rendered useless for the culture of food crops as a consequence of desertification and attendant soil deterioration. The recent famine in Ethiopia may have been precipitated by a drought, but it has been prolonged by desertification brought on by overuse of land incapable of supporting concentrated human or animal populations. The Sahel region of Africa, just south of the Sahara Desert, is a classic example of desertification.

Though desertification makes news most often in the context of Third World countries, the process is ongoing in many other places. In the United States, a large portion of the country is potentially vulnerable: Much of the western half of the country can be classified as semiarid on the basis of its low precipitation. An estimated one-million-plus square miles of land—more than a third of this low-rainfall area—has undergone severe desertification, characterized by loss of desirable native vegetation, seriously increased erosion, and reduced crop yields. The problems are caused or aggravated by intensive use of surface water and overgrazing, and the population of these areas—especially the Sun Belt—is growing.

Why, then, so little apparent concern in this country? The main reason is that so far, the impact on humans has been

Figure 9.29

Degradation—loss of soil, water resources, or biodiversity—has reduced productivity over about 70% of the world's "drylands," and the rest are vulnerable. Over 100 nations are potentially affected. (Black areas are deserts already.)

Base map from UNEP Dryland Degradation Map, 1993. Reprinted by permission Naomi Poulton, Head, Publishing Unit UNEP.

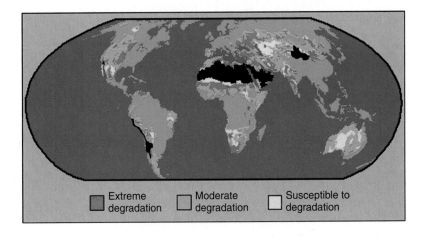

| Extreme degradation | Moderate degradation | Susceptible to degradation |

modest. Other parts of the country can supply needed food. The erosion and associated sediment-redistribution problems are not yet highly visible to most people. Groundwater supplies are being tapped both for water supplies and for irrigation, creating the illusion of adequate water; but as will be seen in chapter 11, groundwater users may be living on borrowed time, or at least borrowed water that is destined to run out. As that happens, it will become painfully obvious that too much pressure from human use was brought to bear on dry lands.

Finally, global climate change threatens to increase the extent of vulnerable arid lands. Higher temperatures and changing precipitation patterns may shift some now-temperate regions in that direction. Chapter 10 explores more fully the evidence for, and possible consequences of, such climate change.

Summary

Glaciers past and present have sculptured the landscape not only in mountainous regions but over wide areas of the continents. They leave behind U-shaped valleys, striated rocks, piles of poorly sorted sediment (till) in a variety of landforms (moraines), and outwash. Most glaciers currently are alpine glaciers, and at present, the majority of these are retreating. The two major ice sheets remaining are in Greenland and Antarctica. Meltwater from glaciers supplies surface and ground water. Therefore, the loss of ice mass and associated water-storage capacity can locally have serious water-supply implications.

Wind moves material much as flowing water does, but less forcefully. As an agent of erosion, wind is less effective than water but may have significant effects in dry, exposed areas, such as beaches, deserts, or farmland. Wind action creates well-sorted sediment deposits, as dunes or in blankets of fine loess. The latter can improve the quality of soil for agriculture. However, loess typically makes a poor base for construction. Features created by wind are most obvious in deserts, which are sparsely vegetated and support little life. The extent of unproductive desert and arid lands is increasing through desertification brought on by intensive human use of these fragile lands.

Key Terms and Concepts

ablation 194	deflation 204	equilibrium line 194	rain shadow 208
abrasion 196	desert 206	glacier 192	slip face 205
alpine glacier 193	desertification 209	loess 205	striations 196
calving 194	drift 196	moraine 197	till 196
continental glacier 193	dune 204	outwash 196	ventifact 203

Exercises

Questions for Review

1. Discuss ways in which glaciers might be manipulated for use as a source of water.

2. Briefly describe the formation and annual cycle of an alpine glacier.

3. What is a moraine? How can moraines be used to reconstruct past glacial extent and movements?

4. What is an ice age? Choose any two proposed causes of past ice ages and evaluate the plausibility of each. (Is the effect on global climate likely to have been large enough? long enough? Is there any geologic evidence to support the proposal? Is the mechanism specific to one particular ice age, or applicable to any?)

5. Describe two ways in which glaciers store water. How can the retreat of alpine glaciers cause both springtime flooding and summer drought?

6. How is sunlight falling on the earth's surface a factor in wind circulation?

7. By what two principal processes does wind erosion occur?

8. Briefly describe the process by which dunes form and migrate.

9. What is loess? Must the sediment invariably be of glacial derivation, as much U.S. loess appears to be?

10. Assess the significance of loess to (a) farming and (b) construction.

11. What is desertification? Describe two ways in which human activities contribute to the process.

Exploring Further in Numbers and Pictures

1. During the Pleistocene glaciation, a large fraction of the earth's land surface may have been covered by ice. Assume that 10% of the 149 million square kilometers was covered by ice averaging 1 kilometer thick. How much would sea level have been depressed over the 361 million square kilometers of oceans? You may find it interesting to examine a bathymetric (depth) chart of the oceans to see how much new land this would have exposed.

2. It is estimated that since 1960, sea level has risen about 30 millimeters due to added water from melting glaciers. Using ocean-area data from problem 1, estimate the volume of ice melted.

3. After the next windstorm, observe the patterns of dust distribution (or, if you are in a snowy climate and season, snow distribution), and try to relate them to the distribution of obstacles that may have altered wind velocity.

4. Information on the U.S. Geological Survey's monitoring of the glaciers in Glacier National Park is at www.nrmsc.usgs.gov/research/glaciers.htm
    Investigate the four different techniques the researchers are using; follow the repeat-photography link and examine the changes shown.

Climate—
Past, Present, and Future

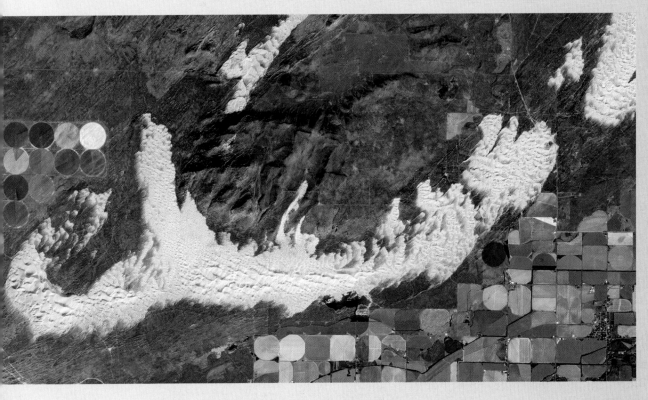

CHAPTER 10

We all experience and deal with weather on a daily basis—sun and clouds, wind, rain, snow—and that weather can vary considerably from day to day or even hour to hour. *Climate* is, in a sense, weather averaged over longer periods of time; as such, it is less variable on a human timescale. But it certainly can and does change over the long span of earth history, as the geologic record makes abundantly clear. Ice ages are just one example of natural climate fluctuation.

Both weather and regional climate can affect the operation or intensity of the surface processes examined in the previous four chapters, and thus the impact they have on human activities. While we can't do much about a day's weather, it has become increasingly evident that humans *can* influence climate on a regional or global scale. This chapter explores various influences on and effects of climate, evidence for climate change, and possible consequences of those changes. Such information will be relevant to discussions of such diverse topics as energy sources, air pollution, and environmental law and policy in future chapters.

Idaho's St. Anthony Dunes reflect climate change. The sediments were deposited in lakes when the region's climate was cooler and wetter, 10,000 years ago at the end of the last ice-sheet advance. The ice retreated, the climate warmed and lakes dried up, and the sand was blown northeast until the winds were blocked by the remains of extinct volcanoes in the Snake River Plain, where the sediments piled up.

NASA Earth Observatory image created by Jesse Allen and Robert Simmon, using EO-1 ALI data provided courtesy of the NASA EO-1 team.

Major Controls on Global Climate; The Greenhouse Effect

Climate is the result of the interplay of a number of factors. The main source of energy input to the earth is sunlight, which warms the land surface, which, in turn, radiates heat into the atmosphere. Globally, how much surface heating occurs is related to the sun's energy output and how much of the sunlight falling on earth actually reaches the surface. Incoming sunlight may be blocked by cloud cover or, as we have previously seen, by dust and sulfuric acid droplets from volcanic eruptions. In turn, heat (infrared rays) radiating outward from earth's surface may or may not be trapped by certain atmospheric gases, a phenomenon known as the **greenhouse effect**.

On a sunny day, it is much warmer inside a greenhouse than outside it. Light enters through the glass and is absorbed by the ground, plants, and pots inside. They, in turn, radiate heat: infrared radiation, not visible light. Infrared rays, with longer wavelengths than visible light, cannot readily escape through the glass panes; the rays are trapped, and the air inside the greenhouse warms up. The same effect can be observed in a closed car on a bright day.

In the atmosphere, molecules of various gases, especially water vapor and carbon dioxide, act similarly to the greenhouse's glass. Light reaches the earth's surface, warming it, and the earth radiates infrared rays back. But the longer-wavelength infrared rays are trapped by these gas molecules, and a portion of the radiated heat is thus trapped in the atmosphere. Hence the term "greenhouse effect." (See figure 10.1.) As a result of the greenhouse effect, the atmosphere stays warmer than it would if that heat radiated freely back out into space. In moderation, the greenhouse effect makes life as we know it possible: without it, average global temperature would be closer to $-17\ ^\circ C$ (about $1^\circ F$) than the roughly $15^\circ C$ ($59^\circ F$) it now is. However, one can have too much of a good thing.

The evolution of a technological society has meant rapidly increasing energy consumption. Historically, we have relied most heavily on carbon-rich fuels—wood, coal, oil, and natural gas—to supply that energy. These probably will continue to be important energy sources for several decades at least. One combustion by-product that all of these fuels have in common is carbon dioxide gas (CO_2).

A "greenhouse gas" in general is any gas that traps infrared rays and thus promotes atmospheric warming. Water vapor is, in fact, the most abundant greenhouse gas in the earth's atmosphere, but human activities do not substantially affect its abundance, and it is in equilibrium with surface water and oceans. Excess water in the atmosphere readily falls out as rain or snow. Some of the excess carbon dioxide is removed by geologic processes (see chapter 18), but since the start of the so-called Industrial Age in the mid-nineteenth century, the amount of carbon dioxide in the air has increased by over 40%—and its concentration continues to climb (figure 10.2). If the heat trapped by carbon dioxide were proportional to the concentration of carbon dioxide in the air, the increased carbon dioxide would by now have caused sharply increased greenhouse-effect heating of the earth's atmosphere.

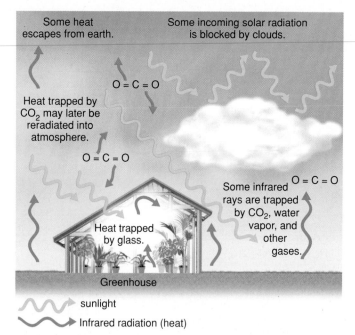

Figure 10.1

The "greenhouse effect" (schematic). Both glass and air are transparent to visible light. Like greenhouse glass, CO_2 and other "greenhouse gases" in the atmosphere trap infrared rays radiating back from the sun-warmed surface of the earth.

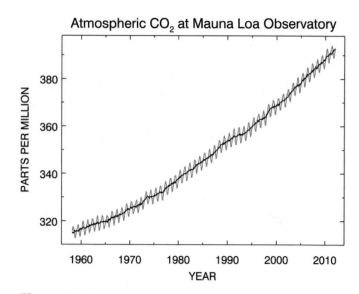

Figure 10.2

Rise in atmospheric CO_2 over past half-century is clear. (Zigzag pattern reflects seasonal variations in local uptake by plants.) Preindustrial levels in ice cores were about 280 ppm (parts per million). One ppm is 0.0001%.

Data from Ralph Keeling, Scripps Institution of Oceanography, and Pieter Tans, NOAA Earth Systems Research Laboratory (ESRL); graph courtesy NOAA/ESRL.

So far, the actual temperature rise has been much more moderate; the earth's climate system is more complex. Indeed, measuring that increase directly, and separating it from the

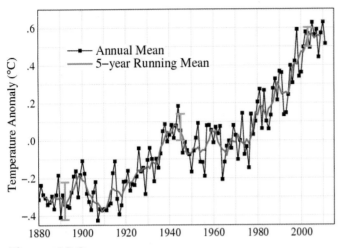

Figure 10.3

Global temperature rise emerges from background noise; green bars indicate uncertainties. (Vertical axis measures temperature deviation, in °C, from 1951–1980 average.)

Image by NASA Goddard Institute for Space Studies

"background noise" of local weather variations, has historically been difficult. Locally, at any one time, a wide variety of local climates can obviously exist simultaneously—icy glaciers, hot deserts, steamy rain forests, temperate regions—and a given region may also be subject to wide seasonal variations in temperature and rainfall. One of the challenges in determining global temperature trends, then, is deciding just how to measure global temperature at any given time. Commonly, several different kinds of data are used, including air temperatures, over land and sea, at various altitudes, and sea surface temperatures. Satellites help greatly by making it possible to survey large areas quickly; see Case Study 10.

Altogether, since the start of the Industrial Age, global surface temperature has risen about 0.85°C (1.5°F) (figure 10.3). Trivial as this may sound, it is already having obvious and profound impacts in many parts of the world. Moreover, the warming is not uniform everywhere; some places are particularly strongly affected. A further concern is what the future may hold.

Climate and Ice Revisited

Early discussions of climate change related to increasing greenhouse-gas concentrations in the atmosphere tended to focus heavily on the prospect of global warming and the resultant melting of earth's reserves of ice, especially the remaining ice sheets. If *all* the ice melted, sea level could rise by over 75 meters (about 250 feet) from the added water alone, leaving aside thermal expansion of that water. About 20% of the world's land area would be submerged. Many millions, perhaps billions, of people now living in coastal or low-lying areas would be displaced, since a large fraction of major population centers grew up along coastlines and rivers. The Statue of Liberty would be up to her neck in water. The consequences to the continental

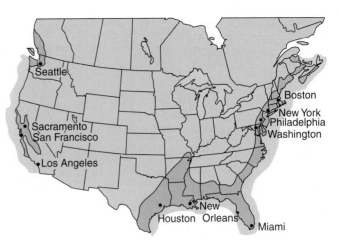

Figure 10.4

The flooding (blue-shaded area) that would result from a 75-meter rise in sea level would flood a large fraction of major cities in the United States and elsewhere.

United States and southern Canada are illustrated in figure 10.4. Also, raising of the base levels of streams draining into the ocean would alter the stream channels and could cause significant flooding along major rivers, and this would not require nearly such drastic sea-level rise.

Such large-scale melting of ice sheets would take time, perhaps several thousand years. On a shorter timescale, however, the problem could still be significant, as some areas are particularly vulnerable (figure 10.5). Complete melting of the

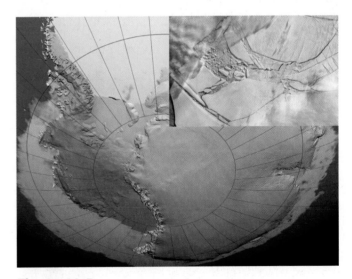

Figure 10.5

From 1957 to 2006, average Antarctic temperature rose about 0.6°C (1.1°F); in West Antarctica, 0.85°C (1.55°F). The West Antarctic ice sheet, grounded below sea level and bordered by floating ice shelves, is especially vulnerable to melting and breakup. Inset shows 2009 breakup of the Wilkins Ice Shelf along the Antarctic Peninsula.

Large image by Trent Schindler, NASA/GSFC Scientific Visualization Studio; inset by Jesse Allen using data courtesy of NASA/GSFC/METI/ERSDAC/JAROS and the U.S./Japan ASTER Science Team.

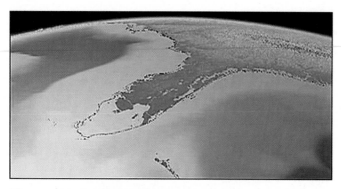

Figure 10.6

Impact on Florida of melting either the West Antarctic or the Greenland ice sheet, which would result in a sea-level rise of about 5 to 6 meters (16 to 20 feet).

Image courtesy William Haxby/NASA

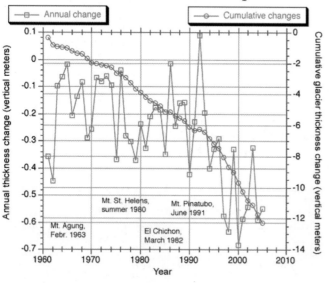

Figure 10.7

Monitoring of hundreds of alpine glaciers worldwide confirms that while thickness, like snowfall, varies from year to year, the net trend is clearly downward. Combining this information with data on glacial retreat indicates a loss of more than 5000 cubic km of ice from alpine glaciers over this period. Note the short-term impact of cooling from major volcanic eruptions.

Image by Mark Durgerov, Institute of Arctic and Alpine Research, University of Colorado at Boulder, courtesy National Snow and Ice Data Center.

West Antarctic ice sheet could occur within a few hundred years. The resulting 5- to 6-meter rise in sea level, though it sounds small, would nevertheless be enough to flood many coastal cities and ports, along with most of the world's beaches; see, for example, figure 10.6. This would be both inconvenient and extremely expensive. (The only consolation is that the displaced inhabitants would have decades over which to adjust to the changes.) Meanwhile, the smaller alpine glaciers are clearly dwindling (figure 10.7).

Arctic polar ice is also demonstrably shrinking, in both area (figure 10.8) and thickness. It is particularly vulnerable because, instead of sitting on long-chilled rock, it floats on the circulating, warming ocean. The good news is that, just as melting ice in a glass does not cause the liquid to overflow, melting sea ice will not cause a rise of sea level. However, concerns about Arctic ecosystems are being raised; for example, polar bears that hunt seals on the ice in Hudson Bay suffer when a shorter ice season reduces their food intake.

The Hidden Ice: Permafrost

Warming is affecting more than the visible ice of cold regions. In alpine climates, winters are so cold, and the ground freezes to such a great depth, that the soil does not completely thaw even in the summer. The permanently frozen zone is **permafrost** (figure 10.9). The meltwater from the thawed layer cannot infiltrate the frozen ground below, so the terrain is often marshy (figure 10.10). Structures built on or in permafrost may be very stable while it stays frozen but sink into the muck when it thaws; vehicles become mired in the sodden ground, so travel may be impossible when the upper soil layers thaw.

Studies have shown that permafrost is also being lost as the world warms. Where some permafrost remains, the thaw penetrates deeper, and the frozen-ground season is shorter. Some areas are losing their permafrost altogether, and lakes drain abruptly when the last of the ice below melts away. All these changes are having effects on ecosystems, animal migrations, and the traditional lifestyles of some native peoples. They are even affecting oil exploration and the Trans-Alaska Pipeline, as we will see in later chapters.

Oceans and Climate

One reason that there is no simple correlation between atmospheric greenhouse-gas concentrations and land surface temperatures is the oceans. That huge volume of water represents a much larger thermal reservoir than the tenuous atmosphere. Coastal regions may have relatively mild climates for this reason: As air temperatures drop in winter, the ocean is there to release some heat to the atmosphere; as air warms in summer, the ocean water absorbs some of the heat. The oceans store and transport a tremendous amount of heat around the globe. A thorough discussion of the role of the oceans in global and local climate is beyond the scope of this chapter, but it is important to appreciate the magnitude of that role.

Recent studies of the balance between the energy received from the sun and the energy radiated back into space have confirmed a net excess of energy absorbed, an average of 0.85 watts per square meter per year. Such a rate of energy absorption, maintained for a thousand years, could melt enough ice to raise sea level by a hundred meters, or raise the temperature of the ocean's surface-water layer by 10°C (nearly 20°F). Moreover, much of this absorbed energy is, for now, being

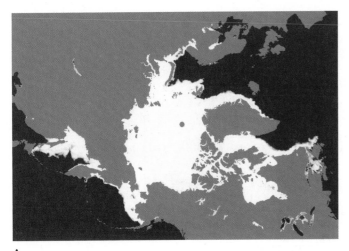

A

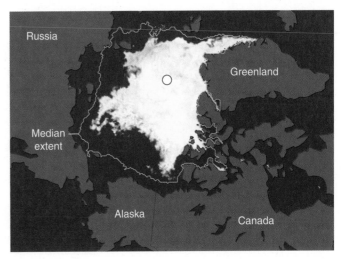

B

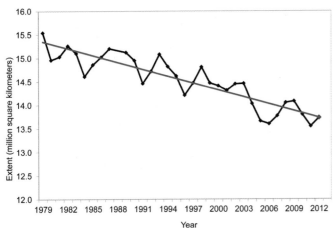

C

Figure 10.8

Average maximum Arctic Sea ice extent (A) has declined only moderately (here, March 2011 as compared to the median for 1979–2000, indicated by the yellow line), but the annual minimum (B) has declined sharply (here, September 2011 compared to the 1979–2000 median, indicated by the yellow line). Thus, on average, annual ice extent (area) has been declining (C). So has the ice thickness, meaning a net loss of large volumes of ice in the Arctic.

(A) and (B) images by Jesse Allen using AMSR-E data provided courtesy National Snow and Ice Data Center. (C) Graph courtesy National Snow and Ice Data Center.

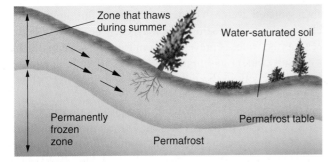

Figure 10.9

In cold climates, a permanently frozen soil zone, permafrost, may be found beneath the land surface.

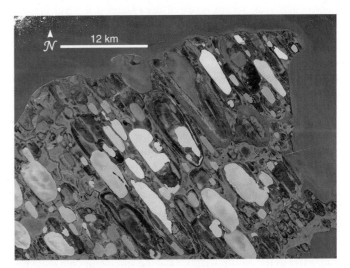

Figure 10.10

A portion of the north coast of the National Petroleum Reserve in Alaska. Abundant lakes reflect soggy soil over ice. Note elongated topographic features, a result of ice-sheet flow during the last Ice Age. Differences in suspended sediment content cause differences in water color in the many lakes.

Image courtesy NASA/GSFC/METI/ERSDAC/JAROS, and the U.S./Japan ASTER Science Team.

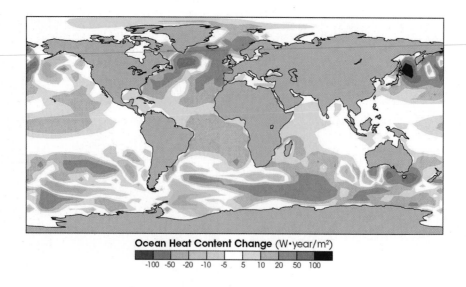

Ocean Heat Content Change (W·year/m²)

-100 -50 -20 -10 -5   5   10   20   50   100

Figure 10.11

Modeled increase in the heat content of the upper 70 meters of the oceans, based on observations from 1993 to 2003. (A watt-year is the amount of energy supplied by 1 watt of power for a year; it equals 8.76 kilowatt-hours.)

NASA image courtesy Jim Hansen et al., Goddard Institute for Space Studies

stored in the oceans (figure 10.11). So it has not yet been reflected in measured global temperature increases.

We have already seen that the oceans supply most of the water vapor to make rain and snow. Water absorbs heat when vaporizing, and more water will evaporate from warmer surface water. Heat from sea surface water supplies the energy to spawn hurricanes and other tropical storms. While meteorologists have not yet seen a systematic increase in the *number* of U.S. hurricanes each year, recent research does suggest a trend toward increasing *intensity* of hurricanes—not good news for residents of the Gulf Coast.

The Thermohaline Circulation

Broadly, oceanic circulation is driven by a combination of winds, which push surface currents, and differences in density within the oceans' waters, related to temperature and salinity. Cold water is denser than warm, and at a given temperature, density increases with increasing salinity. The roles of temperature and salinity are reflected in the name **thermohaline circulation** (remember, "halite" is sodium chloride, salt, the most abundant dissolved material in the oceans) given to the large-scale circulation of the oceans (figure 10.12). Among the notable features are the Gulf Stream, carrying warmed equatorial water to the North Atlantic, and the sinking of cold water in the polar regions. The informal name of "ocean conveyor" is sometimes given to this circulation system in recognition of its heat-transport role.

In the climate records described later in the chapter, scientists have found evidence of episodes of "abrupt climate change" in some locations. One proposed mechanism for sudden cooling of the North Atlantic is disruption of the thermohaline circulation to this region, reducing the influx of

warm water. For instance, extensive melting of the Greenland ice sheet could create a pool of cool, fresh surface water that would block the incoming warm flow; or, increased water evaporation around a warmer equator could increase the salinity of the surface water so much that it would sink rather than flowing northward. Much research is needed to test such hypotheses. In any case, to a geologist, "abrupt" climate change means change on a timescale of decades to centuries—not days to weeks as in recent movie fiction!

El Niño

Most of the vigorous circulation of the oceans is confined to the near-surface waters. Only the shallowest waters, within 100 to 200 meters of the surface, are well mixed by waves, currents, and winds, and warmed and lighted by the sun. The average temperature of this layer is about 15°C (60°F).

Below the surface layer, temperatures decrease rapidly to about 5°C (40°F) at 500 to 1000 meters below the surface. Below this is the so-called *deep layer* of cold, slow-moving, rather isolated water. The temperature of this bottommost water is close to freezing and may even be slightly below freezing (the water is prevented from freezing solid by its dissolved salt content and high pressure). This cold, deep layer originates largely in the polar regions and flows very slowly toward the equator.

When winds blow offshore, they push the warm surface waters away from the coastline also. This, in turn, creates a region of low pressure and may result in *upwelling* of deep waters to replace the displaced surface waters (figure 10.13). The deeper waters are relatively enriched in dissolved nutrients, in part because few organisms live in the cold, dark depths to consume those nutrients. When the nutrient-laden waters rise into

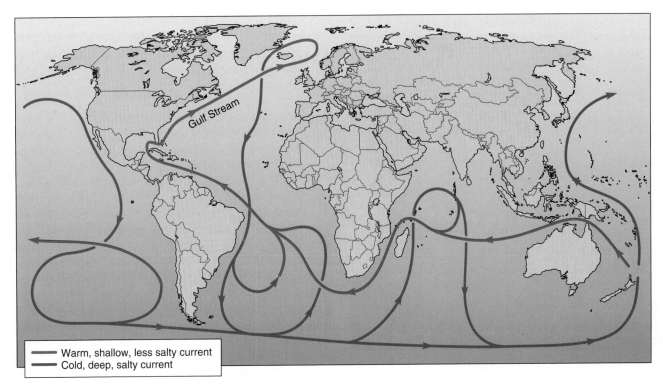

Figure 10.12

The thermohaline circulation, moving water—and thus heat—around the globe.

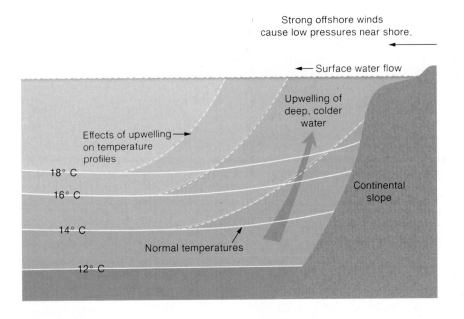

Figure 10.13

Warm surface water normally overlies colder seawater at depth. When offshore winds blow warm waters away from the South American shore and create local low pressure, upwelling of colder deep waters may occur (dashed lines). During an El Niño episode, the winds die down, and upwelling is suppressed (solid lines), so warm surface waters extend to the coast.

the warm, sunlit zone near the surface, they can support abundant plant life and, in turn, animal life that feeds on the plants. Many rich fishing grounds are located in zones of coastal upwelling. The west coasts of North and South America and of Africa are subject to especially frequent upwelling events.

From time to time, however, for reasons not precisely known, the upwelling is suppressed for a period of weeks or longer, as warm waters from the western South Pacific extend eastward to South America (figure 10.14A). Abatement of coastal winds, for one, reduces the pressure gradient driving the

MAR 1 2010

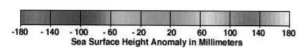

-180 -140 -100 -60 -20 20 60 100 140 180
Sea Surface Height Anomaly in Millimeters

A

upwelling. The reduction in upwelling of the fertile cold waters has a catastrophic effect on the Peruvian anchoveta industry. Such an event is called *El Niño* ("the [Christ] Child") by the fishermen because it commonly occurs in winter, near the Christmas season.

At one time, El Niño events were believed to be rather sporadic and isolated, local events. Now it is recognized that they are cyclic in nature, occurring every four to seven years as part of the El Niño–Southern Oscillation (referring to the southern oceans). The opposite of an El Niño, a situation with unusually cold surface waters off western South America, has been named *La Niña* (figure 10.14B). Extensive shifts in seawater surface temperatures, in turn, cause large-scale changes in evaporation, and thus precipitation, as well as wind-circulation patterns (figure 10.15). Climatic changes associated with El Niño/La Niña events generally include changes in frequency, intensity, and paths of Pacific storms, short-term droughts and floods in various regions of the world, and changes in the timing and intensity of the monsoon season in India. Heavy rains associated with El Niño events have been blamed for major landslides along the Pacific coast of the United States. With respect to global climate change, one concern is that warmer oceans may mean more frequent, intense, and/or prolonged El Niño episodes, with especially devastating consequences to those nations whose much-needed rainy season will be disrupted.

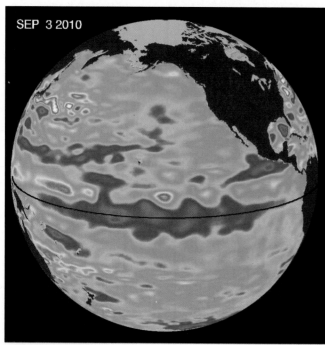

SEP 3 2010

B

Figure 10.14

(A) El Niño: Unusually warm water appears as higher sea-level elevation; as indicated by key; warm water expands. (B) La Niña: Lower sea-level elevations show areas of cooler surface waters.

Images courtesy of JPL/NASA.

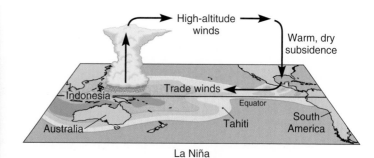

High-altitude winds

Warm, dry subsidence

Trade winds

Indonesia

Equator

Tahiti

South America

Australia

La Niña

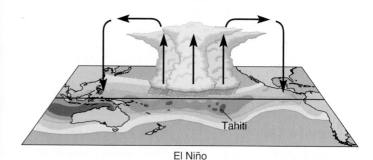

Tahiti

El Niño

Figure 10.15

Under La Niña conditions, warm, wet air over the western Pacific dumps monsoon rains on southeast Asia as it rises and cools; when the air sinks and warms again, over the eastern Pacific, it is drying. During El Niño, surface waters in the eastern Pacific are warmer, which shifts the patterns of evaporation and precipitation; the monsoon season is much drier, while more rain falls on the western United States.

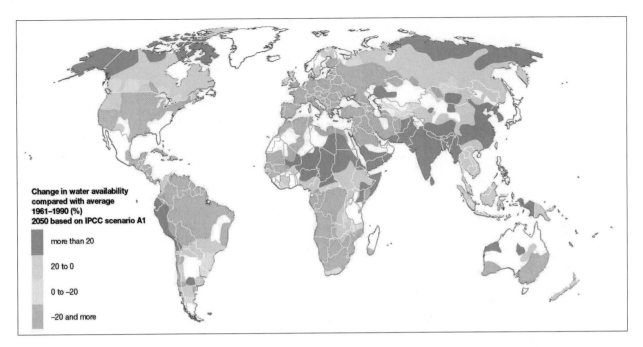

Figure 10.16

Climate modeling based on one scenario seen as likely by the Intergovernmental Panel on Climate Change reveals potentially large drops in water availability by mid-century. Areas such as southern Australia, the Middle East, and the southwestern United States can ill afford this.

Image by Philippe Rekacewicz after Arnell (2004), from UNEP/GRID-Arendal maps and graphics library 2009 http://maps.grida.no/go/graphic/the-contribution-of-climate-change-to-declining-water-availability

Reprinted with permission UNEP. http://www.grida.no/publications/vg/water2/page/3294.aspx

About 1996, investigators first recognized that underlying the El Niño/La Niña is a longer-term phenomenon, dubbed the Pacific Decadal Oscillation (PDO), with a cycle of fluctuating Pacific Ocean surface temperatures over a period of 20–30 years. The PDO, too, affects patterns of evaporation, precipitation, and wind. Still more recently, an Atlantic Multidecadal Oscillation (AMO) has been identified and named. It is entirely possible that more such cycles important to our understanding of climate/ocean relationships remain to be recognized.

Other Aspects of Global Change

As we have seen in previous sections, "global warming" means more than just a little warmer weather everywhere. There are consequences for ice amount and distribution, oceanic circulation, patterns of storms, and more, as will be explored below. It is very important to bear in mind, too, that "global" climate changes do not affect all parts of the world equally or in the same ways.

Changes in wind-flow patterns and amounts and distribution of precipitation will cause differential impacts in different areas, not all of which will be equally resilient (figure 10.16). A region of temperate climate and ample rainfall may not be seriously harmed by a temperature change of a few degrees,

one way or the other, or several inches more or less rainfall. A modest temperature rise in colder areas may actually increase plant vigor and the length of the growing season (figure 10.17), and some dry areas may enjoy increased productivity with increased rainfall.

Warming and moisture redistribution also have potential downsides, of course. In chapter 9, we noted the impact of the loss of alpine glaciers on water supplies. Figure 10.16 suggests potential water shortages as well in areas not relying now on glacial meltwater. In many parts of the world, agriculture is already only marginally possible because of hot climate and little rain. A temperature rise of only a few degrees, or loss of a few inches of rain a year, could easily make living and farming in these areas impossible. Recent projections suggest that summer soil-moisture levels in the Northern Hemisphere could drop by up to 40% with a doubling in atmospheric CO_2. This is a sobering prospect for farmers in areas where rain is barely adequate now, for in many places, irrigation is not an option. Recent evidence and computer modeling also suggest that global warming and associated climate changes may produce more extremes—catastrophic flooding from torrential storms, devastating droughts, "killer heat waves" like that costing hundreds of lives in Chicago in 1995, and the combination of heat and drought that led to deadly wildfires in southern Australia in early 2009 (figure 10.18) and to deaths, wildfires, and crop failures across much of the United States in 2012.

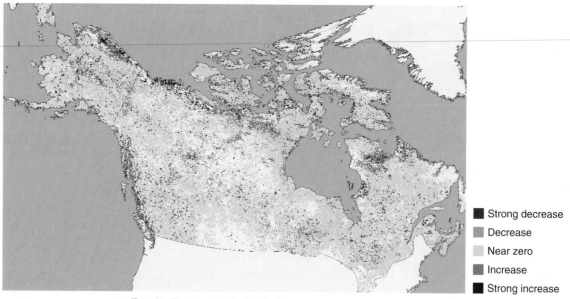

■	Strong decrease
▨	Decrease
▢	Near zero
▨	Increase
■	Strong increase

Trend in Photosynthetic Activity (1982–2003)

Figure 10.17

Plant activity in the northernmost reaches of Alaska and Canada is picking up with global warming, but note that it is declining over a much larger area farther south.

Image courtesy NASA, based on data from Scott Goetz, Woods Hole Research Center

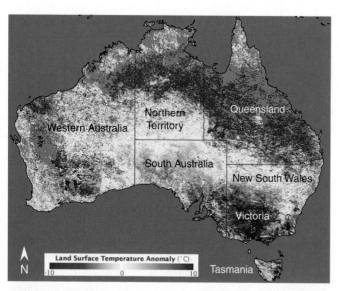

Figure 10.18

Unprecedented heat in southern Australia on 28–31 January and 6–8 February 2009, following six years of drought in 2001–2007, set the stage for raging wildfires that killed over 200 people. Local temperatures reached as high as 48.8°C (120°F), and many longstanding high-temperature records were broken.

Map by Jesse Allen, based on MODIS land surface temperature data.

It has been argued that rising atmospheric CO_2 levels should be beneficial in promoting plant growth. After all, photosynthesis involves using CO_2 and water and solar energy to manufacture more complex compounds and build the plant's structure. However, controlled experiments have shown that not all types of plants grow significantly more vigorously given higher CO_2 concentrations in their air, and even those that do may show enhanced growth only for a limited period of time. Moreover, the productivity of phytoplankton (microscopic plants) in the warming oceans has shown sharp declines (figure 10.19). The reason may be that as the surface water warms and becomes less dense, the warm-water layer becomes more firmly stabilized at the surface, suppressing upwelling of the nutrient-rich cold water below over much of the ocean and mixing of the cold and warm waters. The effect is like that of El Niño, but much more widely distributed. Because some whales and fish depend on the phytoplankton for food, a decline in phytoplankton will have repercussions for those populations also, as well as for humans who may consume larger fish in that food web.

Rising atmospheric CO_2 affects oceans in other ways too. With more CO_2 in the air, more CO_2 dissolves in the oceans, making more carbonic acid. This acidification makes it more difficult for corals to build their calcium-carbonate skeletons. The combination of acidification and warming waters puts severe stress on the coral reefs that support thousands of species in marine ecosystems.

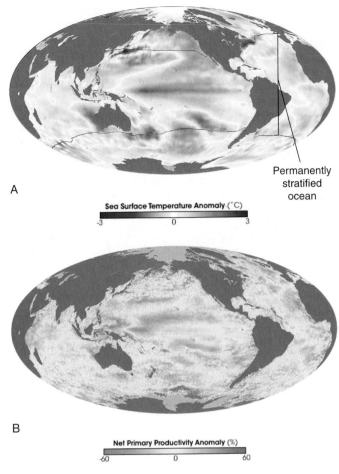

A

Sea Surface Temperature Anomaly (°C)
-3   0   3

Permanently
stratified
ocean

B

Net Primary Productivity Anomaly (%)
-60   0   60

Figure 10.19

(A) Changes in sea surface temperature, 2000–2004. In the "permanently stratified" areas, the oceans show a distinct upper, warmer layer year-round. (B) In general, areas in which phytoplankton productivity declined from 2000 to 2004 are those where the water warmed over the same period.

NASA images by Jesse Allen, based on data provided by Robert O'Malley, Oregon State University

Public-health professionals have identified yet another threat related to global change. Climate influences the occurrence and distribution of a number of diseases. Mosquito-borne diseases, such as malaria and dengue, will be influenced by conditions that broaden or reduce the range of their particular carrier species; the incidence of both has increased in recent years and is currently expected to expand further (figure 10.20). Waterborne parasites and bacteria can thrive in areas of increased rainfall. Efforts to anticipate, and plan responses to, the climate-enhanced spread of disease are growing.

United Nations studies further suggest that less-developed nations, with fewer resources, may be much less able to cope with the stresses of climate change. Yet many of these nations are suffering the most immediate impacts: The very existence of some low-lying Pacific island nations is threatened by

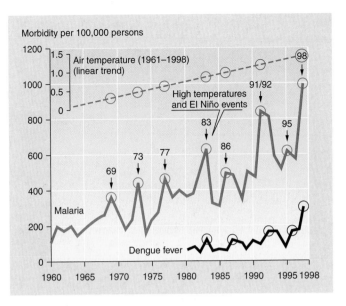

Figure 10.20

Occurrence of malaria and dengue fever in Colombia is correlated with temperature, spiking in warmer El Niño years, but also trending upward overall as temperatures rise.

Graph from United Nations Environment Programme/GRID-Arendal

sea-level rise; reduced soil moisture and higher temperature quickly affect the world's "drylands," where many poorer countries are located; the tropical diseases are spreading most obviously in the Third World where medical care is limited. Much of Africa is vulnerable to multiple threats associated with climate change (figure 10.21). These differential impacts may have implications for global political stability as well.

Evidence of Climates Past

An understanding of past climatic fluctuations is helpful in developing models of possible future climate change. A special challenge in reconstructing past climate is that we do not have direct measurements of ancient temperatures. These must be determined indirectly using various methods, sometimes described as "proxies" for direct temperature records.

Aspects of local climate change may often be deduced from the geologic record, especially surficial deposits and the sedimentary-rock record. For example, the now-vegetated dunes of the Nebraska Sand Hills (figure 10.22) are remnants of an arid "sand sea" in the region 18,000 years ago, when conditions must have been much drier than they are now. See also the chapter-opening image. Much of North Africa was fertile, wheat-growing land in the time of the Roman Empire; small reductions in rainfall, and the resultant crop failures, may have contributed to the Empire's fall. We have noted that

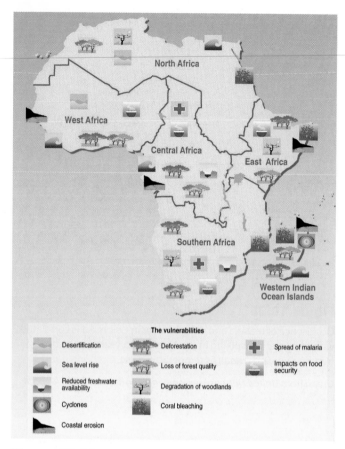

Figure 10.21

Different parts of Africa are vulnerable to different consequences of climate change. ("Coral bleaching" reflects loss of symbiotic algae that give corals their color; it indicates environmental stress and can lead to the death of the coral.)

Image by Delphine Digout after Anna Balance (2002), from UNEP/ GRID-Arendal maps and graphics library, 2012. http://maps.grida.no/ go/graphic/climate-change-vulnerability-in-africa.

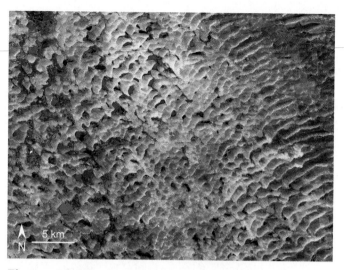

Figure 10.22

The Nebraska Sand Hills formed during the Ice Age: sediment ground off the Rocky Mountains by the ice sheets became outwash in the plains. The dunes developed in dry conditions; now, many are vegetated, and small lakes have accumulated in low spots.

Image courtesy NASA/GSFC/METI/ERSDAC/JAROS, and U.S./Japan ASTER Science Team

glacial deposits further back in the rock record may be recognized in now-tropical regions, and this was part of the evidence for continental drift/plate tectonics. When such deposits were widespread globally, they indicate an ice age at the corresponding time. Conversely, coal deposits indicate warm, wet conditions, conducive to lush plant growth, perhaps in a swampy setting, and widespread coals reflect a warmer Earth. Distinctively warm- or cold-climate animals and plants identified in the fossil record likewise provide evidence of the local environmental conditions at the time they lived.

Other proxies involve the chemistry of sediments, seawater, or snow and ice. For example, from marine sediments comes evidence of ocean-temperature variations. The proportion of calcium carbonate ($CaCO_3$) in Pacific Ocean sediments in a general way reflects water temperature, because the solubility of calcium carbonate is strongly temperature-related: it is more soluble in cold water than in warm. A good idea of global climate at a given time may be gained from compiling such data from many proxies and many places. From such information, we find that climate has, in fact, varied quite a lot throughout earth history, at different times being somewhat cooler or substantially warmer than at present. Particularly when the climatic shifts were relatively large and (geologically speaking) rapid, they have been a factor in the natural selection that is a major force in biological evolution.

A particularly powerful tool involves variations in the proportions of oxygen isotopes in geological materials. By far the most abundant oxygen isotope in nature is oxygen-16 (^{16}O); the heaviest is oxygen-18 (^{18}O). Because they differ in mass, so do molecules containing them, and the effect is especially pronounced when oxygen makes up most of the mass, as in H_2O. Certain natural processes, including evaporation and precipitation, produce fractionation between ^{16}O and ^{18}O, meaning that the relative abundances of the two isotopes will differ between two substances, such as ice and water, or water and water vapor. See figure 10.23. As water evaporates, the lighter $H_2^{16}O$ evaporates preferentially, and the water vapor will then be isotopically "lighter" (richer in ^{16}O, poorer in ^{18}O) than the residual water. Conversely, as rain or snow condenses and falls, the precipitation will be relatively enriched in the heavier $H_2^{18}O$. As water vapor evaporated from equatorial oceans drifts toward the poles, depositing $H_2^{18}O$-enriched precipitation, the remaining

A

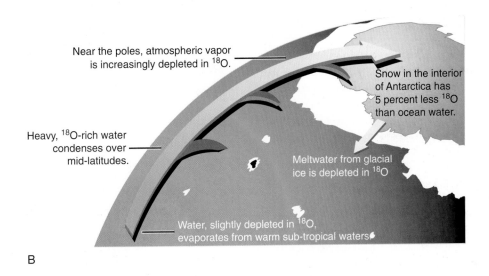

B

Figure 10.23

(A) Mass differences between ^{18}O and ^{16}O lead to different $^{18}O/^{16}O$ ratios in coexisting water and water vapor; the size of the difference is a function of temperature. (B) As the water vapor evaporated near the equator moves poleward and rain and snow precipitate, the remaining water vapor becomes isotopically lighter.

(B) Image by Robert Simmon, courtesy NASA

water vapor becomes progressively lighter isotopically, and so does subsequent precipitation; snow falling near the poles has a much lower $^{18}O/^{16}O$ ratio than tropical rain, or seawater (figure 10.23B).

The fractionation between ^{18}O and ^{16}O in coexisting water vapor and rain or snow is temperature-dependent, more pronounced at lower temperatures. Therefore, at a given latitude, variations in the $^{18}O/^{16}O$ ratio of precipitation reflect variations in temperature. Similarly, oxygen isotopes fractionate between water and minerals precipitated from it. This fractionation, too, is temperature-dependent. Thus, variations in the $^{18}O/^{16}O$ ratio in the oxygen-rich carbonate ($CaCO_3$) or silica (SiO_2) skeletons of marine microorganisms give evidence of variations in

the temperature of the near-surface seawater from which these skeletons precipitated, and they have been used to track past El Niño cycles.

The longevity of the massive continental glaciers—sometimes hundreds of thousands of years—makes them useful in preserving evidence of both air temperature and atmospheric composition in the (geologically recent) past. Tiny bubbles of contemporary air may be trapped as the snow is converted to dense glacial ice. The oxygen-isotope composition of the ice reflects seawater O-isotope composition and the air temperature at the time the parent snow fell. If those temperatures can be correlated with other evidence, such as the carbon-dioxide content of the air bubbles or the presence of volcanic ash layers

Taking Earth's Temperature

Suppose that you were given the assignment of determining the earth's temperature so as to look for evidence of climate change. How would you do it?

Our main concern is clearly temperature at or near the surface, where we and other organisms reside. Thus we can disregard the earth's interior—fortunately, as we can't probe very far into it, and even deep-crustal drill holes are costly. It is also helpful that we focus on trends in temperature over time rather than instantaneous temperatures. We know from experience that, especially in the mid-latitudes, temperatures can fluctuate wildly over short time periods; they can be very different over short distances at the same moment; and they can be quite different from year to year on any given date. If we are looking at global or regional trends, we will be more concerned with averages and how they may be shifting.

Land surface temperatures are the simplest to address. Many weather stations and other fixed observation points have been in place for over a century. Moreover, research has shown that temperature anomalies and trends tend to be consistent across broad regions. (For instance, if your city is experiencing a heat wave or acute cold spell, so are other cities and less-populated areas in the region.) So, data from a modest number of land-based sites should suffice to characterize land surface temperatures.

The oceans obviously present more of a challenge. There have been shipboard temperature measurements for centuries, but relatively far fewer, and typically not repeated at the same locations year after year. There are now stationary buoys that monitor ocean temperatures, but relatively few. Satellites have come into play here. For some decades, sea surface temperatures have been estimated using, as a proxy, radar measurements of sea surface elevation, as in figure 10.14. This is actually a reflection of the temperature through the uppermost layer of seawater, as it is expansion and contraction of that layer as it warms or cools that produces the surface-elevation anomalies. More recently, a more sophisticated approach has been implemented (figure 1). NASA's MODIS detector not only measures directly the infrared radiation (heat) coming from the sea surface. It also measures the concentration of atmospheric water vapor through which the radiated heat is passing—important because water vapor, as a greenhouse gas, will absorb some of that heat on its way to the

satellite, a particular problem in the humid tropics. MODIS can correct for this, and also for interference from such sources as clouds and dust. The result is exceptionally accurate measurement of sea-surface temperatures around the globe.

The average temperature for a given point over the course of a year can then be determined, and those data compared over longer periods of time, or averaged over the earth or a region of it.

Sea Surface Temperature (°C)
-2   35

Figure 1

Sea surface temperatures, 1–8 January 2001, as measured by the Moderate-Resolution Imaging Spectroradiometer (MODIS) aboard NASA's Terra satellite.

NASA image by Jesse Allen based on data provided by the MODIS Ocean Team and the University of Miami Rosenstiel School of Marine and Atmospheric Science Remote Sensing Group

in the ice from prehistoric explosive eruptions, much information can be gained about possible causes of past climatic variability (figure 10.24), and the climate-prediction models can be refined accordingly.

The longest ice-core records come from Antarctica (figure 10.25); they span over 800,000 years. Records from the Greenland ice sheet go back about 160,000 years, and over that

period, show a pattern very similar to the Antarctic results, indicating that the cores do capture global climate trends. The data show substantial temperature variations, spanning a range of about 10°C (20°F), some very rapid warming trends, and obvious correlations between temperature and the concentrations of carbon dioxide (CO_2) and methane (CH_4). Correlation does not show a cause-and-effect relationship, however. The warming

Each year, NASA publishes the resultant trend of global average temperatures (as in figure 10.3) and a map of temperature anomalies around the world (figure 2). The baseline used for reference is the average for the period 1951–1980, a period recent enough that plentiful data are available, long enough to smooth out a good deal of short-term variability, and long enough ago to allow us to look at trends in the last several decades. Clearly, global temperature is rising. In 2011, it was 0.5°C (0.9°F) above that baseline. Though the magnitude of the average increase may not yet look impressive, the anomalies are far from uniform around the globe; the much greater warming of the Arctic has caused particular concern.

Is this really due to the greenhouse effect? Still other measurements suggest that it is. Satellites can also measure atmospheric temperatures at various altitudes. Over the past few decades, temperatures in the troposphere (the lowest 10 km of the atmosphere) have been rising, while those in the stratosphere above it have been falling (figure 3). The greenhouse gases concentrate in the troposphere. Therefore, increasing greenhouse-gas

concentrations in the atmosphere will mean more heat trapped in the troposphere, and less escaping to warm the stratosphere (or radiate into space). The data are thus consistent with surface temperatures rising as a consequence of the observed increase in greenhouse gases.

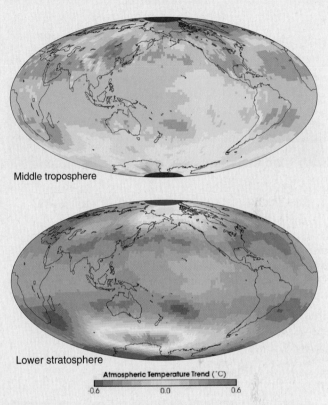

Middle troposphere

Lower stratosphere

Atmospheric Temperature Trend (°C)
-0.6   0.0   0.6

Figure 3

Temperature changes from January 1979 to December 2005 in the middle troposphere, about 5 kilometers (3 miles) above the surface, and in the lower stratosphere, 18 kilometers (11 miles) above the surface. The apparent stratospheric temperature anomalies near the poles may be related to special meteorological conditions of polar regions.

NASA image created by Jesse Allen, using data provided courtesy of Remote Sensing Systems.

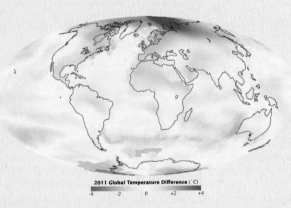

2011 Global Temperature Difference (°C)
-4   -2   0   +2   +4

Figure 2

2011 global surface temperature anomalies, relative to 1951–1980 averages.

NASA image by Jesse Allen, Robert Schmunk, and Robert Simmon, based on data from the Goddard Institute for Space Studies.

might be a result of sharp increases in greenhouse gases in the atmosphere, or the cause (see discussion of climate feedbacks in the next section). One fact is clear: The *highest* greenhouse-gas concentrations seen in the ice cores are about the levels in our atmosphere just prior to the start of the Industrial Age. Since then, we have pushed CO_2 concentrations about 100 ppm higher still, and the numbers continue to climb.

The geologic record does indicate far higher atmospheric CO_2 levels during the much warmer periods of earth's history, long before the time represented by the ice cores. But those occurred tens of millions of years or more before humans occupied the planet, with distinctive ecosystems that presumably adapted to those warmer conditions as global temperatures rose. Modern CO_2 levels are unprecedented in the history of modern humans.

A

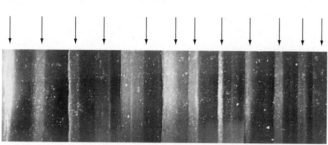

B

Figure 10.24

Ice coring in Greenland and Antarctica supports climate-trend investigations. (A) Cores must be kept frozen to preserve layered structure and trapped gases, so researchers must work in the cold. (B) This 19-cm (7.5-in.) section of Greenland core shows 11 annual layers, defined by summer dust blown onto ice. Core sections are dated by counting these layers.

(A) Photograph by Kendrick Taylor, DRI, University of Nevada–Reno; (B) by A. Gow, U.S. Army Corps of Engineers; both courtesy NOAA/ National Geophysical Data Center.

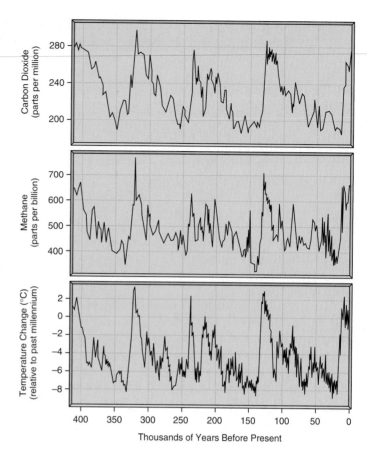

Figure 10.25

Ice-core data from Greenland and Antarctica reveal strong correlations between concentrations of the greenhouse gases carbon dioxide and methane, and temperature changes as indicated by oxygen isotopes.

Whither for the Future? Climate Feedbacks, Predictive Uncertainty

We have already seen some of the many short-term responses of earth systems to global warming associated with increasing greenhouse gases. Trying to project the longer-term trends is complicated, in part, by the fact that some changes tend to reinforce a particular trend, while others tend to counter it and reverse climatic direction. A few examples will illustrate the challenge.

We see that as temperatures rise, one consequence is loss of glacial cover. This decreases the average *albedo* (reflectivity) of the earth's surface, which should mean more heating of the land by sunlight, and more infrared radiation back out into the atmosphere to be captured by greenhouse gases, and thus more heating, more ice melting, and so on. This is an example of positive feedback, a warming trend reinforced.

But wait—if air temperatures rise, evaporation of water from the oceans would tend to increase. Water vapor is a greenhouse gas, so that could further increase heating. However, more water in the air also means more clouds. More clouds, in turn, would reflect more sunlight back into space before it could reach and heat the surface, and that effect would be likely to be more significant. This is an example of negative feedback, in which warming induces changes that tend to cause cooling. Some speculate that over centuries, the world might then cool off, perhaps even enough to initiate another ice advance or ice age far in the future. In fact, NASA at one time identified cloud cover as the single largest source of uncertainty in projections of greenhouse-effect heating.

Now consider the oceans. Atmospheric carbon dioxide can dissolve in seawater, and more dissolves in colder water. So if global warming warms sea surface water, it may release some dissolved CO_2 back into the atmosphere, where it can trap more heat to further warm the surface waters (positive feedback). Or, will melting of polar and glacier ice put so much cool, fresh water on the sea surface that there will be more solution of atmospheric CO_2, removing it from the atmosphere and reducing the heating (negative feedback)? The recent observation that phytoplankton productivity declines in warmer surface waters indicates another positive-feedback scenario: Phytoplankton are plants, and photosynthesis consumes CO_2. Sharply reduced phytoplankton abundance, then, means less CO_2 removed from the atmosphere, and more left there to absorb heat.

Still more complexities exist. For instance, as permafrost regions thaw, large quantities of now-frozen organic matter in the soil can begin to decompose, adding CO_2 and/or CH_4 to the atmosphere—another positive-feedback scenario. If overall, the positive feedbacks dominate, one could eventually have a "runaway greenhouse" effect such as exists on the surface of Venus, where the CO_2-rich atmosphere keeps the planet's surface hot enough to melt lead.

Beyond water vapor, CO_2, and CH_4, there are still more greenhouse gases produced by human activity, including nitrous oxide (N_2O), a product of internal-combustion engines, and the synthetic chlorofluorocarbons. Many of these are even *more* effective at greenhouse-effect heating of the atmosphere than is CO_2. So is methane. Methane is produced naturally during some kinds of decay of organic matter and during the digestive processes of many animals. Human activities that increase atmospheric methane include extraction and use of fossil fuels, raising of domestic livestock (cattle, dairy cows, goats, camels, pigs, and others), and the growing of rice in rice paddies, where bacterial decay produces methane in the standing water. Methane concentrations have also risen, but currently, CH_4 is still far less abundant than CO_2; its concentration in the atmosphere is below 2 parts per million (ppm). If the amounts of minor but potent greenhouse gases continue to rise, they will complicate climate predictions tied primarily to CO_2 projections.

In the near term, the dominant feedbacks are positive, for we see continued warming as CO_2 levels continue to rise. A 2007 report by the Intergovernmental Panel on Climate Change projected further global temperature increases of 1.1 to 6.4°C (2.0 to 11.5°F) by 2100, depending on model assumptions. Considering the effects of the much smaller temperature rise of the last century or so, the implications are substantial. There are real questions about the extent to which humans and other organisms can cope with climate change as geologically rapid as this. We have already noted some of the observed impacts. One recent study has even suggested a possible link between thawing in arctic regions and the occurrence of large ice/rock avalanches.

Some scientists are convinced that drastic action is needed immediately, and they advocate a variety of "geoengineering" solutions, such as reducing incoming sunlight by sending millions of tons of sulfur or soot into the stratosphere each year, to mimic the shading caused by large, explosive volcanic eruptions, or by putting reflective screens into space above the planet. Such strategies would involve substantial costs, and it is not at all clear that we understand the climate system well enough yet to anticipate possible unintended negative consequences of such schemes. This spurs interest in other approaches: energy sources that do not produce CO_2 (chapter 15), research aimed at better understanding of the global carbon cycle and ways to moderate atmospheric CO_2 (chapter 18), and international agreements to limit greenhouse-gas emissions (chapter 19).

Summary

Modern burning of fossil fuels has been increasing the amount of carbon dioxide in the air. The resultant greenhouse-effect heating is melting alpine glaciers, parts of the Greenland and Antarctic ice sheets, Arctic sea ice, and permafrost, and contributing to the thermal expansion of seawater, causing a rise in global sea level. If this continues, flooding of many coastal areas could result. Global warming may influence the oceans' thermohaline circulation and the occurrence of El Niño events. Other probable consequences involve changes in global weather patterns, including the amount and distribution of precipitation, which may have serious implications for agriculture and water resources; more episodes of extreme temperature; and more-intense storms. Depressed productivity of phytoplankton in warmer seawater affects both the marine food web and atmospheric CO_2; ocean acidification and warming put stress on coral reefs. Warming appears to be expanding the occurrence of certain diseases, thus affecting human health.

Cores taken from ice sheets provide data on past temperature and greenhouse-gas fluctuations over 800,000 years. They document a correlation between the two, and the fact that current CO_2 levels in the atmosphere are far higher than at any other time during this period.

Climatic projections for the future are complicated by the number and variety of factors to be considered, including other greenhouse gases, a variety of positive and negative climate feedback mechanisms, and complex interactions between oceans and atmosphere. The current trend is clearly warming. The rock record shows that earth has been subject to many climatic changes. Global temperatures have been more extreme than at present, and other instances of geologically rapid climate change occurred long before human influences. However, such changes have inevitably affected biological communities and, more recently, human societies also. We already observe significant impacts from the warming and associated climatic changes of recent decades, and note that these impacts are much more pronounced in some parts of the globe; some of the results are likely to be irreversible.

Key Terms and Concepts

greenhouse effect 212 permafrost 214 thermohaline circulation 216

| Table 11.1 | The Water in the Hydrosphere | | |

Reservoir	Percentage of Total Water*	Percentage of Fresh Water†	Percentage of Unfrozen Fresh Water
oceans	96.5	—	—
ice, permanent snow	1.74	50.4	—
ground water	1.70	49.3	99.1
lakes and streams salt	0.007	—	—
fresh	0.007	0.20	0.40
soil moisture	0.001	0.03	0.06
atmosphere	0.001	0.03	0.06

Source: Data from NASA and U.S. Geological Survey.

*These figures account for over 99.9% of the water. Organisms (the biosphere) contain only another .0001% of the total water.

†This assumes that all ground water is more or less fresh, since it is not all readily accessible to be tested and classified. In fact, less than half of ground water is believed to be relatively fresh.

Fluid Storage and Mobility: Porosity and Permeability

Porosity and permeability involve the ability of rocks or other mineral materials (sediments, soils) to contain fluids and to allow fluids to pass through them. **Porosity** is the proportion of void space in the material—holes or cracks, unfilled by solid material, within or between individual mineral grains—and is a measure of how much fluid the material can store. Porosity may be expressed either as a percentage (for example, 1.5%) or as an equivalent decimal fraction (0.015). The pore spaces may be occupied by gas, liquid, or a combination of the two. **Permeability** is a measure of how readily fluids pass through the material. It is related to the extent to which pores or cracks are interconnected, and to their size—larger pores have a lower surface-to-volume ratio so there is less frictional drag to slow the fluids down. Porosity and permeability of geologic materials are both influenced by the shapes of mineral grains or rock fragments in the material, the range of grain sizes present, and the way in which the grains fit together (figure 11.1).

Igneous and metamorphic rocks consist of tightly interlocking crystals, and they usually have low porosity and permeability unless they have been broken up by fracturing or weathering. The same is true of many of the chemical sedimentary rocks, unless cavities have been produced in them by slow dissolution. Clastic sediments, however, accumulate as layers of loosely piled particles. Even after compaction or cementation into rock, they may contain more open pore space. Well-rounded equidimensional grains of similar size can produce sediments of quite high porosity and permeability. (This can be illustrated by pouring water through a pile of marbles.) Many sandstones have these characteristics; consider, for example, how rapidly water drains into beach sands. In materials

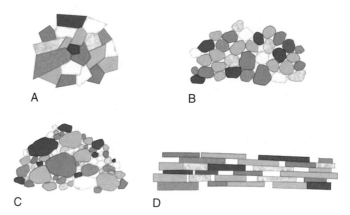

Figure 11.1

Porosity and permeability vary with grain shapes and the way grains fit together. (A) Low porosity in an igneous rock. (B) A sandstone made of rounded grains similar in size has more pore space. (C) Poorly sorted sediment: fine grains fill pores between coarse ones. (D) Packing of plates of clay in a shale may result in high porosity but low permeability. In all cases, fracturing or weathering may increase porosity and permeability beyond that resulting from grain shapes and packing.

containing a wide range of grain sizes, finer materials can fill the gaps between coarser grains. Porosity can thereby be reduced, though permeability may remain high. Clastic sediments consisting predominantly of flat, platelike grains, such as clay minerals or micas, may be porous, but because the flat grains can be packed closely together parallel to the plates, these sediments may not be very permeable, especially in the direction perpendicular to the plates. Shale is a rock commonly made from such sediments. The low permeability of clays can readily be demonstrated by pouring water onto a slab of artists' clay.

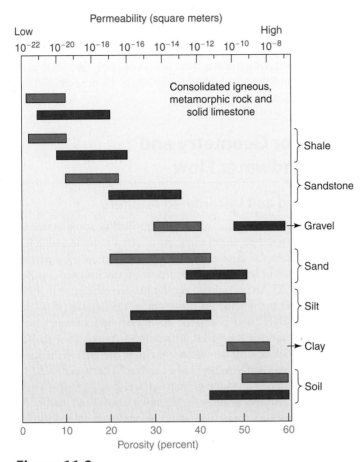

Figure 11.2
Typical ranges of porosities (blue bars) and permeabilities (brown) of various geological materials.

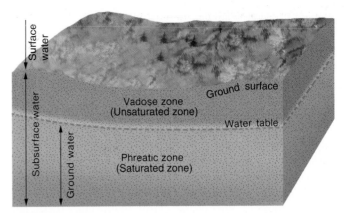

Figure 11.3
Nomenclature of surface and subsurface waters. Ground water is water in the saturated zone, below the water table.

Figure 11.2 gives some representative values of porosity and permeability. These properties are relevant not only to discussions of groundwater availability and use, but also to issues such as stream flooding, petroleum resources, water pollution, and waste disposal.

Subsurface Waters

If soil on which precipitation falls is sufficiently permeable, infiltration occurs. Gravity continues to draw the water downward until an impermeable rock or soil layer is reached, and the water begins to accumulate above it. (The impermeable material itself may or may not contain significant pore water.) Immediately above the impermeable material is a volume of rock or soil that is water-saturated, in which water fills all the accessible pore space, the **saturated zone,** or **phreatic zone. Ground water** is the water in the saturated zone. Usually, ground water is found, at most, a few kilometers into the crust. In the deep crust and below, pressures on rocks are so great that compression closes up any pores that water might fill. Above the saturated zone is rock or soil in which the pore spaces are filled partly with water, partly with air: the **unsaturated zone,** or

vadose zone. The water in unsaturated soil is **soil moisture;** although volumetrically relatively minor (recall table 11.1), it is often an important factor in agricultural productivity. All of the water occupying pore space below the ground surface is, logically, called *subsurface water,* a term that thus includes ground water, soil moisture, and water in unsaturated rocks. The **water table** is the top surface of the saturated zone, where the saturated zone is not confined by overlying impermeable rocks. (See also the discussion of aquifer geometry in the next section.) These relationships are illustrated in figure 11.3.

The water table is not always below the ground surface. Where the water table locally intersects the ground surface, the result may be a lake, stream, spring, or wetland; the water's surface is the water table. The water table below ground is not flat like a tabletop, either. It may undulate with the surface topography and with the changing distribution of permeable and impermeable rocks underground. The height of the water table varies, too. It is highest when the ratio of input water to water removed is greatest, typically in the spring, when rain is heavy or snow and ice accumulations melt. In dry seasons, or when local human use of ground water is intensive, the water table drops, and the amount of available ground water remaining decreases. Ground water can flow laterally through permeable soil and rock, from higher elevations to lower, from areas of higher pressure or potential (see below) to lower, from areas of abundant infiltration to drier ones, or from areas of little groundwater use toward areas of heavy use. Ground water may contribute to streamflow, or water from a stream may replenish ground water (figure 11.4). The processes of infiltration and migration or percolation by which ground water is replaced are collectively called **recharge.** Groundwater **discharge** occurs where ground water flows into a stream, escapes at the surface in a spring, or otherwise exits the aquifer. The term can also apply to the water moving through an aquifer; see the discussion of Darcy's Law later in the chapter.

If ground water drawn from a well is to be used as a source of water supply, the porosity and permeability of the surrounding rocks are critical. The porosity controls the total amount of water available. In most places, there are not vast pools of open water underground to be tapped, just the water dispersed in the rocks' pore spaces. This pore water amounts, in most cases, to a few percent of the rocks' volume at best, although some sands and gravels may occasionally have porosities of 50% or more. The permeability, in turn, governs both the rate at which water can be withdrawn and the rate at which recharge can occur. All the pore water in the world would be of no practical use if it were so tightly locked in the rocks that it could not be pumped out. A rock or soil that holds enough water and transmits it rapidly enough to be useful as a source of water is an **aquifer.** Many of the best aquifers are sandstones or other coarse clastic sedimentary rocks, but any other type of rock may serve if it is sufficiently porous

and permeable—a porous or fractured limestone, fractured basalt, or weathered granite, for instance. Conversely, an **aquitard** is a rock that may store a considerable quantity of water, but in which water flow is slowed, or retarded; that is, its permeability is very low, regardless of its porosity. Shales are common aquitards.

Aquifer Geometry and Groundwater Flow

Confined and Unconfined Aquifers

The behavior of ground water is controlled to some extent by the geology and geometry of the particular aquifer in which it is found. When the aquifer is directly overlain only by permeable rocks and soil, it is described as an **unconfined aquifer** (figure 11.5). An unconfined aquifer may be recharged by infiltration over the whole area underlain by that aquifer, if no impermeable layers above stop the downward flow of water from surface to aquifer. If a well is drilled into an unconfined aquifer, whether rock or soil, the water will rise in the well to the same height as the water table in the aquifer. The water must be actively pumped up to the ground surface.

A **confined aquifer** is bounded above and below by low-permeability rocks (aquitards). Water in a confined aquifer may be under considerable pressure as a consequence of lateral differences in elevation within the aquifer: Water at a higher elevation exerts pressure on water deeper (lower) in the same

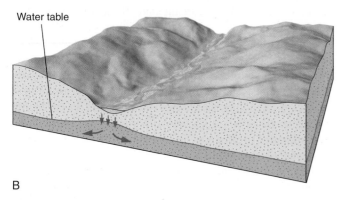

A

B

Figure 11.4

(A) In this natural system, the stream drains water from the region—not only surface runoff, but ground water from adjacent saturated rocks and soil that slowly drains into the stream. Such slow groundwater influx explains why many streams can keep flowing even when there has been no rain for weeks. The stream represents a site of groundwater discharge. (B) Here, water from the stream channel percolates downward to add to the groundwater supply.

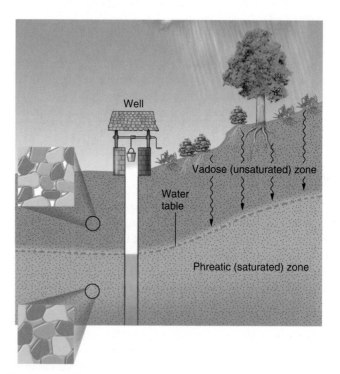

Figure 11.5

An unconfined aquifer. Water rises in an unpumped well just to the height of the water table.

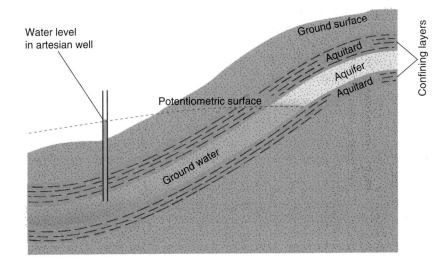

Water level in artesian well

Ground surface

Aquitard

Aquifer

Aquitard

Confining layers

Potentiometric surface

Ground water

Figure 11.6
Natural internal pressure in a confined aquifer system creates artesian conditions, in which water may rise above the apparent (confined) local water table. The aquifer here is a sandstone; the confining layers, shale.

aquifer. The confining layers prevent the free flow of water to relieve this pressure. At some places within the aquifer, the water level in the saturated zone may be held above or below the level it would assume if the aquifer were unconfined and the water allowed to spread freely (figure 11.6).

If a well is drilled into a confined aquifer, the water can rise above its level in the aquifer because of this extra hydrostatic (fluid) pressure. This is called an **artesian system.** The water in an artesian system may or may not rise all the way to the ground surface; some pumping may still be necessary to bring it to the surface for use. In such a system, rather than describing the height of the water table, geologists refer to the height of the **potentiometric surface,** which represents the height to which the water's pressure would raise the water if the water were unconfined. This level will be somewhat higher than the top of the confined aquifer where its rocks are saturated, and it may be above the ground surface, as shown in figure 11.6.

Advertising claims notwithstanding, this is all that "artesian water" means—artesian water is no different chemically from, and no purer, better-tasting, or more wholesome than any other ground water; it is just under natural pressure. Water towers create the same effect artificially: After water has been pumped into a high tower, gravity increases the fluid pressure in the water-delivery system so that water rises up in the pipes of individual users' homes and businesses without the need for many pumping stations. The same phenomenon can be demonstrated very simply by holding up a water-filled length of rubber tubing or hose at a steep angle. The water is confined by the hose, and the water in the low end is under pressure from the weight of the water above it. Poke a small hole in the top side of the lower end of the hose, and a stream of water will shoot up to the level of the water in the hose.

The overlying aquitard also means that a confined aquifer cannot be recharged by infiltration from above. Water extracted can be replenished only by lateral flow from elsewhere in the aquifer, and modern recharge to the aquifer overall may be limited or nonexistent. This has significant implications for water use from confined-aquifer systems.

Darcy's Law and Groundwater Flow

How readily ground water can move through rocks and soil is governed by permeability, but where and how rapidly it actually does flow is also influenced by differences in **hydraulic head** (potential energy) from place to place. Ground water flows from areas of higher hydraulic head to lower. The height of the water table (in an unconfined aquifer) or of the potentiometric surface (confined aquifer) reflects the hydraulic head at each point in the aquifer. All else being equal, the greater the difference in hydraulic head between two points, the faster ground water will flow between them. This is a key result of **Darcy's Law,** which can be expressed this way:

$$Q = K \cdot A \cdot \frac{\Delta h}{\Delta l}$$

where Q = discharge; A = cross-sectional area; K is a parameter known as *hydraulic conductivity* that takes into account both the permeability of the rock or soil and the viscosity and density of the flowing fluid; and ($\Delta h/\Delta l$) is the *hydraulic gradient,* the difference in hydraulic head between two points (Δh) divided by the distance between them (Δl). Hydraulic conductivities, like permeabilities, vary widely among geologic materials; for example, K may exceed 1000 ft/day in a well-sorted gravel but be less than .00001 ft/day in a fresh limestone or unfractured basalt. The hydraulic gradient of ground water is analogous to a stream's gradient. (See figure 11.7.)

Note that this relationship works not only for water, but for other fluids, such as oil; the value of K will change with changes in viscosity, but the underlying behavior is the same. Also, Darcy's Law somewhat resembles the expression for stream discharge, with flow velocity represented by ($K \frac{\Delta h}{\Delta l}$).

Other Factors in Water Availability

More-complex local geologic conditions can make it difficult to determine the availability of ground water without thorough study. For example, locally occurring lenses or patches of relatively

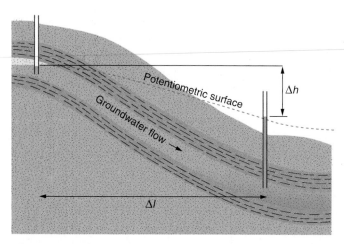

Figure 11.7

Darcy's Law relates groundwater flow rate, and thus discharge, to hydraulic gradient. Height of water in wells reflects relative hydraulic head.

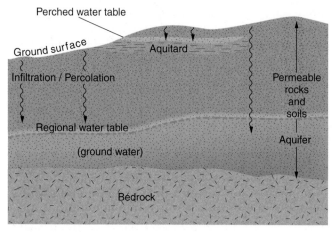

Figure 11.8

A perched water table may create the illusion of a shallow regional water table.

impermeable rocks within otherwise permeable ones may result in a *perched water table* (figure 11.8). Immediately above the aquitard is a local saturated zone, far above the true regional water table. Someone drilling a well above this area could be deceived about the apparent depth to the water table and might also find that the perched saturated zone contains very little total water. The quantity of water available would be especially sensitive to local precipitation levels and fluctuations. Using ground water over the longer term might well require drilling down to the regional water table, at a much greater cost.

As noted earlier, ground water flows (albeit often slowly). Both flow paths (directions) and locations of recharge zones may have to be identified in assessing water availability. For example, if water is being consumed very close to the recharge area and consumption rate exceeds recharge rate, the stored water may be exhausted rather quickly.

If the point of extraction is far down the flow path from the recharge zone, a larger reserve of stored water may be available to draw upon. Like stream systems, aquifer systems also have divides from which ground water flows in different directions, which determine how large a section of the aquifer system can be tapped at any point. Groundwater divides may also separate polluted from unpolluted waters within the same aquifer system.

Consequences of Groundwater Withdrawal

Lowering the Water Table

When ground water must be pumped from an aquifer, the rate at which water flows in from surrounding rock to replace that which is extracted is generally slower than the rate at which water is taken out. In an unconfined aquifer, the result is a circular lowering of the water table immediately around the well, which is called a **cone of depression** (figure 11.9A). A similar feature is produced on the surface of a liquid in a drinking glass when it is sipped hard through a straw. When there are many closely spaced wells, the cones of depression of adjacent wells may overlap, further lowering the water table between wells (figure 11.9B). If over a period of time, groundwater withdrawal rates consistently exceed recharge rates, the regional water table may drop. A clue that this is happening is the need for wells throughout a region to be drilled deeper periodically to keep the water flowing. Cones of depression can likewise develop in potentiometric surfaces. When artesian ground water is withdrawn at a rate exceeding the recharge rate, the potentiometric surface can be lowered.

These should be warning signs, for the process of deepening wells to reach water cannot continue indefinitely, nor can a potentiometric surface be lowered without limit. In many areas, impermeable rocks are reached at very shallow depths. In an individual aquifer system, the lower confining layer may be far above the bedrock depth. Therefore, there is a "bottom" to the groundwater supply, though without drilling, it is not always possible to know just where it is. Groundwater flow rates are highly variable, but in many aquifers, they are of the order of only meters or tens of meters per year. Recharge of significant amounts of ground water, especially to confined aquifers with limited recharge areas, can thus require decades or centuries.

Where ground water is being depleted by too much withdrawal too fast, one can speak of "mining" ground water (figure 11.10). The idea is not necessarily that the water will never be recharged but that the rate is so slow on the human timescale as to be insignificant. From the human point of view, we may indeed use up ground water in some heavy-use areas. Also, as we will see in the next section, human activities may themselves reduce natural recharge, so ground water consumed may not be replaced, even slowly.

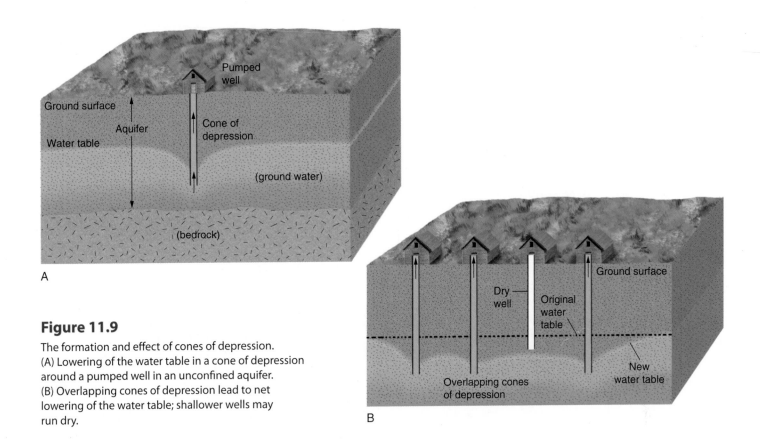

Figure 11.9

The formation and effect of cones of depression. (A) Lowering of the water table in a cone of depression around a pumped well in an unconfined aquifer. (B) Overlapping cones of depression lead to net lowering of the water table; shallower wells may run dry.

Compaction and Surface Subsidence

Lowering of the water table may thus have secondary consequences. The water in saturated rocks and soils provides buoyancy. Aquifer rocks no longer saturated with water may become compacted from the weight of overlying rocks. This decreases their porosity, permanently reducing their water-holding capacity, and may also decrease their permeability. At the same time, as the rocks below compact and settle, the ground surface itself may subside. Where water depletion is extreme—particularly if the water has been extracted from saturated soils, which would typically be more compressible than rocks—the surface subsidence may be several meters. Lowering of the water table/potentiometric surface also may contribute to *sinkhole* formation, as described in a later section.

At high elevations or in inland areas, this subsidence causes only structural problems as building foundations are disrupted. In low-elevation coastal regions, the subsidence may lead to extensive flooding, as well as to increased rates of coastal erosion. The city of Venice, Italy, is in one such slowly drowning coastal area. Many of its historical, architectural, and artistic treasures are threatened by the combined effects of the gradual rise in worldwide sea levels, the tectonic sinking of the Adriatic coast, and surface subsidence from extensive groundwater withdrawal. Drastic and expensive engineering efforts will be needed to save them. The groundwater withdrawal began about 1950. Since 1969, no new wells have been drilled, groundwater use has declined, and subsidence from this cause has apparently stopped, but the city remains awash. Closer to home, the Houston/Galveston Bay area has suffered subsidence of up to several meters from a combination of groundwater withdrawal and, locally, petroleum extraction. Some 80 sq. km (over 30 sq. mi.) are permanently flooded, and important coastal wetlands have been lost. In the San Joaquin Valley in California, half a century of groundwater extraction, much of it for irrigation, produced as much as 9 meters of surface subsidence (figure 11.11).

In such areas, simple solutions, such as pumping water back underground, are unlikely to work. The rocks may have been permanently compacted: This is the case in Venice, where subsidence may have been halted but rebound is not expected because clayey rocks in the aquifer system are irreversibly compacted, and compaction has occurred in the San Joaquin Valley, too. Also, what water is to be employed? If groundwater use is heavy, it may well be because there simply is no great supply of fresh surface water. Pumping in salt water, in a coastal area, will in time make the rest of the water in the aquifer salty, too. And, presumably, a supply of fresh water is still needed for local use, so groundwater withdrawal often cannot be easily or conveniently curtailed.

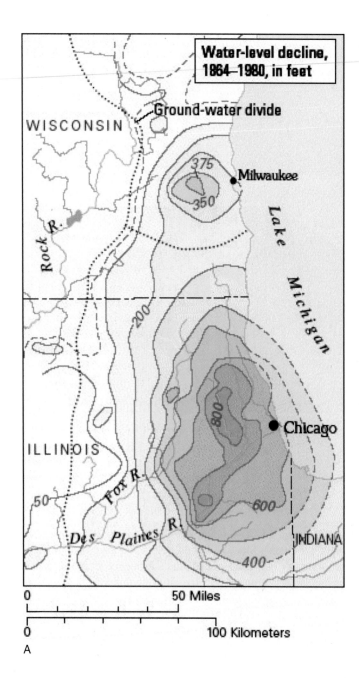

Water-level decline, 1864–1980, in feet

Ground-water divide

WISCONSIN

375

350

Milwaukee

Lake Michigan

Rock R.

200

800

Chicago

ILLINOIS

Fox R.

50

Des Plaines R.

600

INDIANA

400

0	50 Miles

0	100 Kilometers

A

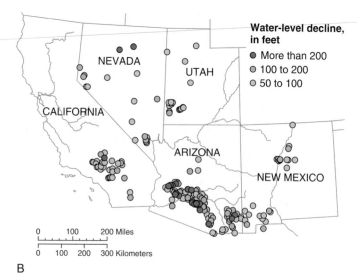

Water-level decline, in feet
- More than 200
- 100 to 200
- 50 to 100

NEVADA

UTAH

CALIFORNIA

ARIZONA

NEW MEXICO

0	100	200 Miles

0	100	200	300 Kilometers

B

2.5 km
N

C

Figure 11.10

(A) Major cities' groundwater consumption can create huge cones of depression. (B) Groundwater "mining" is clearly demonstrated by steep declines in water levels in wells of the dry southwestern U.S., where modern recharge is limited. (C) Ground water in Libya was discovered by accident in the course of oil exploration. In this satellite image taken near the city of Suluq, red represents green vegetation. Modern recharge here is nonexistent; the "fossil water" being used here for irrigation was recharged as much as 40,000 years ago.

(A) and (B) From USGS Fact Sheet 103-03; (A) after USGS Circular 1186. (C) Image by Jesse Allen, courtesy NASA.

Saltwater Intrusion

A further problem arising from groundwater use in coastal regions, aside from the possibility of surface subsidence, is **saltwater intrusion** (figure 11.12). When rain falls into the ocean, the fresh water promptly mixes with the salt water. However, fresh water falling on land does not mix so readily with saline ground water at depth because water in the pore spaces in rock or soil is not vigorously churned by currents or wave action. Fresh water is also less dense than salt water. So the fresh water accumulates in a lens, which floats above the denser salt water. If water use approximately equals the rate of recharge, the freshwater lens stays about the same thickness.

However, if consumption of fresh ground water is more rapid, the freshwater lens thins, and the denser saline ground water, laden with dissolved sodium chloride, moves up to fill in

pores emptied by removal of fresh water. *Upconing* of salt water below cones of depression in the freshwater lens may also occur. Wells that had been tapping the freshwater lens may begin pumping unwanted salt water instead, as the limited freshwater supply gradually decreases. Moreover, once a section of an aquifer becomes tainted with salt, it cannot readily be "made fresh" again. Saltwater intrusion destroyed useful aquifers beneath Brooklyn, New York, in the 1930s and is a significant problem in many

Figure 11.11

Subsidence of as much as 9 meters (almost 30 feet) occurred between 1925 and 1977 in the San Joaquin Valley, California, as a consequence of groundwater withdrawal. Signs indicate former ground surface elevation. Photograph was taken in December 1976.

Photograph courtesy Richard O. Ireland, U.S. Geological Survey

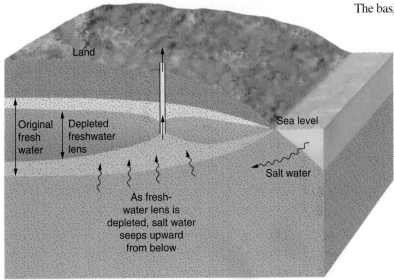

coastal areas nationwide: the southeastern and Gulf coastal states (including the vicinity of Cape Canaveral, Florida), some densely populated parts of California, and Cape Cod in Massachusetts.

Impacts of Urbanization on Groundwater Recharge

Obviously, an increasing concentration of people means an increased demand for water. We saw in chapter 6 that urbanization may involve extensive modification of surface-water runoff patterns and stream channels. Insofar as it modifies surface runoff and the ratio of runoff to infiltration, urbanization also influences groundwater hydrology.

Impermeable cover—buildings, asphalt and concrete roads, sidewalks, parking lots, airport runways—over one part of a broad area underlain by an unconfined aquifer has relatively little impact on that aquifer's recharge. Infiltration will continue over most of the region. In the case of a confined aquifer, however, the available recharge area may be very limited, since the overlying confining layer prevents direct downward infiltration in most places (figure 11.13). If impermeable cover is built over the recharge area of a confined aquifer, then, recharge can be considerably reduced, thus aggravating the water-supply situation. Recent research has led to the development of permeable concrete designed to help address this problem, but a great deal of impermeable cover is typically already in place in urban areas.

Filling in wetlands is a common way to provide more land for construction. This practice, too, can interfere with recharge, especially if surface runoff is rapid elsewhere in the area. A marsh or swamp in a recharge area, holding water for long periods, can be a major source of infiltration and recharge. Filling it in so water no longer accumulates there, and worse yet, topping the fill with impermeable cover, again may greatly reduce local groundwater recharge.

In steeply sloping areas or those with low-permeability soils, well-planned construction that includes artificial recharge basins can aid in increasing groundwater recharge (figure 11.14A). The basin acts similarly to a flood-control retention pond in that it

Figure 11.12

Saltwater intrusion in a coastal zone. If groundwater withdrawal exceeds recharge, the lens of fresh water thins, and salt water flows into more of the aquifer system from below. "Upconing" of saline water also occurs below a cone of depression. Similar effects may occur in an inland setting where water is drawn from a fresh groundwater zone underlain by more saline waters.

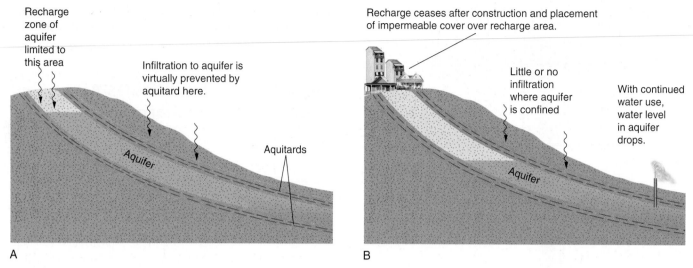

A

B

Figure 11.13

Recharge to a confined aquifer. (A) The recharge area of this confined aquifer is limited to the area where permeable rocks intersect the surface. (B) Recharge to the confined aquifer may be reduced by placement of impermable cover over the limited recharge area. (Vertical scale exaggerated.)

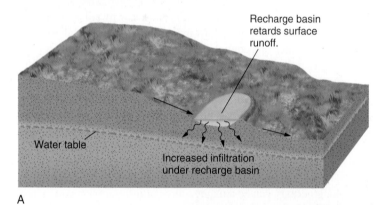

A

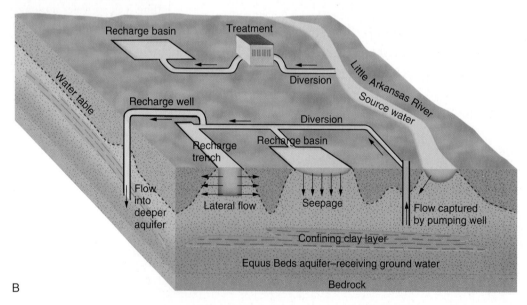

B

Figure 11.14

(A) Artificial recharge basins can aid recharge by slowing surface runoff. (B) USGS artificial-recharge demonstration project near Wichita, Kansas, is designed to stockpile water to meet future needs.

(B) After U.S. Geological Survey

is designed to catch some of the surface runoff during high-runoff events (heavy rain or snowmelt). Trapping the water allows more time for infiltration and thus more recharge in an area from which the fresh water might otherwise be quickly lost to streams and carried away.

Recharge basins are a partial solution to the problem of areas where groundwater use exceeds natural recharge rate, but, of course, they are effective only where there is surface runoff to catch, and they rely on precipitation, an intermittent water source. Increasingly, artificial recharge involves diverting streams, as shown in figure 11.14B.

Karst and Sinkholes

Broad areas of the contiguous United States are underlain by carbonate rocks (limestone and dolomite) or beds of rock salt or gypsum, chemical sediments deposited in shallow seas (figure 11.15). One characteristic that these rock types have in common is that they are extremely soluble in fresh water. Dissolution of these rocks by subsurface water, and occasional collapse or subsidence of the ground surface into the resultant cavities, creates a distinctive terrain known as **karst** (figure 11.16).

Underground, the solution process forms extensive channels, voids, and even large caverns (figure 11.17). This creates large volumes of space for water storage and, where the voids are well interconnected, makes rapid groundwater movement possible. Karst aquifers, then, can be sources of plentiful water; in fact, 40% of the ground water used in the United States for drinking water comes from karst aquifers. However, the news in this regard is not all good. First, the irregular distribution of large voids and channels makes it more difficult to estimate available water volume or to determine groundwater flow rates and paths. Second, because the water flow is typically much more rapid than in non-karst aquifers, recharge and discharge occur on shorter timescales, and water supply may be less predictable. Contaminants can spread rapidly to pollute the water in karst aquifer systems, and the natural filtering that occurs in many non-karst aquifers with finer pores and passages may not occur in karst systems.

Over long periods of time, underground water dissolving large volumes of soluble rocks, slowly enlarging underground caverns, can also erode support for the land above. There may be no obvious evidence at the surface of what is taking place until the ground collapses abruptly into the void, producing a **sinkhole** (figure 11.18A).

The collapse of a sinkhole may be triggered by a drop in the water table as a result of drought or water use that leaves rocks and soil that were previously buoyed up by water pressure unsupported. Failure may also be caused by the rapid input of large quantities of water from heavy rains that wash overlying soil down into the cavern, as happened in late 2004 when hurricanes drenched Florida, or by an increase in subsurface water flow rates.

Sinkholes come in many sizes. The larger ones are quite capable of swallowing many houses at a time: They may be over 50 meters deep and cover several tens of acres. If one occurs in a developed area, a single sinkhole can cause millions of dollars in property damage. Sudden sinkhole collapses beneath bridges, roads, and railways have occasionally caused accidents and even deaths.

Sinkholes are rarely isolated phenomena. As noted, limestone, gypsum, and salt beds deposited in shallow seas can extend over broad areas. Therefore, where there is one sinkhole in a region underlain by such soluble rocks, there are likely to be others. An abundance of sinkholes in an area is a strong hint that more can be expected. Often, the regional topography will give a clue (figure 11.18B). Clearly, in such an area, the subsurface situation should be investigated before buying or building a home or business. Circular patterns of cracks on the ground or conical depressions in the ground surface may be early signs of trouble developing below.

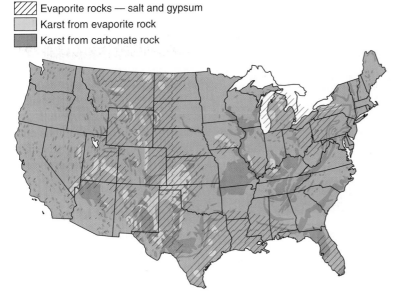

Evaporite rocks — salt and gypsum

Karst from evaporite rock

Karst from carbonate rock

Figure 11.15

Highly soluble rocks underlie more than half of the contiguous United States.

From USGS Fact Sheet 165-00

and certain other ions from water in exchange for added sodium ions. The sodium ions are replenished from the salt (sodium chloride) supply in the water softener. (While softened water containing sodium ions in moderate concentration is unobjectionable in taste or household use, it may be of concern to those on diets involving restricted sodium intake.) The "active ingredient" in water softeners is a group of hydrous silicate minerals known as *zeolites*. The zeolites have an unusual capacity for *ion exchange*, a process in which ions loosely bound in the crystal structure can be exchanged for other ions in solution.

Overall, groundwater quality is highly variable. It may be nearly as pure as rainwater or saltier than the oceans. This adds complexity to the practice of relying on ground water for water supply.

Water Use, Water Supply

General U.S. Water Use

Inspection of the U.S. water budget overall would suggest that ample water is available for use (figure 11.19). Some 4200 billion gallons of precipitation fall on this country each day; subtracting 2750 billion gallons per day lost to evapotranspiration

still leaves a net of 1450 billion gallons per day for streamflow and groundwater recharge. Water-supply problems arise, in part, because the areas of greatest water availability do not always coincide with the areas of concentrated population or greatest demand, and also because a portion of the added fresh water quickly becomes polluted by mixing with impure or contaminated water.

People in the United States use a large amount of water. Biologically, humans require about a gallon of water a day per person, or, in the United States, about 300 million gallons per day for the country. Yet, Americans divert, or "withdraw," about 400 *billion* gallons of water each day—about 1350 gallons per person—for cooking, washing, and other household uses, for industrial processes and power generation, and for livestock and irrigation, a wide range of *"offstream"* water uses. Another several trillion gallons of water are used each day to power hydroelectric plants (*"instream"* use). Of the total water withdrawn, more than 100 billion gallons per day are *consumed*, meaning that the water is not returned as wastewater. Most of the consumed water is lost to evaporation; some is lost in transport (for example, through piping systems).

It might seem easier to use surface waters rather than subsurface waters for water supplies. Why, then, worry about using ground water at all? One basic reason is that, in many dry

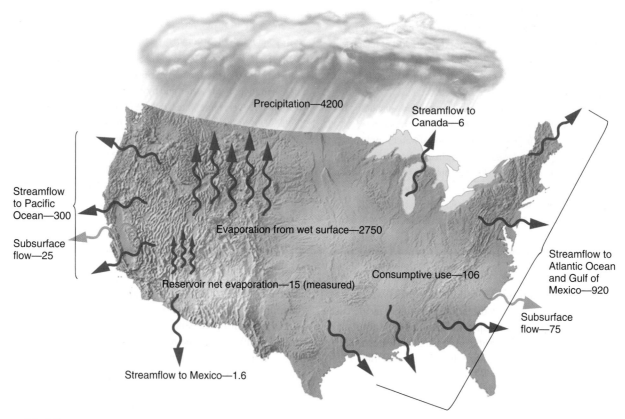

Figure 11.19

In terms of gross water flow, U.S. water budget seems ample. Figures are in billions of gallons per day.

Source: Data from The Nation's Water Resources 1975–2000, *U.S. Water Resources Council*

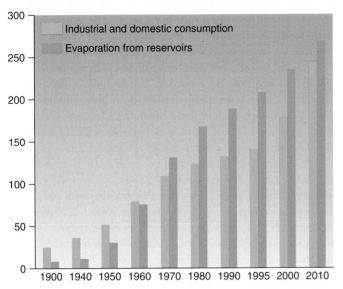

km³ per year

Industrial and domestic consumption
Evaporation from reservoirs

Figure 11.20

As dam/reservoir complexes have proliferated worldwide, evaporation from these reservoirs has outstripped consumption for industrial and domestic uses, despite population growth driving increased consumption. (1 km³ is close to 300 billion gallons)

Graphic by Philippe Rekacewicz, UNEP/GRID-Arendal Maps and Graphics Library, 2009. http://maps.grida.no/go/graphic/more-water-evaporates-from-reservoirs-than-is-consumed-by-humans

areas, there is little or no surface water available, while there may be a substantial supply of water deep underground. Tapping the latter supply allows us to live and farm in otherwise uninhabitable areas.

Then, too, streamflow varies seasonally. During dry seasons, the water supply may be inadequate. Dams and reservoirs allow a reserve of water to be accumulated during wet seasons for use in dry times, but we have already seen some of the negative consequences of reservoir construction (chapter 6). Furthermore, if a region is so dry at some times that dams and reservoirs are necessary, then the rate of water evaporation from the broad, still surface of the reservoir may itself represent a considerable water loss and aggravate the water-supply problem. Globally, in fact, evaporation from reservoirs already exceeds domestic and industrial water consumption combined (figure 11.20)!

Precipitation (the prime source of abundant surface runoff) varies widely geographically (figure 11.21), as does population density. In many areas, the concentration of people far exceeds what can be supported by available local surface waters, even during the wettest season.

Also, streams and large lakes have historically been used as disposal sites for untreated wastewater and sewage, which makes the surface waters decidedly less appealing as drinking waters. A lake, in particular, may remain polluted for decades after the input of pollutants has stopped if there is no place for those pollutants to go or if there is a limited input of fresh water to flush them out. Ground water from the saturated zone, on the

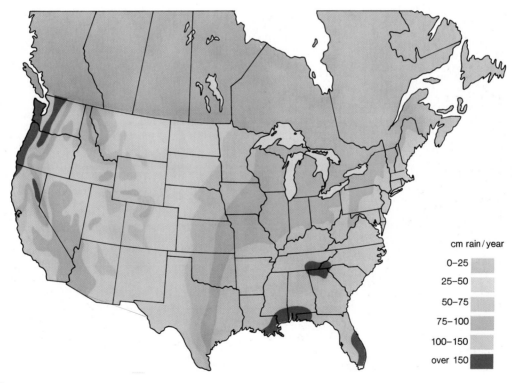

cm rain / year

| 0–25 |
| 25–50 |
| 50–75 |
| 75–100 |
| 100–150 |
| over 150 |

Figure 11.21

Average annual precipitation in the contiguous United States. (One inch equals 2.5 cm.)

Source: U.S. Water Resources Council

other hand, has passed through the rock of an aquifer and has been naturally filtered to remove some impurities—soil or sediment particles and even the larger bacteria—although it can still contain many dissolved chemicals (and in fact, once polluted, can remain so for some time, a problem explored further in chapter 17).

Finally, ground water is by far the largest reservoir of unfrozen fresh water. For a variety of reasons, then, underground waters may be preferred as a supplementary or even sole water source.

Regional Variations in Water Use

Water withdrawal varies greatly by region (figure 11.22), as does water consumption. These two quantities are not necessarily directly related, for the fraction of water withdrawn that is actually consumed depends, in part, on the principal purposes for which it is withdrawn. This, in turn, influences the extent to which local water use may deplete groundwater supplies. Of the fresh water withdrawn, only about 25% is ground water; but many of the areas of heaviest groundwater use rely on that water for irrigation, which consumes a large fraction of the water withdrawn.

Aside from hydropower generation, several principal categories of water use can be identified: municipal (public) supplies (home use and some industrial use in urban and suburban areas), rural use (supplying domestic needs for rural homes and watering livestock), irrigation (a form of rural use, too, but worthy of special consideration for reasons to be noted shortly), self-supplied industrial use (use other than for electricity generation, by industries for which water supplies are separate from municipal sources, and for activities such as mining), and thermoelectric power generation (using fuel—coal, nuclear power, etc.—to generate heat). The quantities withdrawn for and consumed by each of these categories of users are summarized in figure 11.23. The sources and disposition of water for the four major water uses are shown in table 11.2.

A point that quickly becomes apparent from table 11.2 is that while power generation may be the major water *user*, agriculture is the big water *consumer*. Power generators and industrial users account for more than half the water withdrawn, but nearly all their wastewater is returned as liquid water at or near the point in the hydrologic cycle from which it was taken. Most of these users are diverting surface waters and dumping wastewaters (suitably treated, one hopes!) back into the same lake or stream. Together, industrial users and power generation *consume* only about 10 billion gallons per day, or 10% of the total.

A much higher fraction of irrigation water is consumed: lost to evaporation, lost through transpiration from plants, or lost because of leakage from ditches and pipes (figure 11.24). Moreover, more than 40% of the water used for irrigation is ground water. Most of the water lost to evaporation drifts out of

Figure 11.22

U.S. variations in water withdrawals by state. Compare regional patterns with figure 11.21.

Source: U.S. Geological Survey, Estimated Use of Water in the United States in 2005 (USGS Circular 1344).

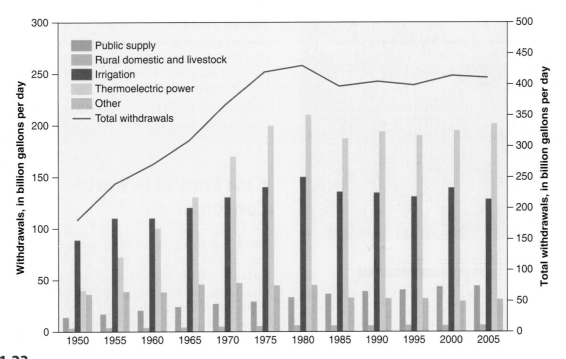

Figure 11.23

Water withdrawal by sector, 1950–2005.

Source: U.S. Geological Survey, Estimated Use of Water in the United States in 2005 *(USGS Circular 1344)*

Table 11.2	Source and Disposition of Water by Use, 1990			
	Thermoelectric Power Generation	**Irrigation**	**Industrial**	**Domestic**
Withdrawal, millions of gallons per day	195,000	137,000	27,800	25,300
Source				
Surface water	100%	63%	67%	1%
Ground water		37%	14%	13%
Public supply			19%	86%
Disposition				
Consumption	2%	56%	15%	23%
Conveyance loss		20%		
Return flow	98%	24%	85%	77%

Data from U.S. Geological Survey. In subsequent reports, consumptive use has not been reported, as it is difficult to determine precisely, but the general patterns shown here should remain consistent.

the area to come down as rain or snow somewhere far removed from the irrigation site. It then does not contribute to the recharge of aquifers or to runoff to the streams from which the water was drawn.

The regional implications of all this can be examined through figures 11.22 and 11.25. The former indicates the relative volumes of water withdrawn, by state. The latter figure shows that many states with high overall water withdrawal use their water extensively for irrigation. Where irrigation use of

water is heavy, water tables have dropped by tens or hundreds of meters (recall figure 11.10) and streams have been drained nearly dry, while we have become increasingly dependent on the crops.

Nor is water supply an issue only in the western United States, by any means. It is an international problem as well. Large areas of China and India, the world's two most populous countries, face severe, imminent shortages of fresh water. Water becomes a political issue between nations when it crosses

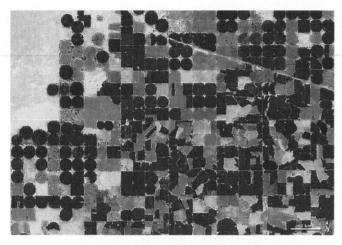

Figure 11.24

Satellites can monitor evapotranspiration, as here in Idaho's Snake River Plain. The natural scrublands (upper left) are using, and losing, much less water than the actively growing crops on irrigated fields (dark blue squares and circles).

NASA image by Robert Simmon, based on data from the Idaho Department of Water Resources

Evapotranspiration (mm/day)

0   4.5   9

Case Studies in Water Consumption

There is no lack of examples of water-supply problems; they may involve lakes, streams, or ground water. Problems of surface-water supply may be either aggravated or alleviated (depending on location) as precipitation patterns shift in reflection of normal climate cycles and/or global change induced by human activities. Where ground water is at issue, however, it is often true that relief through recharge is not in sight, as modern recharge may be insignificant. Many areas are drawing on "fossil" groundwater recharged long ago—for example, the upper Midwest uses ground water recharged over 10,000 years ago as continental ice sheets melted; recall also figure 11.10C. Water availability and population

national boundaries (Table 11.3). In most of the cases cited, the central issues relate either to water availability or to the quality of the available water. One country's water-use preferences and practices can severely disadvantage another, especially if they share a common water source (figures 11.26 and 11.27).

Water withdrawals in million gallons per day

- 0 to 200
- 200 to 1,000
- 1,000 to 5,000
- 5,000 to 15,000
- 15,000 to 25,000

Figure 11.25

Water use for irrigation, by state.

Source: U.S. Geological Survey, Estimated Use of Water in the United States, 2005 (USGS Circular 1344).

| Table 11.3 | A Sampling of Historic International Water Disputes |

River	Countries in Dispute	Issues
Nile	Egypt, Ethiopia, Sudan	Siltation, flooding, water flow/diversion
Euphrates, Tigris	Iraq, Syria, Turkey	Reduced water flow, salinization
Jordan, Yarmuk, Litani, West Bank aquifer	Israel, Jordan, Syria, Lebanon	Water flow/diversion
Indus, Sutlei	India, Pakistan	Irrigation
Ganges-Brahmaputra	Bangladesh, India	Siltation, flooding, water flow
Salween	Myanmar, China	Siltation, flooding
Mekong	Cambodia, Laos, Thailand, Vietnam	Water flow, flooding
Paraná	Argentina, Brazil	Dam, land inundation
Lauca	Bolivia, Chile	Dam, salinization
Rio Grande, Colorado	Mexico, United States	Salinization, water flow, agrochemical pollution
Rhine	France, Netherlands, Switzerland, Germany	Industrial pollution
Maas, Schelde	Belgium, Netherlands	Salinization, industrial pollution
Elbe	Czechoslovakia, Germany	Industrial pollution
Szamos	Hungary, Romania	Industrial pollution

Note that in most cases, central issues to the disputes involve adequacy and/or quality of water supply. In the near future, water-supply disputes are likely to be most intense in the Middle East.

Source: Michael Renner, National Security: The Economic and Environmental Dimensions, *Worldwatch Paper 89, p. 32. Copyright © 1989 Worldwatch Institute, Washington, D.C. www.worldwatch.org. Reprinted by permission.*

distribution and growth together paint a sobering picture of coming decades in many parts of the world (figure 11.28).

The Colorado River Basin

The Colorado River's drainage basin drains portions of seven western states (figure 11.29). Many of these states have extremely dry climates, and it was recognized decades ago that some agreement would have to be reached about which region was entitled to how much of that water. Intense negotiations during the early 1900s led in 1922 to the adoption of the Colorado River Compact, which apportioned 7.5 million acre-feet of water per year each to the Upper Basin (Colorado, New Mexico, Utah, and Wyoming) and the Lower Basin (Arizona, California, and Nevada). (One acre-foot is the amount of water required to cover an area of 1 acre to a depth of 1 foot; it is more than 300,000 gallons.) No provision was made for Mexico, into which the river ultimately flows.

Rapid development occurred throughout the region. Huge dams impounded enormous reservoirs of water for irrigation or for transport out of the basin. In 1944, Mexico was awarded by treaty some 1.5 million acre-feet of water per year, but with no

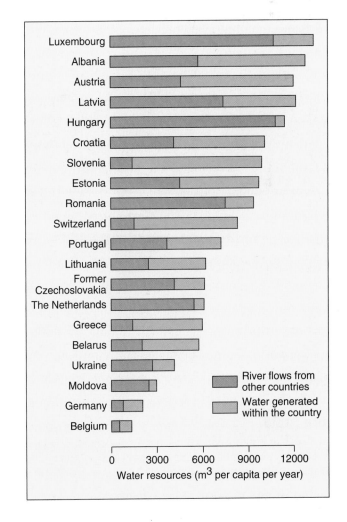

Figure 11.26

"Internal" and "external" contributions to renewable water resources in selected European countries. While some (e.g., Switzerland) control most of their own water resources, others (e.g., Hungary) do not.

Source: The Dobris Report, U.N. Environment Programme, figure 5.4

Figure 11.27

Merowe Dam, Sudan, was built for hydropower generation, but has water-supply implications for many other nations that rely on Nile River water. The dam was completed in 2008; the reservoir is still filling in this image. In this desert climate, considerable water will be lost to evaporation.

NASA image courtesy Image Science and Analysis Laboratory, Johnson Space Center.

stipulation concerning water quality. Mexico's share was to come from the surplus above the allocations within the United States or, if that was inadequate, equally from the Upper and Lower Basins.

Heavy water use has led to a reduction in both water flow and water quality in the Colorado River. The reduced flow results not only from diversion of the water for use but also from large evaporation losses from the numerous reservoirs in the system. The reduced water quality is partly a consequence of that same evaporation, concentrating dissolved minerals, and of selective removal of fresh water. Also, many of the streams flow through soluble rocks, which then are partially dissolved, increasing the dissolved mineral load. By 1961, the water delivered to Mexico contained up to 2700 ppm TDS, and partial crop failures resulted from the use of such saline water for irrigation. In response to protests from the Mexican government, the United States agreed in 1974 to build a desalination plant at the U.S.-Mexican border to reduce the salinity of Mexico's small share of the Colorado.

It has now been shown that sufficient water flow simply does not exist in the Colorado River basin even to supply the full allocations to the U.S. users. The streamflow measurements made in the early 1900s were grossly inaccurate and apparently also were made during an unusually wet period. Estimates made in 1990 put the total flow at only 15.6 million acre-feet, and of that, an estimated 2 million acre-feet are lost to evaporation, especially from reservoirs. The 1922 agreement and Mexican treaties still work tolerably well only because the Upper Basin states are using little more than half their allotted water, but that consumptive use is increasing, so flow to the lower basin is decreasing at about 0.5 million

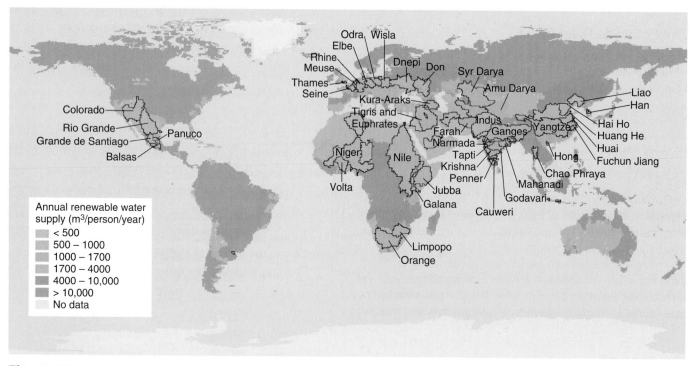

Figure 11.28

World Resources Institute projections suggest inadequate water for many by 2025. Outlined river basins have populations over 10 million. An adequate supply of water is considered to be at least 1700 cubic meters per person per year. Where renewable supplies are less than this, the population may be draining nonrenewable supplies—or suffering without.

Source: Earth Trends 2001, *World Resources Institute*

Figure 11.29

The Colorado River Basin.

Source: U.S. Geological Survey Water Supply Circular 1001

acre-feet per decade. The dry Lower Basin continues to need ever more water as its population grows. Arizona has already gone to court against California over the issue of which state is entitled to how much water; the consequence is that California has been compelled to *reduce* its use of Colorado River water. And California has grown very dependent on that water: The Colorado River Aqueduct, completed in 1941, alone carries over 1.2 million acre-feet of water per year from Lake Havasu to supply Los Angeles and San Diego, and California's 4.4 million acre-feet is already the lion's share of the Lower Basin's allocation.

Moreover, the possibility of greatly increased future water needs looms for another reason: energy. The western states contain a variety of energy resources, conventional and new. Chief among these are coal and oil shale. Extraction of fuel from oil shale is, as we will see in chapter 14, a water-intensive process. In addition, both the coal and oil shale are very likely to be strip-mined, and current laws require reclamation of strip-mined land. Abundant water will be needed to reestablish vegetation on the reclaimed land. Already, energy interests have water-rights claims to more than 1 million acre-feet of water in the Colorado oil shale area alone. The streamflow there, as in many other parts of the Colorado River basin, is already over-appropriated. The crunch will get worse if more claims are fully utilized and additional claims filed with the development of these energy resources. The unconventional fuels have so far remained uneconomic to exploit; but the potential (and associated potential water demand) is still there.

The High Plains (Ogallala) Aquifer System

The Ogallala Formation, a sedimentary aquifer, underlies most of Nebraska and sizeable portions of Colorado, Kansas, and the Texas and Oklahoma panhandles (figure 11.30). The most productive units of the aquifer are sandstones and gravels. Farming

in the region accounts for about 25% of U.S. feed-grain exports and 40% of wheat, flour, and cotton exports. More than 14 million acres of land are irrigated with water pumped from the Ogallala, and with good reason: yields on irrigated land may be triple the yields on similar land cultivated by dry farming (no irrigation).

The Ogallala's water was, for the most part, stored during the retreat of the Pleistocene continental ice sheets. Present recharge is negligible over most of the region. The original groundwater reserve in the Ogallala is estimated to have been approximately 2 billion acre-feet. But each year, farmers draw from the Ogallala more water than the entire flow of the Colorado River. Estimated depletion from predevelopment times through 2005 is over 250 million acre-feet. Recent studies put groundwater consumption at around 10 million acre-feet per year. What is more significant is that the impact is not uniform over the whole aquifer.

Soon after development in the plains began, around 1940, the water table began to drop noticeably. In areas of heavy use—Texas, southwestern Kansas, the Oklahoma panhandle—it had dropped over 100 feet by 1980, and over 150 feet by 2005. The remaining saturated thickness is less than 200 feet over two-thirds of the aquifer's area, and less than 100 feet over much of this southern region. Recent drought in the area has exacerbated the problem (figure 11.31).

Reversion to dry farming, where possible at all, would greatly diminish yields. Reduced vigor of vegetation could lead to a partial return to preirrigation, Dust-Bowl-type conditions. Alternative local sources of water are in many places not at all apparent.

It is noteworthy that there has been little incentive in the short term for the exercise of restraint in water use, and that this is due in large measure to federal policies. Price supports encourage the growing of crops like cotton that require irrigation in the southern part of the region. The government cost-shares in soil-conservation programs and provides for crop-disaster payments, thus giving compensation for the natural consequences of water depletion. And, in fact, federal tax policy provides for groundwater depletion allowances (tax breaks) for High Plains area farmers using pumped ground water, with larger breaks for heavier groundwater use. The results of all these policies aren't surprising.

Still, water-resource professionals are convincing more and more of those dependent on the Ogallala to conserve. Federal, state, and local water-resource specialists are working with farmers and communities to slow water depletion. More-efficient irrigation technology and timing have somewhat reduced water use. Some Ogallala users are trying more creative measures: For example, some farmers in Kansas and Nebraska pump fresh ground water from their land for local towns, which return their treated wastewater to irrigate the fields. Depletion has greatly slowed in the northern part of the region, aided, in part, by unusually heavy precipitation in the late 1990s. Over the longer term, further emphasis on conservation, plus modifications of existing laws that promote groundwater use, could substantially extend the useful life of the Ogallala aquifer.

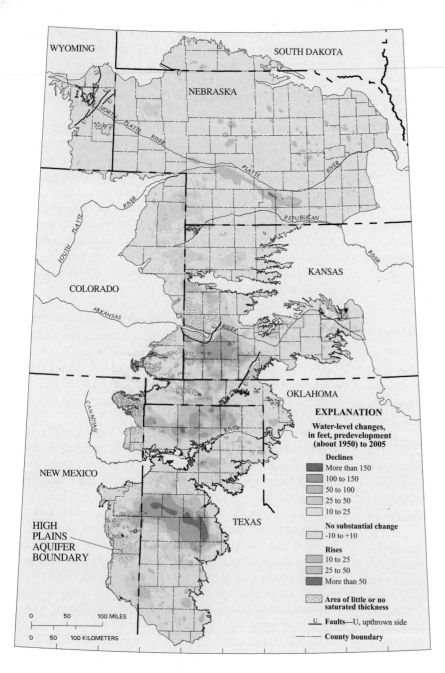

Figure 11.30

Changes in water levels in the Ogallala Formation, predevelopment to 2005. Only in a few places, such as along the Platte River, has the water table risen significantly, in response to short-term-precipitation increases; the overall long-term trend is clearly down.

From V. L. McGuire, Water-level changes in the High Plains Aquifer, predevelopment to 2005 and 2003 to 2005, *U.S. Geological Survey Special Investigations Report 2006-5324.*

The Aral Sea

The Aral Sea lies on the border of Kazakhstan and Uzbekistan. For decades, water from rivers draining into the Aral Sea has been diverted to irrigate land for growing rice and cotton. From 1973 to 1987, the Aral Sea dropped from fourth to sixth largest lake in the world. Its area shrank by more than half, its water volume by over 60%. Its salinity increased from 10% to over 23%, and a local fishing industry failed; the fish could not adapt to the rapid rise in salt content. By now, the South Aral Sea is all but gone (figure 11.32). As the lake has shrunk, former lake-bed sediments—some containing residues of pesticides washed off of the farmland—and salt have been exposed. Meanwhile, the climate has become drier, and dust storms now regularly scatter

the salt and sediment over the region. Ironically, the salt deposited on the fields by the winds is reducing crop yields. Respiratory ailments are also widespread as the residents breathe the pesticide-laden dust. With the loss of the large lake to provide some thermal mass to stabilize the local climate, the winters have become colder, the summers hotter. Here, then, there are serious issues even beyond those of water supply, which are acute enough.

Lake Chad

Lake Chad, on the edge of the Sahara Desert in West Africa, is another example of a disappearing lake. It was once one of the largest bodies of fresh water in Africa, close in area to Lake

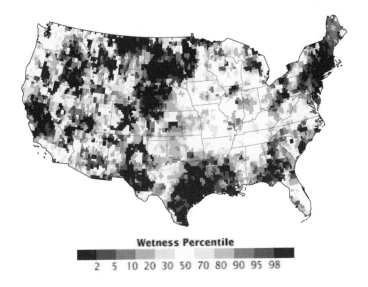

Figure 11.31

Groundwater storage in 2011 compared with the average for 1948–2011. Map is based on data from sensitive satellites that can detect changes in water storage by resultant small changes in earth's gravitational field.

Image by Chris Poulsen, National Drought Mitigation Center, University of Nebraska–Lincoln, based on data from Matt Rudell, NASA Goddard Space Flight Center, and the GRACE science team.

Erie. But like Lake Erie, it was relatively shallow—Lake Chad, indeed, was only 5 to 8 meters deep before the depletion of recent decades—so the volume of water was modest from the outset. A combination of skyrocketing demand for irrigation water from the four countries bordering the lake (Chad, Niger, Nigeria, and Cameroon), plus several decades of declining rainfall, have shrunk Lake Chad to about one-twentieth its former size (figure 11.33). While it still supports vegetation in its former lake bed, the volume of useable water remaining in Lake Chad is severely limited.

Extending the Water Supply

Conservation

The most basic approach to improving the U.S. water-supply situation is conservation. Sometimes that means reducing loss: In response to soaring water demand in the 1980s, officials in Boston focused on fixing leaks in the water-supply system, together with promoting low-flow plumbing fixtures and improving efficiency of industrial processes—and municipal water use declined over 30% in the next two decades. Water is wasted in home use every day—by long showers; inefficient plumbing; insistence on lush, green lawns even in the heat of summer; and in dozens of other ways that individual users could address. Raising livestock for meat requires far more water per pound of protein than growing vegetables for protein. Still, municipal

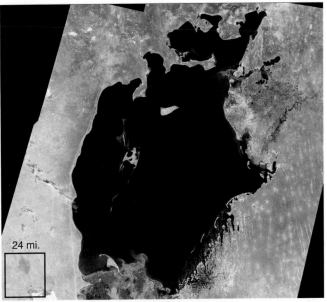

24 mi.

A

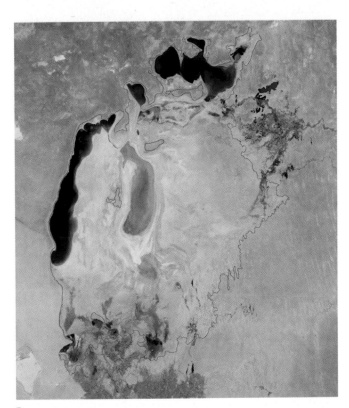

B

Figure 11.32

The shrinking of the Aral Sea: Satellite images, taken in (A) 1973 and (B) 2010. Grey outline on (B) indicates approximate shoreline in 1960.

(A) Image courtesy U.S. Geological Survey; (B) NASA image by Jesse Allen, using data from the Land Processes Distributed Active Archive Center.

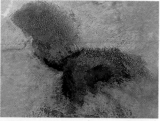

1973   1987

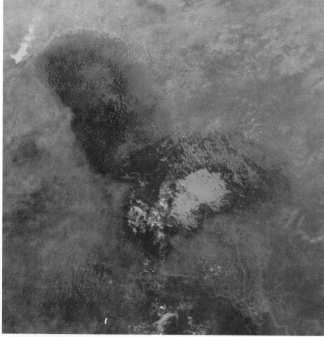

2002

Figure 11.33

Africa's Lake Chad, vanishing right before our eyes. The two older images are false-color composites, in which the red area around the (blue) lake is vegetation. Over time, vegetation has become established on the dry lakebed areas. The 2002 image is natural color, in which the remaining water appears as a brownish puddle at lower right. It is really about 60 km (38 mi) across, which shows what a large expanse of lake has been lost. The exposed lakebed is slowly drying out too, and windblown desert sand is encroaching on remaining vegetation.

Images for 1973 and 1987 courtesy NASA/Goddard Space Flight Center Scientific Visualization Studio; 2002 image by Jesse Allen, based on data from the MODIS Rapid Response Team at NASA/GSFC.

and rural water uses (excluding irrigation) together account for only about 10% of total U.S. water consumption.

The big water drain is plainly irrigation, and that use must be moderated if the depletion rate of water supplies is to be reduced appreciably. For example, the raising of crops that require a great deal of water could be shifted, in some cases at least, to areas where natural rainfall is adequate to support them. In some cases, irrigation methods can also be made more efficient so that far less water is lost by evaporation. This can be done, for instance, by various processes described collectively

as *microirrigation*. Instead of spraying irrigation water over whole fields from which evaporation loss is high, the water can be distributed selectively to just a portion of the land. In an orchard, for instance, pipes can carry water to individual trees or groups of trees and release it slowly into the ground close to the plant roots, with much less waste. The more efficient methods are typically considerably more expensive, too. They have become more attractive as water prices have been driven up by shortages, but they are simply not practical for low-priced field-grown crops for which land is plowed up each year, such as corn, wheat, or soybeans. As noted in the discussion of the High Plains Aquifer System, too, changes in government water policy could provide incentives to conserve, especially with respect to ground water.

Domestic use can be reduced in a variety of ways. For example, lawns can be watered morning or evening when evaporation is less rapid than at midday; or one can forgo traditional lawns altogether in favor of ground covers that don't need watering. Stormwater can be directed into recharge basins rather than dumped into sewer systems. Increasingly, municipalities in dry areas are looking to recycle their wastewater (discussed further in chapter 16).

Interbasin Water Transfer

In the short term, conservation alone will not resolve the imbalance between demand and supply. New sources of supply are needed. Part of the supply problem, of course, is purely local. For example, people persist in settling and farming in areas that may not be especially well supplied with fresh water, while other areas with abundant water go undeveloped. If the people cannot be persuaded to be more practical, perhaps the water can be redirected. This is the idea behind interbasin transfers—moving surface waters from one stream system's drainage basin to another's where demand is higher.

California pioneered the idea with the Los Angeles Aqueduct. The aqueduct was completed in 1913 and carried nearly 150 million gallons of water per day from the eastern slopes of the Sierra Nevada to Los Angeles. In 1958, the system was expanded to bring water from northern California to the southern part of the state. More and larger projects have been undertaken since. Bringing water from the Colorado River to southern coastal California, for example, required the construction of over 300 kilometers (200 miles) of tunnels and canals. Other water projects have transported water over whole mountain ranges. It should be emphasized, too, that such projects are not confined to the drier west: for example, New York City draws on several reservoirs in upstate New York. If population density or other sources of water demand are high, a local supply shortfall can occur even in an area thought of as quite moist.

Dozens of interbasin transfers of surface water have been proposed. Political problems are common even when the transfer involves diverting water from one part of a single state to another. In 1982, it was proposed to expand the aqueduct system to carry water from northern California to the south; 60% of the voters in the southern part of the state were

Figure 11.34

Diversion of surface water that formerly flowed into Mono Lake, California, to supply water to Los Angeles caused lake level to drop, exposing sedimentary formations originally built underwater by reaction of lake water with spring water seeping up from below. The good news is that steps have recently been taken to reduce diversion, and the lake level is beginning to rise again, but slowly.

in favor, but 90% of those in the north were opposed, and the proposition lost. The opposition often increases when transfers among several states are considered. In the 1990s, officials in states around the Great Lakes objected to a suggestion to divert some lake water to states in the southern and southwestern United States. The problems may be far greater when transfers between nations are involved. Various proposals have been made to transfer water from little-developed areas of Canada to high-demand areas in the United States and Mexico. Such proposals, which could involve transporting water over distances of thousands of kilometers, are not only expensive (one such scheme, the North American Water and Power Alliance, had a projected price of $100 billion), they also presume a continued willingness on the part of other nations to share their water. Sometimes, too, the diversion, in turn, causes problems in the region from which the water is drawn (figure 11.34).

Desalination

Another alternative for extending the water supply is to improve the quality of waters not now used, purifying them sufficiently to make them usable. Desalination of seawater, in particular, would allow parched coastal regions to tap the vast ocean reservoirs. Also, some ground waters are not presently used for water supplies because they contain excessive concentrations of dissolved materials. There are two basic methods used to purify water of dissolved minerals: filtration and distillation (figure 11.35).

In a filtration system, the water is passed through fine filters or membranes to screen out dissolved impurities. An advantage of this method is that it can rapidly filter great quantities of water. A large municipal filtration operation may produce several billion gallons of purified water per day. A disadvantage is that the method works best on water not containing very high levels of dissolved minerals. Pumping anything as salty as seawater through the system quickly clogs the filters. This method, then, is most useful for cleaning up only moderately saline ground waters or lake or stream water.

Distillation involves heating or boiling water full of dissolved minerals. The water vapor driven off is pure water, while the minerals stay behind in what remains of the liquid. Because this is true regardless of how concentrated the dissolved minerals are, the method works fine on seawater as well as on less saline waters.

A difficulty, however, is the nature of the necessary heat source. Furnaces fired by coal, gas, or other fuels can be used, but any fuel may be costly in large quantity, and many conventional fuels are becoming scarce. The sun is an alternative possible heat source. Sunlight is free and inexhaustible, and some solar desalination facilities already exist. Their efficiency is limited by the fact that solar heat is low-intensity

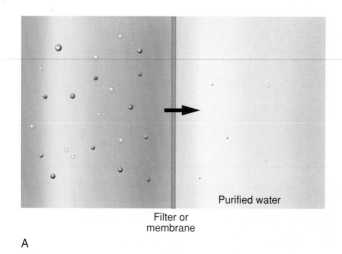

A

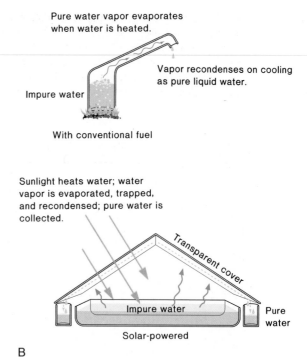

B

Figure 11.35

Methods of desalination. (A) Filtration (simplified schematic): Dissolved and suspended material (represented by dots at left) is screened out by very fine filters. (B) Distillation: As water is heated, pure water is evaporated, then recondensed for use. Dissolved and suspended materials stay behind.

heat. If a large quantity of desalinated water is required rapidly, the water to be heated must be spread out shallowly over a large area, or the rate of water output will be slow. A large city might need a solar desalination facility covering thousands of square kilometers to provide adequate water, and construction on such a scale would be prohibitively expensive even if the space were available.

Desalinated water may be five to ten times more costly to deliver than water pumped straight from a stream or aquifer. For most homeowners, the water bill is a relatively minor expense, so a jump in water costs, if necessitated by the use of desalinated water, would not be a great hardship. Water for irrigation, however, must be both plentiful and cheap if the farmer is to compete with others here and abroad who need not irrigate and if the cost of food production is to be held down. Desalinated water in most areas is simply too expensive for large-scale irrigation use. Unless ways can be found to reduce drastically the cost of desalinated water, agriculture will continue to drain

limited and dwindling surface- and groundwater supplies. Water from the ocean will not be a viable alternative for large-scale irrigation in the United States for some time.

In areas such as the arid Middle East, the more-acute need for irrigation water has made desalinated seawater a viable agricultural option economically. Ironically, a recent study has suggested that, in fact, care must be taken to adjust the chemistry of the resultant water, not only for human consumption, but for plants. A massive Israeli desalination plant began providing desalinated water from the Mediterranean Sea for domestic and agricultural use in late 2005. The water's reduced sulfate concentrations were fine for people, but inadequate for some crops. Calcium, but not magnesium, had been added back to this desalinated water for human health; irrigated plants began to show magnesium-deficiency symptoms. Boron had not been removed, which was fine for people, but toxic for some of the crops. These and other observations demonstrate that using desalinated water may require more than just getting the salt out.

Summary

Most of the water in the hydrosphere at any given time is in the oceans; most of the remaining fresh water is stored in ice sheets. Relatively little water is found in lakes and streams. A portion of the fresher waters on the continents—both surface water and ground water—is diverted for human use. The availability of ground water is influenced by such factors as the presence of suitable aquifers, water quality, and rate of recharge relative to rate of water use. Where ground water is plentiful and subsurface rocks are soluble, dissolution may create caves below ground and sinkholes at the surface, producing a distinctive landscape (karst).

Karst aquifers may be productive, with rapid water flow, but the supply may be erratic and easily polluted.

In the United States, more than half of offstream water use (withdrawal) is for industrial purposes and thermoelectric power generation. However, these industrial applications generally consume very little water. Of the water actually consumed in the United States, approximately 80% is as water loss associated with irrigation. While three-fourths of the water *withdrawn* for use is surface water, close to half of the water *consumed* is ground water, which, in many cases, is being consumed faster than recharge can replace it. Adverse consequences of such rapid consumption of ground water include lowering water tables, surface subsidence, and, in coastal areas, saltwater intrusion. Conservation, interbasin transfers of surface water, and desalination are possible ways to extend the water supplies of high-demand regions. Desalinated water is presently too expensive to use for irrigation on a large scale except in very water-poor countries, which means that it is unlikely to have a significant impact on the largest consumptive U.S. water use for some time to come. The chemistry of water is critical to the health of crops as well as people; with desalinated water, key nutrients may need to be added back to the water before use.

Key Terms and Concepts

aquifer 232
aquitard 232
artesian system 233
cone of depression 234
confined aquifer 232
Darcy's Law 233

discharge 231
ground water 231
hard water 243
hydraulic head 233
karst 239
permeability 230

phreatic zone 231
porosity 230
potentiometric surface 233
recharge 231
saltwater intrusion 236
saturated zone 231

sinkhole 239
soil moisture 231
unconfined aquifer 232
unsaturated zone 231
vadose zone 231
water table 231

Exercises

Questions for Review

1. Explain the importance of porosity and of permeability to groundwater availability for use.

2. Define the following terms: *ground water, water table,* and *potentiometric surface.*

3. How is an artesian aquifer system formed?

4. What is Darcy's Law, and what does it tell us about groundwater flow?

5. Explain how sinkholes develop. What name is given to a landscape in which sinkholes are common?

6. In what ways do karst aquifers differ from non-karst aquifers, in the context of water supply?

7. Explain three characteristics used to describe water quality.

8. What is *hard water,* and why is it often considered undesirable?

9. Describe two possible consequences of groundwater withdrawal exceeding recharge.

10. Explain the process of saltwater intrusion.

11. In what ways may urbanization affect groundwater recharge?

12. Industry is the big water *user,* but agriculture is the big water *consumer.* Explain.

13. Compare and contrast filtration and distillation as desalination methods, noting advantages and drawbacks of each.

14. What factor presently limits the potential of desalination to alleviate agricultural water shortages in many nations?

Exploring Further

1. Where does your water come from? What is its quality? How is it treated, if at all, before it is used? Is a long-term shortage likely in your area? If so, what plans are being made to avert it?

2. Compare figures 11.21 and 11.22. Speculate on the reasons for some of the extremes of water withdrawal shown in figure 11.22, and consider the likelihood that this withdrawal is predominantly *consumption* in the highest-use states.

3. Some seemingly minor activities are commonly cited as great water-wasters. One is letting the tap run while brushing one's teeth. Try plugging the drain while doing this; next, measure the volume of water accumulated during that single toothbrushing (a measuring cup and bucket may be helpful). Consider the implications for water use, bearing in mind the roughly 300 million persons in the United States. Read your home water meter before and after watering the lawn or washing the car, and make similar projections.

4. The USGS now maintains a groundwater site analogous to the "waterwatch" site monitoring streams in real time: groundwaterwatch.usgs.gov/. Examine the site for areas of unusually high or low groundwater levels, and compare with streamflow data for the same areas. Consider what factors might cause groundwater and streamflow levels to be correlated (both high or both low), or not. You may find areas in which some wells show abnormally high levels, and some show abnormally low levels; how can this be?

5. Consider a confined sandstone aquifer with a hydraulic gradient of 0.1 and hydraulic conductivity of 10^{-5} cm/sec, and porosity of 10%. Use Darcy's Law to determine the discharge, in liters per day, through a cross section of 1 square meter. (Watch units as you work this problem!)

CHAPTER 12

Weathering, Erosion, and Soil Resources

oil may not, at first glance, strike many people as a resource requiring special care for its preservation. In most places, even where soil erosion is active, a substantial quantity seems to remain underfoot. Associated problems, such as loss of soil fertility and sediment pollution of surface waters, are even less obvious to the untutored eye and may be too subtle to be noticed readily. Nevertheless, soil is an essential resource on which we depend for the production of the major portion of our food. Soils vary in their suitability not only for agriculture but also for construction and other purposes. Unfortunately, soil erosion is a significant and expensive problem in an increasing number of places as human activities disturb more and more land. In this chapter, we will examine the nature of soil, its formation and properties, aspects of the soil erosion problem, and some strategies for reducing erosion.

The limited plant growth in deserts generally reflects lack of water, not necessarily lack of soil nutrients; the weathering of rocks that contributes to soil formation certainly occurs in deserts. Here in the Mojave Desert, granite outcrops like that at left rear fracture into rectangular blocks. These are gradually transformed by weathering into smoothly rounded boulders (foreground).

Soil Formation

Soil is defined in different ways for different purposes. Engineering geologists define soil very broadly to include all unconsolidated material overlying bedrock. Soil scientists restrict the term *soil* to those materials capable of supporting plant growth and distinguish it from *regolith,* which encompasses all unconsolidated material at the surface, fertile or not. (The loose material on the lunar surface is described as regolith.) This means that in addition to rock and mineral fragments, soils generally must contain organic matter. Conventionally, the term *soil* implies little transportation away from the site at which the soil formed, while the term *sediment* indicates matter that has been transported and redeposited by wind, water, or ice.

Soil is produced by *weathering,* a term that encompasses a variety of chemical, physical, and biological processes acting to break down rocks and minerals. It may be formed directly from bedrock, or from further breakdown of transported sediment such as glacial till. The relative importance of the different kinds of weathering processes is largely determined by climate. Climate, topography, the composition of the material from which the soil is formed, the activity of organisms, and time govern a soil's final composition.

Soil-Forming Processes: Weathering

Mechanical weathering, also called *physical weathering,* is the physical breakup of rocks without changes in the rocks' composition. In a cold climate, with temperatures that fluctuate above and below freezing, water in cracks repeatedly freezes and expands, forcing rocks apart, then thaws and contracts or flows away. Crystallizing salts in cracks may have the same wedging effect. Whatever the cause, the principal effect of mechanical weathering is the breakup of large chunks of rock into smaller ones. In the process, the total exposed surface area of the particles is increased (figure 12.1).

Chemical weathering involves the breakdown of minerals by chemical reaction with water, with other chemicals dissolved in water, or with gases in the air. Minerals differ in the kinds of chemical reactions they undergo. Calcite (calcium carbonate) tends to dissolve completely, leaving no other minerals behind in its place. Calcite dissolves rather slowly in pure water but more rapidly in acidic water. Many natural waters are slightly acidic; acid rainfall or acid runoff from coal strip mines (see chapter 14) is more so and causes more rapid dissolution. This is, in fact, becoming a serious problem where limestone and its metamorphic equivalent—marble—are widely used for outdoor sculptures and building stone (figure 12.2). Calcite dissolution is gradually destroying delicate sculptural features and eating away at the very fabric of many buildings in urban areas and where acid rain is common.

Silicates tend to be somewhat less susceptible to chemical weathering and leave other minerals behind when they are attacked. Feldspars principally weather into clay minerals, an important component of many soils. Ferromagnesian silicates

leave behind insoluble iron oxides and hydroxides and some clays, with other chemical components being dissolved away. Those residual iron compounds are responsible for the reddish or yellowish colors of many soils (figure 12.3). In most climates, quartz is extremely resistant to chemical weathering, dissolving only slightly. Representative weathering reactions are shown in table 12.1.

The susceptibility of many silicates to chemical weathering can be inferred from the conditions under which the silicates formed. Given several silicates that have crystallized from the same magma, those that formed at the highest temperatures tend to be the least stable, or most easily weathered, and vice versa, probably because the lower-temperature silicates have structures with more interlocking of silica tetrahedra in two and three dimensions. A rock's tendency to weather chemically is determined by its mineral composition. For example, a gabbro (the coarsely crystalline equivalent of basalt), formed at high temperatures and rich in ferromagnesian

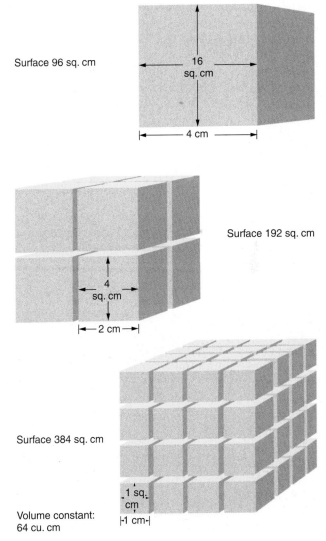

Surface 96 sq. cm

16 sq. cm

4 cm

Surface 192 sq. cm

4 sq. cm

2 cm

Surface 384 sq. cm

1 sq. cm

1 cm

Volume constant: 64 cu. cm

Figure 12.1

Mechanical breakup increases surface area and surface-to-volume ratio.

Figure 12.2

These two monuments commemorating the Battle of Lexington and Concord were erected less than forty years apart, but the marble slab (A) has eroded far more severely than the granite (B).

Figure 12.3

This vivid red Hawaiian soil is derived from weathering of iron-rich basalt flows, exposed here in Waimea Canyon.

minerals, generally weathers more readily than a granite rich in quartz and low-temperature feldspars.

Climate plays a major role in the intensity of chemical weathering. Most of the relevant chemical reactions involve water. All else being equal, then, the more water, the more chemical weathering. Also, most chemical reactions proceed more rapidly at high temperatures than at low ones. Therefore, warm climates are more conducive to chemical weathering than cold ones.

The rates of chemical and mechanical weathering are interrelated. Chemical weathering may speed up the mechanical breakup of rocks if the minerals being dissolved are holding the rock together by cementing the mineral grains, as in some sedimentary rocks. Increased mechanical weathering may, in turn, accelerate chemical weathering through the increase in exposed surface area, because it is only at grain surfaces that minerals, air, and water interact. The higher the ratio of surface area to volume—that is, the smaller the particles—the more rapid the chemical weathering. The vulnerability of surfaces to attack is

Table 12.1   **Some Chemical Weathering Reactions**

Solution of calcite (no solid residue)

$CaCO_3 + 2\,H^+ = Ca^{2+} + H_2O + CO_2$ (gas)

Breakdown of ferromagnesians (possible mineral residues include iron compounds and clays)

$FeMgSiO_4$ (olivine) $+ 2\,H^+ = Mg^{2+} + Fe(OH)_2 + SiO_2{}^*$

$2\,KMg_2FeAlSi_3O_{10}(OH)_2$ (biotite) $+ 10\,H^+ + \frac{1}{2}\,O_2$ (gas) $= 2\,Fe(OH)_3 + Al_2Si_2O_5(OH)_4$ (kaolinite, a clay) $+ 4\,SiO_2{}^* + 2\,K^+ + 4\,Mg^{2+} + 2\,H_2O$

Breakdown of feldspar (clay is the common residue)

$2\,NaAlSi_3O_8$ (sodium feldspar) $+ 2\,H^+ + H_2O = Al_2Si_2O_5(OH)_4 + 4\,SiO_2{}^* + 2\,Na^+$

Solution of pyrite (making dissolved sulfuric acid, H_2SO_4)

$2\,FeS_2 + 5\,H_2O + {}^{15}/_2\,O_2$ (gas) $= 4\,H_2SO_4 + Fe_2O_3 \cdot H_2O$

Notes: Hundreds of possible reactions could be written; the above are only examples of the common kinds of processes involved.

All ions (charged species) are dissolved in solution; all other substances, except water, are solid unless specified otherwise.

Commonly, the source of the H^+ ions for solution of calcite and weathering of silicates is carbonic acid, H_2CO_3, formed by solution of atmospheric CO_2.

*Silica is commonly removed in solution.

---

also shown by the tendency of angular fragments to become rounded by weathering (figure 12.4).

The organic component of soil is particularly important to its fertility and influences other physical properties as well. Biological processes are also important to the formation of soil. Biological weathering effects can be either mechanical or chemical. Among the mechanical effects is the action of tree roots in working into cracks to split rocks apart (figure 12.5). Chemically, many organisms produce compounds that may react with and dissolve or break down minerals. Plants, animals, and microorganisms develop more abundantly and in greater variety in warm, wet climates. Mechanical weathering is generally the dominant process only in areas where climatic conditions have limited the impact of chemical weathering and biological effects—that is, in cold or dry areas.

Finally, airborne chemicals and sediments may *add* components to soil—for example, sulfate from acid rain, salts from sea spray in coastal areas, clay minerals and other fine-grained minerals blown on the wind.

Soil Profiles, Soil Horizons

The result of mechanical, chemical, and biological weathering, together with the accumulation of decaying remains from organisms living on the land and any input from the atmosphere, is the formation of a blanket of soil between bedrock and atmosphere. A cross section of this soil blanket usually reveals a series of zones of different colors, compositions, and physical properties. The number of recognizable zones and the thickness of each vary. A basic, generalized soil profile as developed directly over bedrock is shown in figure 12.6A.

Figure 12.5

Tree roots working into cracks in this granite break up the rock mass. Granite in Yosemite National Park.

Figure 12.4

Angular fragments are rounded by weathering. Corners are attacked on three surfaces, edges on two. In time, fragments become rounded as a result, as shown in the chapter-opening image.

Soil Horizons

O — Organic matter

A — Organic matter mixed with rock and mineral fragments

E — **Zone of leaching** dissolved or suspended materials are carried downward by water

B — **Zone of accumulation** accumulation of iron, aluminum, and clay leached down from the E horizon; contains soluble minerals like calcite in drier climates

C — Weathered parent material, partially broken down

(Bedrock)

A

B

C

Figure 12.6

(A) A generalized soil profile. Individual horizons can vary in thickness. Some may be locally absent, or additional horizons or subhorizons may be identifiable. (B) This Colorado roadcut illustrates well-developed soil horizons; note the dark color of the organic-rich layer at the top— and how thin it is. (C) Soil developed over this sandstone along the California coast shows a thicker A horizon.

At the very top is the **O horizon,** consisting wholly of organic matter, whether living or decomposed—growing plants, decaying leaves, and so on. Below that is the **A horizon.** It consists of the most intensively weathered rock material, being the zone most exposed to surface processes, mixed with organic debris from above. Unless the local water table is exceptionally high, precipitation infiltrates down through the A horizon and below. In so doing, the water may dissolve soluble minerals and carry them away with it. This process is known as **leaching,** and it may be especially intense just below the A horizon, as acids produced by the decay of organic matter seep downward with percolating water. The **E horizon,** below the A horizon, is therefore also known

as the **zone of leaching.** Fine-grained minerals, such as clays, may also be washed downward through this zone.

Many of the minerals leached or extracted from the E horizon accumulate in the layer below, the **B horizon,** also known as the **zone of accumulation.** Soil in the B horizon has been somewhat protected from surface processes. Organic matter from the surface is largely absent from the B horizon. It may contain relatively high concentrations of iron and aluminum oxides, clay minerals, and, in drier climates, even soluble minerals such as calcite. Below the B horizon is a zone consisting principally of very coarsely broken-up bedrock and little else. This is the **C horizon,** which does not resemble our usual idea

Figure 12.7

These vertical cliffs and steep, rapidly eroding talus slopes in western Colorado provide no stable base on which soil can develop and accumulate.

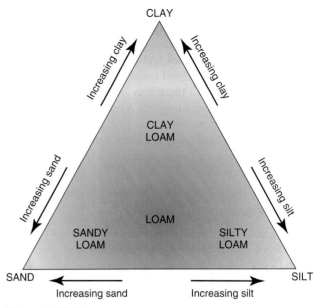

Figure 12.8

Soil-texture terminology reflects particle size, not mineralogy. Diagram simplified from U.S. Department of Agriculture classification scheme.

of soil at all. Similar zonation—though without bedrock at the base of the soil profile—is found in soils developed on transported sediment.

The boundaries between adjacent soil horizons may be sharp or indistinct. In some instances, one horizon may be divided into several recognizable subhorizons. Subhorizons also may exist that are gradational between A and B or B and C. It is also possible for one or more horizons locally to be absent from the soil profile. Some of the diversity of soil types is illustrated in the next section. All variations in the soil profile arise from the different mix of soil-forming processes and starting materials found from place to place. The overall total thickness of soil is partly a function of the local rate of soil formation and partly a function of the rate of soil erosion. The latter reflects the work of wind and water, the topography, and often the extent and kinds of human activities. Where slopes are too steep and erosion too rapid, soil formation may not occur at all (figure 12.7).

Chemical and Physical Properties of Soils

Mechanical weathering merely breaks up rock without changing its composition. In the absence of pollution, wind and rainwater rarely add many chemicals, and runoff water may carry away some leached chemicals in solution. Thus, chemical weathering tends to involve a net subtraction of elements from rock or soil. One control on a soil's composition, then, is the composition of the material from which it is formed. If the bedrock or parent sediment is low in certain critical plant nutrients, the soil produced from it will also be low in those nutrients, and chemical fertilizers may be needed to grow particular crops on that soil even if the soil has never been farmed before. The extent and balance of weathering processes involved in soil formation then determine the extent of further depletion of the soil in various elements. The weathering processes also influence the mineralogy of the soil, the compounds in which its elements

occur. The physical properties of the soil are affected by its mineralogy, the texture of the mineral grains (coarse or fine, well or poorly sorted, rounded or angular, and so on), and any organic matter present.

Color, Texture, and Structure of Soils

Soil *color* tends to reflect compositional characteristics. Soils rich in organic matter tend to be black or brown, while those poor in organic matter are paler in color, often white or gray. When iron is present and has been oxidized by reaction with oxygen in air or water, it adds a yellow or red color (recall figure 12.3). Rust on iron or steel is also produced by oxidation of iron.

Soil *texture* is related to the sizes of fragments in the soil. The U.S. Department of Agriculture recognizes three size components: sand (grain diameters 2–0.05 mm), silt (0.05–0.002 mm), and clay (less than .002 mm). Soils are named on the basis of the dominant grain size(s) present (figure 12.8). An additional term, *loam,* describes a soil that is a mixture of all three particle sizes in similar proportions (10 to 30% clay, the balance nearly equal amounts of sand and silt). Soils consisting of a mix of two particle sizes are named accordingly: for example, "silty clay" would be a soil consisting of about half silt-sized particles, half clay-sized particles. The significance of soil texture is primarily the way it influences drainage. Sandy soils, with high permeability (like many sandstone aquifers), drain quickly. In an agricultural setting, this may mean a need for frequent watering or irrigation. In a flood-control context, it means relatively rapid infiltration. Clay-rich soils, by contrast, may hold a great deal of water but be relatively slow to drain, by virtue of their much lower permeability, a characteristic shared by their lithified counterpart, shale.

Soil *structure* relates to the soil's tendency to form lumps or clods of soil particles. These clumps are technically

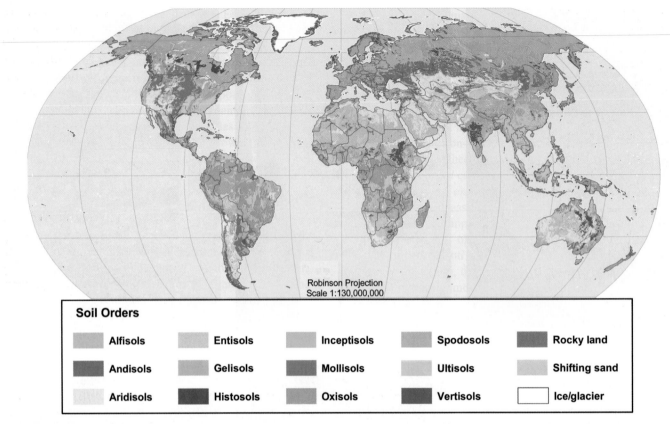

Figure 12.10

Distribution of major soil types worldwide.

Map from USDA Natural Resources Conservation Service

Soils and Human Activities

Soils and their characteristics are not only critical to agricultural activities; their properties influence the stability of construction projects. In the area of soil erosion, human activities may aggravate or moderate the problems—and sometimes may be influenced by the erosion in turn. This section samples several relationships between soil and human activities.

Lateritic Soil

Lateritic soil is common to many less-developed nations and poses special agricultural challenges. A **laterite** may be regarded as an extreme kind of pedalfer. Lateritic soils develop in tropical climates with high temperatures and heavy rainfall, so they are severely leached. Even quartz may have been dissolved out of the soil under these conditions. Lateritic soil may contain very little besides the insoluble aluminum and iron compounds. (Indeed, lateritic weathering has produced useful mineral deposits, as described in Chapter 13.) Soils of the lush tropical rain forests are commonly lateritic, which seems to suggest that lateritic soils have great farmland potential. Surprisingly, however, the opposite is true, for two reasons.

The highly leached character of lateritic soils is one reason. Even where the vegetation is dense, the soil itself has few soluble nutrients left in it. The forest holds a huge reserve of nutrients, but there is no corresponding reserve in the soil. Further growth is supported by the decay of earlier vegetation. As one plant dies and decomposes, the nutrients it contained are quickly taken up by other plants or leached away. If the forest is cleared to plant crops, most of the nutrients are cleared away with it, leaving little in the soil to nourish the crops. Many natives of tropical climates practice a slash-and-burn agriculture, cutting and burning the jungle to clear the land (figure 12.11). Some of the nutrients in the burned vegetation settle into the topsoil temporarily, but relentless leaching by the warm rains makes the soil nutrient-poor and infertile within a few growing seasons. Nutrients could, in principle, be added through synthetic chemical fertilizers. However, many of the nations in regions of lateritic soil are among the poorer developing countries, and vast expenditures for fertilizer are simply impractical.

Even with fertilizers available, a second problem with lateritic soils remains. A clue to the problem is found in the term *laterite* itself, which is derived from the Latin for "brick." A lush rain forest shields lateritic soil from the drying and baking effects of the sun, while vigorous root action helps to

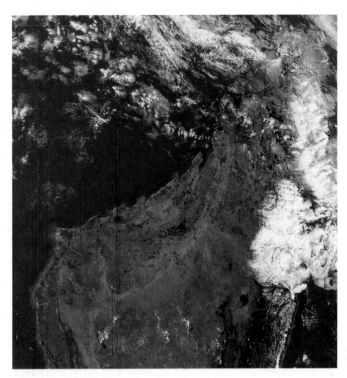

Figure 12.11

Northern Madagascar. Originally, most of Madagascar was covered by jungle and lush vegetation, as along the coast at right. Deforestation from slash-and-burn agriculture has bared most of the island, exposing its predominantly lateritic soil. Note red sediment washing into the ocean along the northwestern coast.

Image by Brian Montgomery, Robert Simmon, and Reto Stöckli, courtesy NASA

keep the soil well broken up. When its vegetative cover is cleared and it is exposed to the baking tropical sun, lateritic soil can quickly harden to a solid, bricklike consistency that resists infiltration by water or penetration by crops' roots. What crops can be grown provide little protection for the soil and do not slow the hardening process very much. Within five years or less, a freshly cleared field may become completely uncultivable. The tendency of laterite to bake into brick has been used to advantage in some tropical regions, where blocks of hardened laterite have served as building materials. Still, the problem of how to maintain crops remains. Often, the only recourse is to abandon each field after a few years and clear a replacement. This results, over time, in the destruction of vast tracts of rain forest where only a moderate expanse of farmland is needed. Also, once the soil has hardened and efforts to farm it have been abandoned, it may revegetate only very slowly, if at all.

In Indochina, in what is now Cambodian jungle, are the remains of the Khmer civilization that flourished from the ninth to the sixteenth centuries. There is no clear evidence of why that civilization vanished. A major reason may have been the difficulty of agriculture in the region's lateritic soil. (It is clear that the phenomenon of laterite solidification was occurring at that

time: Chunks of hardened laterite were used to construct temples at Angkor Wat and elsewhere.) Perhaps the Mayas, too, moved north into Mexico to escape the problems of lateritic soils. In Sierra Leone in west Africa, increased clearing of forests for firewood, followed by deterioration of the exposed soil, has reduced the estimated carrying capacity of the land to 25 persons per square kilometer; the actual population is already over 40 persons per square kilometer. Part of the concern over deforestation associated with harvesting of hardwoods in South America relates to loss of habitat and possible extinction of unique species (loss of biodiversity), but in part it relates to the fact that what is left behind is barren lateritic soil. Today, some countries with lateritic soil achieve agricultural success only because frequent floods deposit fresh, nutrient-rich soil over the depleted laterite. The floods, however, cause enough problems that flood-control efforts are in progress or under serious consideration in many of these areas of Africa and Asia. Unfortunately, successful flood control could create an agricultural disaster for such regions.

Wetland Soils

The lateritic soils are leached, oxidized, and inclined to bake hard. The soils of wetlands, by contrast, tend to be rich in accumulated organic matter, the decay of which consumes dissolved oxygen, and soft. (Recall figure 12.9F.) Wetlands provide vital habitat for waterbirds and distinctive ecological niches for other organisms and can act as natural retention ponds for floodwaters. Some act as settling ponds, reducing sediment pollution and also water pollution from contaminants carried on the sediment in water passing through them. Unfortunately, wetlands have not always been properly appreciated; many of these swampy areas have been drained for farmland or for development, or simply to provide water wanted elsewhere for irrigation. In drier climates, reduced precipitation can exacerbate wetland loss, as in the case of Lake Chad, described in chapter 11. As the value of wetlands has become more widely recognized, concerted regulatory efforts to preserve, protect, or restore wetlands have expanded in the United States (figure 12.12) and elsewhere. Direct human impacts aside, however, there is an added threat to many coastal wetlands: rising sea level, as noted in chapter 8.

Soil Erosion

Weathering is the breakdown of rock or mineral materials in place; *erosion* involves physical removal of material from one place to another. Soil erosion is caused by the action of water and wind. Rain striking the ground helps to break soil particles loose (figure 12.13), and the harder it falls, the more soil can be dislodged. Surface runoff and wind together carry away loosened soil. The faster the wind and water travel, the larger the particles and the greater the load they move. Therefore, high winds cause more erosion than calmer ones, and fast-flowing surface runoff moves more soil than slow runoff. This, in turn, suggests that, all else being equal, steep and unobstructed

Figure 12.12

Even where undeveloped land is plentiful, wetlands can serve special roles. This wetland, near Anchorage, Alaska, is protected as a wildlife refuge. A raised wooden boardwalk (at right) allows visitors to view and photograph from the perimeter without disturbing birds or animals.

Figure 12.13

The impact of a single raindrop loosens many soil particles.

Photograph courtesy of USDA Soil Conservation Service

slopes are particularly susceptible to erosion by water, for surface runoff flows more rapidly over them. Flat, exposed land is correspondingly more vulnerable to wind erosion, as surface runoff is slower and obstacles to deflect the wind are lacking. The physical properties of the soil also influence its vulnerability to erosion.

Rates of soil erosion can be estimated in a variety of ways. Over a large area, erosion due to surface runoff may be judged by estimating the sediment load of streams draining the area. On small plots, runoff can be collected and its sediment load measured. Controlled laboratory experiments can be used to simulate wind and water erosion and measure their impact. Because wind is harder to monitor comprehensively, especially over a range of altitudes, the extent of natural wind erosion is more difficult to estimate. Generally, it is less significant than erosion through surface-water runoff, except under drought conditions (for example, during the Dust Bowl era).

Figure 12.14

A 1930s dust storm in eastern Colorado.

Photograph courtesy of USDA Soil Conservation Service

The Dust Bowl area proper, although never exactly defined, comprised close to 100 million acres of southeastern Colorado, northeastern New Mexico, western Kansas, and the Texas and Oklahoma panhandles. The farming crisis there during the 1930s resulted from an unfortunate combination of factors: clearing or close grazing of natural vegetation, which had been better adapted to the local climate than were many of the crops; drought that then destroyed those crops; sustained winds; and poor farming practices.

Once the crops had died, there was nothing to hold down the soil. The action of the wind was most dramatically illustrated during the fierce dust storms that began in 1932. The storms were described as "black blizzards" that blotted out the sun (figure 12.14). Black rain fell in New York, black snow in Vermont, as windblown dust moved eastward. People choked on the dust, some dying of suffocation or of a "dust pneumonia" similar to the silicosis miners develop from breathing rock dust.

As the dust from Kansas and Oklahoma settled on their desks, politicians in Washington, D.C., realized that something had to be done. In April 1935, a permanent Soil Conservation Service was established to advise farmers on land use, drainage, erosion control, and other matters. By the late 1930s, concerted efforts by individuals and state and federal government agencies to improve farming practices to reduce wind erosion—together with a return to more normal rainfall—had considerably reduced the problems. However, renewed episodes of drought and/or high winds in the 1950s, the winter of 1965, and the mid-1970s led to similar, if less extreme, dust storms and erosion. In recent years, the extent of wind damage has actually been limited somewhat by the widespread use of irrigation to maintain crops in dry areas and dry times. But, as sources of that irrigation water are depleted, future spells of combined drought and wind may yet produce more scenes like that in figure 12.14. Even now, erosion by both wind and water, while declining, remains a significant problem over much of the United States (figure 12.15).

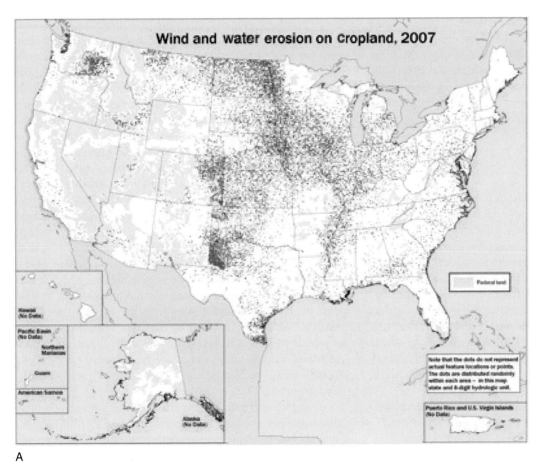

Wind and water erosion on cropland, 2007

A

**Erosion exceeding the soil loss tolerance rate
on cropland, 2007**

B

Figure 12.15

(A) Each blue dot represents 100,000 tons per year of soil lost to erosion by surface water; each red dot, 100,000 tons per year lost to wind erosion. Not surprisingly, wind erosion is more severe in the drier plains of the West. (Grey-shaded areas in the United States are federal lands.) (B) The USDA Natural Resources Conservation Service defines the maximum soil loss tolerance rate as the "maximum rate of annual soil loss that will permit crop productivity to be sustained economically and indefinitely on a given soil." Red dots represent areas where that limit is being exceeded by 100,000 tons per year on highly erodible cropland, meaning that soil quality is being degraded and productivity is threatened. In areas denoted by green dots, erosion losses also exceed 100,000 tons per year above that limit, but the cropland is not highly erodible; the soil loss is spread over a greater area.

Maps after USDA NRCS National Resources Inventory 2007.

If all the soil in an area is lost, farming clearly becomes impossible. Long before that point, however, erosion becomes a cause for concern. The topsoil of the A horizon, with its higher content of organic matter and nutrients, is especially fertile and suitable for agriculture, and it is the topsoil that is lost first as the soil erodes. Fertilizer may need to be added, at substantial expense, to compensate for loss of fertile topsoil. In addition, the organic-matter-rich topsoil usually has the best structure for agriculture—it is more permeable, more readily infiltrated by water, and retains moisture better than deeper soil layers.

Soil erosion from cropland leads to reduced crop quality and reduced agricultural income. Six inches of topsoil loss in western Tennessee has reduced corn yields by over 40%, and such relationships are not uncommon. Even when the nutrients required for adequate crop growth are added through fertilizers, other chemicals that contribute to the nutritional quality of the food grown may be lacking; in other words, the food itself may be less healthful. Also, the soil eroded from one place is deposited, sooner or later, somewhere else. If a large quantity is moved to other farmland, the crops on which it is deposited may be stunted or destroyed, although small additions of fresh topsoil may enrich cropland and make it more productive.

A subtle consequence of soil erosion in some places has been increased persistence of toxic residues of herbicides and pesticides in the soil. The loss of nutrients and organic matter through topsoil erosion may decrease the activity of soil microorganisms that normally speed the breakdown of some toxic agricultural chemicals. Many of these chemicals, which contribute significantly to water pollution, also pollute soils.

Another major problem related to soil erosion is sediment pollution (figure 12.16). In the United States, about 750 million tons per year of eroded sediment end up in lakes and streams. This decreases the water quality and may harm wildlife. The problem is still more acute when those sediments contain toxic chemical residues, as from agricultural herbicides and pesticides: The sediment is then both a physical and a potential chemical pollutant. A secondary consequence of this sediment load is the infilling of stream channels and reservoirs, restricting navigation and decreasing the volume of reservoirs and thus their usefulness for their intended purposes, whether for water supply, hydropower, or flood control. Tens of millions of dollars are spent each year to dredge and remove unwanted sediment deposits.

In recent years, there has been an increase in several kinds of activities that may increase soil erosion in places other than farms and cities. One is strip mining, which leaves behind readily erodable piles of soil and broken rock, at least until the land is reclaimed. Another is the use of off-road recreational vehicles (ORVs). This last is a special problem in dry areas where vegetation is not very vigorous or easily reestablished. Fragile plants can easily be destroyed by a passing ORV, leaving the land bare and vulnerable to intensified erosion (figure 12.17).

Figure 12.18 shows that erosion during active urbanization (highway and building construction and so forth) is considerably more severe than erosion on any sort of undeveloped land. Erosion during highway construction in the Washington, D.C., area, for example, was measured at rates up to 75 tons/acre/year. However, the total U.S. area undergoing active development is only about 1.5 million acres each year, as compared to nearly 400 million devoted to cropland. Also, construction disturbance is of relatively short duration. Once the land is covered by buildings, pavement, or established

Figure 12.16

Post-storm runoff carries a heavy sediment load from this Tennessee farm field to a drainage ditch leading to a nearby stream.

Photograph by Tim McCabe, courtesy USDA Natural Resources Conservation Service.

Figure 12.17

Fragile vegetation on this Oregon dune does not survive disturbance by off-road vehicles, leaving exposed sand more vulnerable to erosion.

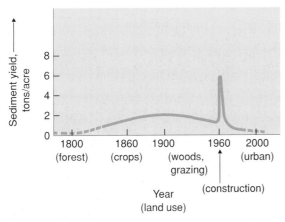

Figure 12.18

The impact of urbanization and other human activities on soil erosion rates. Construction may cause even more intensive erosion than farming, but far more land is affected by farming, and for a longer time. In this study of the Washington, D.C. area, erosion rates rose sharply during a highway construction boom around 1960; once construction was completed, erosion became negligible.

From "A Cycle of Sedimentation and Erosion in Urban River Channels," M. G. Wolman from Geografiska Annaler, *Vol. 49, series A, figure 1, by permission of Scandinavian University Press*

landscaping, erosion rates typically drop to below even natural, predevelopment levels. (Whether this represents an optimum land use is another issue.) The problems of erosion during construction, therefore, may be very intensive but are typically localized and short-lived.

Soil Erosion versus Soil Formation

Estimates of the total amount of soil erosion in the United States vary widely, but estimates by the U.S. Department of Agriculture Natural Resources Conservation Service put the figure at several billion tons per year. To a great extent, human activities, such as construction and farming, account for the magnitude of soil loss (figure 12.19). On cropland, average erosion losses from wind and water together have been estimated at 5.6 tons per acre per year. In very round numbers, that is about a 0.04 centimeter (0.02 inch) average thickness of soil removed each year. The rate of erosion from land under construction might be triple that. Erosion rates from other intensely disturbed lands, such as unreclaimed strip mines, might be at least as high as those for construction sites.

How this compares with the rate of soil *formation* is difficult to generalize because the rate of soil formation is so sensitive to climate, the nature of the parent rock material, and other factors. Upper limits on soil formation rates, however, can be determined by looking at soils in glaciated areas of the northern United States last scraped clean by the glaciers tens of thousands of years ago. Where the parent material was glacial till, which should weather more easily than solid rock, perhaps

A

B

C

Figure 12.19

Examples of water erosion of soil. (A) Gully erosion of soil bank exposed during construction. (B) Gullying caused by surface runoff downhill along furrows on Iowa farmland. (C) This severely wind-eroded field dramatically illustrates the role plant roots can play in holding soil in place.

Photographs (A) and (B) by Lynn Betts, all courtesy of USDA Natural Resources Conservation Service

1 meter (about 3 feet) of soil has formed in 15,000 years in the temperate, fairly humid climate of the upper Midwest. The corresponding average rate of 0.006 centimeters (0.002 inches) of soil formed per year is an order of magnitude less than the average cropland erosion rate. Furthermore, soil formation in many areas and over more resistant bedrocks is slower still. In some places in the Midwest and Canada, virtually no soil has formed on glaciated bedrocks, and in the drier southwest, soil formation is also likely to be very much slower. It seems clear that erosion is stripping away farmland far faster than the soil is being replaced. In other words, soil in general—and high-quality farmland—is also a nonrenewable resource in many populated areas.

Strategies for Reducing Erosion

The wide variety of approaches for reducing erosion on farmland basically involve either reducing the velocity of an eroding agent or protecting the soil from its effects. Under the latter heading come such practices as leaving stubble in the fields after a crop has been harvested rather than clearing and tilling the field, and planting cover crops in the off-season between cash crops. In all such cases, the plants' roots help to hold the soil in place, and the plants themselves, to some extent, shield the soil from wind and rain (figure 12.20).

The lower the wind or water runoff velocity, the less material carried. Surface runoff may be slowed on moderate slopes by contour plowing (figure 12.21A). Plowing rows parallel to the contours of the hill, perpendicular to the direction of water flow, creates a ridged land surface so that water does not rush downhill as readily (contrast with figure 12.19B). Other slopes

A

B

C

Figure 12.21

Decreasing slope of land, or breaking up the slope, decreases erosion by surface runoff. (A) Contour-plowed field in northern Iowa. (B) Terracing creates "steps" of shallower slope from one long, steeper slope. (C) Terraces become essential on very steep slopes, as with these terraces carved out of jungle.

(A) Photograph by Tim McCabe; (A) and (B) courtesy of USDA Natural Resources Conservation Service; (C) © The McGraw-Hill Companies, Inc./Barry Barker, photographer

Figure 12.20

On this Iowa cornfield, stubble has been left in the field over winter; soil is then cultivated and planted in one step in the spring, so it is never left exposed between harvest and planting. This is described as "no-till" agriculture.

Photograph by Lynn Betts, courtesy of USDA Natural Resources Conservation Service

A

B

C

Figure 12.22

Creating near-ground irregularities decreases wind velocity.
(A) Windbreaks create wind barriers on this otherwise-flat North Dakota farmland. (B) Strip cropping also breaks up the surface topography, decreasing wind velocity. Allamakee County, Iowa. (C) Contour strip cropping of corn and alfalfa on the Iowa/Minnesota border.

(A) Photograph by Erwin Cole; (B) and (C) by Tim McCabe; all courtesy USDA Natural Resources Conservation Service

may require terracing (figure 12.21B), whereby a single slope is terraced into a series of shallower slopes, or even steps that slant backward into the hill. Again, surface runoff makes its way down the shallower slope more slowly, if at all, and therefore carries far less sediment with it. Grass is typically left on the steeper slopes at the edges of the terraces to protect them from erosion. Terracing has, in fact, been practiced since ancient times. Both terracing and contour plowing, by slowing surface runoff, increase infiltration and enhance water conservation as well as soil conservation. Often, terracing and contouring are used in conjunction (figures 12.21B,C).

Wind can be slowed down by planting hedges or rows of trees as windbreaks along field borders or in rows perpendicular to the dominant wind direction (figure 12.22A) or by erecting low fences, like snow fences, similarly arrayed. This does not altogether stop soil movement, as shown by the ridges of soil that sometimes pile up along the windbreaks. However, it does reduce the distance over which soil is transported, and some of the soil caught along the windbreaks can be redistributed over the fields. Strip cropping (figure 12.22B), alternating crops of different

heights, slows near-ground wind by making the land surface more irregular. Combining strip cropping and contouring (figure 12.22C) may help reduce both wind and water erosion.

A major obstacle to erosion control on farmland is cost. Many of the recommended measures—for example, terracing and planting cover crops—can be expensive, especially on a large scale. Even though the long-term benefits of reduced erosion are obvious and may involve long-term savings or increased income, the effort may not be made if short-term costs seem too high. The benefits of a particular soil conservation measure may be counterbalanced by substantial drawbacks, such as reduced crop yields. Also, if strict erosion-control standards are imposed selectively on farmers of one state (as by a state environmental agency, for instance), those farmers are at a competitive disadvantage with respect to farmers elsewhere who are not faced with equivalent expenses. Even meeting the same standards does not cost all farmers everywhere equally. Increasingly, the government is becoming involved in cost-sharing for erosion-reduction programs so that the financial burden does not fall too heavily on individual farmers. Since

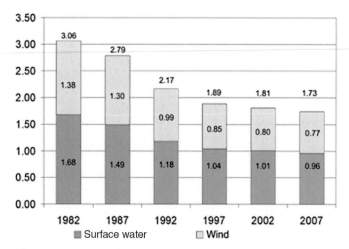

Figure 12.23

Erosion on cropland, in billions of tons; includes land removed from cultivation under the Conservation Reserve Program.

From USDA National Resources Conservation Service, National Resources Inventory 2007

the Dust Bowl era, over $20 billion in federal funds have been spent on soil conservation efforts.

One simple approach to erosion reduction on farmland is to take the most susceptible land out of cultivation altogether. The USDA's Conservation Reserve Program pays farmers to plant grass or trees on some seriously erodible land, requiring that they agree to leave the land uncultivated for 10 to 15 years. However, with the surge in interest in ethanol fuels produced from corn, there is new economic pressure for farmers to opt out of the program when their contracts expire, and a corresponding surge in erosion may result.

The USDA Natural Resources Conservation Service reports that a total of 1.73 billion tons of soil was lost from cropland in 2007. Clear progress has been made in erosion reduction over the past few decades (figure 12.23), though the rate of progress slowed after 1997. All agricultural areas of the United States show reduced erosion rates since 1982, but rates still remain high in some areas, notably the southern plains and the mountain region in the West (figure 12.24).

Other strategies can minimize erosion in nonfarm areas. For example, in areas where vegetation is sparse, rainfall limited, and erosion a potential problem, ORVs should be restricted to prescribed trails. In the case of urban construction projects, one reason for the severity of the resulting erosion is that it is common practice to clear a whole area, such as an entire housing-project site, at the beginning of the work, even though only a small portion of the site is actively worked at any given time. Clearing the land in stages, as needed, minimizes the length of time the soil is exposed and reduces urban erosion. Mulch or temporary ground covers can protect soil that must be left exposed for longer periods. Stricter mining regulations requiring reclamation of strip-mined land (see

Figure 12.24

Erosion rates vary significantly by region, as does the proportion of land that is planted with crops. In general, higher erosion rates are seen in drier regions.

After USDA NRCS National Resources Inventory 2007.

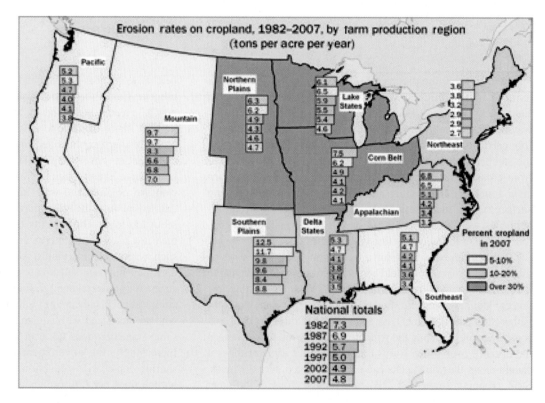

Plantations in Paradise—Unintended Consequences

Earlier in the chapter, we focused on the characteristics of tropical soils and how they change with farming. Problems associated with agricultural activities in such areas are by no means limited to Third World countries, or to the soil itself. Broad areas of the Hawaiian Islands are devoted to pineapple and sugarcane crops. Practices associated with the latter are particularly controversial.

When the sugarcane is ready to be harvested, it is set afire. The leaves are consumed; only the charred canes remain. This serves several purposes: Like slash-and-burn methods of clearing land, it restores nutrients to the soil; it reduces the volume and weight of material that has to be hauled to the sugar mill and processed, by burning off the unwanted leaves in the field; and the heat kills insects and their eggs, fungi, and other organisms that might otherwise linger to attack the next planting. However, it also contributes to air pollution, producing an acrid-sweet brown smoke that is unsightly (figure 1A)

and occasionally disrupts traffic when visibility on nearby roads becomes too poor.

The cane fields are at least generally on fairly level ground, where the huge cane-hauling trucks can operate, so sediment runoff between crops is reduced. Some of the pineapple plantations occupy much steeper terrain, and sediment runoff has created problems offshore, along beaches and among coral-reef communities. Corals require clear (as well as warm) waters to thrive. A major reason that Australia's Great Barrier Reef could grow so spectacular is that relatively few rivers carry sediment off the east side of the continent. Increased sediment runoff makes the water murkier, and corals die (figure 2B). The World Resources Institute estimates that between 20% and 25% of reefs around the world are threated by inland pollution and sediment erosion, and another 30% by coastal development, which can also increase sediment runoff.

A

B

Figure 1

The brown, smoky haze from burning sugar cane is distinctive and very visible (A); the process clears the cane field and restores nutrients to the tropical soil for the next crop (B).

A

B

Figure 2

(A) In clear tropical waters, colorful corals and associated reef communities can thrive. (B) Sediment runoff clouds the water, and the coral reef and ecosystem suffer.

A   B

Figure 12.25

(A) Salt buildup on land surface in Colorado, due to irrigation: salty water wicks to the drier surface and evaporates, leaving salty crust.
(B) Kesterson Wildlife Refuge, with selenium-contaminated ponds. These would be an attractive habitat for waterfowl, were it not for the toxicity of the waters.

(A) Photograph by Tim McCabe, courtesy of USDA Natural Resources Conservation Service; (B) courtesy of USGS Photo Library, Denver, CO.

chapter 13) are already significantly reducing soil erosion and related problems in these areas.

Where erosion cannot be eliminated, sediment pollution may nevertheless be reduced. "Soil fences" that act somewhat like snow fences, and hay-lined soil traps through which runoff water must flow, may reduce soil runoff from construction sites. On either construction sites or farmland, surface runoff water may be trapped and held in ponds to allow suspended sediment to settle out in the still ponds before the clarified water is released; such settling ponds are illustrated in chapter 17.

Irrigation and Soil Chemistry

We have noted leaching as one way in which soil chemistry is modified. Application of fertilizers, herbicides, and pesticides—whether on agricultural land or just one's own backyard—is another way that soil composition is changed, in this case by the addition of a variety of compounds through human activities. Less-obvious additions can occur when irrigation water redistributes soluble minerals.

In dry climates, for example, irrigation water may dissolve salts in pedocal soils; as the water evaporates near the surface, it can redeposit those salts in near-surface soil from which they had previously been leached away. As the process continues, enough salt may be deposited that plant growth suffers (figure 12.25A).

A different kind of chemical-toxicity problem traceable to irrigation was recognized in California's San Joaquin Valley in the mid-1980s. Wildlife managers in the Kesterson Wildlife Refuge noticed sharply higher incidence of deaths, deformities, and reproductive failures in waterfowl nesting in and near the artificial wetlands created by holding ponds for

irrigation runoff water (figure 12.25B). In the 1960s and 1970s, that drainage system had been built to carry excess runoff from the extensively irrigated cropland in the valley, precisely to avoid the kind of salinity buildup just described. The runoff was collected in ponds at Kesterson and seemed to create beneficial habitat. What had not initially been realized is that the irrigated soils were also rich in selenium.

Selenium is an element present in trace amounts in many soils. Small quantities are essential to the health of humans (note that a recommended daily allowance is shown on your multivitamin/mineral-supplement package!), livestock, and other organisms. As with many essential nutrients, however, large quantities can be toxic; and selenium in particular tends to accumulate in organisms that ingest it and to increase in concentration up a food chain. From selenium-rich rocks in the area, selenium-rich soils were derived. This was no particular problem until the soluble selenium was dissolved in the irrigation water, drained into the ponds at Kesterson, and concentrated there by evaporation in the ponds, where it built up through the algae→insects→fish→birds food chain. In time, it reached levels obviously toxic to the birds.

The short-term solution at Kesterson was to fill in the ponds. The episode underscored the more general issue of the need to consider the effects of adding water to soils which, in dry climates, may contain quantities of soluble minerals that, when mobilized in solution, can present a new hazard.

The Soil Resource—The Global View

Soil degradation is a concern worldwide. Desertification, erosion, deterioration of lateritic soil, contamination from pollution, and other chemical modification by human activity all

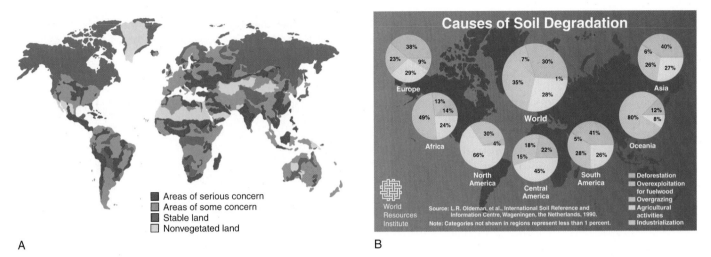

Figure 12.26

Soil degradation of various kinds is a global problem. (A) Six continents are significantly affected by soil degradation; on four (Africa, Australia, Europe, and South America), little land is considered stably vegetated. (B) The causes of soil degradation, however, vary around the world.

Sources: (A) data from Global Resource Information Database of U.N. Environment Programme; (B) from World Resources Institute

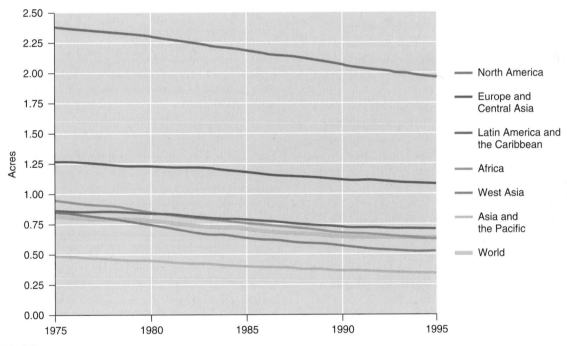

Figure 12.27

Acres of arable land per capita, for the world and various regions of it. The declines may be due to population growth, the spread of cities and suburbs onto arable land, or loss of farmland through degradation, depending on the region in question. Although the most rapid drop is occurring in North America, this region still enjoys far more available farmland per capita than most of the rest of the world; this resource is especially scarce in Asia. In some places, the amount of arable land could be increased by irrigation, but as we saw in the last chapter, increased reliance on irrigation may simply aggravate water-supply problems.

Data from the U.N. Environment Programme Global Resource Information Database

contribute to reduced soil quality, fertility, and productivity. The problems are widespread (figure 12.26A); the contributing causes vary regionally (figure 12.26B).

Land area is finite. Differences in population density as well as soil quality create great disparities in per-capita arable land by region of the globe, and the numbers decline as population grows and as arable land is covered up by development (figure 12.27). Soil degradation simply further diminishes the availability of the farmland needed to feed the world.

Soils and Suspects

Forensic geology may be said to have its roots in the Sherlock Holmes stories of Sir Arthur Conan Doyle. Holmes was a new sort of investigator, one who used careful observation of myriad small details in identifying criminals and unraveling mysteries. The concept of using traces of soils to track where a person or object had been was quite novel at the time. Since then, geologic evidence of many kinds has figured in the solution of real-life crimes.

Some are white-collar crimes, involving fraud or theft. Scam artists selling gullible investors shares in a mining prospect may "salt" samples of the supposed ore, adding bits of gold or other valuable mineral to make it seem richer than it is. (The stakes can be very high: A massive fraud of this kind involving a mine in Borneo cost investors billions of dollars.) Investigation can reveal the additions, which may, by their composition or texture, be obviously incompatible with the host ore material. In a famous theft case, a shipment of electronics that had been sent from Texas to Argentina via Miami disappeared en route, replaced by concrete blocks. Study of the sand in the concrete, and comparison with local sands at each of the shipment's stops, indicated that the substitution was made in Florida, which narrowed the search for the culprits and led to their eventual arrest. Protected cacti stolen from federal land have been linked to their source by the soil clinging to their roots.

Some of the most spectacular cases involve murder or kidnaping. A coverup originally complicated the investigation of the death in Mexico of an agent of the U.S. Drug Enforcement Agency. The body was moved from where it was originally buried to a farm unconnected with the crime. But the source location of a small sample of soil attached to the body was subsequently identified, the coverup revealed, and the murderers prosecuted.

Not all cases are straightforward. Soils and sediments differ in many ways—color, texture, mineralogy, sorting, and more (figure 1). However, not every sample is unique to a small, specific area. In the above-mentioned murder case, for example, the soil contained rhyolitic ash with a distinctive mineralogy, found only in one small volcanic area in Mexico. Also, a very small sample may not be representative of the average composition of the soil or sediment from

Figure 1

Three distinctive soils (clockwise from lower left): a quartz-rich sand, a commercial potting soil rich in vermiculite and organic matter, and a poorly sorted soil with very fine clay and rocky gravel together. With a spoonful-sized sample you could characterize each fairly well; with only a few grains, it would be much more difficult to compare your sample accurately with others.

which it comes. Comparison of soil samples rarely involves the level of precision of, say, DNA matching. Still, it is often sufficient to indicate, for instance, that a suspect was in an area where that person claims never to have been, and thereby break an alibi, or to associate the suspect with a person, area, or object of interest to an investigation. Analytical techniques are also becoming more sophisticated over time, making geologic evidence more powerful.

For an introduction to forensic geology and some of the cases it has helped to solve, the interested reader is referred to *Evidence from the Earth,* a book written for the nonspecialist by experienced forensic geologist Raymond C. Murray.

Summary

Soil forms from the breakdown of rock materials through mechanical and chemical weathering, often accelerated by biological activity. Weathering rates are closely related to climate, with chemical weathering tending to dominate in warmer, wetter regions, and mechanical weathering in cooler, dryer areas. The character of the soil reflects the nature of the parent material and the kinds and intensities of weathering processes that formed it. The wetter the climate, the more leached the soil. The lateritic soil of tropical climates is particularly unsuitable for agriculture: Not only is it highly leached of nutrients, but when exposed, it may harden to a bricklike solidity. The texture of soil is a major determinant of its drainage characteristics; soil structure is related to its suitability for agriculture. Agricultural practices may also alter soil chemistry, whether through direct application of additional chemicals or through irrigation-induced changes in natural cycles.

Soil erosion by wind and water is a natural part of the rock cycle. Where accelerated by human activity, however, it can also be a serious problem, especially on farmland or, locally, in areas subject to construction or strip mining. Erosion rates far exceed inferred rates of soil formation in many places. A secondary problem is the resultant sediment pollution of lakes, streams, and nearshore ocean waters. Strategies to reduce soil erosion on farmland include terracing, contour plowing, planting or erecting windbreaks, the use of cover crops, strip cropping, and minimum-tillage farming. Leaving the most erodible land uncultivated also reduces erosion overall; but growing interest in biofuels is encouraging increased farming on marginal lands. Elsewhere, restriction of ORVs, more selective clearing of land during construction, the use of sediment traps and settling ponds, and careful reclamation of strip-mined areas could all help to minimize soil erosion. Globally, arable land per capita is declining, as potential farmland is lost to development and degradation and population grows.

Key Terms and Concepts

A horizon 262
B horizon 262
chemical weathering 259
C horizon 262

E horizon 262
laterite 266
leaching 262
mechanical weathering 259

O horizon 262
pedalfer 264
pedocal 264
soil 259

zone of accumulation 262
zone of leaching 262

Exercises

Questions for Review

1. Briefly explain how the rate of chemical weathering is related to (a) the amount of precipitation, (b) the temperature, and (c) the amount of mechanical weathering.

2. Sketch a generalized soil profile and indicate the A horizon, B horizon, C horizon, E horizon, O horizon, zone of leaching, and zone of accumulation. Is such a profile always present?

3. How do pedalfer and pedocal types of soil differ, and in what kind of climate is each more common?

4. The lateritic soil of the tropical jungle is poor soil for cultivation. Explain.

5. Soil erosion during active urbanization is far more rapid than it is on cultivated farmland, and yet the majority of soil-conservation efforts are concentrated on farmland. Why?

6. Cite and briefly describe three strategies for reducing cropland erosion. How is interest in biofuels such as ethanol related to soil erosion?

7. Irrigation can, over time, cause harmful changes in soil chemistry that reduce crop yields or create toxic conditions in runoff water. Explain briefly.

8. Describe three factors that are reducing the amount of available farmland per capita around the world.

Exploring Further

1. Choose any major agricultural region of the United States and investigate the severity of any soil erosion problems there. What efforts, if any, are being made to control that erosion, and what is the cost of those efforts?

2. Visit the site of an active construction project; examine it for evidence of erosion, and erosion-control activities.

3. How does your state fare with respect to soil erosion? What factors can you identify that contribute to the extent—or the lack—of soil erosion? (The National Resources Inventory 2007 may be a place to start; the Natural Resources Conservation Service or your local geological survey may be additional resources.)

4. Assume that solid rock can be eroded at a rate comparable to that at which it can be converted to soil, and use the rate of 0.006 centimeter per year as representative. How long would it take to erode away a mountain 1500 meters (close to 5000 feet) high?

5. Take a sample of local soil; examine it closely with a magnifying lens and describe it as fully as you can. What features of it (if any) might be relatively unusual? Does it resemble any of the samples in Case Study 12.2?

6. Consider the slopes of the trend lines for North America, Asia, and the world in figure 12.27, approximating each by a constant value. Project into the future and estimate the dates at which the North American trend would intersect each of the other two. Consider what factors might change these slopes for each area, and thus your projected dates.

Mineral and Rock Resources

Though we may not often think about it, virtually everything we build or create or use in modern life involves rock, mineral, or fuel resources. In some cases, this is obvious: We can easily see the steel and aluminum used in vehicles and appliances; the marble or granite of a countertop or floor; the gold, silver, and gemstones in jewelry; the copper and nickel in coins. But our dependence on these resources is much more pervasive. Asphalt highways (which themselves use petroleum resources) are built on crushed-rock roadbeds. A wooden house is built on a slab or foundation of concrete (which requires sand and limestone to make), held together with metal nails, finished inside with wallboard (made with gypsum), wired with copper or aluminum, given a view by windows (glass made from quartz-rich sand); the

bricks of a brick house are made from clay. Your toothpaste cleans your teeth using mineral abrasives (such as quartz, calcite, corundum, or phosphate minerals), reduces cavities with fluoride (from the mineral fluorite), is probably colored white with mineral-derived titanium dioxide (the same pigment as in white paint), and, if it sparkles, may contain mica flakes. The computers and other electronic devices on which we increasingly rely would not exist without metals for circuits and semiconductors.

The bulk of the earth's crust is composed of fewer than a dozen elements. In fact, as noted in table 1.2, eight chemical elements make up more than 98% of the crust. Many of the elements *not* found in abundance in the earth's crust, including industrial and precious metals, essential components of chemical

The Carajás Mine in northeastern Brazil is one of the world's largest iron mines, containing an estimated 18 billion tons of iron ore. Additional metals that can be recovered from the ore include copper, nickel, manganese, and gold. For scale, cleared area at center is about 1 mile square. Deforestation outside the mine property is visible at top of image.

NASA image created by Jesse Allen, using EO-1 ALI data provided courtesy of the NASA EO-1 team

fertilizers, and elements like uranium that serve as energy sources, are vitally important to society. Some of these are found in very minute amounts in the average rock of the continental crust: copper, 100 ppm; tin, 2 ppm; gold, 4 ppb. Clearly, then, many useful elements must be mined from very atypical rocks. This chapter looks at the occurrences of a variety of rock and mineral resources and examines the U.S. and world supply-and-demand picture.

Resources, Reserves, and Ore Deposits

In a broad sense, *resources* are all those things that are necessary or important to human life and civilization, that have some value to individuals and/or to society. What constitutes a resource may thus change with time or social context. Many of the fuels, building materials, metals, and other substances important to modern technological civilization (which are thus resources in a modern context) were of no importance to the earliest cave dwellers. Even today, some societies or nations may have no use or need for minerals or fuels that highly technological societies find essential, so the global resource-availability picture may differ greatly from that for a single country.

In discussing supply/demand questions at any scale, distinctions are made between quantities of reserves and various other categories of resources, as summarized in figure 13.1. The **reserves** are that quantity of a given geologic material that has been found and that can be recovered economically with existing technology. Usually, the term is used only for material not already consumed, as distinguished from **cumulative reserves,** which include the quantity of the material already used up in addition to the remaining (unused) reserves. The reserves represent the most conservative estimate of how much of a given metal, mineral, or fuel remains unused. Beyond that, several additional categories of resources can be identified.

Those deposits that have already been found but that cannot presently be profitably exploited are the **subeconomic resources** (also known as **conditional resources**). Some may be exploitable with existing technology but contain ore that is too low-grade (see below) or fuel that is too dispersed to produce a profit at current prices. Others may require further advances in technology before they can be exploited. Still, these deposits have at least been located, and the quantity of material they represent can be estimated fairly well.

Then there are the undiscovered resources. These are sometimes subdivided into **hypothetical resources,** additional deposits expected to be found in areas in which some deposits of the material of interest have already been found, and the **speculative resources,** those deposits that might be found in explored or unexplored regions where deposits of the material are not already known to occur. Estimates of undiscovered resources are extremely rough by nature, and it would obviously be unwise to count too heavily on deposits that have not even been found yet, especially for the near future.

An **ore** is a rock in which a valuable or useful metal occurs at a concentration sufficiently high, relative to average rocks, that it may be economically worth mining. A given ore deposit may be described in terms of the enrichment or **concentration factor** of a metal of interest:

$$\text{Concentration factor of a given metal in an ore} = \frac{\text{Concentration of the metal in the ore deposit}}{\text{Concentration of the metal in average continental crust}}$$

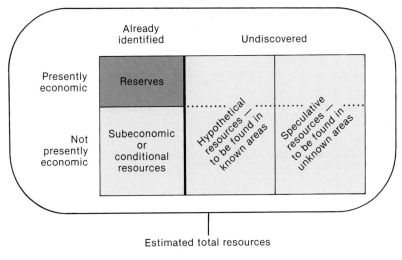

Figure 13.1

Different categories of reserves and resources.

The higher the concentration factor, the richer the ore, and the less of it needs to be mined to extract a given amount of the metal.

In general, the minimum concentration factor required for profitable mining is inversely proportional to the average crustal concentration: If ordinary rocks are already fairly rich in the metal, it need not be concentrated much further to make mining it economic, and vice versa. Metals like iron or aluminum, which make up about 6% and 8% of average continental crust, respectively, need only be concentrated by a factor of 4 to 5 times for mining to be profitable. Copper must be enriched about 100 times relative to average rock, while mercury (average concentration 80 ppb) must be enriched to about 25,000 times its average concentration before the ore is rich enough to mine profitably. The exceptions to this rule are those few extremely valuable elements, such as gold, that are so high-priced that a small quantity justifies a considerable amount of mining. Even at several hundred dollars an ounce, gold need be found in concentrations only a few thousand times its average 4 ppb to be worth mining. This inverse relationship between average concentration and the concentration factor required for profitable mining further suggests that economic ore deposits of the relatively abundant metals (such as iron and aluminum) might be more plentiful than economic deposits of rarer metals, and this is indeed commonly the case.

The value of the mineral or metal extracted and its concentration in a particular deposit, then, are major factors determining the profitability of mining a specific deposit. The economics are naturally sensitive to world demand. If demand climbs and prices rise in response, additional, not-so-rich ore deposits may be opened up; a fall in price causes economically marginal mines to close. The discovery of a new, rich deposit of a given ore may make other mines with poorer ores noncompetitive and thus uneconomic in the changing market. The practicality of mining a specific ore body may also depend on the mineral(s) in which a metal of interest is found, because this affects the cost to extract the pure metal. Three iron deposits containing equal concentrations of iron are not equally economic if the iron is mainly in oxides in one, in silicates in another, and in sulfides in the third. The relative densities of the iron-bearing minerals, compared to other minerals in the rock, will affect how readily those iron-rich minerals can be separated from the rest; the nature and strengths of the chemical bonds in the iron-rich phases will influence the energy costs to break them down to extract the iron, and the complexity of chemical processes required; the concentration of iron in the iron-rich minerals (rather than in the rock as a whole) will determine the yield of iron from those minerals, after they are separated. Different types of mines, too, involve different costs. These factors and others influence the economics of mining. Given the definition of reserves, fluctuating economic factors will affect whether some ore deposits are counted as reserves or put in the "subeconomic resources" category—and for a specific deposit, that can change over time.

By definition, ores are somewhat unusual rocks. Therefore, it is not surprising that the known economic mineral deposits are very unevenly distributed around the world. (See, for example, figure 13.2.) The United States controls about 27% of the world's known molybdenum reserves. But although the United States is the major world consumer of aluminum, using over 10% of the total produced, it has virtually no domestic aluminum ore deposits; Australia, Guinea, and Brazil together control about 60% of the world's aluminum reserves. Congo (Kinshasa, formerly Zaire) accounts for nearly half the known economic cobalt deposits, Chile for almost 30% of the copper, and China, Indonesia, and Brazil for, respectively, 31%, 17%, and 13% of the tin. South Africa controls over 40% of the chromium, and about 95% of reserves of the platinum-group metals. These great disparities in mineral wealth among nations are relevant to later discussions of world supply/demand projections.

Types of Mineral Deposits

Deposits of economically valuable rocks and minerals form in a variety of ways. This section does not attempt to review them all but describes some of the more important processes involved.

Igneous Rocks and Magmatic Deposits

Magmatic activity gives rise to several different kinds of deposits. Certain igneous rocks, just by virtue of their compositions, contain high concentrations of useful silicate or other minerals. The deposits may be especially valuable if the rocks are coarse-grained, so that the mineral(s) of interest occur as large, easily separated crystals. **Pegmatite** is the term given to unusually coarse-grained igneous intrusions (figure 13.3). In some pegmatites, single crystals may be over 10 meters (30 feet) long. Feldspars, which provide raw materials for the ceramics industry, are common in pegmatites. Many pegmatites also are enriched in uncommon elements. Examples of rarer minerals mined from pegmatites include tourmaline (used both as a gemstone and for crystals in radio equipment) and beryl (mined for the metal beryllium when crystals are of poor quality, or used as the gemstones aquamarine and emerald when found as clear, well-colored crystals).

Other useful minerals may be concentrated within a cooling magma chamber by gravity. If they are more or less dense than the magma, they may sink or float as they crystallize, instead of remaining suspended in the solidifying silicate mush, and accumulate in thick layers that are easily mined (figure 13.4). Chromite, an oxide of the metal chromium, and magnetite, one of the iron oxides, are both quite dense. In a magma of suitable bulk composition, rich concentrations of these minerals may form in the lower part of the magma chamber during crystallization of the melt. The dense precious metals, such as gold and platinum, may also be concentrated during magmatic crystallization. Even where disseminated throughout an igneous rock body, a very valuable mineral commodity may be worth mining. This is true, for example, of diamonds. One reason for the rarity of diamonds is that they must

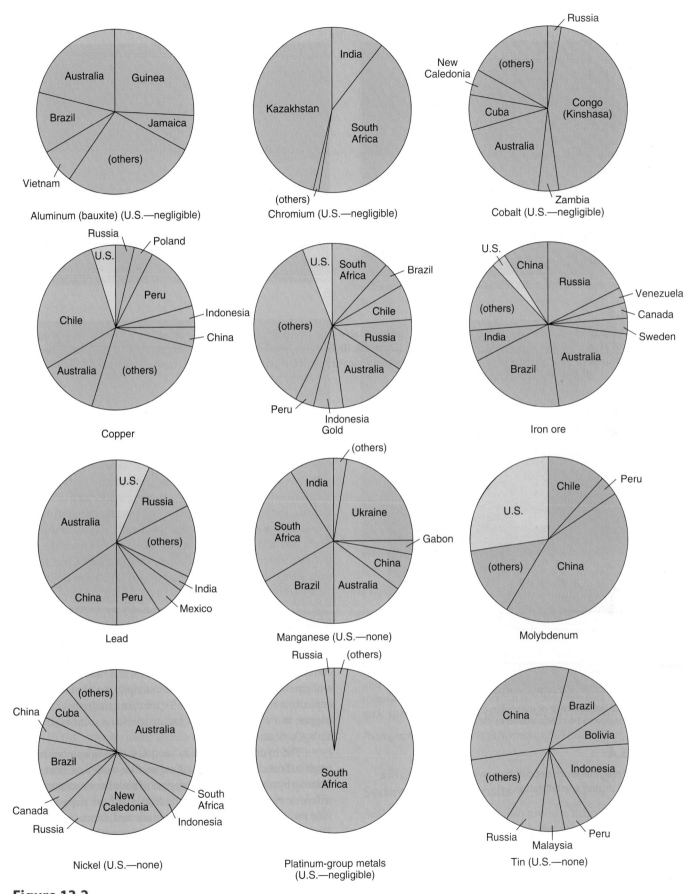

Figure 13.2

Proportions of world reserves of some nonfuel minerals controlled by various nations. Although the United States is a major consumer of most metals, it is a major producer of very few. *Source: Data from* Mineral Commodity Summaries 2012, *U.S. Geological Survey*

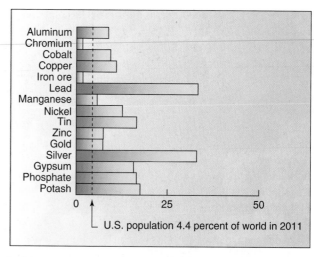

Figure 13.12

United States share of global consumption of selected materials. Assumes that world consumption of each commodity approximately equals world production.

Source: Data from Mineral Commodity Summaries 2012, *U.S. Geological Survey*

A

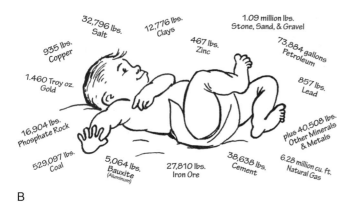

B

Figure 13.11

(A) Per-capita consumption of selected mineral and rock resources in the United States, 2011 (1 kg = 2.2 lb). Many of these numbers are down from prior years due to economic slowdown. (B) "The American Baby" of the Mineral Information Institute; MII takes similar data, together with average U.S. life expectancy, to calculate the average total lifetime consumption by that individual of various rock, mineral, and fuel resources. Note that the total is about 3 million pounds!

Source: (A) Data from Mineral Commodity Summaries 2012, *U.S. Geological Survey (B) from Mineral Information Institute, an affiliate of the SME Foundation.*

U.S. Mineral Production and Consumption

All told, the United States consumes huge quantities of mineral and rock materials, relative to both its population and its production of most minerals. Per-capita consumption of many materials is shown in figure 13.11. Table 13.1 and figure 13.12 give a global perspective by showing 2011 U.S. primary production and consumption data and the percentages of world totals that these figures represent. Keep in mind when looking at the consumption figures that only about 4.4% of the people in the world live in the United States; and for many materials, 2011 demand was substantially lower than previous years' as a consequence of the economic downturn.

The relationship of domestic consumption to domestic production can be complex. "Primary production" is production from new ore, excluding production from recycling or from partially processed ore imported from elsewhere; total production includes the latter kinds of production as well. (This accounts for differences in figures for metals in the last two columns of table 13.1.) Sometimes, the United States even imports materials of which it has ample domestic supplies, perhaps to stay on good trade terms with countries having other, more critical commodities that the United States also wishes to import or because the cost of complying with U.S. environmental regulations makes it cheaper to import the commodity from another nation. For example, in the case of molybdenum, in 2011, U.S. primary production was 64,000 metric tons, against consumption of 50,000 tons—yet we imported 18,000 tons; in the same year, we produced 415 metric tons of gold (including recycling) and consumed 200, but imported another 670 tons.

Of course, where production falls short of consumption, importing is necessary to meet domestic demand. Moreover, low

Material	U.S. Primary Production	U.S. Primary Production as % of World	U.S. Consumption	U.S. Consumption as % of World[‡]	U.S. Production as % of U.S. Consumption[■]	
					Primary	Total
Metals						
aluminum[x]	1990	4.5	3900	8.8	51	87
chromium	0	0	400	1.7	0	40
cobalt	0	0	8.7	9.4	0	25
copper	1120	7.0	1780	11.0	62	70
iron ore	54,000	1.9	49,000	1.8	110	110
lead	117	7.7	1500	33.3	7.8	88
manganese	0	0	810	5.8	0	0
nickel	0	0	228	12.7	0	0
tin	0	0	42.1	16.6	0	4.9
zinc	117	6.1	924	7.5	12.7	27
gold[†]	237	8.8	200	7.4	118	208
silver[†]	1160	4.9	7850	33.0	14.5	32
platinum group[†]	16.2	4.0	ø	ø	ø	ø
Nonmetals						
clays	25,900	*	21,900	*	118	118
gypsum	9400	6.4	23,400	15.8	40	53
phosphate	28,400	14.9	31,800	16.6	89	89
potash	1100	3.0	6500	17.6	17	17
salt	44,000	15.2	57,000	19.7	77	77
sulfur	8800	12.8	11,600	16.8	76	76
Construction						
sand and gravel, construction	790,000	*	790,000	*	100	100
stone, crushed	1,110,000	*	1,150,000	*	96	96

[†]All production and consumption figures in thousands of metric tons, except for gold, silver, and platinum-group metals, for which figures are in metric tons. One metric ton = 1000 kg = 2200 lb.

[‡]Assumes that overall, world production approximates world consumption. This may not be accurate if, for example, recycling is extensive.

[■]"Primary" and "Total" headings refer to U.S. production; total production includes production from recycled scrap.

*World data not available

ø Consumption data not available

[x]Partly from bauxite imports

Source: Mineral Commodity Summaries 2012, U.S. Geological Survey.

domestic production may not necessarily mean that the United States is conserving its own resources, but that it has none to produce or, at least, no economic deposits. This is true, for example, of chromium, an essential ingredient of stainless steel, and manganese, which is currently irreplaceable in the manufacture of steel. Currently, for materials such as these, the United States is wholly or heavily dependent on imports (figure 13.13).

Leaving aside the unanswerable political question of how long any country can count on an uninterrupted flow of necessary mineral imports, how well supplied is the world, overall, with minerals? How imminent are the shortages? We will address these questions below.

World Mineral Supply and Demand

Certain assumptions must underlie any projections of "lifetimes" of mineral reserves. One is future demand. For reasons already outlined, demand can fluctuate both up and down. Furthermore, not all commodities follow the same trends. Making the assumption that world production annually approximates demand, at least for "new" mine production, we find that just over one recent decade, annual demand for iron ore remained approximately constant; for copper and zinc, it increased by 10% and 20%, respectively; for lead, it declined by 7%. Rather than attempt to make precise demand estimates for the future, the following discussion assumes demand equal to primary production at constant 2011 levels.

Table 13.2 shows projections of the lifetimes of selected world mineral reserves, assuming that constant demand. Note that, for most of the metals, the present reserves are projected to last only a few decades, even at these constant consumption levels. Consider the implications if consumption resumes the rapid growth rates of the mid-twentieth century.

Table 13.2	World Production and Reserves Statistics, 2011*			
Material	Production	Reserves	Projected Lifetime of Reserves (years)	Estimated Resources
bauxite	220,000	29,000,000	132	55,000,000–75,000,000
chromium	24,000	480,000°	20	12,000,000
cobalt	98.0	7500	76	15,000
copper	16,100	690,000	43	3,700,000[†]
iron ore	2,800,000	170,000,000	61	800,000,000
lead	4500	85,000	19	1,500,000
manganese	14,000	630,000	45	**
nickel	1800	80,000	44	130,000
tin	253	4800	19	**
zinc	12,400	250,000	20	1,900,000
gold	2.7	51	19	n.a.
silver	23.8	530	19	n.a.
platinum group	0.40	66	165	100
gypsum	148,000	**	**	**
phosphate	191,000	71,000,000	372	300,000,000
potash	37,000	9,500,000	257	250,000,000

*All production, reserve, and resource figures in thousands of metric tons.

**Resources and reserves estimated as "large" but not quantified.

°Some additional world reserve figures not available, so projected "lifetime" is lower limit.

[†]Includes 0.7 billion tons copper estimated to occur in manganese nodules; "extensive" resources of nickel are also projected to occur in these nodules.

Source: Mineral Commodity Summaries 2012, U.S. Geological Survey.

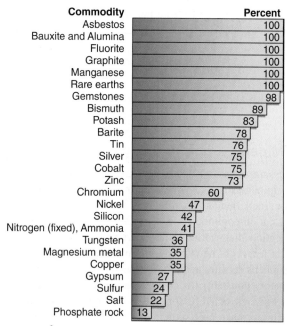

Commodity	Percent	Major Sources (2007–2010)
Asbestos	100	Canada, Zimbabwe
Bauxite and Alumina	100	Jamaica, Brazil, Guinea, Australia
Fluorite	100	Mexico, China, South Africa, Mongolia
Graphite	100	China, Mexico, Canada, Brazil
Manganese	100	South Africa, Gabon, China, Australia
Rare earths	100	China, France, Estonia, Japan
Gemstones	98	Israel, India, Belgium, South Africa
Bismuth	89	China, Belgium, United Kingdom
Potash	83	Canada, Belarus, Russia
Barite	78	China, India
Tin	76	Peru, Bolivia, Indonesia, China
Silver	75	Mexico, Canada, Peru, Chile
Cobalt	75	China, Norway, Russia, Canada
Zinc	73	Canada, Peru, Mexico, Ireland
Chromium	60	South Africa, Kazakhstan, Russia, China
Nickel	47	Canada, Russia, Australia, Norway
Silicon	42	China, Russia, Venezuela, Canada
Nitrogen (fixed), Ammonia	41	Trinidad and Tobago, Russia, Canada, Ukraine
Tungsten	36	China, Bolivia, Canada, Germany
Magnesium metal	35	Israel, Canada, China
Copper	35	Chile, Canada, Peru, Mexico
Gypsum	27	Canada, Mexico, Spain
Sulfur	24	Canada, Mexico, Venezuela
Salt	22	Canada, Chile, Mexico, The Bahamas
Phosphate rock	13	Morocco, Peru

Figure 13.13

Proportions of U.S. needs for selected minerals supplied by imports as a percentage of apparent material consumption, 2011. Principal sources of imports indicated at right; note the variety of nations represented.

Source: Data from Mineral Commodity Summaries 2012, U.S. Geological Survey

Table 13.3	Projected Lifetimes of U.S. Mineral Reserves (assuming complete reliance on domestic reserves)	
Material	Reserves*	Projected Lifetime (years)
bauxite[†]	20,000	7.5
chromium	620	1.5
cobalt	33,000	3.8
copper	35,000	20
iron ore	6,900,000	144
lead	6100	4
manganese	0	0
nickel	0	0
tin	0	0
zinc	12,000	13
gold	3	15
silver	25	3.2
platinum group	0.9	**
gypsum	700,000	30
phosphate	1,400,000	44
potash	130,000	20

*Reserves in thousands of metric tons.

**Accurate consumption data not available.

[†]Note that bauxite consumption is only a partial measure of total aluminum consumption; additional aluminum is consumed as refined aluminum metal, of which there are no reserves.

Source: Mineral Commodity Summaries 2012, U.S. Geological Survey.

Such projections also presume unrestricted distribution of minerals so that the minerals can be used as needed and where needed, regardless of political boundaries or such economic factors as nations' purchasing power. It is interesting to compare global projections with corresponding data for the United States only (table 13.3). Even at constant 2011 consumption rates, the United States has only a decade or two of reserves of most of these materials. Of some metals, including chromium, manganese, nickel, and tin, the United States has essentially no reserves at all. In some cases, too, we may have considerable domestic reserves that we are simply not mining because the particular elements cannot be extracted as cheaply in this country as they can elsewhere in the world. We may thus have comfortable long-term supplies of such mineral resources; but because mines and mineral-processing facilities take years to bring into production, we could face critical near-term shortages if our imported supply is abruptly cut off. See Case Study 13.1.

Some relief in the overall supply picture can be anticipated from the economic component of the definition of reserves. For example, as currently identified reserves are depleted, the law of supply and demand will drive up minerals' prices. This, in turn, will make some of what are presently subeconomic resources profitable to mine; some of those deposits will effectively be reclassified as reserves. Improvements in mineral-processing technology can accomplish the same result.

In addition, continued exploration can be expected to locate additional economic deposits, particularly as exploration methods become more sophisticated, as described later in the chapter.

The technological advances necessary for the development of some of the subeconomic resources, however, may not be achieved rapidly enough to help substantially. Also, if the move is toward developing lower- and lower-grade ore, then by definition, more and more rock will have to be processed and more and more land disturbed to extract a given quantity of mineral or metal. Some of the consequences of mining activities, which will have increasing impact as more mines are developed, are also discussed later in this chapter.

What other prospects exist for averting future mineral-resource shortages?

Minerals for the Future: Some Options Considered

One way to extend nonrenewable resources, in theory, would be to reduce consumption rates, or at least hold them steady. In practice, this seems unlikely to occur, even in the developed nations where resource consumption is already high.

The United States has the world's highest rate of car ownership: In 2009, 59% of U.S. households owned two or more cars, and 23%, three or more, for total household vehicle ownership of about 680 vehicles per 1000 people. Many factors—trends toward smaller households, necessity, more leisure for travel, desires for convenience or status—put upward pressure even on those U.S. car-ownership numbers. Or consider the impact of new technologies: for example, microwave ovens, personal computers, or cellular phones (increasing in number so fast that telephone companies in urban areas scrambled to create new area codes to accommodate them). In many cases, the new devices are additions to, rather than replacements for, other devices already owned (conventional ovens, typewriters, landline phones). All represent consumption of materials (as well as energy, to manufacture and often to use). While it is perhaps perfectly natural for people to want to acquire goods that increase comfort, convenience, efficiency, or perceived quality of life, the result may be increases in already high resource-consumption rates. In the United States, from 1950 to 1990, the population grew by about 65%, while materials consumption increased twice as fast, by more than 130%. Most of the increase was in plastics and industrial minerals.

Furthermore, even if the technologically advanced nations were to restrain their appetite for minerals, they represent a minority of the world's people. If the less-developed nations are to achieve a standard of living for their citizens that is even remotely comparable to that in the most industrialized nations, vast quantities of minerals and energy will be needed. For example, 2008 data put total motor vehicles at only 36 per 1000 people in China and 13 per 1000 in India, while the global average was about 240 total vehicles per 1000 persons, and for the United States, 842 vehicles per thousand. Historically, China has even fallen below the norm for its GDP among its neighbors (figure 13.14). However,

The Not-So-Rare Rare Earths

As we noted in chapter 2, certain groups of elements within the periodic table have collective names—for example, the alkali metals, those elements in the first column. The group known as the **rare-earth elements** (REE) is the set of elements with atomic numbers from 57 (lanthanum) to 71 (lutetium); see figure 2.2. Their names and chemical symbols are listed in table 1. They are similar both in ionic size and in ionic charge (nearly always +3), so they behave very similarly geochemically and tend to be found together in the same minerals.

Geologically speaking, the REE are actually not all that rare. Their abundances in average continental crust are about a thousand times that of gold, and comparable to the abundances of common metals like lead or tin. They were long considered "rare" mainly because they are seldom found as the major cations in minerals; instead, they tend to substitute for common cations in silicates and other minerals, so their overall abundances were not realized until geologists had made detailed chemical analyses of many rocks and minerals. Even then, their usefulness was not immediately recognized.

Modern technology has changed that. REE are used in a tremendous variety of applications: as catalysts; in making and polishing glass; as components in alloys, lasers, optical fibers, permanent magnets, fluorescent lights, color TV tubes and flat-panel screens, superconductors, rechargeable batteries (both small ones and those used in hybird and electric vehicles), and much more. The REE have become essential to a great many industries.

The United States began mining REE in the mid-1960s, primarily at Mountain Pass, California. In the 1980s, China began REE mining, and its share of production rose while that of the United States and the rest of the world declined (figure 1). In 2002, owing to

Table 1	Chemical Symbols and Names of the Rare-Earth Elements		
La	lanthanum	Tb	terbium
Ce	cerium	Dy	dysprosium
Pr	praseodymium	Ho	holmium
Nd	neodymium	Er	erbium
Pm	prometheum	Tm	thulium
Sm	samarium	Yb	ytterbium
Eu	europium	Lu	lutetium
Gd	gadolinium		

economic conditions and environmental issues, the Mountain Pass mine ceased operation. By 2011, China accounted for over 95% of the world's REE production.

This had been a concern for some time, but suddenly became a major global issue in fall of 2010. First, in a political dispute with Japan, China suspended REE exports to that country; within weeks, the suspension was extended to the United States and Europe. REE prices began to rise sharply. Exports to the United States and Europe were resumed within weeks, and those to Japan within months, but concerns about supply stability and cost outside China remained acute. During all this, prices of REE on world markets had soared, some rising by a factor of 30 or more, and although prices

---

China's economy is growing rapidly, with vehicle ownership quadrupling in a decade. If China's population of over 1.3 billion persons were to increase its car-ownership rate just to the global average, that would mean more than 250 million additional vehicles. Consider the resource implications if such populous nations as China and India were to increase their car ownership rates just tenfold (leaving them still well below the U.S. level) and correspondingly increased their consumption of the consumer goods plentiful in the most developed countries.

If demand cannot realistically be cut, supplies must be increased or extended. We must develop new sources of minerals, in either traditional or nontraditional areas, or we must conserve our minerals better so that they will last longer.

New Methods in Mineral Exploration

Most of the "easy ores," or near-surface mineral deposits in readily accessible places, have probably already been discovered, at least in developed and/or well-explored regions. Fortunately, a variety of new methods are being applied to the search. Geophysics provides some assistance. Rocks and minerals vary in density, and in magnetic and electrical properties, so changes in rock types or distribution below the earth's surface can cause small variations in the gravitational or magnetic field measured at the surface, as well as in the electrical conductivity of the rocks. Some ore deposits may be detected in this way. As a simple example, because many iron deposits are strongly magnetic, large magnetic anomalies may indicate the presence and the extent of a subsurface body of iron ore. Radioactivity is another readily detected property; uranium deposits may be located with the aid of a Geiger counter.

Geochemical prospecting is another approach with increasing applications. It may take a variety of forms. Some studies are based on the recognition that soils reflect, to a degree, the chemistry of the materials from which they formed. The soil over a copper-rich ore body, for instance, may itself be enriched in copper relative to surrounding soils. Surveys of soil chemistry over a wide area can help to pinpoint likely spots to dig in search of ores.

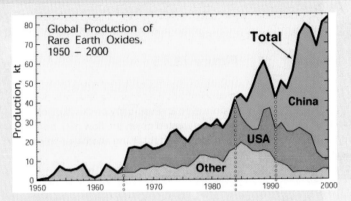

Figure 1

For decades, the United States was self-sufficient with respect to REE; now, like most nations, we depend entirely on imports from China. (kt = kilotons = thousands of tons)

After U.S. Geological Survey Fact Sheet 087-02.

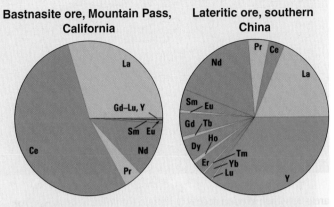

Figure 2

Different REE ores contain different proportions of individual REE. Yttrium (Y) is just above La in the periodic table and behaves chemically so much like it that Y is sometimes considered part of the REE. Bastnäsite is a rare mineral, an REE carbonate-fluoride.

From U.S. Geological Survey Fact Sheet 087-02.

did drop somewhat once exports resumed, they remain far above 2009 levels. (The applications differ for different REE, so their prices do also. At the end of 2011, prices ranged from about $62/kg for lanthanum to $3850/kg for europium oxide; europium is uniquely useful as the red phosphor in liquid-crystal displays, among other applications.)

All this has motivated the United States and other technological nations to rethink their dependence on REE imports. Because the REE substitute readily in the crystal structures of many minerals, they seldom become highly concentrated in ores. The United States is

actually fortunate to have about 12% of the world's known REE reserves (China controls 50%). In 2011, the first steps were taken toward reopening the Mountain Pass mine, and other possible locations for REE mining are being identified. The relative abundances of the various REE differ between deposits (figure 2), so one nation's REE reserves may not match its demands precisely, but in gross numbers, U.S. reserves amount to about 100 times 2011 world REE production, so we should be able to develop substantial domestic REE sources. Other nations, such as Japan, are not so favorably placed.

Occasionally, plants can be sampled instead of soil: Certain plants tend to concentrate particular metals, making the plants very sensitive indicators of locally high concentrations of those metals. Ground water may contain traces of metals leached from ore deposits along its flow path (figure 13.15). Even soil gases can supply clues to ore deposits. Mercury is a very volatile metal, and high concentrations of mercury vapor have been found in gases filling soil pore spaces over mercury ore bodies. Geochemical prospecting has also been used in petroleum exploration.

Remote sensing methods are becoming increasingly sophisticated and valuable in mineral exploration. These methods rely on detection, recording, and analysis of wave-transmitted energy, such as visible light and infrared radiation, rather than on direct physical contact and sampling. Aerial photography is one example, satellite imagery another. Remote sensing, especially using satellites, is a quick and efficient way to scan broad areas, to examine regions having such rugged topography or hostile climate that they cannot easily be explored on foot or with surface-based vehicles, and to view

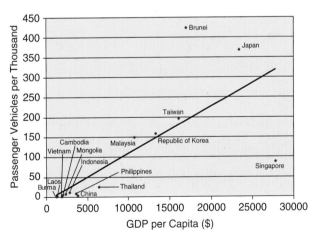

Figure 13.14

There is a generally positive correlation between GDP and car ownership in east and southeast Asia. Data for 2002.

From USGS Open-File Report 2004-1374.

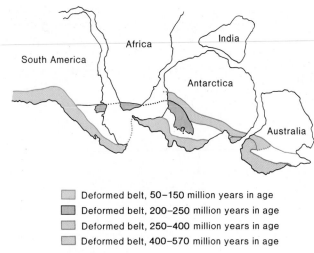

Deformed belt, 50–150 million years in age
Deformed belt, 200–250 million years in age
Deformed belt, 250–400 million years in age
Deformed belt, 400–570 million years in age

Figure 13.18

Pre-continental-drift reassembly of landmasses suggests locations of possible ore deposits in unexplored regions by extrapolation from known deposits on other continents.

Source: Data from C. Craddock et al., Geological Maps of Antarctica, Antarctic Map Folio Series, folio 12, copyright 1970 by the American Geographical Society

ore may occur in the matching mountain belt. Such reasoning is partly responsible for the supposition that economically valuable ore deposits may exist on Antarctica. (Some of the legal/political consequences of such speculations are explored in chapter 19; fundamental concerns about the wisdom of mineral exploration in the fragile Antarctic environment have motivated international agreements to restrict resource exploration there.)

Marine Mineral Resources

Increased consideration is also being given to seeking mineral wealth in unconventional places. In particular, the sea may provide partial solutions to some mineral shortages. Seawater itself contains dissolved in it virtually every chemical element. However, most of what it contains is dissolved halite, or sodium chloride. Most needed metals occur in far lower concentrations. Vast volumes of seawater would have to be processed to extract small quantities of such metals as copper and gold. The costs would be very high, especially since existing technology is not adequate to extract, selectively and efficiently, a few specific metals of interest. Several types of underwater mineral deposits have greater potential.

During the last Ice Age, when a great deal of water was locked in ice sheets, worldwide sea levels were lower than at present. Much of the area of the now-submerged continental shelves was dry land, and streams running off the continents flowed out across the exposed shelves to reach the sea. Where the streams' drainage basins included appropriate source rocks, these streams might have produced valuable placer deposits. With the melting of the ice and rising sea levels, any placers on the continental shelves were submerged. Finding and mining these deposits may be costly and difficult, but as land deposits are exhausted, placer deposits on the continental shelves may

well be worth seeking. Already, more than 10% of world tin production is from such marine placers; potentially significant amounts of titanium, zirconium, and chromium could also be mined from them if the economics were right.

The hydrothermal ore deposits forming along some seafloor spreading ridges are another possible source of needed metals. In many places, the quantity of material being deposited is too small, and the depth of water above would make recovery prohibitively expensive. However, the metal-rich muds of the Red Sea contain sufficient concentrations of such metals as copper, lead, and zinc that some exploratory dredging is underway, and several companies are interested in the possibility of mining those sediments. Along a section of the Juan de Fuca ridge, off the coast of Oregon and Washington, hundreds of thousands of tons of zinc- and silver-rich sulfides may already have been deposited, and the hydrothermal activity continues. A Canadian and a U.K.-based mining company have each secured exploration licenses for substantial areas of the territorial waters of New Zealand, Fiji, Tonga, and other parts of the southwest Pacific basin, and hope to be developing hydrothermal sulfides associated with plate boundaries in this region within a decade.

Perhaps the most widespread undersea mineral resource is **manganese nodules.** These are lumps up to about 10 centimeters in diameter, composed mostly of manganese minerals. They also contain lesser but potentially valuable amounts of copper, nickel, cobalt, platinum, and other metals. Indeed, the value of the minor metals may be a greater motive for mining manganese nodules than the manganese itself. The nodules are found over much of the deep-sea floor, in regions where sedimentation rates are slow enough not to bury them (figure 13.19A). At present, the costs of recovering these nodules would be high compared to the costs of mining the same metals on land, and the technical problems associated with recovering the nodules from beneath several kilometers of seawater remain to be worked out. Still, manganese nodules represent a sufficiently large metal resource that they have become a subject of International Law of the Sea conferences (see chapter 19). With any ocean resources outside territorial waters— manganese nodules, hydrothermal deposits at spreading ridges, fish, whales, or whatever—questions of ownership, or of who has the right to exploit these resources, inevitably lead to heated debate. Potential environmental impacts of recovering such marine resources have also not been fully analyzed.

Conservation of Mineral Resources

Overall need for resources is likely to increase, but for selected individual materials that are particularly scarce, perhaps demand can be moderated. One way is by making substitutions. One might, for certain applications, replace a very rare metal with a more abundant one. The extent to which this is likely to succeed is limited, since reserves of most metals are quite limited, as was clear from tables 13.2 and 13.3. Substituting one metal for another reduces demand for the second while increasing demand for the first; the focus of the shortage problem shifts, but the problem persists. Nonmetals are replacing metals in other applications. Unfortunately, many of these nonmetals

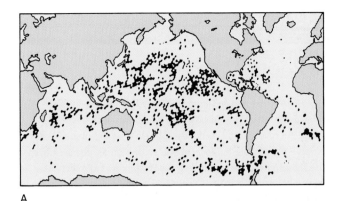

A

B

Figure 13.19

Manganese nodules. (A) Manganese-nodule distribution on the sea floor. (B) Manganese nodules on the floor of the northeast Pacific Ocean. A nodule grows by only about 1 centimeter in diameter over 1 million years, so they are exposed only where sedimentation rates are very low.

(A) Data from G. R. Heath, "Manganese Nodules: Unanswered Questions," Oceanus Vol. 25, No. 3, pp. 37–41. 1982. Copyright © Woods Hole Oceanographic Institute. (B) Photograph © Institute of Oceanographic Sciences/NERC/Photo Researchers

Table 13.4	Metal Recycling in the United States, 1986–2011 (recycled scrap as percentage of consumption)					
Metal	**1986**	**1992**	**1999**	**2005**	**2008**	**2011**
aluminum	15	30	20	16	30	36
chromium	21	26	20	29	32	40
cobalt	15	25	34	27	20	24
copper	22	42	33	30	31	35
lead	50	62	68	72	74	83
manganese	0	0	0	0	0	0
nickel	25	30	38	39	38	43

Source: Mineral Commodity Summaries 1990, 1993, 2000, 2006, 2009, and 2012, U.S. Geological Survey.

need to be extracted from mines, reducing—or at least moderating increases in—primary production. Some metals are already extensively recycled, at least in the United States (table 13.4 lists a few examples). Worldwide, recycling is less widely practiced, in part because the less-industrialized countries have accumulated fewer manufactured products from which materials can be retrieved. Still, given the limited U.S. reserves of many metals, recycling can be a very important means of extending those reserves. Additional benefits of recycling include a reduction in the volume of the waste-disposal problem and a decrease in the extent to which more land must be disturbed by new mining activities.

Unfortunately, not all materials lend themselves equally well to recycling. Among those that work out best are metals that are used in pure form in sizeable pieces—copper in pipes and wiring, lead in batteries, and aluminum in beverage cans. The individual metals are relatively easy to recover and require minimal effort to purify for reuse. Recycling can be more energy-efficient than producing new metal by mining and refining new ore, which results in the dual benefits of saving energy and cutting costs: The U.S. Environmental Protection Agency estimates that recycling just one ton of aluminum cans saves the energy equivalent of 1665 gallons of gasoline. (And of course, this means a reduction in greenhouse-gas emissions as well, yet another benefit.)

Where different materials are intermingled in complex manufactured objects, it is more difficult and costly to extract individual metals. Consider trying to separate the various metals from a refrigerator, a lawn mower, or a computer. The effort required to do that, even if it were technically possible, would make most of the materials recovered far too costly and thus noncompetitive with new production. Only in a few rare cases are metals valuable enough that the recycling effort may be worthwhile; for example, there is interest in recovering the platinum from catalytic converters in exhaust systems.

Alloys present special problems. The United States uses tens of millions of tons of steel each year, and some of it is indeed recycled scrap. However, steel is not a single chemical substance. It is iron alloyed with one or more other elements, and, often, the composition is very specific to the application. Chromium is added to make stainless steel; titanium, molybdenum, or tungsten steels are high-strength steels; other alloys are

are plastics or other materials derived from petroleum, which has simply contributed to petroleum consumption. As we will see in chapter 14, supplies of that resource are limited, too. For some applications, ceramics or fiber products can be substituted; however, sometimes the physical, electrical, or other properties required for the particular application demand the use of metals. There are alternatives to the REE for most applications, but they generally do not work as well.

The most effective way to extend mineral reserves, for some metals at least, may be through recycling. Overall use of the metals may increase, but if ways can be found to reuse a significant fraction of these metals repeatedly, proportionately less new metal will

Mining Your Cell Phone?

Consumer electronics make up a fast-growing segment of the U.S. waste stream. Unlike many consumer products that are discarded only when they cease to function, many types of electronics, such as cell phones and computers, may be replaced for other reasons—because they have become obsolete or because consumers want new features that are only available on new models, for instance. Many of the replaced items end up in landfills. This not only increases the challenge of finding adequate space for solid-waste disposal; it adds a variety of toxic metals, from antimony and arsenic to mercury and selenium, to those landfills. From there, the toxins may escape to contaminate air and water, as will be explored in later chapters.

However, viewed and handled differently, these products can provide us resources. The individual cell phone may not seem a rich treasure, but collectively, cell phones represent a significant source of recyclable metal (figure 1). There are over 325 million cell phones actively in use in the United States at present. The Environmental Protection Agency (EPA) estimates that each year, over 140 million cell phones are taken out of service, whether discarded, recycled, or set aside. Hundreds of millions more are believed already to be "in storage"—stuck into closets or desk drawers or otherwise tucked away, no longer being used but not actually having been discarded. Are all those phones worth anything? Consider just five of the metals they typically contain:

Metal	Ounces per Cell Phone*	Commodity Price**	Weight in 325 Million Cell Phones	Value in 325 Million Cell Phones
Copper	0.563	$ 4.05/lb	5700 tons	$ 46.2 million
Silver	.0113	$34.50/oz	3.6 million oz	124 million
Gold	.0011	$ 1600/oz	350,000 oz	560 million
Palladium	.000484	$ 730/oz	160,000 oz	117 million
Platinum	.0000110	$1720/oz	3500 oz	6.0 million

*From USGS Fact Sheet 2006-3097
**From Mineral Commodity Summaries 2012, U.S. Geological Survey
Except for copper, all weights and prices reported in ounces are troy ounces, as is customary for precious metals.

This table shows that just the cell phones in active use represent about $853 million worth of these metals alone; those "in storage" would constitute a substantial resource ready to be tapped. Granted, the figures do not take into account the costs of recycling, but in any event it is clear that there is value here, as well as potential for mineral-resource conservation (and reduced waste disposal and pollution into the bargain). Yet the EPA estimates that only 10% of the 140 million cell phones "retired" each year are currently recycled. The figures for personal computers and their components and peripherals are not much better: Though they contain not only recyclable metals but recyclable glass and plastic as well, only about 18% of discarded computer products are being recycled.

A major reason given for the low recycling rate on such products is that many people lack access to, or are simply unaware of, programs to recycle their electronic items. To help, the EPA now not only provides online information about what is sometimes called "eCycling," but provides links to other organizations that can put consumers in touch with local electronics-recycling programs. If you want to investigate the options for recycling your own used electronics, see the EPA's eCycling home page:

www.epa.gov/epawaste/conserve/materials/ecycling/index.htm

Figure 1

Manufactured goods such as cell phones may not suggest the concept of "resource" as obviously as a chunk of banded iron ore, but many can be recycled to yield useful materials.

used when other properties are important. If the composition of the alloy is critical to a particular application, clearly one cannot just toss any old steel scrap into the recycling pot. Each type of steel would have to be recycled individually. Facilities for separate collection and recycling of many different compositions of steel are rare. Progress is nevertheless being made: The U.S. automotive industry in 2005 recycled over 17 million vehicles by means of 200⁺ car shredders, and contributed over 13 million tons of shredded scrap steel to the steel industry, which, in turn, produces 100 million tons of new steel each year. As with aluminum, it takes much less energy to recycle old steel scrap than to manufacture new steel.

Some materials are not used in discrete objects at all. The potash and phosphorus in fertilizers are strewn across the land and cannot be recovered. Road salt washes off streets and into the soil and storm sewers. These things clearly cannot be recovered and reused.

For the foregoing and other reasons, it is unrealistic to expect that all minerals can ever be wholly recycled. However, more than half of the U.S. consumption of certain metals already is being recovered from old scrap. Where this is possible, the benefits are substantial.

Impacts of Mining-Related Activities

Mining and mineral-processing activities can modify the environment in a variety of ways, and few areas in the United States are untouched by mining activities (figure 13.20). Most obvious

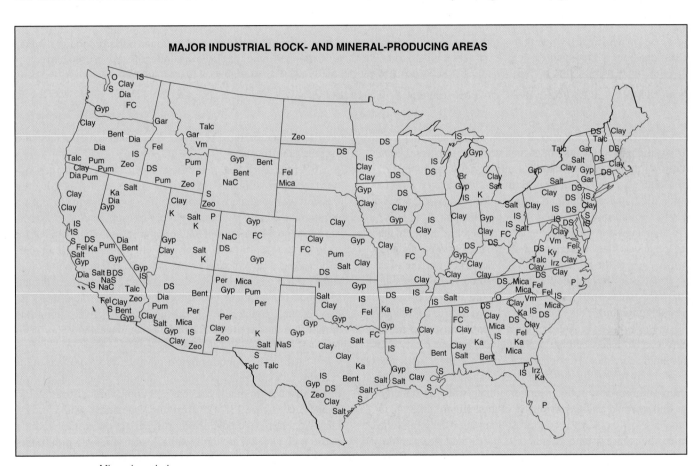

MAJOR INDUSTRIAL ROCK- AND MINERAL-PRODUCING AREAS

Mineral symbols

Bent	Bentonite	Fel	Feldspar	Ka	Kaolin
B	Borates	Gar	Garnet	Ky	Kyanite
Br	Bromine	Gyp	Gypsum	Mica	Mica
Clay	Common clay	Irz	Ilmenite, rutile,	O	Olivine
Dia	Diatomite		and zircon	Per	Perlite
DS	Dimension stone	IS	Industrial sand	P	Phosphate
FC	Fire clay	I	Iodine	K	Potash

Pum Pumice and pumicite
Salt Salt
NaC Soda ash
NaS Sodium sulfate
S Sulfur
Talc Talc
Vm Vermiculite
Zeo Zeolites

Figure 13.20

This map shows only sites of mining for industrial rock and mineral resources, but clearly shows that these activities are widespread in the United States. Still more places are affected by mining for metal ores.

Source: Mineral Commodity Summaries 2003, U.S. *Bureau of Mines*

A   B

Figure 13.21

Subsurface mining activities sometimes affect the earth's surface. (A) Subsidence pits and troughs over abandoned underground coal mines in Sheridan County, Wyoming. (B) Remains of Lake Emma, Sunnyside Mine, Colorado. Mining too shallowly under the lake caused it to collapse into the mine workings in 1978. Mining—for gold, silver, and various metal sulfides—continues at the site today.

Photograph (A) by C. R. Dunrud, (B) by P. Carrara; both courtesy USGS Photo Library.

is the presence of the mine itself. Both underground mines and surface mines have their own sets of associated impacts. Though safety is improving, the hazards of mining activities to the miners should also not be overlooked: mining is a dangerous occupation. Both underground and surface mines can also be sources of water pollution, as explored further in chapters 14 and 17.

Underground mines are generally much less apparent than surface mines. They disturb a relatively small area of the land's surface close to the principal shaft(s). Waste rock dug out of the mine may be piled close to the mine's entrance, but in most underground mines, the tunnels follow the ore body as closely as possible to minimize the amount of non-ore rock to be removed and, thus, mining costs. When mining activities are complete, the shafts can be sealed, and the area often returns very nearly to its pre-mining condition. However, near-surface underground mines occasionally have collapsed years after abandonment, when supporting timbers have rotted away or ground water has enlarged the underground cavities through solution (figure 13.21A). In some cases, the collapse occurred so long after the mining had ended that the existence of the old mines had been entirely forgotten. On the other hand, misjudgment occasionally results in collapse of an active mine (figure 13.21B).

Surface-mining activities consist of either open-pit mining (including quarrying) or strip-mining. Quarrying extracts rock to be used either intact (for building or facing stone) or crushed (as for cement-making, roadbed, etc.), and is also commonly used for unconsolidated sands and gravels; see figure 13.22. Open-pit mining is practical when a large, three-dimensional ore body is located near the surface (figure 13.23 and chapter-opening photograph). Most of the material in the pit contains the valuable commodity and is extracted for processing. Both procedures permanently change the topography, leaving a large hole. The exposed rock may begin to weather and, depending

on the nature of the ore body, may release pollutants into surface runoff water.

As noted earlier, many metal ores are hydrothermal deposits rich in sulfides. Pyrite (iron sulfide, FeS_2) is not only frequently associated with sulfides of rarer metals in ores; it is a common trace mineral in many types of rock (and often associated with coal beds as well). Where pyrite and other sulfides are exposed to reaction with water and with oxygen in the air, sulfuric acid is a typical weathering product, and the result is acid runoff water. Such acid mine drainage can be a problem, in fact, with both surface and underground mines. The acid drainage can continue for decades or centuries after active mining has stopped, as the exposed sulfides continue to weather. This problem will be explored further in chapter 17.

Strip-mining, more often used to extract coal than mineral resources, is practiced most commonly when the material of interest occurs in a layer near and approximately parallel to the surface (figure 13.24). Overlying vegetation, soil, and rock are stripped off, the coal or other material is removed, and the waste rock and soil are dumped back as a series of **spoil banks** (figure 13.25A). Until fairly recently, that was all that was done. The broken-up material of the spoil banks, with its high surface area, is very susceptible to both erosion and chemical weathering. Chemical and sediment pollution of runoff from spoil banks was common. Vegetation reappeared on the steep, unstable slopes only gradually, if at all (figure 13.25B). Now, much stricter laws govern strip-mining activities, and reclamation of new strip mines is required. Reclamation usually involves regrading the area to level the spoil banks and to provide a more gently sloping land surface (figure 13.25C); restoring the soil; replanting grass, shrubs, or other vegetation; and, where necessary, fertilizing and/or watering the area to help establish vegetation. The result, when successful, can be land on which

A

B

Figure 13.22

(A) Granite quarrying at the Rock of Ages quarry, Barre, Vermont.

(B) Quarrying sand and gravel, Wasatch Front, Utah.

Photograph (A) The McGraw-Hill Companies, Inc./Doug Sherman, photographer; (B) The McGraw-Hill Companies/John A. Karachewski, photographer

Figure 13.23

The world's largest open-pit mine: Bingham Canyon, Utah. Over $20 billion worth of minerals have been extracted from this one mine over more than a century of operation.

Courtesy of Kennecott

Figure 13.24

Strip-mining for manganese in South Africa.

Courtesy USGS Photo Library, Denver, CO.

evidence of mining activities has essentially been obliterated (figure 13.25D).

Naturally, reclamation is much more costly to the mining company than just leaving spoil banks behind. Costs can reach several thousand dollars per acre. Also, even when conscientious reclamation efforts are undertaken, the consequences may not be as anticipated. For example, the inevitable changes in topography after mining and reclamation may alter the drainage patterns of regional streams and change the proportion of runoff flowing to each. In dry parts of the western United States, where every drop of water in some streams is already spoken for, the consequences may be serious for landowners downstream. In such areas, the water needed to sustain the replanted vegetation may also be in very short supply. An additional problem is

A

B

C

D

Figure 13.25

Strip-mining and land reclamation. (A) Spoil banks, Rainbow coal strip mine, Sweetwater County, Wyoming. (B) Revegetation of ungraded spoils at Indian Head mine is slow. (C) Grading of spoils at Indian Head coal strip mine, Mercer County, North Dakota. (D) Reclaimed portion of Indian Head mine one year after seeding.

Photographs by H. E. Malde, USGS Photo Library, Denver, CO.

A

B

Figure 13.26

As the Bingham Canyon mine (figure 13.23) is approached, huge tailings piles stand out against the mountains (A). Closer inspection of the tailings (B) shows obvious erosion by surface runoff. Local residents report metallic-tasting water, attributed to copper leached from tailings.

presented by old abandoned, unreclaimed mines that were operated by now-defunct companies: Who is responsible for reclamation, and who pays? (Superfund, discussed in chapter 16, was created in part to address a similar problem with toxic-waste dumps.)

Mineral processing to extract a specific metal from an ore can also cause serious environmental problems. Processing generally involves finely crushing or grinding the ore. The fine waste materials, or **tailings,** that are left over may end up heaped around the processing plant to weather and wash away, much like spoil banks (figure 13.26). Generally, traces of the ore are also left behind in the tailings. Depending on the nature of the ore, rapid weathering of the tailings may produce acid drainage and leach out harmful elements, such as mercury, arsenic, cadmium, and uranium, to contaminate surface and ground waters. Uncontrolled usage of uranium-bearing tailings as fill and in concrete in Grand Junction, Colorado, resulted in radioactive buildings that had to be extensively torn up and rebuilt before they were safe for occupation.

The chemicals used in processing are often hazardous also. For example, cyanide is commonly used to extract gold from its ore. Water pollution from poorly controlled runoff of cyanide solutions can be a significant problem. Additional concerns arise as attention turns to trying to recover additional gold from old tailings by "heap-leaching," using cyanide solutions percolating through the tailings piles to dissolve out the gold. Gold is valuable, and heap-leaching old tailings is relatively inexpensive, especially in comparison to mining new ore. But unless the process is very carefully controlled, both metals and cyanide can contaminate local surface and ground waters. Smelting to extract metals from ores may, depending on the ores involved and on emission controls, release arsenic, lead, mercury, and other potentially toxic elements along with exhaust gases and ash. Sulfide-ore processing also releases the sulfur oxide gases that are implicated in the production of acid rain (chapter 18). These same gases, at high concentrations, can destroy nearby vegetation: During the twentieth century, one large smelter in British Columbia destroyed all conifers within *19 kilometers* (about 12 miles) of the operation. More-stringent pollution controls are curbing such problems today. Still, mineral processing can be a huge potential source of pollutants if not carefully executed. Worldwide, too, not all nations are equally sensitive to the hazards; residents of areas with weak pollution controls may face significant health risks.

Summary

Economically valuable mineral deposits occur in a variety of specialized geologic settings—igneous, sedimentary, and metamorphic. Both the occurrence of and the demand for minerals are very unevenly distributed worldwide. Projections for mineral use, even with conservative estimates of consumption levels, suggest that present reserves of most metals and other minerals could be exhausted within decades. As shortages arise, some currently subeconomic resources will be added to the reserves, but the quantities are unlikely to be sufficient to extend supplies greatly. Other strategies for averting further mineral shortages include applying new exploration methods to find more ores, looking to undersea mineral deposits not exploited in the past, and recycling metals to reduce the demand for newly mined material. Decreasing mining activity would minimize the negative impacts of mining, such as disturbance of the land surface and the release of harmful chemicals through accelerated weathering of pulverized rock or as a consequence of mineral-processing activities.

Key Terms and Concepts

banded iron formation 285	hydrothermal ores 284	pegmatite 282	speculative resources 281
concentration factor 281	hypothetical resources 281	placer 286	spoil banks 302
conditional resources 281	kimberlites 284	rare-earth elements 294	subeconomic resources 281
cumulative reserves 281	manganese nodules 298	remote sensing 295	tailings 305
evaporite 286	ore 281	reserves 281	

Exercises

Questions for Review

1. Explain how economics and concentration factor relate to the definition of an ore.

2. Describe two examples of magmatic ore deposits.

3. Hydrothermal ore deposits tend to be associated with plate boundaries. Why?

4. What is an evaporite? Give an example of a common evaporite mineral.

5. Explain how stream action may lead to the formation of placer deposits. Why is there interest in exploring for placers on the continental shelves?

6. What is the distinction between the quantities of *reserves* and of *resources*? What are *conditional resources*?

7. As mineral reserves are exhausted, some resources may be reclassified as reserves. Explain.

8. How might plate-tectonic theory contribute to the search for new ore deposits?

9. How are satellites contributing to ore prospecting? Note two advantages of such remote-sensing methods.

10. What metallic mineral resource is found over much of the deep-sea floor, and what political problem arises in connection with it?

11. Why are aluminum and lead comparatively easy to recycle, while steel is less so?

12. Describe one hazard associated with underground mining.

13. What water-pollution problem is particularly associated with mining of sulfide ore deposits?

14. What steps are involved in strip-mine reclamation? Can land always be fully restored to its pre-mining condition with sincere effort? Explain.

15. Why are tailings from mineral processing a potential environmental concern?

Exploring Further

1. Choose an area that has been subjected to extensive surface mining (examples include the coal country of eastern Montana or Appalachia and the iron ranges of Upper Michigan). Investigate the history of mining activities in this area and of legislation relating to mine reclamation. Assess the impact of the legislation.

2. Select one metallic mineral resource and investigate its occurrence, distribution, consumption, and reserves. Evaluate the impact of its customary mining and extraction methods. (How and where is it mined? How much energy is used in processing it? What special pollution problems, if any, are associated with the processing? Can this metal be recycled, and if so, from what sorts of material? Are there practical substitutes for it?) You might start at the USGS website **minerals.usgs.gov/**.

3. Choose an everyday object, and investigate what mineral materials go into it and where they come from. (The Mineral Information Institute's website might be a place to start.)

4. Select one or more recyclable metals found in a computer, and perform an analysis like that of Case Study 13.2. (The EPA's eCycling site plus USGS publications from the online Suggested Readings list should provide the necessary data.)

Energy Resources— Fossil Fuels

The earliest and, at first, only energy source used by primitive people was the food they ate. Then, wood-fueled fires produced energy for cooking, heat, light, and protection from predators. As hunting societies became agricultural, people used energy provided by animals—horsepower and ox power. The world's total energy demands were quite small and could readily be met by these sources.

Gradually, animals' labor was replaced by that of machines, new and more sophisticated technologies were developed, and demand for manufactured goods rose. These factors all greatly increased the need for energy and spurred the search for new energy sources (figure 14.1). It was first wood, then primarily the fossil fuels that eventually met those needs, and they continue to dominate our energy consumption (figure 14.2). They also pose certain environmental problems. All contribute to the carbon-dioxide pollution in the atmosphere, and several can present other significant hazards such as toxic spills and land subsidence (oil), or sulfur pollution, acid runoff, and mine collapse (coal).

The term *fossil* refers to any remains or evidence of ancient life. Energy is stored in the chemical bonds of the organic compounds of living organisms. The ultimate source of that energy is the sun, which drives photosynthesis in plants. The **fossil fuels** are those energy sources that formed from the remains of once-living organisms. These include oil, natural gas, coal, and fuels derived from oil shale and tar sand. When we burn them, we are using that stored energy. The differences in the physical properties among the various fossil fuels arise from differences in the starting materials from which the fuels formed and in what happened to those materials after the organisms died and were buried within the earth. What the fossil fuels all have in common is that they are **nonrenewable** energy sources, meaning that the processes by which they form are so slow that they are not replaceable on a human timescale. Whatever we consume is, from our perspective, gone for good.

Oil slick from the *Deepwater Horizon* drilling rig, almost a month after the accident. The Mississippi Delta is west of the slick. The rig was located near the center of the brightest-gray part of the slick. Currents in the Gulf of Mexico have swept part of the slick south/southeast.

NASA image by Jeff Schmaltz, MODIS Rapid Response Team.

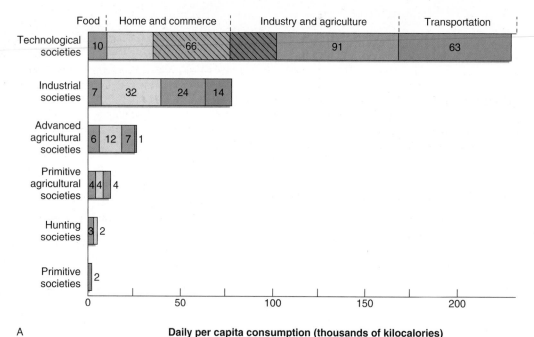

Food | Home and commerce | Industry and agriculture | Transportation

Technological societies: 10 | 66 | 91 | 63

Industrial societies: 7 | 32 | 24 | 14

Advanced agricultural societies: 6 | 12 | 7 | 1

Primitive agricultural societies: 4 | 4 | 4

Hunting societies: 3 | 2

Primitive societies: 2

A

Daily per capita consumption (thousands of kilocalories)

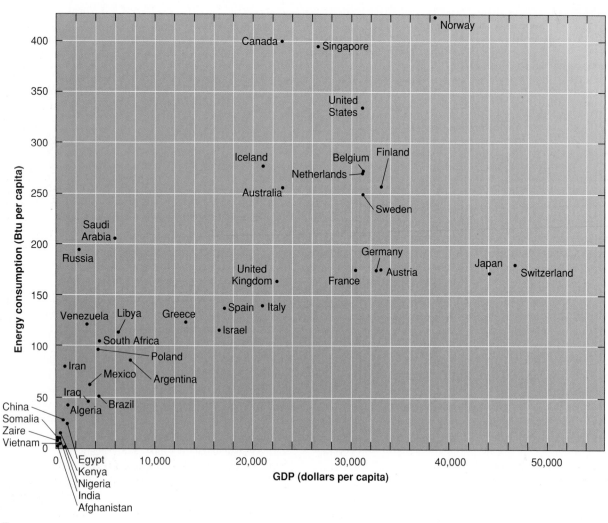

B

Figure 14.1

(A) As societies evolve from primitive to agricultural to industrialized, the daily per capita consumption of energy increases sharply. The body needs, on average, about 2000 kilocalories per day to live (the "kilocalorie," a measure of energy, is the food calorie that dieters count). However, in the United States in 2007, per capita energy consumption was 230,000 kilocalories per day: 40% for residential/commercial uses, 32% for industry, 29% for transportation. Of the total, about 40% was consumed in the generation of electricity. Note that, as societies advance, even the energy associated with the food individuals eat increases. Members of modern societies do not consume 10,000 kilocalories of food energy directly, but behind the 2,000 or so actually consumed are large additional energy expenditures associated with agriculture, food processing, and preservation, and even the conversion of plants to meat by animals, which in turn are consumed as food. The shaded area in the top bar of the graph is electricity usage. (B) There is a variable but generally positive correlation between GDP and energy consumption: The more energy consumed, the higher the value of goods and services produced, and generally, the higher the standard of living and the level of technological development as well.

(A) Source: Graph after Earl Cook, "The Flow of Energy in an Industrialized Society," Scientific American, © *1971.*

(B) Data from International Energy Annual 2001, *U.S. Energy Information Administration, and 2002* World Population Data Sheet, *Population Reference Bureau.*

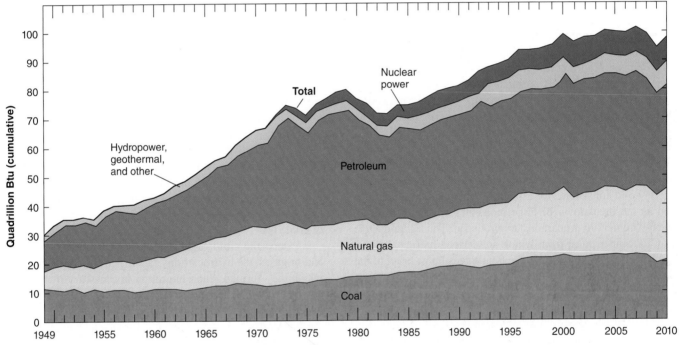

Figure 14.2

U.S. energy consumption, 1949–2010. The trend is clearly upward, except following the Arab oil embargo of the early 1970s, and in the early 1980s, when high prices and a sluggish economy combined to depress energy consumption; leveling-off since 2005 is also attributed to economic factors. Note that figures are expressed in terms of energy derived from each source so that the contributions can be more readily compared. For energy equivalents of the several fossil fuels, see the table of units.

Source: Annual Energy Review 2010, *U.S. Energy Information Administration, Department of Energy.*

Formation of Oil and Natural Gas Deposits

Petroleum is not a single chemical compound. Liquid petroleum, or **oil,** comprises a variety of liquid hydrocarbon compounds (compounds made up of different proportions of the elements carbon and hydrogen). There are also gaseous hydrocarbons (**natural gas**), of which the compound methane (CH_4) is the most common. How organic matter is transformed into liquid and gaseous hydrocarbons is not fully understood, but the main features of the process are outlined below.

The production of a large deposit of any fossil fuel requires a large initial accumulation of organic matter, which is rich in carbon and hydrogen. Another requirement is that the organic debris be buried quickly to protect it from the air so that decay by biological means or reaction with oxygen will not destroy it.

Microscopic life is abundant over much of the oceans. When these organisms die, their remains can settle to the sea floor. There are also underwater areas near shore, such as on many continental shelves, where sediments derived from continental erosion accumulate rapidly. In such a setting, the starting requirements for the formation of oil are satisfied: There is an abundance of organic matter rapidly buried by sediment. Oil and much natural gas are believed to form from such accumulated marine microorganisms. Continental oil fields often reflect the presence of marine sedimentary rocks below the surface. Additional natural gas deposits not associated with oil may form from deposits of plant material in sediment on land.

As burial continues, the organic matter begins to change. Pressures increase with the weight of the overlying sediment or rock; temperatures increase with depth in the earth; and slowly, over long periods of time, chemical reactions take place. These reactions break down the large, complex organic molecules into simpler, smaller hydrocarbon molecules. The nature of the hydrocarbons changes with time and continued heat and pressure. In the early stages of petroleum formation in a marine deposit, the deposit may consist mainly of larger hydrocarbon molecules ("heavy" hydrocarbons), which have the thick, nearly solid consistency of asphalt. As the petroleum matures, and as the breakdown of large molecules continues, successively "lighter" hydrocarbons are produced. Thick liquids give way to thinner ones, from which are derived lubricating oils, heating oils, and gasoline. In the final stages, most or all of the petroleum is further broken down into very simple, light, gaseous molecules—natural gas. Most of the maturation process occurs in the temperature range of 50° to 100°C (approximately 120° to 210°F). Above these temperatures, the remaining hydrocarbon is almost wholly methane; with further temperature

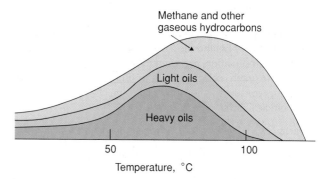

Figure 14.3

The process of petroleum maturation in a marine deposit, in simplified form.

Source: After J. Watson, Geology and Man, 1983, *Allen and Unwin, Inc.*

increases, methane can be broken down and destroyed in turn (see figure 14.3).

A given oil field yields crude oil containing a distinctive mix of hydrocarbon compounds, depending on the history of the material. The refining process separates the different types of hydrocarbons for different uses (see table 14.1). Some of the heavier hydrocarbons may also be broken up during refining into smaller, lighter molecules through a process called *cracking,* which allows some of the lighter compounds such as gasoline to be produced as needed from heavier components of crude oil.

Once the solid organic matter is converted to liquids and/or gases, the hydrocarbons can migrate out of the rocks in which they formed. Such migration is necessary if the oil or gas is to be collected into an economically valuable and practically usable deposit. The majority of petroleum source rocks are fine-grained clastic sedimentary rocks of low permeability, from which it would be difficult to extract large quantities

Table 14.1	Fuels Derived from Liquid Petroleum and Gas	
	Material	**Principal Uses**
Heavier hydrocarbons	waxes (for example, paraffin)	candles
	heavy (residual) oils	heavy fuel oils for ships, power plants, and industrial boilers
	medium oils	kerosene, diesel fuels, aviation (jet) fuels, power plants, and domestic and industrial boilers
	light oils	gasoline, benzene, and aviation fuels for propeller-driven aircraft
	"bottled gas" (mainly butane, C_4H_{10})	primarily domestic use
Lighter hydrocarbons	natural gas (mostly methane, CH_4)	domestic/industrial use and power plants

Source: Data from J. Watson, Geology and Man, *copyright © 1983 Allen and Unwin, Inc.*

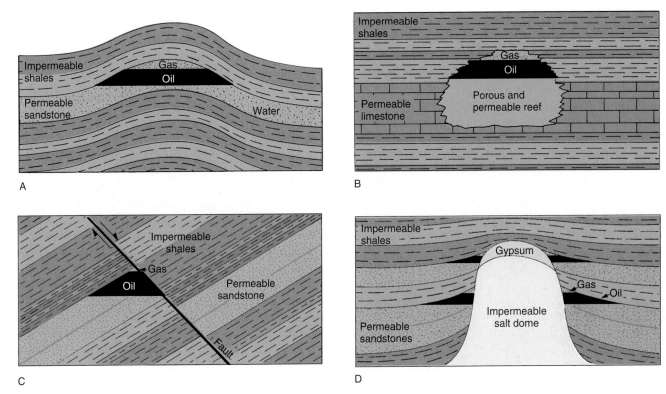

Figure 14.4

Types of petroleum traps. (A) A simple fold trap. (B) Petroleum accumulated in a fossilized ancient coral reef. (C) A fault trap. (D) Petroleum trapped against an impermeable salt dome, which has risen up from a buried evaporite deposit.

of oil or gas quickly. Despite the low permeabilities, oil and gas are able to migrate out of their source rocks and through more permeable rocks over the long spans of geologic time. The pores, holes, and cracks in rocks in which fluids can be trapped are commonly full of water. Most oils and all natural gases are less dense than water, so they tend to rise as well as to migrate laterally through the water-filled pores of permeable rocks.

Unless stopped by impermeable rocks, oil and gas may keep rising right up to the earth's surface. At many known oil and gas seeps, these substances escape into the air or the oceans or flow out onto the ground. These natural seeps, which are one of nature's own pollution sources, are not very efficient sources of hydrocarbons for fuel if compared with present drilling and extraction processes, although asphalt from seeps in the Middle East was used in construction five thousand years ago.

Commercially, the most valuable deposits are those in which a large quantity of oil and/or gas has been concentrated and confined (trapped) by impermeable rocks in geologic structures (see figure 14.4). The reservoir rocks in which the oil or gas has accumulated should be relatively porous if a large quantity of petroleum is to be found in a small volume of rock and should also be relatively permeable so that the oil or gas flows out readily once a well is drilled into the reservoir. If the reservoir rocks are not naturally very permeable, it may be possible to fracture them artificially with explosives or with water or gas

under high pressure to increase the rate at which oil or gas flows through them.

The amount of time required for oil and gas to form is not known precisely. Since virtually no petroleum is found in rocks younger than 1 to 2 million years old, geologists infer that the process is comparatively slow. Even if it took only a few tens of thousands of years (a geologically short period), the world's oil and gas is being used up far faster than significant new supplies could be produced. Therefore, as noted earlier, oil and natural gas are among the nonrenewable energy sources. We have an essentially finite supply with which to work.

Supply and Demand for Oil and Natural Gas

As with minerals, the most conservative estimate of the supply of an energy source is the amount of known reserves, "proven" accumulations that can be produced economically with existing technology. A more optimistic estimate is total resources, which include reserves, plus known accumulations that are technologically impractical or too expensive to tap at present, plus some quantity of the substance that is expected to be found and extractable. As with minerals, estimates of energy reserves are thus sensitive both to price fluctuations and to technological advances.

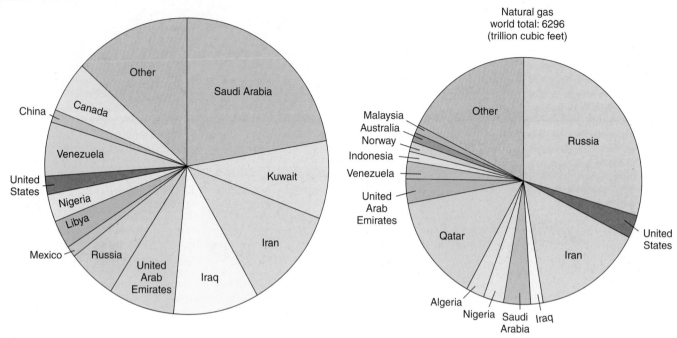

Figure 14.5

Estimated proven world reserves of crude oil and natural gas. Note that inclusion of the fuel in Canada's oil sands (discussed later in the chapter) would increase world oil reserves by more than 150 billion barrels, and make Canada's oil reserves second only to those of Saudi Arabia.

Source: Country proportions are averages of estimates of BP Statistical Review, Oil and Gas Journal and World Oil, as summarized in International Energy Annual 2006, U.S. Energy Information Administration. World totals are updated per data released by Energy Information Administration, 2009.

Oil

Oil is commonly discussed in units of barrels (1 barrel = 42 gallons). Worldwide, over one trillion barrels of oil have been consumed, according to estimates by the U.S. Geological Survey. The estimated remaining reserves are about 1.26 trillion barrels (figure 14.5). That does not sound too ominous until one realizes that close to half of the consumption has occurred in the last quarter-century or so. Also, global demand continues to increase as more countries advance technologically.

Oil supply and demand are very unevenly distributed around the world. Some low-population, low-technology, oil-rich countries (Libya, for example) may be producing fifty or a hundred times as much oil as they themselves consume. At the other extreme are countries like Japan, highly industrialized and energy-hungry, with no petroleum reserves at all. The United States alone consumes over 20% of the oil used worldwide, more than all of Europe together, and about twice as much as China, the next-highest individual-country consumer. As with mineral resources, the size of petroleum resources of a given nation are not correlated with geographic size. Many have speculated that much of the international interest in the Middle East is motivated as much by concern for its estimated 60% share of world oil reserves as by humanitarian or other motives.

The United States initially had perhaps 10% of all the world's oil resources. Already over 400 billion barrels have been

produced and consumed; remaining U.S. proven *reserves* are only about 22 billion barrels. For more than three decades, the United States has been using up about as much domestic oil each year as the amount of new domestic reserves discovered, or proven, and in many years somewhat more, so the net U.S. oil reserves have generally been decreasing year by year, with a small recent increase as oil prices have risen sharply (see table 14.2). Domestic oil production has likewise been declining (figures 14.6 and 14.7).

Table 14.2	Proven U.S. Reserves Crude Oil and Natural Gas, 1977–2009	
Year	**Crude Oil (billions of barrels)**	**Natural Gas (trillions of cu ft)**
1977	31.8	207.4
1982	29.5	201.5
1987	28.7	187.2
1992	25.0	165.0
1997	23.9	167.2
2002	24.0	186.9
2007	21.0	211.1
2008	20.6	255.0
2009	22.3	283.9

Source: Data from Annual Energy Review 2007, U.S. Energy Information Administration, Department of Energy; 2008 and 2009 data from Energy Information Administration.

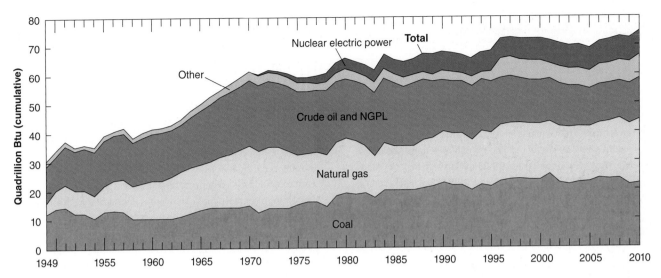

Figure 14.6

U.S. energy production since the mid-twentieth century, by energy source. Note that domestic production remains rather level despite sharp rises in some fossil-fuel prices, and rising consumption. (Natural gas plant liquids (NGPL) are liquid fuels produced during refining/processing of natural gas.)

Source: Annual Energy Review 2010, *U.S. Energy Information Administration.*

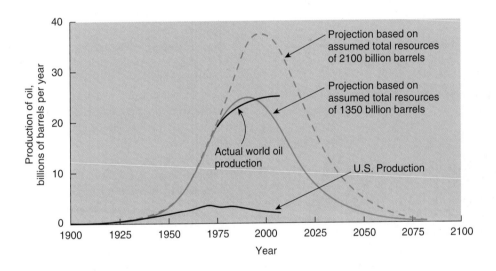

Figure 14.7

When this classic projection by M. King Hubbert was first widely published in the early 1970s, many people were shocked to discover that the total supply of oil appeared so limited. In fact, the total resource estimates had long existed in the geologic literature. The supply problem was made acute by the rapid rise in consumption starting in the 1950s. Hubbert accurately predicted the slower rise in world production of oil after 1975, which to a great extent is the result of depletion of reserves, especially easily accessible ones. U.S. production peaked in 1970 for that reason. The "blip" on the downslope represents the effect of added production from Alaska. Recent projections, using revised USGS estimates of total recoverable world oil resources of about 3000 billion barrels, place the peak of world oil production within about five years of 2020, depending upon model assumptions.

Source: Projections from M. King Hubbert, "The Energy Resources of the Earth," Scientific American, Sept. 1971, p. 69. Production data from Annual Energy Review 2008, U.S. Energy Information Administration.

So, the United States has for many years relied heavily on imported oil to meet part of its oil demand. Without it, U.S. reserves would dwindle more quickly. The United States consumes about 7 billion barrels of oil per year, to supply about 37% of all the energy used. Currently, about half of the oil consumed is imported from other countries. Principal sources currently are Canada, Mexico, Saudi Arabia, Venezuela, and Nigeria, with other nations contributing as well (for example, in 2010, the United States was importing 400,000 barrels a day from Iraq). Simple arithmetic demonstrates that the rate of oil

The Arctic National Wildlife Refuge: To Drill or Not to Drill?

The Prudhoe Bay field on the North Slope of Alaska is the largest U.S. oil field ever found. It has yielded over 11 billion barrels of oil so far, with an estimated remaining reserve of 2 billion barrels. Its richness justified the $8 billion cost of building the 800-mile Trans-Alaska pipeline. West of Prudhoe Bay lies the 23-million-acre tract now known as the National Petroleum Reserve–Alaska (NPRA), first established as Naval Petroleum Reserve No. 4 by President Warren Harding in 1923. East of Prudhoe Bay is the 19-million-acre Arctic National Wildlife Refuge (ANWR), established in 1980 (figure 1).

Geologic characteristics of, and similarities among, the three areas suggested that oil and gas might very well be found in both NPRA and ANWR. Thus when establishing ANWR, Congress deferred a final decision on the future of the coastal portion, some 1.5 million acres known as the "1002 Area" after the section of the act that addressed it.

Little petroleum exploration occurred in NPRA until the late 1990s. In 1996, oil was found in the area between NPRA and Prudhoe Bay. Then, beginning in 2000, exploration wells began to find oil and gas in NPRA.

Estimating potential oil and gas reserves in a new region is a challenge. Drilling and seismic and other data constrain the types and distribution of subsurface rocks. Samples from drill cores, and comparisons with geologically similar areas, allow projection of quantities of oil and gas that may be present. Only a portion of the in-place resources may be technologically recoverable—perhaps 30–50% of the oil, 60–70% of the gas. And of course, the market price of each commodity will determine how much is economically recoverable. Combining all of these considerations resulted in a range of estimates of oil and gas reserves in NPRA and the ANWR 1002 Area (figure 2). At crude-oil prices of $40 per barrel, the analysis suggested a mean probability of close to 7 billion barrels of economically recoverable oil each in NPRA and in the ANWR 1002 Area. Moreover, it is believed that the oil accumulations in the ANWR 1002 Area are, on average, larger, meaning that they could be developed more efficiently. (Development of the gas reserves would require another pipeline.)

A few years after this USGS analysis was done, oil prices began to rise sharply. One would naturally expect that that would increase the projected volume of reserves (though note that the difference between estimated reserves at $30 per barrel and at $40 per barrel in the original analysis was rather modest). In mid-2008, in fact, oil prices briefly exceeded $130 per barrel—and then dropped precipitously, falling below $40 per barrel by early 2009. As this is written, oil is back around $100/barrel, but there is no assurance of how long it will stay there. Such volatility dictates caution in making reserve projections based on assumptions of very high oil prices, and it certainly prompts oil companies to be cautious in moving forward to develop the more marginal deposits.

For one thing, drilling costs continue to rise, particularly in challenging areas like Alaska. Permafrost is a delicate foundation for construction, as will be explored further in chapter 20, and has become more so with Arctic warming. For this and other reasons, costs of drilling onshore in Alaska rose from $283 per foot drilled in 2000 to $1880 per foot in 2005, and offshore costs are higher.

Furthermore, development of new oil fields is never quick, and this is likely to be especially true in ANWR. A 2008 analysis by the U.S. Energy Information Administration (EIA) conservatively suggested a lag of *ten years* from the time oil leases would be authorized by the federal government to the first oil production. This breaks down as follows:

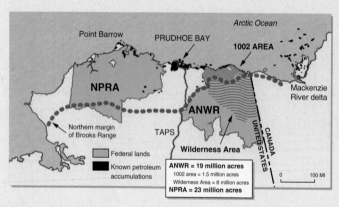

Figure 1

The NPRA and ANWR flank the petroleum-rich Prudhoe Bay area. The future of the "1002 Area" of the ANWR is currently being hotly debated.

After USGS Fact Sheet 040-98.

consumption in the United States is very high compared even to the estimated total U.S. resources available. Some of those resources have yet to be found, too, and will require time to discover and develop. Both U.S. and world oil supplies could be nearly exhausted within decades (figure 14.7), especially considering the probable acceleration of world energy demands. On occasion, explorationists do find the rare, very large concentrations of petroleum—the deposits on Alaska's North Slope and beneath Europe's North Sea are examples. Yet even these make only a modest difference in the long-term picture. The Prudhoe Bay oil field, for instance, represented reserves of only about 13 billion barrels, a great deal for a single region, but less than two

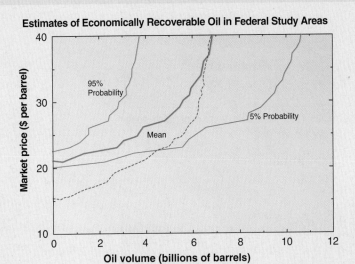

Estimates of Economically Recoverable Oil in Federal Study Areas

95% Probability

Mean

5% Probability

Market price ($ per barrel)

Oil volume (billions of barrels)

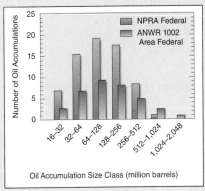

Number of Oil Accumulations

NPRA Federal

ANWR 1002 Area Federal

Oil Accumulation Size Class (million barrels)

Market Price ($/barrel)	Economically Recoverable Oil (billion barrels)	
	NPRA	ANWR
15	0	0
–	–	–
20	0	3.2
21	0.4	4.0
22	1.3	4.4
23	2.8	5.0
24	3.1	5.2
25	3.7	5.6
26	4.0	5.8
27	4.8	6.0
28	5.1	6.2
29	5.4	6.3
30	5.6	6.3
–	–	–
35	6.4	6.6
40	6.9	6.8

Table of mean values

Figure 2

U.S. Geological Survey estimates of economically recoverable oil in the federally owned portions of NPRA and the ANWR 1002 Area (other areas are Native lands), and size distribution of individual oil accumulations (inset graph). Red lines are for NPRA; dashed blue line is mean for ANWR 1002 Area.

After USGS Fact Sheet 045-02.

2–3 years for writing and acceptance of an Environmental Impact Statement, preliminary analysis of exploration data by interested companies, and a lease auction

2–3 years for the drilling of *one* exploratory well (a slow process partly because of the location, and partly because exploratory wells involve extensive data collection, not just drilling)

1–2 years for the company to formulate a plan for developing its oil field, and for the Bureau of Land Management (the responsible federal agency here) to approve that plan

3–4 years to build the necessary oil-processing facilities and feeder pipelines to connect the new oil to the Trans-Alaska Pipeline, and to drill development wells to extract the oil

Thus, a company would need confidence in the price of oil ten or more years in the future before being certain of the economic viability of a particular deposit. Indeed, the EIA concluded that even prices prevailing at the time of its analysis (over $120 per barrel) would not motivate development of smaller or more technically difficult oil fields in ANWR until after 2030; the richer and more accessible deposits already identified as reserves by USGS would certainly be developed first. As oil prices have dropped since the EIA assessment, speculation about reserves beyond those identified in the USGS study is more complicated. A further complicating factor is the rapid rise in natural-gas reserves in the last few years due to new discoveries and the development of "shale gas," discussed later in the chapter, which has driven down natural-gas prices and made gas a more attractive fuel than oil for many applications.

Those who favor drilling in the ANWR 1002 Area point to the current massive U.S. dependence on foreign oil, the financial benefits to Alaskans of additional oil production, the fact that there are tens of millions of acres of other wilderness. They also note that oil production technology has become more sophisticated so as to leave less of a "footprint"—for example, one can drill one well at the surface that splays out in multiple directions at depth rather than having to drill multiple wells from the surface to tap the same deposits. Those opposed to drilling note the extreme environmental fragility of alpine ecosystems in general, the biological richness of the 1002 Area in particular (the ANWR has been called the Alaskan Serengeti), and the fact that the United States consumes over 7 billion barrels of crude oil in just a year (so, why risk the environment for so relatively little oil?). The debate continues.

years' worth of U.S. consumption. U.S. oil production peaked at 9.64 million barrels per day in 1970, and has declined since then; it was 5.5 million barrels per day in 2010. Concern about dependence on imported oil led to the establishment of the Strategic Petroleum Reserve in 1977, but the amount of oil in the SPR stayed at about 550 million barrels from 1980 to 2000,

while the number of days' worth of imports that it represents declined from a high of 115 in 1985 to 50 in 2001 (figure 14.8). Only renewed addition of imported oil reversed the decline in the number of days of imports held in the SPR, but that figure was still only 73 in 2010. Such facts increase pressure to develop new domestic oil sources; see Case Study 14.1.

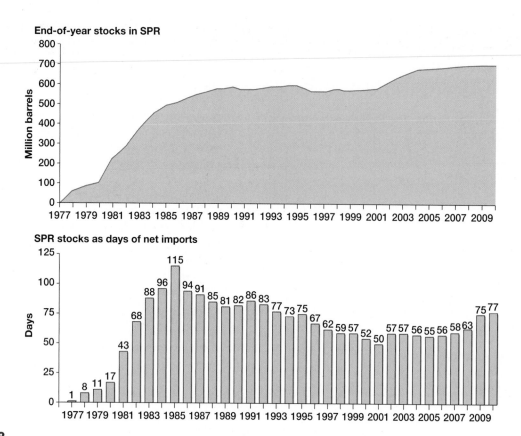

Figure 14.8

Strategic Petroleum Reserve (SPR) was established to provide a "cushion" in the event of disruption of imports of oil. The length of time the SPR will last declines as consumption of imported oil rises. Only significant infusions of oil after 9/11, plus some decline in consumption due to economic conditions, have reversed the decline in number of days of imports in SPR.

Source: Annual Energy Review 2010, *U.S. Energy Information Administration, Department of Energy.*

Natural Gas

The supply/demand picture for conventional natural gas is similar to that for oil (see figures 14.2 and 14.6). Natural gas presently supplies about 25% of the energy used in the United States. The United States has proven natural gas reserves of nearly 300 trillion cubic feet. However, roughly 20 trillion cubic feet of these reserves are consumed per year, and most years, less has been found in new domestic reserves than the quantity consumed. The net result has again been generally declining reserves—at least until recently (table 14.2). Prior to the mid-1980s, U.S. natural-gas production roughly equaled consumption. Since then, we have been importing significant amounts of natural gas; imports now account for about 10% of consumption. That may be changing, however. While the U.S. supply of conventional natural gas, like oil, could be used up in a matter of decades, technology has given a boost to our gas reserves, by allowing development of novel natural-gas sources that previously could not be tapped economically.

One is **coal-bed methane.** During coal formation, quantities of methane-rich gas are also produced. Historically, this methane has been treated as a hazardous nuisance, as it is toxic and potentially explosive, and its incidental release during coal

mining has contributed to rising methane concentrations in the atmosphere. However, technology has improved to the point that much of it is now economically recoverable. Where it can be extracted for use rather than wasted, it contributes significantly to U.S. gas reserves: by 2010, an estimated 20 trillion cubic feet of natural gas reserves occurring as coal-bed methane had been identified. And we know where to find it because U.S. coal beds are already well mapped, as noted in the later discussion of coal. The major technological obstacle to recovery of coal-bed methane is the need to extract and dispose of the water that typically pervades coal beds, water that is often saline and requires careful disposal to avoid polluting local surface or ground water. However, coal-bed methane is already being produced in the Wasatch Plateau area of Utah and the Powder River Basin in Wyoming. As extraction technology is improved, increased development of this resource can be expected.

An even larger resource is **shale gas.** The fine sediments that are later lithified to become shale are often rich in organic matter. Over time, heat and pressure may break this down into natural gas. Most shales are low in permeability, so extracting this gas economically was not possible. However, the technology to increase permeability by hydraulic fracturing, or **fracking**

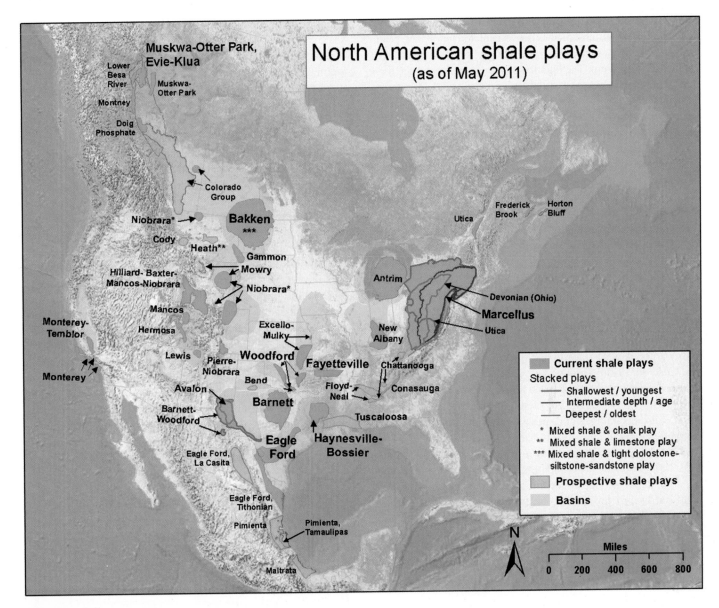

Figure 14.9

Shale-gas "plays" (sites of gas-bearing shales) in North America. Names on map are geological formation names that identify the rock units in which the shale gas is found at each location.

Map by U.S. Energy Information Administration.

as it is commonly called, and to drill horizontally along shale beds to facilitate fracking and gas extraction, has improved to the point that the gas in many shales can indeed be counted in our reserves and added to our production. By 2010, shale-gas reserves were estimated at over 60 trillion cubic feet, and the number continues to rise. See figure 14.9 for locations of gas-bearing shales that may be future sources of natural gas. (It should be noted that the rapid expansion of fracking activity has raised concerns about possible induced seismicity and groundwater pollution. As of this writing, such problems have not yet been documented, though there is evidence of minor induced seismicity related to disposal of waste fluids, as noted in chapter 16.)

Coal-bed methane and shale gas both illustrate how technological advances can shift resources into the "reserve" category. Prior to 2007, production of both was negligible. In 2010, they accounted for 7% and 16% of U.S. natural-gas production respectively, and the gap between U.S. production and consumption of natural gas is narrowing.

Future Prospects for Oil and Gas

Many people tend to assume that, as oil and natural gas supplies dwindle and prices rise, there is increased exploration and discovery of new reserves so that we will not really run out of these fuels. But the supplies are ultimately finite, after all.

Moreover, while decreasing reserves have prompted exploration of more areas, most regions not yet explored for oil have been neglected precisely because they are unlikely to yield appreciable amounts of petroleum. The high temperatures involved in the formation of igneous and most metamorphic rocks would destroy organic matter, so oil would not have formed or been preserved in these rocks. Nor do these rock types tend to be very porous or permeable, so they generally make poor reservoir rocks as well, unless fractured. The large portions of the continents underlain predominantly by igneous and metamorphic rocks are, for the most part, very unpromising places to look for oil. Some such areas are now being explored, but with limited success. Other, more promising areas may be protected or environmentally sensitive.

The costs of exploration have gone up, and not simply due to inflation. In the United States, drilling for oil or gas in 2006 cost an average of $324 *per foot,* and as noted in Case Study 14.1, drilling costs in some areas of future resource potential, notably Alaska, are far higher than the average. The average oil or natural gas well drilled was over 6400 feet deep (it was about 3600 feet in 1949); even with increasingly sophisticated methods of exploration to target areas most likely to yield economic hydrocarbon deposits, nearly half of exploratory wells come up dry. Costs for drilling offshore are substantially higher than for land-based drilling, too, and many remaining sites where hydrocarbons might be sought are offshore. Yield from producing wells is also declining, from a peak average of 18.6 barrels of oil per well per day in 1972 to 10.4 barrels in 2010.

Despite a quadrupling in oil prices between 1970 and 1980 (*after* adjustment for inflation), U.S. proven reserves continued to decline. They are declining still, despite oil prices recently at record levels. This is further evidence that higher prices do not automatically lead to proportionate, or even appreciable, increases in fuel supplies. Also, many major oil companies have been branching out into other energy sources beyond oil and natural gas, suggesting an expectation of a shift away from petroleum in the future.

Enhanced Oil Recovery

A few techniques are being developed to increase petroleum production from known deposits. An oil well initially yields its oil with minimal pumping, or even "gushes" on its own, because the oil and any associated gas are under pressure from overlying rocks, or perhaps because the oil is confined like water in an artesian system. Recovery using no techniques beyond pumping is *primary recovery.* When flow falls off, water may be pumped into the reservoir, filling empty pores and buoying up more oil to the well *(secondary recovery).* Primary and secondary recovery together extract an average of one-third of the oil in a given trap, though the figure varies greatly with the physical properties of the oil and host rocks in a given oil field. On average, then, two-thirds of the oil in each deposit has historically been left in the ground. Thus, additional *enhanced recovery* methods have attracted much interest.

Enhanced recovery comprises a variety of methods beyond conventional secondary recovery. Permeability of rocks can be increased by deliberate fracturing, using explosives or even water

under very high pressure (as with the fracking used to extract shale gas). Carbon dioxide gas under pressure can be used to force out more oil. Hot water or steam may be pumped underground to warm thick, viscous oils so that they flow more easily and can be extracted more completely. There have also been experiments using detergents or other substances to break up the oil.

All of these methods add to the cost of oil extraction. That they are now being used at all is largely a consequence of higher oil prices in the last two decades and increased difficulty of locating new deposits. Researchers in the petroleum industry believe that, from a technological standpoint, up to an additional 40% of the oil initially in a reservoir might be extractable by enhanced-recovery methods. This would substantially increase oil reserves. A further positive feature of enhanced-recovery methods is that they can be applied to old oil fields that have already been discovered, developed by conventional methods, and abandoned, as well as to new discoveries. In other words, numerous areas in which these techniques could be used (if the economics were right) are already known, not waiting to be found. It should, however, be kept in mind that enhanced recovery may involve increases in problems such as ground subsidence or groundwater pollution that arise also with conventional recovery methods. Even conventional drilling uses large volumes of drilling mud to cool and lubricate the drill bits. The drilling muds may become contaminated with oil and must be handled carefully to avoid polluting surface or ground water. When water is more extensively used in fracturing rock or warming the oil, the potential for pollution is correspondingly greater.

Unconventional Natural Gas Sources

Major new potential sources of natural gas are **methane hydrates.** Gas (usually methane) hydrates are crystalline solids of gas and water molecules (figure 14.10). These hydrates have

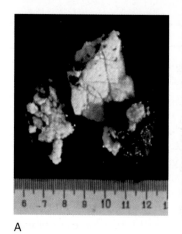
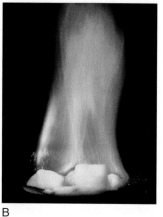

A   B

Figure 14.10

(A) Irregular chunks of gas hydrate from sediments in the Sea of Okhotsk. Scale is in centimeters. (B) Methane released from a piece of hydrate burning.

(A) Photograph courtesy U.S. Geological Survey. (B) Photograph by J. Pinkston and L. Stern , U.S. Geological Survey

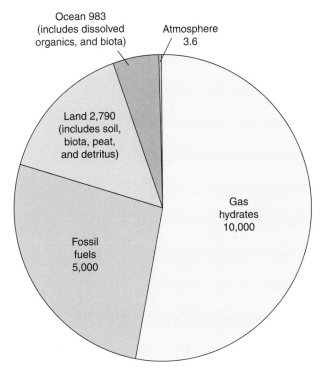

Ocean 983
(includes dissolved
organics, and biota)

Atmosphere
3.6

Land 2,790
(includes soil,
biota, peat,
and detritus)

Gas
hydrates
10,000

Fossil
fuels
5,000

Figure 14.11

The amount of carbon believed to exist in gas hydrates exceeds that in all known fossil fuels and other organic-carbon reservoirs combined. Units are billions of tons of carbon.

Data from U.S. Geological Survey.

been found to be abundant in arctic regions and in marine sediments. The U.S. Geological Survey has estimated that the amount of carbon found in gas hydrates may be twice that in all known fossil fuels on earth (figure 14.11), and a 2008 study suggested a probable 85 trillion cubic feet of (yet-undiscovered) methane in gas hydrates of Alaska's North Slope alone. Methane in methane hydrate thus represents a huge potential natural gas resource. Just how to tap methane hydrates safely for their energy is not yet clear.

In arctic regions, the hydrates are stabilized by the low temperature. If these arctic sediments are warmed enough by the continued warming of the Arctic to break down some of the hydrates, or if they are disturbed in the course of extracting methane, methane gas would be released into the atmosphere, which would further increase greenhouse-effect heating. Considering that methane is a far more efficient greenhouse gas than is CO_2 and that the amount of methane locked in hydrates now is estimated at about 3000 times the methane currently in the atmosphere, the potential impact is large and represents yet another source of uncertainty in global-climate modeling.

Some evidence suggests that additional natural gas reserves exist at very great depths in the crust. Thousands of meters below the surface, conditions are sufficiently hot that any oil would have been broken down to natural gas (recall figure 14.3). This natural gas, under the tremendous pressures exerted by the overlying rock, may be dissolved in the water filling the pore spaces in the rock, much as carbon dioxide is dissolved in a bottle of soda. Pumping this water to the surface is like taking off the bottle cap: The pressure is released, and the gas bubbles out. Enormous quantities of natural gas might exist in such **geopressurized zones;** estimates of the amount of potentially recoverable gas in this form range from 150 to 2000 trillion cubic feet.

Special technological considerations are involved in developing these deposits. It is difficult and very expensive to drill to these deep, high-pressure accumulations. Also, many of the fluids in which the gas is dissolved are very saline brines that cannot be casually released at the surface without damage to the quality of surface water; the most effective means of disposal is to pump the brines back into the ground, but this is costly. On the plus side, however, the hot fluids themselves may represent an additional geothermal energy source (see the discussion of geothermal energy in chapter 15). Geopressurized natural gas may well become an important supplementary fossil fuel in the future, though its total potential and the economics of its development are poorly known. It is unlikely that enough can be found and produced soon enough to solve near-future energy problems.

There is also natural gas in "tight" (low-permeability) sandstones in the Rocky Mountains and elsewhere. The quantity of recoverable gas in these rocks is uncertain. Recent estimates have ranged from 60 to 840 trillion cubic feet. In principle, some modification of the fracking technology now used for shale-gas recovery might be developed for the "tight" sands. However, at present such techniques for these rocks do not generally yield gas at a price competitive with that of gas from other sources.

Conservation

Conservation of oil and gas is, potentially, a very important way to stretch remaining supplies. More and more individuals, businesses, and governments in industrialized societies are practicing energy conservation out of personal concern for the environment, from fear of running out of nonrenewable fuels, or simply for basic economic reasons when energy costs have soared. The combined effects of conservation and economic recession can be seen in the flattening of the U.S. energy-consumption curve in figure 14.2 during the mid-1970s to mid-1980s. However, beginning about 1990, energy consumption resumed rising steadily, reaching an all-time high in 2004 before slackening slightly. Some of the factors in rising energy consumption can be climatic—unusual heat in July and August, unusual cold in November and December—but other consumption increases are discretionary. Low gasoline prices, for example, make fuel-efficient cars and other fuel-conservation methods such as carpooling less urgent priorities for some consumers. Worldwide, too, the United States has long had relatively inexpensive gasoline and consumed it freely as a result (figure 14.12). See also Case Study 14.2.

Conservation can buy some much-needed time to develop alternative energy sources. However, *world* energy consumption will probably not decrease significantly in the near future.

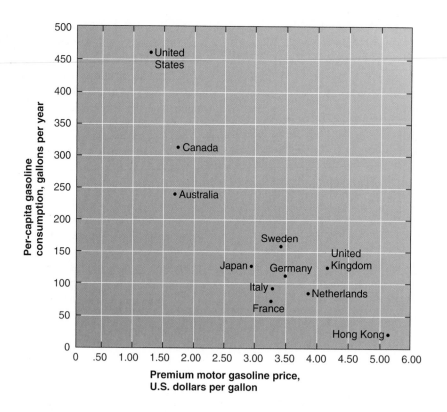

Figure 14.12

Among industrialized nations worldwide, there is an apparent inverse correlation between gasoline price and gasoline consumption. It remains to be seen if recent rises in U.S. gasoline prices will change long-established patterns of high gasoline use; through 2010, U.S. consumption was still about 440 gallons per person, despite fuel prices above $3/gallon.

Data for graph from International Energy Annual 2006, *U.S. Energy Information Administration.*

Even if industrialized countries consistently adopt more energy-efficient practices, demand in the many nonindustrialized countries is expected to continue to rise. Many technologically less-developed countries view industry and technology as keys to better, more prosperous lives for their people. They resent being told to conserve energy by nations whose past voracious energy consumption has largely led to the present squeeze. Assuming that world demand for energy stays at least as high as it is now, we face a global crisis of dwindling petroleum and natural-gas supplies over the next few decades. In the meantime, heavy reliance on oil and gas continues to have some serious environmental consequences, notably oil spills.

Oil Spills

Natural oil seeps are not unknown. The LaBrea Tar Pits of prehistoric times are an example. In fact, it is estimated that oil rising up through permeable rocks escapes into the ocean at the rate of 600,000 tons per year, though many natural seeps seal themselves as leakage decreases remaining pressure. Tankers that flush out their holds at sea continually add to the oil pollution of the oceans and, collectively, are a significant source of such pollution. Oil spills represent the largest negative impacts from the extraction and transportation of petroleum, although as a source

of water pollution, they are less significant volumetrically than petroleum pollution from careless disposal of used oil, amounting to only about 5% of petroleum released into the environment. And although we tend to hear only about the occasional massive, disastrous spill, there are many more that go unreported in the media: The U.S. Coast Guard reports about 10,000 spills in and around U.S. waters each year, totaling 15 to 25 million gallons of oil annually. Natural disasters add to the total: In 2005, damage to oil-drilling and processing facilities in the Gulf of Mexico from Hurricane Katrina caused about 100 spills of a total of 8 million gallons of oil. The oil spills about which most people are concerned, however, are the large, sudden, catastrophic spills that occur in two principal ways: from accidents during drilling of offshore oil wells and from wrecks of oil tankers at sea.

When an oil spill occurs, the oil, being less dense than water, floats. The lightest, most volatile hydrocarbons start to evaporate initially, decreasing the volume of the spill somewhat but polluting the air. Then a slow decomposition process sets in, due to sunlight and bacterial action. After several months, the mass may be reduced to about 15% of the starting quantity, and what is left is mainly thick asphalt lumps. These can persist for many months more. Oil is toxic to marine life, causes water birds to drown when it soaks their feathers, may decimate fish and shellfish populations, and can severely damage the economies of beach resort areas.

A variety of approaches to damage control following an oil spill have been tried. In calm seas, if a spill is small, it may be contained by floating barriers and picked up by specially designed "skimmer ships" that can skim up to fifty barrels of oil per hour off the water surface. Some attempts have been made to soak up oil spills with peat moss, wood shavings, and even chicken feathers. Large spills or spills in rough seas are a greater problem. When the tanker *Torrey Canyon* broke up off Land's End, England, in 1967, several strategies were tried, none very successfully. First, an attempt was made to burn off the spill. By the time this was tried, the more flammable, volatile compounds had evaporated, so aviation fuel was poured over the spill, and bombs were dropped to ignite it! This did not really work very well, and in any event, it would have produced a lot of air pollution. Some French workers used ground chalk to absorb and sink the oil. Sinking agents like chalk, sand, clay, and ash can be effective in removing an oil spill from the sea surface, but the oil is no healthier for marine life on the ocean bottom. Furthermore, the oil may separate out again later and resurface. The British mixed some 2 million gallons of detergent with part of the spill, hoping to break up the spill so that decomposition would work more rapidly. (Just as small mineral grains weather faster than large ones, due to their relatively higher surface area, so small oil droplets decompose faster than large ones.) The detergents, in turn, turned out to be toxic to some organisms, too.

Tanker disasters are potentially becoming larger all the time. The largest supertankers are longer than the Empire State Building is high (over 335 meters) and can carry nearly 2 million barrels of oil (about 80 million gallons). The largest single marine tanker spill ever resulted from the wreck of the *Amoco Cadiz* near Portsall, France, in 1978; the bill for cleaning up what could be recovered of the 1.6 million barrels spilled was more than $50 million, and at that, negative environmental impacts were still detectable seven years later. The largest tanker accident in U.S. waters was that of the *Exxon Valdez,* in 1989.

The port of Valdez lies at the southern end of the Trans-Alaska Pipeline (figure 14.13). Through that pipeline flow 1.6–1.7 million barrels of oil a day. From Valdez, the oil is transported by tanker to refinery facilities elsewhere. Prince William Sound, on which Valdez is located, is home to a great variety of wildlife, from large mammals (including whales, dolphins, sea otters, seals, and sea lions) to birds, shellfish, and fish (of which salmon and herring are particularly important economically).

Early in the morning of 23 March 1989, the tanker *Exxon Valdez,* loaded with 1.2 million barrels (50 million gallons) of crude oil from Alaska's North Slope, ran aground on Bligh Island in Prince William Sound. Response to the spill was delayed for ten to twelve hours after the accident, in part because an oil-containment barge happened to be disabled. More than 10 million gallons of oil escaped, eventually to spread over more than 900 square miles of water. The various cleanup efforts cost Exxon an estimated $2.5 billion.

Skimmers recovered relatively little of the oil, as is typical: Three weeks after the accident, only about 5% of the oil had been recovered. Chemical dispersants were not very successful

(perhaps because their use was delayed by concerns over *their* impact on the environment), nor were attempts to burn the spill. Four days after the accident, the oil had emulsified by interaction with the water into a thick, mousse-like substance, difficult to handle, too thick to skim, impossible to break up or burn effectively. Oil that had washed up on shore had coated many miles of beaches and rocky coast (figure 14.14).

Figure 14.13

The oil tanker port at Valdez, Prince William Sound, Alaska, southern end of the Trans-Alaska Pipeline. (Note classic glacial valley in background.)

Figure 14.14

Beach cleanup workers in oil-splattered raingear mop oil from beach on LaTouche Island, Prince William Sound.

Photograph from Alaska State Archive slide, courtesy Oil Spill Public Information Center.

Energy Prices, Energy Choices

In recent years, consumers have been surprised by rapid swings in fossil-fuel prices. Many factors influence both energy prices and energy consumption. On the price side, the underlying cost of the commodity may vary little except when political events or natural disasters constrain availability (figure 1); speculation in the financial markets may exacerbate the effects of these factors, as in 2008. Short-term supply-and-demand issues are often more significant influences on retail prices of petroleum products. As noted in the chapter, crude oil must be refined into a range of petroleum products, and refiners make projections about the relative amounts of gasoline, heating oil, etc. that will be needed at different times of year. Unexpected developments (such as a fierce, prolonged cold spell in heating-oil country) can drive up prices until more of the needed fuel can be

produced. Likewise with natural gas: Although it's readily delivered by pipeline, it can only be extracted and transported so fast, and if use is unexpectedly heavy for some time, prices will again rise until demand slackens or supply catches up.

Consumers can moderate their own costs through conservation but often feel little incentive to do so if financial pressures are low. Gasoline prices generally fell in the 1980s and 1990s, and consumption tended to rise in complementary fashion (figure 2). In part, this can be related to the rise in popularity of vans, sport-utility vehicles, and other vehicles with notably lower mileage than is mandated for passenger cars by the Environmental Protection Agency (figure 3). The sharp price rise in gasoline in 2008 depressed consumption a bit, but the effect, like the extreme gasoline prices, was short-lived.

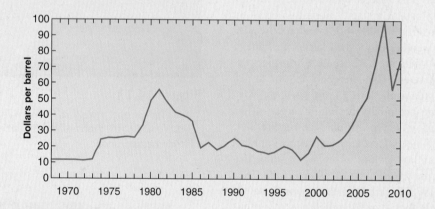

Figure 1

The Arab oil embargo and other events in the Middle East led to a spike in crude-oil prices in the 1980s. Prices then subsided for nearly two decades, but damage to U.S. production by Hurricane Katrina, and wars in the Middle East, combined to push prices higher in the mid-2000s. Speculation drove them sharply higher in 2008, since which time they have fluctuated sharply. The global economic slowdown is expected to depress demand and prices in the near term. Normalized to year 2000 dollars.

Source: U.S. Energy Information Administration.

Thousands of birds and marine mammals died or were sickened by the oil. The annual herring season in Valdez was cancelled, and salmon hatcheries were threatened.

When other methods failed, it seemed that only slow degradation over time would get rid of the oil. Given the size of the spill and the cold Alaskan temperatures, prospects seemed grim; microorganisms that might assist in the breakdown grow and multiply slowly in the cold. It appeared highly likely that the spill would be very persistent. So, as an experiment, Exxon scientists sprayed miles of beaches in the Sound with a fertilizer solution designed to stimulate the growth of naturally occurring microorganisms that are known to "eat" oil, thereby contributing to its decomposition. Within two weeks, treated beaches

were markedly cleaner than untreated ones. Five months later, the treated beaches still showed higher levels of those microorganisms than did untreated beaches. This type of treatment isn't universally successful. The fertilizer solution runs off rocky shores, so this treatment won't work on rocky stretches of oil-contaminated coast; nor is it as effective once the oil has seeped deep into beach sands. But microorganisms may indeed be the future treatment of choice. Examination of the long-term effects of another strategy, hot-water washing of beaches, has suggested that it may actually do more harm than good. The heat can kill organisms that had survived the oil, the pressure of the hot-water blast can drive oil deeper into the sediments where degradation is slower, and the water treatment may wash oil

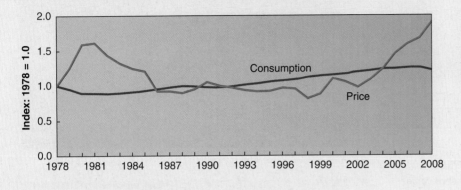

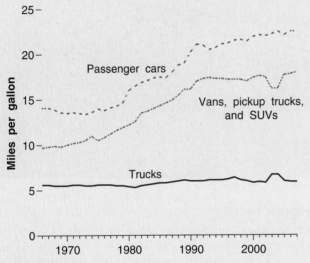

Figure 3

Mileage of passenger cars has risen steadily—largely due to EPA mandates; mileage of vans, SUVs, and pickup trucks has lagged consistently behind, and the gap has widened in recent years.

Source: U.S. Energy Information Administration.

And what about the electric car as an alternative to gasoline? At present, about 55,000 electric cars are in use in the United States; they make up less than 10% of the so-called alternative-fueled vehicles. Unfortunately they neither save energy nor eliminate CO_2 emissions. Their main application at present is in easing *local* pollution problems. The electricity to power them still has to be generated somehow (and a great deal of energy is lost in the process, as will be explored in chapter 15). The electric car may operate cleanly and efficiently, but making that electricity still consumes energy and creates pollution somewhere— and currently, about 70% of our electricity is generated using fossil fuels. More-efficient hybrid-technology cars have also been developed, but it will be a long time before their numbers on the highways are significant. Still, improving the fuel economy of motor vehicles, of whatever sort, could substantially reduce oil consumption, considering that motor gasoline accounts for nearly half the petroleum consumption in the United States.

onto beaches and into tidepools that were unaffected by the original spill.

Precautionary measures can be increased—for instance, double-hulled tankers and tugboat escorts are now mandatory for oil tankers in Prince William Sound, and practice drills in manipulation of oil-containment booms are now regularly held (figure 14.15)—but as long as demand for petroleum remains high, some accidents are, perhaps, inevitable.

The search for solutions continues. In November 2002, the tanker *Prestige* went down off the Iberian coast. The Spanish government's solution was to tow the tanker further offshore and sink it, on the presumption that the estimated 60,000 tons of oil still in the tanker would solidify in the deeper,

colder water. Alas, not so. For months thereafter, oil continued to leak to the surface—about 125 tons a day, which ultimately blanketed more than 900 km of coastline from Portugal to France, costing an estimated $1 billion in cleanup costs and prompting a spate of lawsuits that are ongoing as this is written.

Drilling accidents have been a growing concern as more areas of the continental shelves are opened to drilling. Normally, the drill hole is lined with a steel casing to prevent lateral leakage of the oil, but on occasion, the oil finds an escape route before the casing is complete. Alternatively, drillers may unexpectedly hit a high-pressure pocket that causes a blowout.

This is what happened on 20 April 2010 to the offshore drilling rig *Deepwater Horizon* as it drilled an exploratory well

Figure 14.15

More than twenty years after the grounding of the *Exxon Valdez*, practice with equipment used to contain oil spills keeps Valdez-area residents prepared for any future tanker accidents.

A

B

Figure 14.16

(A) The Coast Guard conducted a controlled burn on 6 May 2010 to destroy some of the oil from the *Deepwater Horizon* rig before it could reach land. Note oil-containment boom in foreground—and air pollution from the burn. (B) Sediment samples from the beach face at East Ship Island, Mississippi, show alternating layers of oil-rich and cleaner sand.

(A) U.S. Navy photo by Mass Communication Specialist 2nd Class Justin Stumberg/Released. (B) Photograph by Shane Stocks, U.S. Geological Survey Mississippi Water Science Center.

about 65 km (40 miles) off the Louisiana coast. The blowout-prevention system failed. The resulting explosion and fire killed eleven workers and injured several more. The half-billion-dollar rig burned for two days before the remains of the structure collapsed and sank to the bottom of the Gulf of Mexico. Meanwhile, oil gushed from the blown-out well, creating both a huge and spreading slick on the surface (see the chapter-opening photograph) and underwater plumes of oil drifting away from the drill site. At risk were many plant and animal species; shrimp, mussel, and clam fisheries; and coastal tourism, businesses, and recreational activities. Oil continued to leak for twelve weeks before efforts to cap (plug) the well finally succeeded. Government estimates put the amount of oil that escaped from the well at over 4.9 million barrels (over 200 million gallons), more than ten times the size of the *Exxon Valdez* spill.

Many of the strategies described earlier were applied to this spill. Some oil was burned in the Gulf to keep it offshore (figure 14.16A), since oil that reached the beaches would be much harder to remove (figure 14.16B) and would negatively affect more people, organisms, and ecosystems. Skimmer ships from around the world were mustered to collect as much of the oil as possible. Dispersants—some 1.8 million gallons—were applied to both the surface slick and the subsurface plumes to break up the oil droplets to speed microbial decomposition.

By late 2010, it was estimated that about a quarter of the oil remained as slicks on the water or as oil and tar on the shores of the Gulf; a quarter had been burned or collected; a quarter had been dispersed as droplets; and a quarter had evaporated or dissolved. Cleanup will continue for some time. Questions remain about potential toxic effects on organisms of the residual oil and dispersants, and whether microbial decomposition of the oil will cause further harm to animal life in the Gulf by depleting dissolved oxygen in the water. (See chapter 17 for discussion of dissolved oxygen and water pollution.)

By April 2011, all fisheries in federal waters that had been closed due to the spill had been reopened, though monitoring for safety continues. A number of federal agencies, includ-ing the EPA and the U.S. Geological Survey, continue to study the aftereffects of the spill. Shortly after the spill, the government instituted a six-month moratorium on issuance of new permits for offshore drilling in the Gulf. Concerned that the *Deepwater Horizon* accident might, in part, have been due to lax oversight, the administration also created a new agency to regulate offshore drilling, the Bureau of Ocean Energy Management, Regulation, and Enforcement. Permitting has now resumed, but slowly, with new, more stringent regulations in place, aimed at averting another accident like this one.

This spill, like many others before it, seems to have confirmed that there is no simple, effective solution to a major oil spill; hence the interest in taking every precaution to try to prevent one!

Coal

Prior to the discovery and widespread exploitation of oil and natural gas, wood was the most commonly used fossil fuel, followed by coal as the industrial age began in earnest in the nineteenth century. However, coal was bulky, cumbersome, and dirty to handle and to burn, so it fell somewhat out of favor, particularly for home use, when the liquid and gaseous fossil fuels became available. Currently, though it supplies only about 21% of total U.S. energy needs, coal plays a key role in the energy picture, for coal-fired power plants account for nearly half of U.S. electric power generation, which, in turn, consumes over 90% of U.S. coal production. With oil and gas supplies dwindling, many energy users are looking again at the potential of coal for other applications also.

Formation of Coal Deposits

Coal is formed not from marine organisms, but from the remains of land plants. A swampy setting, in which plant growth is lush and where there is water to cover fallen trees, dead leaves, and other debris, is especially favorable to the initial stages of coal formation. The process requires **anaerobic** conditions, in which oxygen is absent or nearly so, since reaction with oxygen destroys the organic matter. (A pile of dead leaves in the street does not turn into fuel, and in fact, over time, leaves very little solid residue.)

The first combustible product formed under suitable conditions is *peat*. Peat can form at the earth's surface, and there are places on earth where peat can be seen forming today. Further burial, with more heat, pressure, and time, gradually dehydrates the organic matter and transforms the spongy peat into soft brown coal (**lignite**) and then to the harder coals (**bituminous** and **anthracite**); see figure 14.17. As the coals become harder, their carbon content increases, and so does the amount of heat released by burning a given weight of coal. The hardest, high-carbon coals (especially anthracite), then, are the most desirable as fuels because of their potential energy yield. As with oil, however, the heat to which coals can be subjected is limited: overly high temperatures lead to metamorphism of coal into graphite.

The higher-grade coals, like oil, apparently require formation periods that are long compared to the rate at which coal is being used. Consequently, coal too can be regarded as a nonrenewable resource. However, the world supply of coal represents an energy resource far larger than that of petroleum, and this is also true with regard to the U.S. coal supply. In terms of energy equivalence, U.S. coal reserves represent more than 50 times the energy in the remaining oil reserves and 30 times the energy of remaining U.S. conventional natural-gas reserves.

Coal Reserves and Resources

Coal resource estimates are subject to fewer uncertainties than corresponding estimates for oil and gas. Coal is a solid, so it does not migrate. It is found, therefore, in the sedimentary rocks in which it formed; one need not seek it in igneous or metamor-

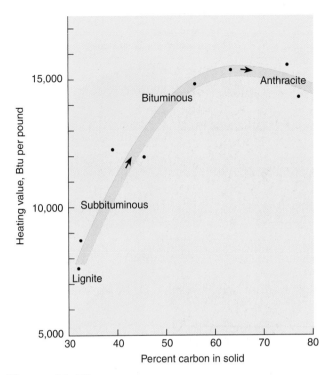

Figure 14.17

Change in character of coal with increasing application of heat and pressure. There is a general trend toward higher carbon content and higher heat value with increasing grade, though heat value declines somewhat as coal tends toward graphite.

phic rocks. It occurs in well-defined beds that are easier to map than underground oil and gas concentrations. And because it formed from land plants, which did not become widespread until 400 million years ago, one need not look for coal in more ancient rocks.

The estimated world reserve of coal is about one trillion tons; total resources are estimated at over 10 trillion tons. The United States is particularly well supplied with coal, possessing about 27% of the world's reserves, over 255 billion tons of recoverable coal (figure 14.18). Total U.S. coal resources may approach ten times that amount, and most of that coal is yet unused and unmined. At present, coal provides between 20 and 25% of the energy used in the United States. While the United States has consumed probably half of its total petroleum resources, it has used up only a few percent of its coal. Even if only the reserves are counted, the U.S. coal supply could satisfy U.S. energy needs for more than 200 years at current levels of energy use, if coal could be used for all purposes. As a supplement to other energy sources, coal could last many centuries. Unfortunately, heavy reliance on coal has drawbacks.

Limitations on Coal Use

A primary limitation of coal use is that solid coal is simply not as versatile a substance as petroleum or natural gas. A major shortcoming is that it cannot be used directly in most forms of modern transportation, such as automobiles and airplanes. (It

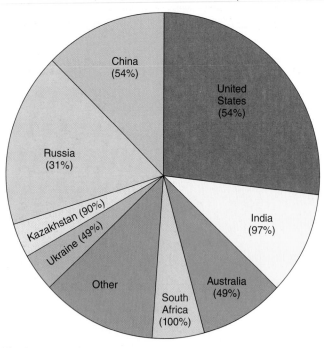

Total
948 billion tons
(47% anthracite and bituminous)

Figure 14.18

World coal reserves, in millions of tons. Sum of anthracite and bituminous for each country in parentheses, indicating the quality of the coal.

Source: Data on country proportions and quality from International Energy Annual 2006 *and* Annual Energy Review 2007; *total world reserve from* International Energy Outlook; *all from U.S. Energy Information Administration.*

was, of course, used to power trains in the past, but coal-fired locomotives quickly fell out of favor when cleaner fuels became available.) Coal is also a dirty and inconvenient fuel for home heating, which is why it was abandoned in favor of oil or natural gas. Therefore, given present technology, coal simply cannot be adopted as a substitute for oil in all applications.

Coal can be converted to liquid or gaseous hydrocarbon fuels—gasoline or natural gas—by causing the coal to react with steam or with hydrogen gas at high temperatures. The conversion processes are called **gasification** (when the product is gaseous) and **liquefaction** (when the product is liquid fuel). Both processes are intended to transform coal into a cleaner-burning, more versatile fuel, thus expanding its range of possible applications.

Commercial coal gasification has existed on some scale for 150 years. Many U.S. cities used the method before inexpensive natural gas became widely available following World War II. Europeans continued to develop the technologies into the 1950s. At present, only experimental coal-gasification plants are operating in the United States.

Current gasification processes yield a gas that is a mixture of carbon monoxide and hydrogen with a little methane. The heat derived from burning this mix is only 15 to 30% of what can be obtained from burning an equivalent volume of natural gas. Because the low heat value makes it uneconomic to transport this gas over long distances, it is typically burned only where produced. Technology exists to produce high-quality gas from coal, equivalent in heat content to natural gas, but it is presently uneconomical when compared to natural gas, and becoming more so with the recent development of shale gas. Research into improved technologies continues.

Pilot projects are also underway to study *in situ* underground coal gasification, through which coal would be gasified in place and the gas extracted directly. The expected environmental benefits of not actually mining the coal include reduced land disturbance, water use, air pollution, and solid waste produced at the surface. Potential drawbacks include possible groundwater pollution and surface subsidence over gasified coal beds. Underground gasification may provide a means of using coals that are too thin to mine economically or that would require excessive land disruption to extract. However, in the near future, simple extraction of coal-bed methane is likely to yield economical gas more readily.

Like gasification, the practice of generating liquid fuel from coal has a longer history than many people realize. The Germans used it to produce gasoline from their then-abundant coal during World War II; South Africa has a liquefaction plant in operation now, producing gasoline and fuel oil. A variety of technologies exist, and several noncommercial pilot plants have operated in the United States. Technological advances made during the 1980s slashed costs by 60% and greatly improved yields (to about 70% of original coal converted to liquid fuel). Even so, their liquid fuel products have not historically been economically competitive with conventional petroleum. The higher gasoline prices rise, of course, the more economically viable coal liquefaction becomes.

Recent renewed interest in U.S. energy independence has revived and intensified discussion of liquefaction, also known as *coal-to-liquids* technology. The most obvious advantage is that it would allow us to use our plentiful coal resources to meet a need now satisfied by petroleum, much of it imported (though of course, it would result in more rapid depletion of the coal supply). However, there are also clear concerns.

The process is actually a multistage one: The coal is first gasified, and some impurities, such as sulfur gases, removed from the resulting gas mix (which does mean that, at least in terms of those pollutants, the eventual fuel can be cleaner-burning than conventional gasoline). The gas is then converted to liquid fuels such as gasoline, diesel, and jet fuel. It is a water-intensive process, in which an estimated 6–10 barrels of water are consumed for every barrel of liquid fuel produced. To the extent that coal mining must be expanded to meet demand for such fuel, the impacts of that mining, described below, are also expanded. And the process of producing liquid fuel from coal generates more than twice the greenhouse-gas emissions of producing gasoline from petroleum. Costs aside, then, there are significant issues to be resolved before liquid fuels derived from coal are likely to be embraced on a commercial scale in the United States.

Environmental Impacts of Coal Use

Gases

A major problem posed by coal is the pollution associated with its mining and use. Like all fossil fuels, coal produces carbon dioxide (CO_2) when burned. In fact, it produces significantly more carbon dioxide per unit energy released than oil or natural gas. (It is the burning of the carbon component—making CO_2—that is the energy-producing reaction when coal is burned; when hydrocarbons are burned, the reactions produce partly CO_2, partly H_2O as their principal by-products, the proportions varying with the ratio of carbon to hydrogen in the particular hydrocarbon(s).) The relationship of carbon dioxide to greenhouse-effect heating was discussed in chapter 10. The additional pollutant that is of special concern with coal is sulfur.

The sulfur content of coal can be more than 3%, some in the form of the iron sulfide mineral pyrite (FeS_2), some bound in the organic matter of the coal itself. When the sulfur is burned along with the coal, sulfur gases, notably sulfur dioxide (SO_2), are produced. These gases are poisonous and are extremely irritating to eyes and lungs. The gases also react with water in the atmosphere to produce sulfuric acid, a very strong acid. This acid then falls to earth as acid rainfall. Acid rain falling into streams and lakes can kill fish and other aquatic life. It can acidify soil, stunting plant growth. It can dissolve rock; in areas where acid rainfall is a severe problem, buildings and monuments are visibly corroding away because of the acid. The geology of acid rain is explored further in chapter 18.

Oil can contain appreciable sulfur derived from organic matter, too, but most of that sulfur can be removed during the refining process, so that burning oil releases only about one-tenth the sulfur gases of burning coal. Some of the sulfur can be removed from coal prior to burning, but the process is expensive and only partially effective, especially with organic sulfur. Alternatively, sulfur gases can be trapped by special devices ("scrubbers") in exhaust stacks, but again, the process is expensive (in terms of both money and energy) and not perfectly efficient. From the standpoint of environmental quality, then, low-sulfur coal (1% sulfur or less) is more desirable than high-sulfur coal for two reasons: first, it poses less of a threat to air quality, and second, if the sulfur must be removed before or after burning, there is less of it to remove, so stricter emissions standards can be met more cheaply. On the other hand, much of the low-sulfur coal in the United States, especially western coal, is also lower-grade coal (see figure 14.19), which means that more of it must be burned to yield the same amount of energy. This dilemma is presently unresolved.

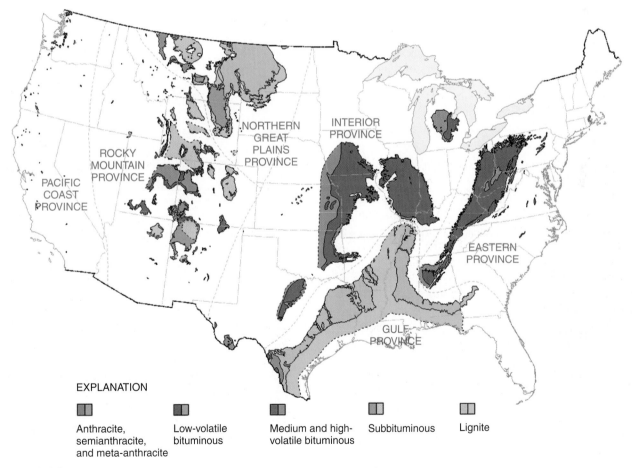

EXPLANATION

Anthracite, semianthracite, and meta-anthracite

Low-volatile bituminous

Medium and high-volatile bituminous

Subbituminous

Lignite

Figure 14.19

Distribution of U.S. coal fields. Within each color, darker shading represents current or likely future reserves; lighter shades are deposits of unknown or lower quality. *Source: U.S. Geological Survey.*

Another toxic substance of growing concern with increasing coal use is mercury. A common trace element in coal, mercury is also a very volatile (easily-vaporized) metal, so it can be released with the waste gases. The consequences are examined in chapter 17.

Ash

Coal use also produces a great deal of solid waste. The ash residue left after coal is burned typically ranges from 5 to 20% of the original volume. The ash, which consists mostly of noncombustible silicate minerals, also contains toxic metals. If released with waste gases, the ash fouls the air. If captured by scrubbers or otherwise confined within the combustion chamber, this ash still must be disposed of. If exposed at the surface, the fine ash, with its proportionately high surface area, may weather very rapidly, and the toxic metals such as selenium and uranium can be leached from it, thus posing a water-pollution threat. Uncontrolled erosion of the ash could likewise cause sediment pollution. The magnitude of this waste-disposal problem should not be underestimated. A single coal-fired electric power plant can produce over a million tons of solid waste a year. There is no obvious safe place to dump all that waste. Some is used to make concrete, and it has been suggested that the ash could actually serve as a useful source of some of the rarer metals concentrated in it. At present, however, much of the ash from coal combustion is deposited in landfills. Between combustion and disposal, the ash may be stored wet, in containment ponds. These, too, can pose a hazard: In December 2008, a containment pond along the Emory River in Tennessee failed, releasing an estimated 5.4 million cubic meters of saturated ash from the Kingston Fossil Plant into the river (figure 14.20). The surge of sludge buried nearly half a square mile of land, damaged many homes, and released such pollutants as arsenic, chromium, and lead into the water. It was originally estimated, optimistically, that cleanup might take 4 to 6 weeks. It now appears that the process will take years, and projections of total cleanup costs range up to $1 billion.

Coal-Mining Hazards and Environmental Impacts

Coal mining poses further problems. Underground mining of coal is notoriously dangerous, as well as expensive. Mines can collapse; miners may contract black lung disease from breathing the dust; there is always danger of explosion from pockets of natural gas that occur in many coal seams. A tragic coal-mine fire in Utah in December 1984 and a methane explosion in a Chinese coal mine in 2004 that killed over 100 people were reminders of the seriousness of the hazards. Recent evidence further suggests that coal miners are exposed to increased cancer risks from breathing the radioactive gas radon, which is produced by natural decay of uranium in rocks surrounding the coal seam.

Even when mining has ceased, fires may start in underground mines (figure 14.21). An underground fire that may have been started by a trash fire in an adjacent dump has been

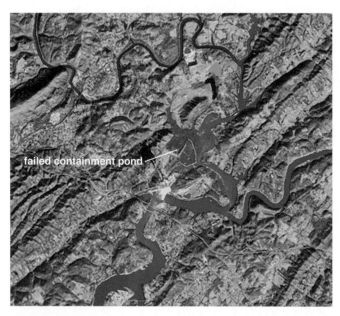

Figure 14.20

In this false-color Landsat image of the December 2008 Tennessee ash spill, vegetation appears in green, bare ground pink or brown, clear water dark blue, and ash-laden water light blue.

Image courtesy Ronald Beck, U.S. Geological Survey, and NASA

Figure 14.21

Fire out of control in abandoned underground coal mine in Sheridan County, Wyoming.

Photograph by C. R. Dunrud, USGS Photo Library, Denver, CO.

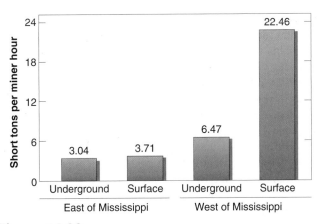

Figure 14.22

Until 1970, virtually all U.S. coal mining occurred east of the Mississippi. Now most mining occurs west of the Mississippi, and the majority of that mining is strip mining.

Source: Annual Energy Review 2007, *U.S. Energy Information Administration*

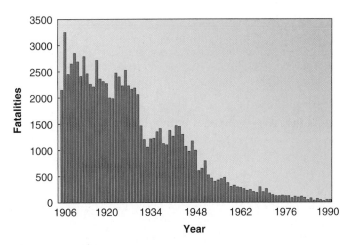

Figure 14.23

Drop in coal-mine fatalities reflects partly safer mining practices, partly the shift from underground mining to strip-mining.

Source: U.S. Bureau of Mines, Minerals Today, *Oct. 1991, p. 19.*

burning under Centralia, Pennsylvania, since 1962; over 1000 residents have been relocated, at a cost of more than $40 million, and collapse and toxic fumes threaten the handful of remaining residents resisting evacuation. And while Centralia's fire may be among the most famous, it is by no means the only such fire in the United States: there are 38 more in Pennsylvania alone. Coal-mine fires are a problem in other nations as well. China relies heavily on coal for its energy; it has been estimated that uncontrolled coal-mine fires there consume up to 20 percent of their coal production, the carbon-dioxide emissions from them are estimated to account for several percent of global CO_2 emissions, and the associated pollution is believed to be aggravating respiratory problems in China. Mining of India's largest coal field began in 1894; the first fires there started in 1916, and fifty years later, the fires could be found across the whole coal field. They burn on today.

The rising costs of providing a safer working environment for underground coal miners, together with the greater technical difficulty (and thus lower productivity) of underground mines, are largely responsible for a steady shift in coal-mining methods in the United States, from about 20% surface mining around 1950 to nearly 70% surface mining by 2010. Surface mining continues to dominate, as demand for low-sulfur western coal increases (figure 14.22). The positive side of this shift in methods is that it has been accompanied by a sharp decline in U.S. coal-mine fatalities (figure 14.23); strip-mining of coal is far safer for the miners.

A significant portion of U.S. coal, particularly in the west, occurs in beds very close to the surface. It is relatively cheap to strip off the vegetation and soil and dig out the coal at the surface, thus making strip-minable coal very attractive economically in spite of land reclamation costs, even when the coal beds are thin (figure 14.24). Still, strip-mining presents its own problems. Strip-mining in general was discussed in chapter 13

Figure 14.24

Rainbow coal strip mine, Sweetwater County, Wyoming. Note the very thin coal beds (black) relative to waste rock (brown).

Photograph by H. E. Malde, USGS Photo Library, Denver, CO.

(recall figure 13.25). A particular problem with strip-mining coal involves, again, the sulfur in the coal. Not every bit of coal is extracted from the surrounding rock. Some coal and its associated sulfur are left behind in the waste rock in spoil banks. Pyrite is also common in the shales with which coal often is interlayered, and these pyritic shales form part of the spoil banks. The sulfur in the spoils can react with water and air to produce runoff water containing sulfuric acid:

$$2\,FeS_2 + 5\,H_2O + \frac{15}{2}\,O_2 = 4\,H_2SO_4 + Fe_2O_3 \cdot H_2O$$

    pyrite   water   oxygen   sulfuric   hydrated iron
                                 acid   oxide

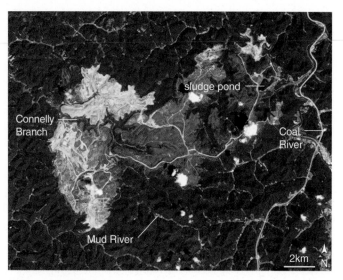

Figure 14.25

Hobet Mine, West Virginia. Some of the surplus spoils have been used to fill in Connelly Branch valley, and filling in more streams has been approved. Note the scale of this mountaintop mine complex. (2 km = 1¼ mi)

NASA Earth Observatory image created by Robert Simmon, using Landsat data provided by the U.S. Geological Survey.

This same reaction is responsible for most of the acid mine drainage discussed in chapter 13. Because plants grow poorly in very acid conditions, this acid slows revegetation of the area by stunting plant growth or even preventing it altogether. The acid runoff can also pollute area ground and surface waters, killing aquatic plants and animals in lakes and streams and contaminating the water supply. Very acidic water also can be particularly effective at leaching or dissolving some toxic elements from soils, further contributing to water pollution. Coal strip mines, like others, can be reclaimed, but in addition to regrading and replanting the land, it is frequently necessary to replace the original topsoil (if it was saved) or to bring in fresh soil so that sulfur-rich rocks are not left exposed at the surface, where weathering is especially rapid. (Even underground coal mines may have associated acid drainage, but water circulation there is generally more restricted.) The question of water availability to support plant regrowth is of special concern in the western United States: 50% of U.S. coal reserves are found there, at least 40% of these reserves would be surface-mined, and most western regions underlain by coal are very dry, as is evident in figure 14.24.

When coal surface mines are reclaimed, efforts are commonly made to restore original topography—although if a thick coal seam has been extracted, this may not be easy. The artificial slopes of the restored landscape may not be altogether stable, and as natural erosional processes begin to sculpture the land surface, natural slope adjustments in the form of slumps and slides may occur. In dry areas, thought must be given to how the reclamation will modify drainage patterns (surface and subsurface). In wetter areas, the runoff may itself be a problem, gullying slopes and contributing to sediment pollution of

surface waters until new vegetation is established. In short, reclamation can be a multifaceted challenge.

This has proved especially true with the mountaintop-removal style of surface mining that has become common since the 1990s in the Appalachian coal country. With mountaintop removal, vegetation is cleared and then, as the name suggests, explosives and earth-moving equipment are used to break up and remove the entire hilltop. Coal is sorted out from waste rock and soil, and then rinsed, producing a sediment-laden sludgy wastewater that must be contained in ponds to prevent surface-water pollution. Other impacts aside, the combination of hilltop removals and valley fills produces a much different, and flatter, topography than the original (figure 14.25). Yet the 1977 Surface Mining Control and Reclamation Act, discussed further in chapter 19, mandates that the mined land be restored to its "approximate original contour." Lawsuits over this and other environmental consequences of such mining are proliferating as demand for coal drives mining expansion.

Oil Shale

Oil shale is very poorly named. The rock, while always sedimentary, need not be a shale, and the hydrocarbon in it is not oil! The potential fuel in oil shale is a sometimes-waxy solid called **kerogen,** which is formed from the remains of plants, algae, and bacteria. Kerogen is not a single compound; the term is a general one describing organic matter that is not readily dissolved either in water or in organic solvents, and the term can also be applied to solid precursor organic compounds of oil and natural gas in other types of rocks. The physical properties of kerogen dictate that the oil shale must be crushed and heated to distill out the hydrocarbon as "shale oil," which then is refined somewhat as crude oil is to produce various liquid petroleum products.

The United States has nearly three-fourths of the world's known oil shale resources. About 70% is in the Green River Formation (figure 14.26), which holds enough hydrocarbon to yield an estimated 1½ trillion barrels of shale oil. For a number of reasons, the United States is not yet using this apparently vast resource to any significant extent, and it may not be able to do so in the near future.

One reason for this is that much of the kerogen is so widely dispersed through the oil shale that huge volumes of rock must be processed to obtain moderate amounts of shale oil. Even the richest oil shale yields only about three barrels of shale oil per ton of rock processed. The cost is not presently competitive with that of conventional petroleum, except when oil prices are extremely high, and to date they have not remained high enough long enough. Nor have large-scale processing facilities yet been built—present plants are strictly experimental. More-efficient and cheaper processing technologies are needed.

Another problem is that a large part of the oil shale is located at or near the surface. At present, the economical way to mine it, therefore, appears to be surface- or strip-mining with its attendant land disturbance. The Green River Formation is

Figure 14.26

Oil shale of the Green River Formation, which is found in parts of Colorado, Utah, and Wyoming.

© *The McGraw-Hill Companies, Inc./John A. Karachewski, photographer.*

located in areas of the western United States that are already chronically short of water. The dry conditions would make revegetation of the land after strip-mining especially difficult. Some consideration has been given to *in situ* extraction of the kerogen, warming the rock in place and using pressure to force out the resulting fluid hydrocarbon, without mining the rock, to minimize land disruption. But the necessary technology is poorly developed, though a pilot project does exist, and the process leaves much of the fuel behind.

The water shortage presents a further problem. Current processing technologies require large amounts of water—on the order of three barrels of water per barrel of shale oil produced. Just as water for reclamation is in short supply in the west, so, too, is the water to process the oil shale.

Finally, since the volume of the rock actually increases during processing, it is possible to end up with a 20 to 30% larger volume of waste rock to dispose of than the original volume of rock mined. Aside from any problems of accelerated weathering of the crushed material, there remains the basic question of where to put it. Because it will not all fit back into the space mined, the topography will inevitably be altered.

Scientists and economists differ widely in the extent to which they see oil shale as a promising alternative to conven-

tional oil and gas. Certainly, the water-shortage, waste-disposal, and land-reclamation problems will have to be solved before shale oil can be used on a large scale. It is unlikely that those problems can be solved within the next few decades, although over the longer term, oil shale may become an important resource, and the higher conventional oil prices rise, the more competitive "shale oil" becomes, at least in that respect. For now however, while we have large oil shale *resources*, they cannot be considered fuel *reserves*.

Oil Sand

Oil sands, also known as tar sands, are sedimentary rocks containing a very thick, semisolid, tarlike petroleum called **bitumen.** The heavy petroleum in oil sands is believed to be formed in the same way and from the same materials as lighter oils. Oil-sand deposits may represent very immature petroleum deposits, in which the breakdown of large molecules has not progressed to the production of the lighter liquid and gaseous hydrocarbons. Alternatively, the lighter compounds may have migrated away, leaving this dense material behind. Either way, the bitumen is too thick to flow out of the rock. Like oil shale, oil sand presently must either be mined, crushed, or heated to extract the petroleum, or, for *in situ* mining, large volumes of water or steam must be pumped underground to reduce the viscosity of the bitumen so it can be pumped out. However it is recovered, the bitumen is then refined into various fuels.

Many of the environmental problems associated with oil shale likewise apply to oil sand. Because the bitumen is disseminated through the rock, large volumes of rock must be mined and processed to extract significant amounts of petroleum. Many oil sands are near-surface deposits, so the mining method used is commonly strip-mining. The processing requires a great deal of water, and the amount of waste rock after processing may be larger than the original volume of oil sand. The negative impact of oil-sand production naturally increases as production increases.

In Canada, the extensive Athabasca oil sand in the province of Alberta is estimated to contain over 170 billion barrels of recoverable bitumen reserves (about 70% of the world's known oil-sand reserves), with total resources ten times that. Indeed, Canadian supplies of "unconventional" petroleum resources far exceed that nation's "conventional" oil deposits, and if they were counted as conventional petroleum, Canada would rank second only to Saudi Arabia in oil reserves.

Initial development was slow in part because of environmental concerns and in part due to the technical challenges. The bitumen makes up only 3 to 18% of the rock volume; on average, 2 tons of oil sand must be processed to yield 1 barrel of oil. Processing is energy-intensive as well as water-intensive. However, as the price of conventional oil has risen, oil-sand development has accelerated. Several open-pit mining operations are functioning in the province. For the deeper tar sands, *in situ* extraction is used. One method, nicknamed the "huff-and-puff" procedure, involves injection of hot steam for several weeks to warm the bitumen, after which the less-viscous fluid can be pumped out for a month or so before it chills and stiffens and reheating is necessary. Altogether, more than a million barrels of oil a day are being produced from Canada's oil sands. The scale of operations is large and growing (figure 14.27), as is the scale of corresponding environmental impacts, including land disturbance by mining; accumulation of oily tailings that leave oil slicks on tailings ponds, which can endanger migrating birds; and the fact that processing oil sand releases two to three times as much in CO_2 emissions as extracting conventional oil via wells, along with sulfur gases, hydrocarbons, and fine particulate pollutants. Those environmental impacts cause serious concern, but the resource is too large and too valuable to ignore.

The United States has virtually no oil-sand resources, so it cannot look to oil sand to solve its domestic energy problems even if the environmental difficulties could be overcome. However, Canada is already the single largest supplier of U.S. oil imports, and oil from the oil sands helps to assure that that situation can continue.

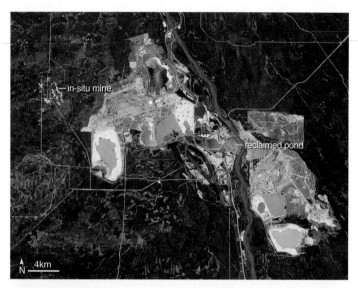

Figure 14.27

A portion of Athabascan oil-sand mining activity. Even *in situ* mining (left) requires forest clearing. Canada's environmental laws require reclamation of the land, but the reclaimed (filled-in) tailings pond at center is not yet showing plant regrowth. (4 km = 2.5 mi)

NASA Earth Observatory image created by Robert Simmon, using Landsat data provided by the U.S. Geological Survey.

Summary

The United States now relies on fossil fuels for about 83% of its energy: About 37% comes from oil, 25% from natural gas, and 21% from coal. All the fossil fuels are nonrenewable energy sources. Known conventional oil and gas supplies have been seriously depleted and may be exhausted within decades, though shale gas and coalbed methane have recently become significant new sources of natural-gas reserves. Oil spills, too, are a subject of concern, especially as much of the remaining U.S. petroleum is offshore or in the Arctic. Remaining coal supplies are much larger than oil and gas reserves, but many hazards and unsolved environmental problems are associated with heavy reliance on coal: acid rain, acid runoff, mine accidents and fires, strip-mine reclamation issues, and more. U.S. oil shale and Canadian oil sand represent significant hydrocarbon resources; however, they share a number of technical and environmental issues of concern: the fact that the hydrocarbon is dispersed in the host rock, the need for strip mining, extraction methods that require large volumes of water and produce large volumes of tailings. Although "shale oil" is likely to remain uneconomic for some time, development of the substantial petroleum reserves in the oil sands is active and expanding. Methane hydrates likewise represent a large resource, but the challenge of how to extract their fuel has not yet been resolved, and carries with it the risk of adding more methane—a potent greenhouse gas—to the atmosphere.

Key Terms and Concepts

anaerobic 325	fossil fuel 307	lignite 325	oil 309
anthracite 325	fracking 316	liquefaction 326	oil sand 331
bitumen 331	gasification 326	methane hydrate 318	oil shale 330
bituminous 325	geopressurized zones 319	natural gas 309	petroleum 309
coal-bed methane 316	kerogen 330	nonrenewable 307	shale gas 316

Exercises

Questions for Review

1. A society's level of technological development strongly influences its per-capita energy consumption. Explain.

2. What are fossil fuels?

3. Briefly describe how conventional oil and gas deposits form and mature.

4. Compare and contrast past and projected U.S. consumption of petroleum and coal.

5. What is enhanced oil recovery, and why is it of interest? Give two examples of the method.

6. What is "fracking," and how has it allowed shale gas to be added to U.S. natural-gas reserves?

7. What advantages does coal-bed methane have over other "unconventional" gas sources? Describe one problem associated with its extraction.

8. Explain the nature of geopressurized natural gas resources, and note at least one obstacle to their exploitation.

9. What are methane hydrates, and where are they found? Why are they both a substantial potential energy resource, and a concern with respect to global climate change?

10. What is coal, and how and from what does it form?

11. What is coal liquefaction, and why is it of interest? Describe one reason it is not widely practiced in the United States.

12. What air-pollution problems are associated particularly with coal, relative to other fossil fuels?

13. List and describe at least three potential negative environmental impacts of coal mining. Why is mountaintop removal particularly controversial?

14. What are oil shales and oil sands?

15. Oil shale and tar sand share a number of technical/environmental drawbacks. Cite and explain several of these.

Exploring Further

1. Select a particular region or major city and investigate its energy consumption. Identify the principal energy sources and the proportion that each contributes; see whether these proportions are significantly different now from what they were ten and twenty-five years ago. To what extent have the types and quantities of energy used been sensitive to economic factors?

2. If possible, visit an area that has been or is being drilled for oil or mined for coal. Observe any visible signs of negative effects on the environment, and note any efforts being made to minimize such impacts.

3. Check the current status of the *Deepwater Horizon* cleanup analysis; see online NetNotes for some possible information sources.

Energy Resources—
Alternative Sources

In the discussion of the (nonrenewable) fossil fuels in chapter 14, the point was made that, with the conspicuous exception of coal, most of the U.S. supply of recoverable fossil fuels could be exhausted within decades. It was further pointed out that coal is not the most environmentally benign of energy sources. Alternative energy sources for the future are thus needed, both to supply essential energy and to spare the environment as much disruption as possible. And this is true worldwide, for globally, too, fossil fuels are the primary energy source (figure 15.1).

The extent to which alternative energy sources are required and how soon they will be needed is directly related to future world energy demand, which is difficult to predict precisely. In general, consumption can be expected to rise as population increases and as standards of living improve. However, the correlation between energy consumption and standard of living is perhaps even less direct than the correlation between mineral consumption and living standards.

By way of a simplified example, consider the addition of central heating to a home. The furnace is likely to contain about the same quantity of material, regardless of its type or efficiency; the drain on mineral reserves is approximately fixed. The very addition of that furnace represents a certain jump in both mineral and energy consumption. However, depending on the efficiency of the unit, the "tightness" of the home's insulation, and the climate, the amount of fuel the furnace must consume to maintain a certain level of heat in the home will vary enormously. The best-insulated modern homes may require as little as one-tenth the energy for heating as the average U.S. home. Also, the quantity of material used to manufacture the furnace

---

High in the West Maui mountains of Hawaii, a chain of wind-power generators sits on a ridge. The islands have no fossil fuels, and transporting materials there is expensive, so generating energy locally from renewable sources is especially attractive.

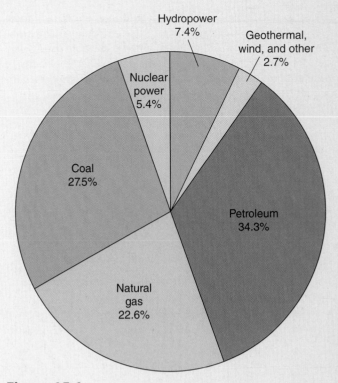

Figure 15.1

World energy production by source, 2008.

Source: Data from International Energy Outlook 2011, *U.S. Energy Information Administration, Department of Energy.*

will not change whether the homeowner maintains the temperature at 60°F or 80°F, but the amount of energy consumed plainly will. And, of course, the climate in which the home is built influences its heating costs. Both cold and heat, taken to extremes, cause increased energy demand—for heating and air conditioning, respectively. Thus there are considerable regional variations in energy consumption in the United States because of varying population density, degree of industrialization, and climatic variations.

In the area of transportation, similar variability arises. As standards of living rise to the point that motorized transportation becomes commonplace, consumption of both materials and energy rises. However, the amount of energy used depends heavily on the mode of transportation chosen and its fuel efficiency. It is more fuel-efficient to transport fifty people in a bus than in a dozen automobiles, but the distribution of people and their travel patterns must be clustered appropriately in order for mass transportation to work in a practical sense. Just among passenger cars, fuel efficiency varies greatly (which has substantial implications for possible energy conservation; see Table 15.1). When projecting future energy use in developing countries, the underlying population base is important, too: An increase of 50 vehicles per thousand persons in China, with population over 1 billion, means far more added automobiles than an increase of 200 vehicles per thousand in South Korea, with a population of only about 50 million, less than one-twentieth of China's.

Thus, in deciding just how much energy various alternative sources must supply, assumptions must be made not only about the rates of increase in standards of living (growth of GNP) but also about the degree of energy efficiency with which the growth is achieved. Other factors that influence both demand and projections include prices of various forms of energy (which recent history suggests is difficult to assess years, let alone decades, into the future) and public policy decisions (to pursue or not to pursue nuclear power, for example) in those nations that are major energy consumers.

Figure 15.2 shows one recent projection of world energy demand to 2030. Whether or not it is perfectly accurate, it clearly

Table 15.1	Fuel Efficiency, Fuel Consumption	
	Passenger Cars	**Vans, Pickup Trucks, and SUVs**
Number of vehicles	135,900,000	101,500,000
Average miles per gallon (est.)	22.6	18.1
Average annual miles driven per vehicle	11,800	10,950
Total annual miles driven by vehicle type	1.60 trillion	1.11 trillion
Total gasoline consumption, gallons	70.8 billion	61.3 billion

If the vans/pickups/SUVs achieved the same average mpg as the passenger cars (22.6), total gasoline consumed by this group would be only 49.1 billion gallons, for a saving of 12.2 billion gallons (290 million barrels) of petroleum per year.

If all of these vehicles met the 2012 EPA average fuel-economy requirement for *new* vehicles (25.2 mpg), the fuel savings would be 7.3 billion gallons for the passenger cars and 17.3 billion gallons for the van/pickup/SUV group, for a total of 24.6 billion gallons (586 million barrels). If all the passenger cars just met the 2012 requirement of 32.8 mpg for *new* passenger cars, this would save 22 billion gallons (524 million barrels) of gasoline.

If the passenger cars achieved the 51 mpg city mileage rating of the most efficient gas-electric hybrid of the 2012 model year, their total gasoline consumption would be just 31.4 billion gallons, saving 39.4 billion gallons (938 million barrels) per year. And if the vans/pickups/SUVs achieved that same mileage on average, the saving would be another 39.5 billion gallons (940 million barrels), for combined savings from both vehicle categories of 1.88 billion barrels per year—about a quarter of total annual U.S. oil consumption!

Data from U.S. EPA and Highway Statistics 2007, *Federal Highway Administration, U.S. Department of Transportation.*

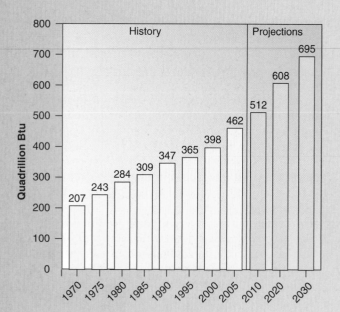

Figure 15.2

World energy consumption, historic and projected to 2030. Projected increase from 2005 to 2030 is about 50%.

Source: International Energy Outlook 2008, *U.S. Energy Information Administration.*

indicates substantial anticipated growth in energy use. If fossil fuels are inadequate in quantity and/or environmentally problematic, one or more alternatives must be far more extensively developed. This chapter surveys some possibilities. Evaluation of each involves issues of technical practicality, environmental consequences, and economic competitiveness.

Nuclear Power—Fission

Fission—Basic Principles

The phrase *nuclear power* actually comprises two different types of processes with different advantages and limitations. **Fission** is the splitting apart of atomic nuclei into smaller ones, with the release of energy. **Fusion** is the combining of smaller nuclei into larger ones, also releasing energy. Currently, only one of these processes is commercially feasible: fission. Fission basics are outlined in figure 15.3. (For a review of the terminology related to atoms, see chapter 2.) Very few isotopes—some 20 out of more than 250 naturally occurring isotopes—can undergo fission spontaneously, and do so in nature. Some additional nuclei can be induced to split apart, and some naturally fissionable nuclei can be made to split up more rapidly, thus increasing the rate of energy release. The fissionable nucleus of most interest in modern nuclear power reactors is the isotope of uranium with 92 protons and 143 neutrons, uranium-235.

A uranium-235 nucleus can be induced to undergo fission by firing another neutron into the nucleus. The nucleus splits into two lighter nuclei (not always the same two) and releases additional neutrons as well as energy. Some of the newly released neutrons can induce fission in other nearby uranium-235 nuclei, which, in turn, release more neutrons and more energy in breaking up, and so the process continues in a **chain reaction.** A controlled chain reaction, with a continuous, moderate release of energy, is the basis for fission-powered reactors (figure 15.4). The energy released heats cooling water that circulates through the reactor's core. The heat removed from the core is transferred through a heat exchanger to a second water loop in which steam is produced. The steam, in turn, is used to run turbines to produce electricity.

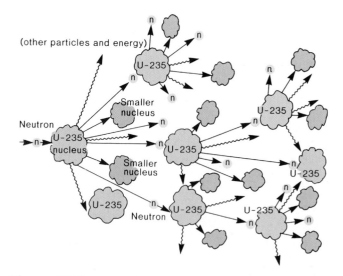

Figure 15.3

Nuclear fission and chain reaction involving uranium-235 (schematic). Neutron capture by uranium-235 causes fission into two smaller nuclei plus additional neutrons, other subatomic particles, and energy. Released neutrons, in turn, cause fission in other uranium-235 nuclei. As U-235 nuclei are used up, reaction rate slows; eventually fresh fuel must replace "spent" fuel.

This scheme is somewhat complicated by the fact that a chain reaction is not sustained by ordinary uranium. Only 0.7% of natural uranium is uranium-235. The material must be processed to increase the concentration of this isotope to several percent of the total to produce reactor-grade uranium. As the reactor operates, the uranium-235 atoms are split and destroyed so that, in time, the fuel is so depleted (spent) in this isotope that it must be replaced with fresh fuel enriched in uranium-235.

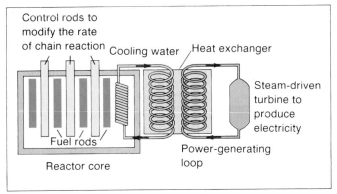

Figure 15.4

Schematic diagram of conventional nuclear fission reactor. Heat is generated by chain reaction; withdrawing or inserting control rods between fuel elements varies rate of reaction, and thus rate of release of heat energy. Cooling water also serves to extract heat for use. Heat is transferred to power loop via heat exchanger, so the cooling water, which may contain radioactive contaminants, is isolated from the power-generating equipment.

The Geology of Uranium Deposits

Worldwide, 95% of known uranium reserves are found in sedimentary or metasedimentary rocks. In the United States, the great majority of deposits are found in sandstone. They were formed by weathering of uranium source rocks, followed by uranium migration in and deposition by ground water.

Minor amounts of uranium are present in many crustal rocks. Granitic rocks and carbonates may be particularly rich in uranium (meaning that its concentration in these may be in the range of ppm to tens of ppm; in most rocks, it is even lower). Uranium is concentrated in carbonate rocks during precipitation of the carbonates from water, including seawater. As uranium-bearing rocks weather under near-surface conditions, that uranium goes into solution: uranium is particularly soluble in an oxygen-rich environment. The uranium-bearing solutions then infiltrate and join the groundwater system. As they percolate through permeable rocks, such as sandstone, they may encounter chemically reducing conditions, created by some factor such as an abundance of carbon-rich organic matter, or sulfide minerals, in shales bounding the sandstone. Under reducing conditions, the solubility of uranium is much lower. The dissolved uranium is then precipitated and concentrated in these reducing zones. Over time, as great quantities of uranium-bearing ground water percolate slowly through such a zone, a large deposit of precipitated uranium ore may form.

World estimates of available uranium are somewhat difficult to obtain, partly because the strategic importance of uranium leads to some secrecy. Even for the United States, the reserve estimates are strongly sensitive to price, as one would expect. Table 15.2 summarizes present reserve and resource estimates at different price levels. With the type of nuclear reactor

Table 15.2	U.S. Uranium Reserve and Resource Estimates	
Recoverable Costs	**Reserves (million pounds of U_3O_8)**	**Resources, Including Reserves (million pounds)**
$30/lb U_3O_8	300	3800
$50/lb U_3O_8	900	6400
$100/lb U_3O_8	1400	9800

Source: Data from Energy Information Administration, U.S. Department of Energy.

Actual uranium prices (in constant 1983 dollars) rose from under $20/lb U_3O_8 in the early 1970s, to over $70/lb on the spot market in 1976, then declined sharply to under $15/lb in 1990, and remained depressed for some years, largely as a result of reduced demand for power-plant fuel. Domestic production virtually ceased in the early 1990s. Peak production of recent years has been below 5 million pounds U_3O_8, while over 55 million pounds were imported in 2010. However, world demand and prices began to rise beginning in 2005, and by late 2010 the price was between $40 and $50 per pound.

currently in commercial operation in the United States, nuclear electricity-generating capacity probably could not be increased to much more than four times present levels by the year 2020 without serious fuel shortages. This increase in production would supply less than 15% of total U.S. energy needs. In other words, the rare isotope uranium-235 is in such short supply that the United States could use up our reserves within several decades, assuming no improvements in reactor technology and no significant uranium imports. Moreover, interest in nuclear power has been rising in the United States as more emphasis is placed on using energy sources that do not increase greenhouse-gas emissions. Worldwide, too, new fission plants continue to be built. So, means of increasing the fuel supply for fission reactors are an issue.

Extending the Nuclear Fuel Supply

Uranium-235 is not the only possible fission-reactor fuel, although it is the most plentiful naturally occurring one. When an atom of the far more abundant uranium-238 absorbs a neutron, it is converted into plutonium-239, which is, in turn, fissionable. Uranium-238 makes up 99.3% of natural uranium and over 90% of reactor-grade enriched uranium. During the chain reaction inside the reactor, as freed neutrons move about, some are captured by uranium-238 atoms, making plutonium. Spent fuel could be *reprocessed* to extract this plutonium, which could be purified into fuel for future reactors, as well as to re-enrich the remaining uranium in uranium-235. How reprocessing would alter the nuclear fuel cycle is shown in figure 15.5. Fuel reprocessing with recovery of both plutonium and uranium could reduce the demand for "new" enriched uranium by an estimated 15%.

A **breeder reactor** can maximize the production of new fuel. Breeder reactors produce useful energy during operation, just as conventional "burners" using up uranium-235 do, by fission in a sustained chain reaction within the reactor core. In addition, they are designed so that surplus neutrons not required to sustain the chain reaction are used to produce more

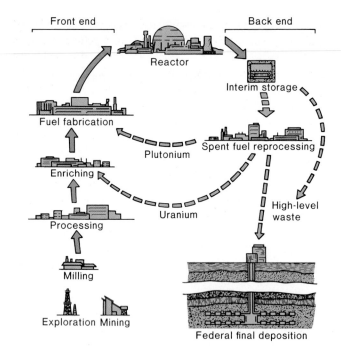

Fuel cycle as it operates currently

Fuel cycle as it would operate with spent fuel reprocessing and federal waste storage

Figure 15.5

The nuclear fuel cycle, as it currently operates and as it would function with fuel reprocessing.

Source: U.S. Energy Information Administration, Department of Energy.

fissionable fuels from suitable materials, like plutonium-239 from uranium-238, or thorium-233, which is derived from the common isotope thorium-232. Breeder reactors could synthesize more nuclear fuel for future use than they would actively consume while generating power.

Breeder-reactor technology is more complex than that of conventional water-cooled burners. The core coolant is liquid metallic sodium; the reactor operates at much higher temperatures. The costs would be substantially higher than for burner reactors, perhaps close to $10 billion per reactor by the time any breeders could be completed in the United States. The breeding process is slow, too: The break-even point after which fuel produced would exceed fuel consumed might be several decades after initial operation. If the nuclear-fission option is to be pursued vigorously in the twenty-first century, the reprocessing of spent fuels and the use of breeder reactors are essential. Yet at present, reprocessing is minimal in the United States, and in early 1985, Congress cancelled funding for the experimental Clinch River breeder reactor, in part because of high estimated costs. No commercial breeders currently exist in the United States, and only a handful of breeder reactors are operating in the world, though some—notably the Superphénix reactor in France—are functioning successfully. Even conventional reactors fell out of favor in the United States for some years, for several reasons.

Concerns Related to Nuclear Reactor Safety

A major concern regarding the use of fission power is reactor safety. In normal operation, nuclear power plants release very minor amounts of radiation, which are believed to be harmless. (A general discussion of radiation and its hazards is presented in chapter 16.) The small but finite risk of damage to nuclear reactors through accident or deliberate sabotage is more worrisome to many.

One of the most serious possibilities is a so-called loss-of-coolant event, in which the flow of cooling water to the reactor core would be interrupted. Resultant overheating of the core might lead to **core meltdown,** in which the fuel and core materials would deteriorate into a molten mass that might or might not melt its way out of the containment building and thus release high levels of radiation into the environment, depending upon the design of the reactor and containment building. A partial loss of coolant, with 35 to 45% meltdown, occurred at Three Mile Island in 1979 (figure 15.6).

No matter how far awry the operation of a commercial power plant might go, and even if there were a complete loss of coolant, the reactor could *not* explode like an atomic bomb. Bomb-grade fuels must be much more highly enriched in the fissionable isotope uranium-235 for the reaction to be that intensive and rapid. Also, the newest reactors have additional safety features designed to reduce the risk of accident. However, an ordinary explosion originating within the reactor (or by saboteurs' use of conventional explosives) could, if large enough, rupture both the containment building and reactor core, and thus release large amounts of radioactive material. The serious accidents at Chernobyl and, more recently, Fukushima reinforced reservations about reactor safety in many people's minds; see Case Study 15.1.

Plant siting is another problem. Siting nuclear plants close to urban areas puts more people potentially at risk in case of accident; placing the plants far from population centers where energy is needed means more transmission loss of

Figure 15.6

Three Mile Island near Harrisburg, Pennsylvania; damaged reactor remains shut down, while others are still operative.

© Doug Sherman/Geofile.

electricity (which already claims nearly 10% of electricity generated). Proximity to water is often important for cooling purposes but makes water pollution in case of mishap more likely. There are also concerns about the structural integrity of nuclear plants located close to fault zones—the Diablo Canyon station in California, the Humboldt Bay nuclear plant, and others; several proposed reactor sites have been rejected when investigation revealed nearby faults. This concern, too, was reinforced in the aftermath of the March 2011 earthquake in Japan.

Concerns Related to Fuel and Waste Handling

The mining and processing of uranium ore are operations that affect relatively few places in the United States (figure 15.7). Nevertheless, they pose hazards because of uranium's natural radioactivity. Miners exposed to the higher radiation levels in uranium mines have experienced higher occurrence rates of some types of cancer. Carelessly handled tailings from processing plants have exposed others to radiation hazards; recall the case of Grand Junction, Colorado, mentioned in chapter 13, where radioactive tailings were unknowingly mixed into concrete used in construction.

The use of reprocessing or breeder reactors to produce and recover plutonium to extend the supply of fissionable fuel

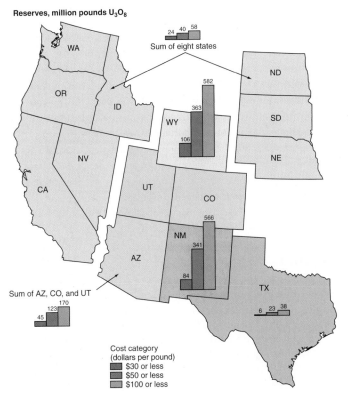

Figure 15.7

Locations of U.S. uranium reserves.

Source: Annual Energy Review 2007, *U.S. Energy Information Administration, Department of Energy.*

poses special problems. Plutonium itself is both radioactive and chemically toxic. Of greater concern to many people is that, as a readily fissionable material, it can also be used to make nuclear weapons. Extensive handling, transport, and use of plutonium would pose a significant security problem and would therefore require very tight inventory control to prevent the material from falling into hostile hands.

The radioactive wastes from the production of fission power are another concern. Radiation hazards and radioactive-waste disposal are considered within the broader general context of waste disposal in chapter 16. Here, two aspects of the problem are highlighted. First, radioactive materials cannot be treated by chemical reaction, heating, and so on to make them nonradioactive. In this respect, they differ from many toxic chemical wastes that can be broken down by appropriate treatment. Second, there has been sufficient indecision about the best method of radioactive-waste disposal and the appropriate site(s) for it that none of the radioactive wastes generated anywhere in the world have been disposed of permanently to date. Currently, the wastes are in temporary storage while various disposal methods are being explored, and many of the temporary waste-holding sites are filled almost to capacity. Clearly, acceptable waste-disposal methods must be identified and adopted, if only to dispose of wastes already accumulated.

Nuclear plants have another unique waste problem. The bombardment of the reactor core and structure by neutrons and other atomic debris from the fission process converts some of the structural materials to radioactive ones and changes the physical properties of others, weakening the structure. At some point, then, the plant faces **decommissioning:** It must be taken out of operation, broken down, and the most radioactive parts delivered to radioactive-waste disposal sites. In late 1982, an electricity-generating plant at Shippingport, Pennsylvania, became the first commercial U.S. fission power plant to face decommissioning, after twenty-five years of operation. Costs of demolition and disposal have exceeded $800 million for a single power plant in the United States, and the process may take a decade or more. Dozens of reactors are being, or have been, decommissioned worldwide. A total of 28 U.S. reactors, including Shippingport, had been retired by the close of 2010. Historically, the nominal lifetime allowed by regulatory agencies was about 30 years; currently, it is 40–50 years; with rigorous plant monitoring and replacement of some key components, it may be possible to operate a fission-powered electricity-generating plant for 70–80 years. Still, sooner or later, decommissioning will take its fiscal and waste-disposal toll. In the United States and worldwide, reactors are aging (figure 15.8).

Risk Assessment, Risk Projection

Much of the debate about energy sources focuses on hazards: what kinds, how many, how large. Many of the risks associated with various energy sources may not be immediately obvious. No energy source is risk-free, but, then, neither is living. The question is, what constitutes "acceptable" risk (and, acceptable to whom)? Underground miners of both coal and uranium face

A Tale of Two Disasters: Chernobyl and Fukushima

On 26 April 1986, an accident occurring at the nuclear power plant at Chernobyl, Ukraine, in the former USSR, dwarfed any U.S. reactor accident in scale and regional impact and reinforced many fears about the safety of the nuclear industry. The accident resulted from a combination of equipment problems and human error, as plant operators conducted an experiment to test system response to unusual conditions. The results were disastrous. A power surge overheated the reactor core; a runaway chain reaction ensued; an explosion and core meltdown occurred. What caused the explosion is not clear. Possibly, leaks allowed water and steam to hit the hot graphite of the reactor core, producing hydrogen gas, which is highly explosive. The explosion would have prevented the insertion of control rods to slow the reaction and disrupted the emergency cooling system.

Quantities of radioactive material escaped from the reactor building, drifting with atmospheric circulation over Scandinavia and eastern Europe (figure 1). The official report was that thirty-one deaths were directly attributable to the accident; 203 persons were hospitalized for acute radiation sickness; 135,000 people were evacuated. Downwind, preventive measures to minimize possible health impacts were suggested: For example, some people consumed large doses of natural iodine to try to saturate the thyroid gland (which concentrates iodine) and thus prevent its absorption of the radioactive iodine in fallout from the reactor. The republics most severely affected were Ukraine and

Byelarus. Radioactive fallout settled on the surrounding land. Ultimately, more than 120,000 people were evacuated from the contaminated area, though some have stayed (figure 2).

Some of the long-range health effects of the accident are still not known, partly because some of these consequences develop long after radiation exposure and partly because the effects of radiation are not always distinguishable from the effects of other agents (see chapter 16). Studies have begun to document increases in medical problems attributable to radiation—thyroid cancer in children, chromosome damage—near the reactor site. Estimates vary widely but put the increased lifetime risk of contracting radiogenic cancer at 0 to 0.02% in Europe, 0.003% in the Northern Hemisphere overall. Over the long term, total cancer cases resulting from the Chernobyl incident are projected at 5000 to 100,000. Unusually high rates of mutations have also been reported in animal populations that have moved into the evacuated region around Chernobyl.

Can a Chernobyl-style accident happen at a commercial nuclear reactor in the United States? Basically, no, because the design of these reactors is fundamentally different from the design of the Chernobyl reactor (though a few research reactors in this country have a core of Chernobyl type). In any reactor, some material must serve as a *moderator*, to slow the neutrons streaming through the core enough that they can interact with nuclei to sustain the chain reaction. In the Chernobyl reactor, the moderator is graphite; the core is built largely of graphite blocks. In commercial U.S. reactors (and most reactors in western nations), the moderator is water. Should the core of a water-moderated reactor overheat, the water would vaporize. Steam is a poor moderator, so the rate of chain reaction would slow down. (Emergency cooling of the core would still be needed, but as the reaction slowed, so would the rate of further heat production.) However, graphite remains solid up to extremely high temperatures, and hot graphite and cold graphite are both effective moderators. In a graphite reactor, therefore, a runaway chain reaction is not stopped by changes in the moderator. At Chernobyl, as the graphite continued to overheat, it began to burn, turning the reactor core into the equivalent of a giant block of

DAY 6

Scale in km

0 100 200   400   600   800   1000   1200

Figure 1

Simulation of the spread of a cloud of radioactive iodine (iodine-131) drifting downwind from the Chernobyl reactor site in the week after the accident. For comparison, maximum *annual* radiation emission permitted at the boundary of a nuclear facility in the United States is 25 rem. (A *rem* is a unit of radiation exposure that takes into account both the energy and nature of the radiation, so that different types of radiation can be compared in terms of their effects on the human body.) Map is as of 0000 Greenwich Mean Time (GMT); the accident began on 26 April 1986 at 1:23 GMT. Day 6 is 1 May.

Source: Data from Mark A. Fischetti, "The Puzzle of Chernobyl" in IEEE Spectrum 23, July 1986, pp. 38–39. Copyright © 1986 IEEE.

6 mi

Cesium-137
curies/sq km
> 40
15–40

Figure 2

Distribution of radioactive fallout from Chernobyl can be estimated from cesium-137 levels. The reactor complex is at the northwest end of the cooling-water pond at lower right of image. In this 1992 false-color Landsat image, cultivated crops are in shades of bright red; abandoned fields that have reverted to native vegetation appear blue-grey. Note that despite the obvious hazard, many areas within the highest-radioactivity zones are still being cultivated. (A *curie* is a measure of radioactivity. Named for Marie Curie, discoverer of radium, it is equal to the activity (radioactive decay rate) of a gram of pure radium: In a sample with activity of one curie, 37 billion atoms would be undergoing radioactive decay every second.)

Image courtesy EROS Data Center, U.S. Geological Survey

radioactive charcoal. (To put it out, workers eventually had to smother it in concrete.) The fuel rods in the burning core ruptured and melted from the extreme heat. The core meltdown, together with the explosion, led to the release of high levels of radioactive materials into the environment. The extent of that release was also far greater for the Chernobyl reactor than it would have been for a U.S. reactor suffering a similar accident because the containment structures of U.S. reactors are much more substantial and effective. The accident focused international attention sharply on the issue of reactor safety and also on the reactor-operator interface and operator training. The incident also emphasized that, like air pollution, radioactive fallout ignores international boundaries.

Japan, with a highly technological society and no fossil-fuel resources, has long embraced nuclear power. The Fukushima accident has caused some lessening of that enthusiasm. (What follows is an abbreviated summary of a very complex series of events and problems; see online references and NetNotes for more complete information.)

As noted in chapter 4, the Fukushima Dai-Ishi plant was designed with both earthquakes and tsunamis in mind. When the 11 March 2011 quake hit, all six reactors were shut down within seconds by automated control-rod insertion; when external power supplies were cut off, a backup generating system to circulate cooling water to the still-hot reactor cores was promptly activated. A tsunami warning did not raise undue concern, as the predicted height was 3 meters and the plant was more than 10 meters above sea level. Unfortunately, the actual tsunami was far higher, estimated later at 14 meters or more. It overwhelmed the protective seawall, surged into the plant, flooded and knocked out eleven of twelve backup generators. Without power, operators had no functioning instruments to monitor the reactors. The company had additional power supplies on trucks that could be brought in in case of emergency—but the scale of this natural disaster was so great that the trucks could not get through the flood of survivors trying to flee the area on badly damaged roads. Plant operators cobbled together car batteries to power their instruments, but the data they obtained were incomplete and, in retrospect, apparently not entirely accurate. Cooling water was boiling to steam and pressure was building inside several reactor buildings; reactor 1 was in the worst condition, with core meltdown beginning. Fire trucks were brought in to pump first fresh water, then (when the fresh ran out) salt water, to try to cool the reactors' cores. Local residents were evacuated to allow workers to vent some of the accumulated (now-radioactive) gas from inside containment vessels to reduce pressure. Plant workers struggled to get control of the situation and bring all reactors to stable shutdown. But on the afternoon of 13 March, a stray spark ignited hydrogen gas that had formed from the hot steam inside reactor 1; the explosion released considerable radiation, and made things both harder and more dangerous for workers by littering the whole complex with rubble and radioactive material. Explosions eventually occurred in units 2, 3, and 4, and meltdowns in units 2 and 3, as well. The site is still being stabilized and cleaned up.

Given that Japan is an island nation, and given the prevailing winds and currents, there was no major radiation threat to other countries. The World Health Organization has also confirmed that radiation in fallout and runoff to the Pacific Ocean would have been diluted quickly enough not to contaminate the region's seafood supplies. But more than 100,000 people were evacuated in connection with the Fukushima accident, many of whom are still not able to return home. As with Chernobyl, there remains a risk of radiation-related health problems, particularly among plant workers and emergency personnel who were exposed to the highest radiation levels, and the full extent of these problems will not be known for decades. A Japanese investigation has produced a number of recommendations for changes that collectively would have averted the Fukushima disaster if implemented beforehand, but Japan and other nations are nevertheless reconsidering the nuclear-power option in the light of this latest incident.

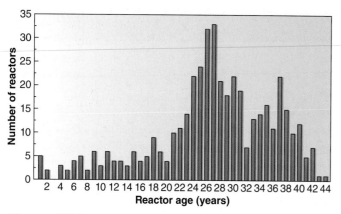

Figure 15.8

Age distribution of nuclear reactors worldwide, December 2011. Few face decommissioning yet, but many may in the next decade or two.

Source: International Atomic Energy Agency.

increased health risks, but the power consumer does not see those risks directly. Those living downstream from a dam are at risk; so are those downwind from a nuclear or fossil-fuel plant.

A power plant of 1-billion-watt capacity serves the electricity needs of about 1 million people. One study in the mid-1970s projected accidental deaths at 0.2 per year for a uranium-fueled plant this size (virtually all associated with mining), compared to an estimated 2.6 to 4 deaths per year for a comparable coal-fired plant, depending upon mining method, including significant risks associated with processing and transporting the very much larger volume of coal involved. By way of comparison, in the U.S. in 2005, per million people, there were 146 deaths from motor-vehicle accidents, 110 from accidental poisoning, 13 from drowning, 66 from falls, and 5 from accidents involving firearms.

A further consideration when comparing risks may be how well-defined the risks are. That is, estimates of risks from fires, automobile accidents, and so on are actuarial, based on considerable actual past experience and hard data. By contrast, projections of risks associated with extensive use of fission power are based on limited past experience, plus educated guesses about future events based, in part, on hard-to-quantify parameters like probability of human errors. In 1975, a report prepared for the U.S. Nuclear Regulatory Commission estimated the risk of core meltdown at one per million reactor-years of operation. To date, total global fission-reactor operations add up to about 15,000 reactor-years, so stastically the accident rate is running far ahead of that estimate. Each such event leads to improvements in design and operation of reactors, but the history has given many people pause.

In the early 1970s, it was predicted that 25% of U.S. energy would be supplied by nuclear power by the year 2000. At the end of 2010, there were 104 nuclear power plants (figure 15.9) operating

Figure 15.9

Distribution of U.S. nuclear power plants. In December 2010 there were 104 operable plants. In addition, four new plants have permits to begin construction, and license applications for 17 more projects were pending with the Nuclear Regulatory Commission, reflecting renewed interest in the use of nuclear power for generating electricity. (Due to space limitations, symbols do not represent actual locations; the number of plants in each state *is* accurate.)

Source: Data from U.S. Energy Information Administration, U.S. Nuclear Regulatory Commission, and International Atomic Energy Agency.

in the United States (down from a 1990 peak of 112), and they accounted for only 9% of U.S. energy production. For some years, cancellations far exceeded new orders, as utilities decided to power their new generating plants with coal, not nuclear fission.

Nuclear plants do have lower fueling and operating costs than coal-fired plants. The small quantity of fuel required for their operation can also be more easily stockpiled against interruption by strikes or transportation delays. However, nuclear plants are more costly to build than coal-fired plants and now require longer to plan, construct, and license (nine to twelve years for U.S. nuclear plants; six to ten for coal). These economic and time factors, together with increasingly vocal public opposition to nuclear power plants (itself a contributing factor in many licensing delays), made the coal option more appealing to many utilities.

Worldwide, reliance on nuclear power varies widely. Some nations have no nuclear-power plants. They account for only 2 to 5% of electricity production in such countries as Mexico, India, Pakistan, and China, but over 30% in much of Europe, and in some countries (Slovakia, Belgium, France), over 50% (figure 15.10). Some nations have turned to nuclear fission for lack of fossil-fuel resources. With the twenty-first century's increasing focus on limiting CO_2 emissions, the clean operation of nuclear fission plants has enhanced their appeal to many. In December 2010 there were 441 operable nuclear fission plants in the world, and 67 more under construction, including 28 in China, 11 in Russia, 6 in India, and 5 in Korea. The number of plants under construction increased more than 60% from 2006 to 2009, reflecting fission's renewed attractiveness.

Different people weigh the pros and cons of nuclear fission power in different ways. For many, the uncertainties and potential risks, including the dangers of reactor accident, outweigh the benefits. For others, the problems associated with using coal (the most readily available alternative for generating electricity) appear at least as great. At present, nonfossil alternatives to nuclear fission are seriously inadequate in scale to replace it, as we will see.

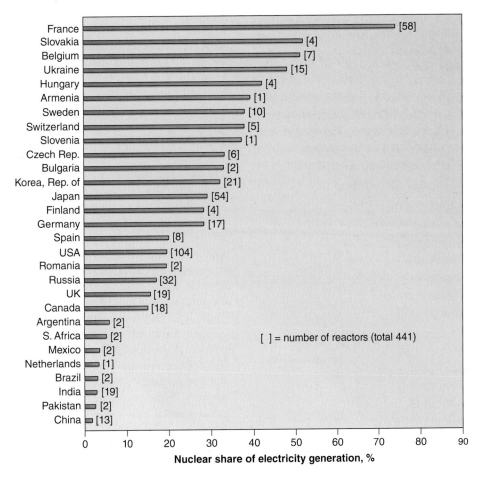

Figure 15.10

Percentage of electricity generated by nuclear fission varies greatly by country, and not simply in proportion to numbers of reactors; where electricity consumption is moderate, a few reactors can account for a large share.

Source: Data for December 2010; from International Atomic Energy Agency.

Nuclear Power—Fusion

Nuclear *fusion* is the opposite of fission. As noted earlier, fusion is the process by which two or more smaller atomic nuclei combine to form a larger one, with an accompanying release of energy. It is the process by which the sun generates its vast amounts of energy. In the sun, simple hydrogen nuclei containing one proton are fused to produce helium. For technical reasons, fusion of the heavier hydrogen isotopes deuterium (nucleus containing one proton and one neutron) and tritium (one proton and two neutrons) would be easier to achieve on earth. That fusion reaction is diagrammed in figure 15.11.

Hydrogen is plentiful because it is a component of water; the oceans contain, in effect, a huge reserve of hydrogen, an essentially inexhaustible supply of the necessary fuel for fusion. This is true even considering the scarcity of deuterium, which is only 0.015% of natural hydrogen. (Tritium is rarer still and would probably have to be produced from the comparatively rare metal lithium. However, the magnitude of energy release from fusion reactions is such that this is not a serious constraint.) The principal product of the projected fusion reactions—helium—is a nontoxic, chemically inert, harmless gas. There could be some mildly radioactive light-isotope by-products of fusion reactors, but they would be much less hazardous than many of the products of fission reactors.

Since fusion is a far "cleaner" form of nuclear power than fission, why not use it? The principal reason is technology, or lack of it. To bring about a fusion reaction, the reacting nuclei must be brought very close together at extremely high temperatures (millions of degrees at least). The natural tendency of hot gases is to expand, not come together, and no known physical material could withstand such temperatures to contain the reacting nuclei. The techniques being tested in laboratory fusion experiments are elaborate and complex, either involving containment of the fusing materials with strong magnetic fields or using lasers to heat frozen pellets of the reactants very rapidly. At best, the experimenters have been able to achieve the necessary conditions for fractions of a second, and the energy required to bring about the fusion reactions has exceeded the energy released thereby. Certainly, the present technology does not permit the construction of commercial fusion reactors in which controlled, sustained fusion could be used as a source of energy.

Scientists in the field estimate that several decades of intensive, expensive research will be needed before fusion can become a commercial reality. In tight economic times, it can be a hard sell. In 2008, the Department of Energy cancelled funding for what would have been the fourth experimental fusion reactor in the United States, in part due to cost overruns. One such reactor being funded by an international consortium and built in France has an estimated cost of $12 billion; its construction is not expected to be complete until 2018, with initial power production projected in 2026. Even when the technology for commercial fusion power is developed, it will be costly. Some estimates suggest that a commercial fusion electricity-generating plant could cost tens of billions of dollars. Nevertheless, the abundance of the fuel supply and the relative cleanness of fusion (as compared to both fission and fossil-fuel power generation) make it an attractive prospect, at least for later in the twenty-first century. Fusion, however, is a means of generating electricity in stationary power plants only. This fact limits its potential contribution toward satisfying total energy needs.

Solar Energy

The earth intercepts only a small fraction of the energy radiated by the sun. Much of that energy is reflected or dissipated in the atmosphere. Even so, the total solar energy reaching the earth's surface far exceeds the energy needs of the world at present and for the foreseeable future. The sun can be expected to go on shining for approximately 5 billion years—in other words, the resource is inexhaustible, which contrasts sharply with nonrenewable sources like uranium or fossil fuels. Sunlight falls on the earth without any mining, drilling, pumping, or disruption of the land. Sunshine is free; it is not under the control of any company or cartel, and it is not subject to embargo or other political disruption of supply. The *use* of solar energy is essentially pollution-free, at least in the sense that the absorption of sunlight for heat or the operation of a solar cell for electricity are very "clean" processes (though solar electricity has other environmental costs; see "Solar Electricity"). It produces no hazardous solid wastes, air or water pollution, or noise. For most current applications, solar energy is used where it falls, thereby avoiding transmission losses. All these features make solar energy an attractive option for the future. Several practical limitations on its use also exist, however, particularly in the short term.

The solar energy reaching the earth is dissipated in various ways. Some is reflected into space; some heats the

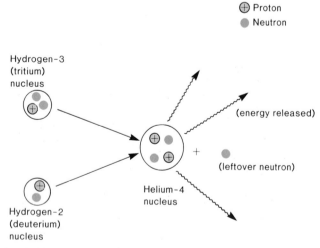

Figure 15.11

Schematic diagram of one nuclear fusion reaction. Other variants are possible.

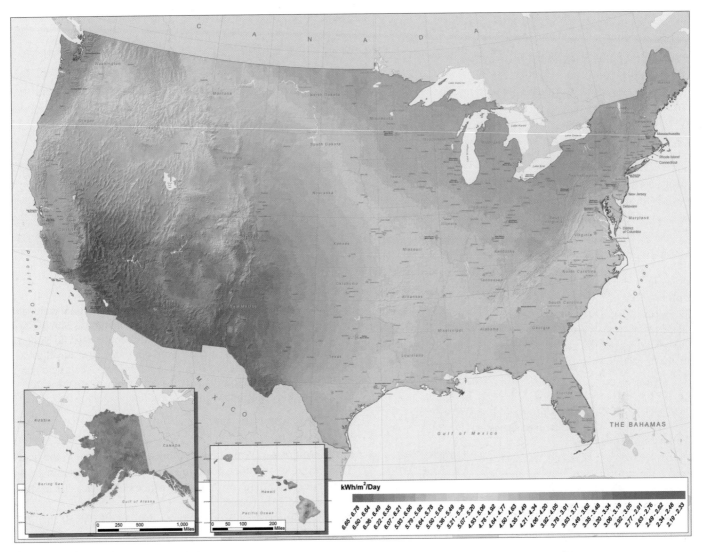

Figure 15.12

Solar-energy potential for electricity generation by photovoltaic (solar) cells. Units of kWh/m^2/day are kilowatt-hours per square meter per day. Clearly, maximum potential in the United States is in the Southwest.

Map by B. Roberts, courtesy National Renewable Energy Laboratory, U.S. Department of Energy.

atmosphere, land, and oceans, driving ocean currents and winds. The sun supplies the energy needed to cause evaporation and thus to keep the hydrologic cycle going; it allows green plants to generate food by photosynthesis. More than ample energy is still left over to provide for all human energy needs, in principle. However, the energy is distributed over the whole surface of the earth. In other words, it is a very dispersed resource: Where large quantities of solar energy are used, solar energy collectors must cover a wide area. Sunlight is also variable in intensity, both from region to region (figure 15.12) and from day to day as weather conditions change. The two areas in which solar energy can make the greatest immediate contribution are in space heating and in the generation of electricity, uses that together account for about two-thirds of U.S. energy consumption.

Solar Heating

Solar space heating typically combines direct use of sunlight for warmth with some provision for collecting and storing additional heat to draw on when the sun is not shining. It is typically employed on the scale of an individual building. The simplest approach is *passive-solar heating,* which does not require mechanical assistance. The building design should allow the maximum amount of light to stream in through south and west windows during the cooler months. This heats the materials inside the house, including the structure itself, and the radiating heat warms indoor air. Media used specifically for storing heat include water—in barrels, tanks, even indoor swimming pools—and the rock, brick, concrete, or other dense solids used in the building's construction (figure 15.13A). These supply a

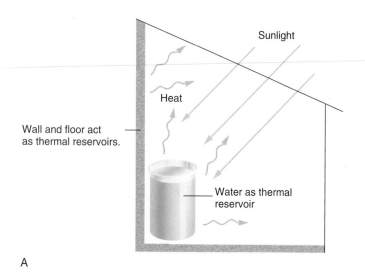

A

Wall and floor act as thermal reservoirs.

Sunlight

Heat

Water as thermal reservoir

In summer, leaves and wide eaves together shade windows, reducing incoming sunlight.

Lower-angle winter sun streams through tree branches and under eaves.

B

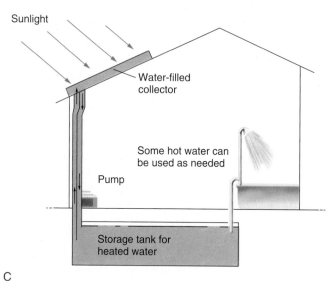

Sunlight

Water-filled collector

Some hot water can be used as needed

Pump

Storage tank for heated water

C

Figure 15.13

(A) Basics of passive-solar heating with water or structural materials as thermal reservoir: Sunlight streams into greenhouse with glass roof and walls, heat is stored for nights and cloudy days. (B) Design features of home and landscaping can optimize use of sun in colder weather, provide protection from it in summer. (C) A common type of active-solar heating system with a pump to circulate the water between the collector and the heat exchanger/storage tank.

thermal mass that radiates heat back when needed, in times of less (or no) sunshine. Additional common features of passive-solar design (figure 15.13B) include broad eaves to block sunshine during hotter months (feasible because the sun is higher in the sky during summer than during winter) and drapes or shutters to help insulate window areas during long winter nights. Other possible variations on passive-solar heating systems are too numerous to describe here.

Active-solar heating systems usually involve the mechanical circulation of solar-heated water (figure 15.13C). The flat solar collectors are water-filled, shallow boxes with a glass surface to admit sunlight and a dark lining to absorb sunlight and help heat the water. The warmed water is circulated either directly into a storage tank or into a heat exchanger through which a tank of water is heated. The solar-heated water can provide both space heat and a hot-water supply. If a building already uses conventional hot-water heat, incorporating solar collectors is not necessarily extremely expensive (especially considering the free "fuel" to be used). With the solar collectors mounted on the roof, an active-solar system does

not require the commitment of any additional land to the heating system—another positive feature. The method can be as practical for urban row houses or office buildings as for widely spaced country homes.

While solar heating may be adequate by itself in mild, sunny climates, in areas subject to prolonged spells of cloudiness or extreme cold, a conventional backup heating system is almost always needed. In the latter areas, then, solar energy can greatly reduce but not wholly eliminate the need for some consumption of conventional fuels. It has been estimated that, in the United States, 40 to 90% of most homes' heating requirements could be supplied by passive-solar heating systems, depending on location. It is usually more economical to design and build passive-solar technology features into a new structure initially than to incorporate them into an existing building later (retrofit). (A few 100%-solar homes have been built, even in such severe climates as northern Canada. However, the variety of solar-powered devices, extra insulation, and other features required can add tens of thousands of dollars to the cost of such homes, especially when an existing home is retrofitted, pricing this option out of the reach of many home buyers. Moreover, superinsulation can aggravate indoor air pollution; see chapter 18.)

Solar Electricity

Direct production of electricity using sunlight is accomplished through **photovoltaic cells,** also called simply "solar cells" (see figure 15.14). In simplest form, they consist of two layers of

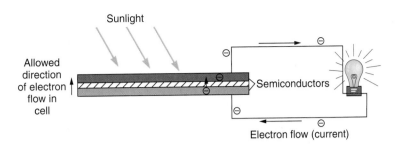

Sunlight

Allowed direction of electron flow in cell

Semiconductors

Electron flow (current)

Figure 15.14

In a photovoltaic cell, incident sunlight dislodges electrons in top layer, which flow through wire as electric current, to return to the other side of the cell. Accumulated electrons move back to upper layer of cell and the cycle continues.

semiconductor material sandwiched together, with a barrier between that allows electrons to flow predominantly in one direction only. Sunlight striking the exposed side can dislodge some electrons, which flow as electric current through a circuit, to return to the cell and continue the cycle.

Solar cells have no moving parts and, like solar heating systems, do not emit pollutants during operation. For many years, they have been the principal power source for satellites and for remote areas difficult to reach with power lines (figure 15.15). A major limitation on solar-cell use has historically been cost, which is several times higher per unit of power-generating capacity than for either fossil-fuel or nuclear-powered generating plants. This has restricted the appeal of home-generated solar electricity. The high cost is partly a matter of technology (present solar cells are not very efficient, though they are being improved) and partly one of scale (the industry is not large enough to enjoy the economies of mass production).

Figure 15.15

Solar electricity is very useful in remote areas. High in the mountains of Denali National Park, at the base camp that is the takeoff point for expeditions to climb Denali (Mount McKinley), solar cells (right) power vital communications equipment.

Currently, low solar-cell efficiency and the diffuse character of sunlight continue to make photovoltaic conversion an inadequate option for energy-intensive applications, such as many industrial and manufacturing operations. Even in the areas of strongest sunlight in the United States, incident radiation is of the order of 250 watts per square meter. Operating with commercial solar cells of about 20% efficiency means power generation of only 50 watts per square meter. In other words, to keep one 100-watt lightbulb burning would require at least 2 square meters of collectors (with the sun always shining). A 100-megawatt power plant would require 2 square kilometers of collectors (nearly one square mile!), and many nuclear or coal-fired generating plants have more than ten times that capacity. Using solar cells at that scale represents a large commitment of both land and the mineral resources from which the collectors are made.

For the collector array alone, a 100-megawatt solar electric plant operating with basic solar cells would use at least an estimated 30,000 to 40,000 tons of steel, 5000 tons of glass, and 200,000 tons of concrete. A nuclear plant, by contrast, would require about 5000 tons of steel and 50,000 tons of concrete; a coal-fired plant, still less. The solar cells present additional resource issues as their use is scaled up.

There are various photovoltaic technologies, but some of the most efficient use materials such as gallium, arsenic, selenium, indium, and tellurium. Most of these are potentially toxic, presenting health hazards in both the mining and manufacturing processes. The United States is essentially 100% dependent on imports for arsenic, gallium, and indium. As demand has increased, prices of some of these elements have risen spectacularly: tellurium went from $13 to $350 per kilogram just over seven years, 2004 to 2011.

Siting a sizable array requires a substantial commitment and disturbance of land. Its presence could alter patterns of evaporation and surface runoff. These considerations could be especially critical in desert areas, which are the most favorable sites for such facilities from the standpoint of intensity and constancy of incident sunlight. Construction could also disturb desert-pavement surfaces and accelerate erosion. It has even been suggested that covering large areas of light-colored, reflective desert sand with sunlight-absorbing solar panels could increase earth's net heating, exacerbating global warming.

Figure 15.16

To make solar electricity without photovoltaic cells, solar heat can be used to make steam to power turbines, but it must first be concentrated, as here in the California desert, where parabolic mirrors focus sunlight on tubes of water.

Photograph © The McGraw-Hill Companies, Inc./Doug Sherman, photographer.

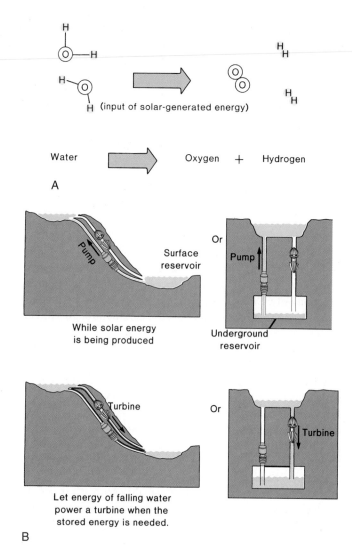

Figure 15.17

Some possible schemes for storing the energy of solar-generated electricity. (A) Use solar electricity to break up water molecules into hydrogen and oxygen; recombine them later (burn the hydrogen) to release energy. (B) Use solar energy to pump water up in elevation; when the energy is needed, let the water fall back and use it to generate hydropower.

Clearly, then, for large-scale applications, it would be important to concentrate the solar energy being used to generate electricity. This can be done using arrays of parabolic mirrors, which can focus sunlight from a broader area onto a small expanse of solar cells. Or, the solar energy can be concentrated to heat water or another medium to run turbines to generate electricity, just as other fuels heat water to make steam for power generation. This approach uses no photovoltaic cells at all (figure 15.16), though large mirror arrays are then needed, and these involve their own mineral-resource issues: They are typically coated with silver, for which we are already 75% dependent on imports and which is becoming more expensive (the price rose from \$6.70 to \$34.50 per troy ounce from 2004 to 2011). Or, they could be made of aluminum—of which we import 100%.

Storing solar electricity is also a more complex matter than storing heat. For individual homeowners, batteries may suffice, but no wholly practical scheme for large-scale storage has been devised, despite advances in battery technology. Some of the proposals are shown in figure 15.17.

Solar-generated electricity could eventually supply perhaps 10 to 15% of U.S. electricity needs, but currently, it accounts for less than one-tenth of a percent. Major improvements in efficiency of generation and in storage technology are needed before the sun can be the principal source of electricity, resource and land-use issues aside. It may be still longer before solar electricity can contribute to the transportation energy budget.

Geothermal Energy

The earth contains a great deal of heat, some of it left over from its early history, some continually generated by decay of radioactive elements in the earth. Slowly, this heat is radiating away and the earth is cooling down, but under normal circumstances, the rate of heat escape at the earth's surface is so slow that we do not even notice it and certainly cannot use it. If the heat escaping at the earth's surface were collected over an average square meter for a year, it would be sufficient only to heat about 2 gallons of water to the boiling point, though local heat flow can be substantially higher, for example in young volcanic areas.

Seasonal variations in surface temperatures do not significantly affect this outward heat flow, and indeed do not penetrate very deeply underground, because rocks and soils conduct heat poorly. This characteristic can easily be demonstrated by turning over a flat, sunbaked rock on a bright but cool day. The surface exposed to the sun may be hot to the touch, but the heat will not have traveled far into the rock; the underside stays cool for some time. So, temperatures tend to remain nearly constant only a few tens of feet underground. This is the key fact that makes possible a type of geothermal technology that can address both heating and cooling needs at the scale of individual buildings: geothermal heat pumps. Over much of the United States, at a depth of 20 feet, temperatures fluctuate only a few degrees over the year, and average 50 to 60°F (10–15°C). Pipes laid in the deep soil can circulate water between this zone of fairly constant temperature and a heat exchanger at the surface, drawing on earth's heat to warm the building in winter when the air is colder, and/or carrying away heat from the building into the soil when surface air is hotter. No emissions are generated in the process. Supplementary heating or cooling may also be required, but the EPA reports that typically, homeowners save 30 to 70% on heating and 20 to 50% on cooling using such "geoexchange" systems, and over 1 million such systems are currently installed in the United States.

Larger-scale applications of geothermal energy require more specialized conditions.

Traditional Geothermal Energy Uses

Magma rising into the crust from the mantle brings unusually hot material nearer the surface. Heat from the cooling magma heats any ground water circulating nearby (figure 15.18). This is the basis for extracting **geothermal energy** on a commercial scale. The magma-warmed waters may escape at the surface in geysers and hot springs, signalling the existence of the shallow heat source below (figure 15.19). More subtle evidence of the presence of hot rock at depth comes from sensitive measurements of heat flow at the surface, the rate at which heat is being conducted out of the ever-cooling earth: High heat flow signals unusually high temperatures at shallow depths. High heat flow and recent (or even current) magmatic activity go together and, in turn, are most often associated with plate boundaries. Therefore, most areas in which geothermal energy is being tapped extensively are along or near plate boundaries (figure 15.20).

Exactly how the geothermal energy is used depends largely on how hot the system is. In some places, the ground water is warmed, but not enough to turn to steam. Still, the

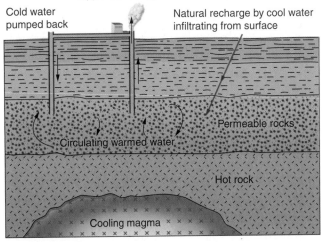

Figure 15.18
Geothermal energy is utilized by tapping circulating warmed ground water.

Figure 15.19
One of the many thermal features in Yellowstone National Park: Lone Star Geyser. Structure is built by deposition of dissolved minerals.

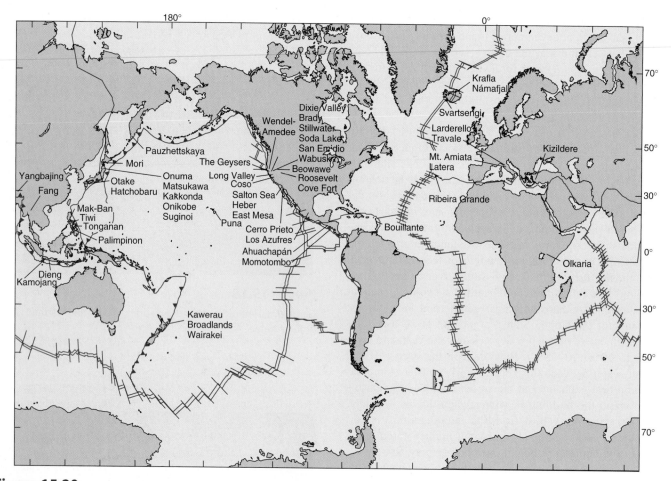

Figure 15.20

Geothermal power plants worldwide.

Source: Figure prepared by L. J. Patrick Muffler and Ellen Lougee, U.S. Geological Survey; plate boundaries supplied by Charles DeMets, University of Wisconsin at Madison.

water temperatures may be comparable to what a home heating unit produces (50 to 90°C, or about 120 to 180°F). Such warm waters can be circulated directly through homes to heat them. This is being done in Iceland and in parts of the former Soviet Union.

Other geothermal areas may be so hot that the water is turned to steam. The steam can be used, like any boiler-generated steam produced using conventional fuels, to run electric generators. The largest U.S. geothermal-electricity operation is The Geysers, in California (figure 15.21), which has operated since 1960 and now has a generating capacity of close to 2 billion watts. In 1989, The Geysers and six smaller geothermal areas together generated close to 10 billion kilowatt-hours of electricity (0.3% of total energy consumed) in this country, though its output has since declined, as explained below. Other steam systems are being used in Larderello, Italy, and in Japan, Mexico, the Philippines, and elsewhere. Altogether, there are about forty sites worldwide where geothermal electricity is actively being developed.

Where most feasible, geothermal power is quite competitive economically with conventional methods of generat-

ing electricity. The use of geothermal steam is also largely pollution-free. Some sulfur gases derived from the magmatic heat source may be mixed with the steam, but these certainly pose no more serious a pollution problem than sulfur from coal burning. Moreover, there are no ash, radioactive-waste, or carbon-dioxide problems as with other fuels. Warm geothermal *waters* may be a somewhat larger problem. They frequently contain large quantities of dissolved chemicals that not only build geyser structures as in figure 15.19, but that can clog or corrode pipes (a potentially significant problem that will increase operational costs) or may pollute local ground or surface waters if allowed to run off freely. Sometimes, there are surface subsidence problems, as at Wairakei (New Zealand), where subsidence of up to 0.4 meters per year has been measured. Subsidence problems may be addressed by reinjecting water, but the long-term success of this strategy is unproven. The fluid injection—whether to counter subsidence or for future heat extraction—may cause its own problems; studies at The Geysers indicate increased seismicity associated with fluid injection, though most of the earthquakes are below magnitude 4.

Figure 15.21

The Geysers geothermal power complex, California, is the largest such facility in the world.

Photograph © The McGraw-Hill Companies, Inc./John A. Karachewski, photographer.

While the environmental difficulties associated with geothermal power are relatively small, three other limitations severely restrict its potential. First, each geothermal field can only be used for a period of time—a few decades, on average—before the rate of heat extraction is seriously reduced. This is a negative consequence of the fact that rocks conduct heat very poorly. As hot water or steam is withdrawn from a geothermal field, it is replaced by cooler water that must be heated before use. Initially, the heating can be rapid, but in time, the permeable rocks become chilled to such an extent that water circulating through them heats too slowly or too little to be useful. The heat of the magma has not been exhausted, but its transmittal into the permeable rocks is slow. Some time must then elapse before the permeable rocks are sufficiently reheated to resume normal operations. Steam pressure at The Geysers has declined rapidly in recent years, forcing the idling of some of its generating capacity: By 1991, despite a capacity of 2 billion watts, electricity production was only 1½ billion watts; power generation had declined to half its 1987 peak within a decade, by 2001 was down to 1.2 billion watts, with the first four installed generating units entirely idle, and by 2010 had further decreased, to 725 million watts.

A second limitation of geothermal power is that not only are geothermal power plants stationary, but so is the resource itself. Oil, coal, or other fuels can be moved to power-hungry population centers. Geothermal power plants must be put where the hot rocks are, and long-distance transmission of the power they generate is not technically practical, or, at best, is ineffi-

cient. Most large cities are far removed from major geothermal resources. Also, of course, geothermal power cannot contribute to such energy uses as transportation.

The total number of sites suitable for geothermal power generation is the third limitation. Clearly, plate boundaries cover only a small part of the earth's surface, and many of them are inaccessible (seafloor spreading ridges, for instance). Not all have abundant circulating subsurface water in the area, either. Even accessible regions that do have adequate subsurface water may not be exploited. Yellowstone National Park has the highest concentration of thermal features of any single geothermal area in the world, but because of its scenic value and uniqueness, the decision was made years ago not to build geothermal power plants there.

Alternative Geothermal Sources

Many areas away from plate boundaries have heat flow somewhat above the normal level, and rocks in which temperatures increase with depth more rapidly than in the average continental crust. The **geothermal gradient** is the rate of increase of temperature with increasing depth in the earth. Even in ordinary crust, the geothermal gradient is about 30°C/kilometer (about 85°F/mile). Where geothermal gradients are at least 40°C/kilometer, even in the absence of much subsurface water, the region can be regarded as a potential geothermal resource of the **hot-dry-rock** type. Deep drilling to reach usefully high temperatures must be combined with induced circulation of water pumped in from the surface (and perhaps artificial fracturing to increase permeability, as in some oil and gas fields) to make use of these hot rocks. The amount of heat extractable from hot-dry-rock fields is estimated at more than ten times that of natural hot-water and steam geothermal fields, just because the former are much more extensive. Most of the regions identified as possible hot-dry-rock geothermal fields in the United States are in thinly populated western states with restricted water supplies (figure 15.22). There is, therefore, a large degree of uncertainty about how much of an energy contribution they could ultimately make. Certainly, hot-dry-rock geothermal energy will be less economical than that of the hot-water or steam fields where circulating water is already present. Although some experimentation with hot-dry-rock fields is underway—Los Alamos National Laboratory has an experimental project in operation in the Jemez Mountains of New Mexico—commercial development in such areas is not expected in the immediate future.

The geopressurized natural gas zones described in chapter 14 represent a third possible type of geothermal resource. As hot, gas-filled fluids are extracted for the gas, the heat from the hot water might, in theory, also be used to generate electricity. Substantial technical problems probably will be associated with developing this resource, as was noted. Moreover, it probably would be economically infeasible to pump the spent water back to such depths for recycling. No geopressurized zones are presently being developed for geothermal power, even on an experimental basis.

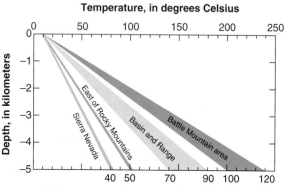

Figure 15.22

The area east of the Rocky Mountains has a geothermal gradient and surface heat flow typical of world average continental crust; selected areas west of the Rockies have more energy potential. The faster temperature increases with depth, the closer to the surface are usefully warm rocks, and—all else being equal—the greater the geothermal-energy potential.

After USGS Circular 1249.

Hydropower

The energy of falling or flowing water has been used for centuries. It is now used primarily to generate electricity. Hydroelectric power has consistently supplied a small percentage of U.S. energy needs for several decades; it currently provides about 2.5% of U.S. energy (about 6% of U.S. electricity). The principal requirements for the generation of substantial amounts of hydroelectric power are a large volume of water and the rapid movement of that water. Nowadays, commercial generation of hydropower typically involves damming up a high-discharge stream, impounding a large volume of water, and releasing it as

desired, rather than operating subject to great seasonal variations in discharge. The requirement of plentiful surface water is reflected in large regional variations in water use for hydropower generation (figure 15.23).

Hydropower is a very clean energy source in that the water is not polluted as it flows through the generating equipment. No chemicals are added to it, nor are any dissolved or airborne pollutants produced. The water itself is not consumed during power generation; it merely passes through the generating equipment. In fact, water use for hydropower in the United States is estimated to be more than 2½ times the average annual surface-water runoff of the nation, which is possible because the same water can pass through multiple hydropower dams along a stream. Hydropower is renewable as long as the streams continue to flow. Its economic competitiveness with other sources is demonstrated by the fact that nearly one-third of U.S. electricity-generating plants are hydropower plants; worldwide, about 7% of all energy consumed is hydropower.

The Federal Power Commission has estimated that the potential energy to be derived from hydropower in the United States, if it were tapped in every possible location, is about triple current hydropower use. In principle, then, hydropower could supply one-fifth of U.S. electricity if consumption remained near present levels. However, development is unlikely to occur on such a scale.

Limitations on Hydropower Development

We have already considered, in chapters 6 and 11, some of the problems posed by dam construction, including silting-up of reservoirs, habitat destruction, water loss by evaporation, and even, sometimes, earthquakes. Evaluation of the risks of various energy sources must also consider the possibility of dam failure. There are over a thousand dams in the United States (not all constructed for hydropower generation). Several dozen have failed within the twentieth century. Aside from age and poor design or construction, the reasons for these failures may include geology itself. Fault zones often occur as topographic lows, and streams thus frequently flow along fault zones. It follows that a dam built across such a stream is built across a fault zone, which may be active or may be reactivated by filling the reservoir. Not all otherwise-suitable sites, then, are safe for hydropower dams.

Indeed, the late 1990s saw some reversals in U.S. hydropower development. For the first time, the Federal Energy Regulatory Commission ordered the dismantling of a dam, over the owner's objections: the Edwards Dam in Maine, the removal of which began in July of 1999. The move was a response to concerns over the cumulative environmental impact of 162 years of the dam's existence, and as it was a small facility (accounting for less than 1% of Maine's electric power), opposition to the dismantling was limited. More-vigorous objections are being made to the proposed removal of four dams on the Snake River in Washington state. Concern for local salmon populations is among the reasons for the removal, but as these dams and reservoirs are more extensively used both

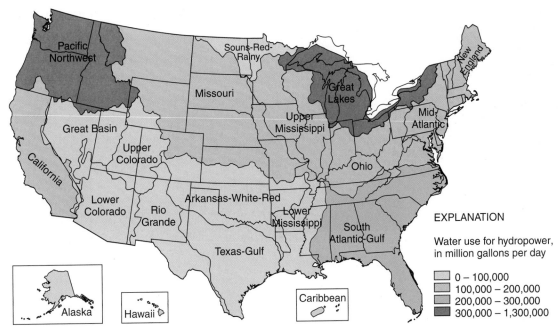

Figure 15.23

Water use for hydropower generation in the United States is naturally concentrated where streamflow is plentiful.

Source: 1995 Estimated Water Use in the United States, *U.S. Geological Survey.*

EXPLANATION

Water use for hydropower, in million gallons per day

- 0 – 100,000
- 100,000 – 200,000
- 200,000 – 300,000
- 300,000 – 1,300,000

for power generation and to supply irrigation water, their destruction would have more negative impact on local people than did the Edwards Dam removal. As of this writing, these dams' ultimate fate is uncertain.

Other sites with considerable power potential may not be available or appropriate for development. Construction might destroy a unique wildlife habitat or threaten an endangered species. It might deface a scenic area or alter its natural character; suggestions for additional power dams along the Colorado River that would have involved backup of reservoir water into the Grand Canyon were met by vigorous protests. Many potential sites—in Alaska, for instance—are just too remote from population centers to be practical unless power transmission efficiency is improved.

An alternative to development of many new hydropower sites would be to add hydroelectric-generating facilities to dams already in place for flood control, recreational purposes, and so on. Although the release of impounded water for power generation alters streamflow patterns, it is likely to have far less negative impact than either the original dam construction or the creation of new dam/reservoir complexes. Still, it is clear that flood control and power generation are somewhat conflicting aims: the former requires considerable reserve storage capacity, while the latter is enhanced by impounding the maximum volume of water.

Like geothermal power, conventional hydropower is also limited by the stationary nature of the resource. In addition, hydropower is more susceptible to natural disruptions than other sources considered so far. Just as torrential precipitation and 100-year floods are rare, so, too, are prolonged droughts—but they do happen. U.S. hydropower generation declined by

about 25% from 1986 to 1988, largely as a consequence of drought. A western drought that began in 1999 dropped water levels in Lake Powell, the reservoir behind Glen Canyon Dam (figure 15.24), by more than 100 feet, reducing water storage by more than 50%. Hydropower generation at the dam declined from a 1997 high of 6.7 billion kilowatt-hours (kwh) to just 3.2 billion kwh in 2005, and it has remained depressed along with the lake level, though water storage and power generation both rebounded somewhat from 2007 to 2012. Heavier reliance on hydropower could thus leave many energy consumers vulnerable to interruption of service in times of extreme weather.

For various reasons, then, it is unlikely that numerous additional hydroelectric power plants will be developed in the United States. This clean, cheap, renewable energy source can continue indefinitely to make a modest contribution to energy use, but it cannot be expected to supply much more energy in the future than it does now. Still, hydropower is an important renewable energy source in the United States (figure 15.25).

Worldwide, future hydropower development is expected to be most rapid in Asia, but opposition exists there, too. The largest such project ever, the Three Gorges Dam project in China went forward despite international protests from environmental groups. Supply disruptions are not unique to the United States, either: Serious droughts in 1999 in Latin and South America sharply reduced generating capacity at existing hydroelectric facilities, forcing Mexico to buy electricity from the United States and causing Chile to look toward developing more fossil-fuel power-generating capacity as an alternative. (See further discussion of dams in chapter 20.)

A

B

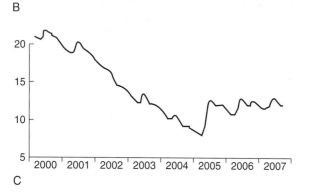

C

Figure 15.24

Glen Canyon Dam hydroelectric project, Arizona. In dry surroundings such as these, evaporation losses from reservoirs are high, and they can exacerbate regional water-supply problems. (A) Photo taken in 2005 shows drought-lowered level of Lake Powell; note the "bathtub ring" behind the dam that indicates the normal level of the reservoir when full. (B) Lake outline in 2006, superimposed on a 2000 image, illustrates shrinkage in surface area. (C) Decline in water storage in the reservoir has both hydropower and water-supply consequences. Water storage (vertical axis) in millions of acre-feet.

Image (B) by Jesse Allen, courtesy NASA; (C) Data courtesy NASA

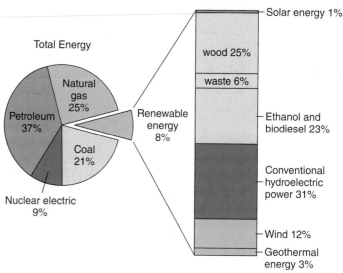

Figure 15.25

Hydropower is currently the dominant single renewable energy source in the United States, though biofuels collectively now surpass it. Renewable sources still provide a relatively small proportion of energy used.

Source: Annual Energy Review 2011, *U.S. Energy Information Administration.*

Energy from the Oceans

Three different approaches exist for extracting energy from Earth's oceans: harnessing the energy of waves or tides, or making use of temperature differences between deep and shallow waters.

Tides represent a great volume of shifting water. In fact, all large bodies of standing water on the earth, including the oceans and large lakes like the Great Lakes, show tides. Why not also harness this moving water as an energy source? Unfortunately, the energy represented by tides is too dispersed in most places to be useful. Average beach tides reflect a difference between high-tide and low-tide water levels of about 1 meter. A commercial tidal-power electricity-generating plant requires at least 5 meters difference between high and low tides for efficient generation of electricity and a bay or inlet with a narrow opening that could be dammed to regulate the water flow in and out (figure 15.26). The proper conditions exist in very few places in the world. Tidal power is being used in a small way at several locations—at the Bay of Fundy in Nova Scotia, near St. Malo, France, in the Netherlands, and in the former Soviet Union. So although tidal power shares the environmental benefits of conventional hydropower, its total potential is limited, estimated at only 2% of the energy potential of conventional hydropower worldwide.

Ocean thermal energy conversion (OTEC) is another clean, renewable technology that is currently in the developmental stages. It exploits the temperature difference between warm

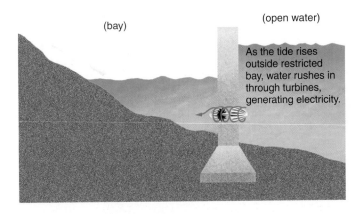

As the tide rises outside restricted bay, water rushes in through turbines, generating electricity.

When tide is out, water confined in bay again moves through turbines to generate power.

Figure 15.26

Tidal-power generation uses flowing water to generate electricity, as with conventional hydropower.

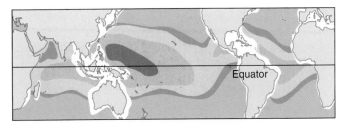

Temperature difference between surface and depth of 1000 m

- ☐ Less than 18°C
- ☐ 18° to 20°C
- ☐ 20° to 22°C
- ☐ 22° to 24°C
- ☐ More than 24°C
- ☐ Depth less than 1000 m

Figure 15.27

Given the thermal and other requirements of OTEC, tropical islands are likely to be the first sites for its development.

After National Renewable Energy Laboratory, U.S. Department of Energy.

surface water and the cold water at depth. Either the warm water is vaporized and used directly to run a turbine, or its heat is used to vaporize a "working fluid" to do so; the vapor is recondensed by chilling with the cold water. No fuel is burned, no emissions released, and the vast scale of the oceans assures a long life for such facilities (in contrast, for instance, to geothermal fields). Where the water is vaporized, the vapor is pure water, which, when recondensed, can be used for water supply—this is essentially a distillation process. The cold seawater can be used for other purposes, such as air conditioning or aquaculture.

However, specialized conditions are needed for OTEC to be a productive energy source. The deep cold water must be accessible near shore; coastlines with broad shelves are unsuitable. The temperature difference between warm and cold seawater must be at least 40°F (22°C) year-round, which is true only near the equator (figure 15.27). Thus, this technology is most suitable for select tropical islands. (An experimental OTEC facility exists on Hawaii, but has yet to produce electricity.) And there is concern about possible damage to reefs and coastal ecosystems from the pipes and other equipment. The near-term potential of OTEC is consequently very limited.

Over much of the oceans, waves ripple incessantly over the surface. The up-and-down motion of the water can be harnessed in various ways to generate electricity. The bobbing water can drive a pump to push water through a turbine, for example; or, in an enclosed chamber that is partially submerged,

the rise and fall of the water surface can produce pulses of compression in the air above, which can drive an air-powered turbine. These systems too are clean and renewable. A small wave-energy system is currently in operation off the coast of Portugal. However, appropriate sites are limited by concerns over the visual impact of the equipment in coastal areas, and over possible disruption of natural sediment-transport patterns. These issues, together with relatively high costs, have so far prevented widespread development of wave energy.

Wind Energy

Because the winds are ultimately powered by the sun, wind energy can be regarded as a variant of solar energy. It is clean and, like sunshine, renewable indefinitely (at least for 5 billion years or so). Wind power has been utilized to some extent for more than two thousand years; the windmills of the Netherlands are probably the best-known historic example. Today, there is considerable interest in making more extensive use of wind power for generating electricity.

Wind energy shares certain limitations with solar energy. It is dispersed, not only in two dimensions but in three: It is spread out through the atmosphere. Wind is also erratic, highly variable in speed both regionally and locally. The regional variations in potential power supply are even more significant than they may appear from average wind velocities because windmill power generation increases as the cube of wind speed. So if wind velocity doubles, power output increases by a factor of 8 ($2 \times 2 \times 2$).

Figure 15.28 shows that most of the windiest places on land in the United States are rather far removed physically from most of the heavily populated and heavily industrialized areas. As with other physically localized power sources, the technological difficulty of long-distance transmission of electricity will limit wind power's near-term contribution in areas of high electricity consumption until transmission efficiencies improve.

Even where average wind velocities are great, strong winds do not always blow. This presents the same storage problem

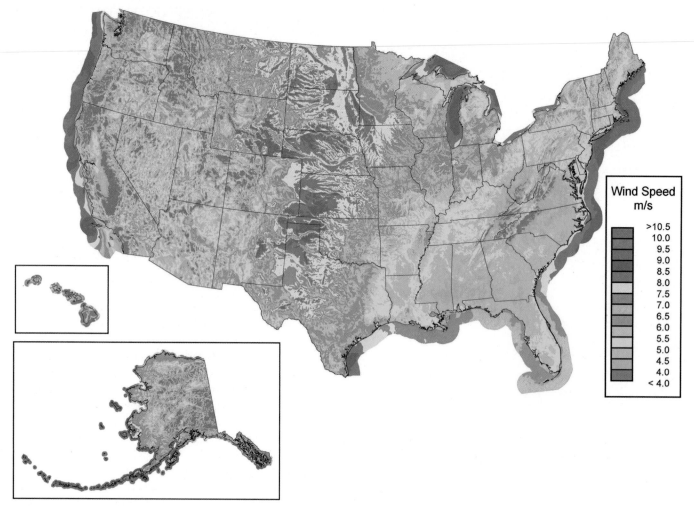

Figure 15.28

Average annual wind speed 80 meters (about 90 ft) above the surface, a common height for wind turbines. Topographic irregularities, including tall vegetation, disrupt wind flow, so the strongest consistent winds are found in the Central Plains and offshore.

Source: National Renewable Energy Laboratory, U.S. Department of Energy.

as does solar electricity, and it has likewise not yet been solved satisfactorily. At present, wind-generated electricity is used most commonly to supplement conventionally generated power when wind conditions are favorable, with conventional plants bearing most or all of the demand load at other times. Still, design improvements have steadily increased the reliability of wind generators; the newest wind turbines are available for some power generation at least 95% of the time.

Winds blow both more strongly and more consistently at high altitudes. Tall wind vanes, perhaps a mile high, may provide more energy with less of a storage problem. However, such tall structures are technically more difficult to build and require substantial quantities of material. They may also be deemed more objectionable visually.

The ultimate potential of wind energy is unclear. Certainly, blowing wind represents far more total energy than we can use, but most of it cannot be harnessed. Most schemes for commercial wind-power generation of electric power involve

"wind farms," concentrations of many wind generators in a few especially favorable windy sites (figure 15.29). The limits to wind-power use then include the area that can be committed to wind-generator arrays, as well as the distance the electricity can be transmitted without excessive loss in the power grid. About 1000 1-megawatt wind generators are required to generate as much power as a sizable conventional coal- or nuclear-powered electric generating plant. The units have to be spread out, or they block each other's wind flow. Spacing them at four windmills per square kilometer requires about 250 square kilometers (100 square miles) of land to produce energy equivalent to a 1000-megawatt power plant. However, the land need not be devoted exclusively to the wind-powered generators. Farming and the grazing of livestock, common activities in the Great Plains area, could continue on the same land concurrently with power generation.

In recent years, interest in generating wind power using offshore turbines has grown. This is in large part because it

Figure 15.29

Wind-turbine array near Palm Springs, California. Different elements of the array can take advantage of various velocities of wind. Some of the smaller turbines turn even in light wind, while the larger turbines require stronger winds to drive them.

© The McGraw-Hill Companies, Inc./Doug Sherman, photographer.

would make wind power more accessible to the many large cities located at or near the coast (including those bordering the Great Lakes, such as Chicago), as is evident from figure 15.28. There are both technical issues and questions of jurisdiction to be addressed (a utility cannot own or lease an area of the ocean as it can a plot of land, and even offshore, a wind-turbine array would cover a substantial area). Aesthetic and practical objections to the presence of offshore turbine arrays exist also. Still, the fact that transmission losses of wind-generated electricity could be greatly reduced through offshore generation makes this a potentially attractive option.

The storage problem remains in any case. Other concerns relating to wind farms, on land or offshore, include the aesthetic impact (the generator arrays must naturally be very exposed); interference with and deaths of migrating birds and bats; the noise associated with a large number of windmills operating;

and disruption of communications—TV, radio, cellular telephones, perhaps even aircraft communications.

At one time, it was projected that the United States might produce 25 to 50% of its electricity from wind power by the year 2000, but the necessary major national program for wind-energy development did not materialize. This percentage would represent a significant fraction of anticipated total energy needs, and it also would contribute to conserving nonrenewable fuels. In fact, in 2010, wind-powered electric generating facilities in the United States accounted for only about 4% of U.S. generating capacity, and less than 3% of actual electricity generation. However, wind-power capacity has been rising sharply in the United States in recent years (figure 15.30), and those in the industry believe that this country could be using wind to supply 20% of our energy by 2030.

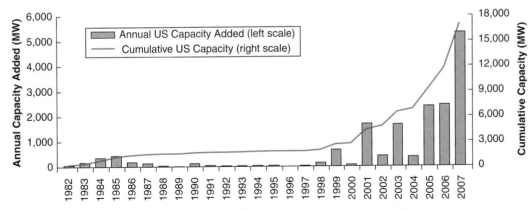

Figure 15.30

U.S. wind power installed capacity, 1981–2007. (MW=megawatts) The erratic pattern of capacity additions after 1999 reflects the presence or absence of federal tax credits for wind-energy development, with the lapse of those credits in 2000, 2002, and 2004 resulting in minimal capacity additions in those years. Federal policy can thus be a powerful influence in energy resource development. From 2008–2011, more than 5,000 MW of capacity were added each year, so that by the end of 2011, total wind-energy generating capacity exceeded 47,000 MW.

Source: National Renewable Energy Laboratory, using data from the American Wind Energy Association.

Electricity's Hidden Energy Costs

Have you ever looked closely at your electric bill? Many utilities now provide information on the sources of their power, and the associated pollution (figure 1). Such considerations are very much part of current discussions about energy sources. At one time, the main concern was availability of the fossil fuels. Now global focus has shifted to the emission of greenhouse gases like CO_2, and the relative pollutant outputs of different energy sources are a significant part of the debate. With nuclear power, however, the issue of waste disposal is also a concern.

A contributing factor in rising U.S. energy use (and associated pollution) is what has been described as the "electrification" of U.S.

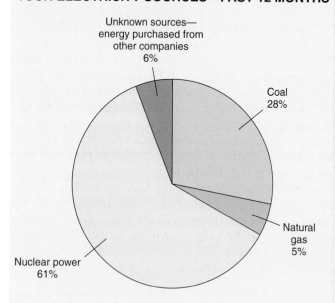

YOUR ELECTRICITY SOURCES—PAST 12 MONTHS

Unknown sources— energy purchased from other companies 6%

Coal 28%

Natural gas 5%

Nuclear power 61%

AVERAGE EMISSIONS and NUCLEAR WASTE per 1000 kilowatt-hours (kWhs) PRODUCED from known sources for the past 12 months	
Carbon dioxide	659.84 lb.
Nitrogen oxides	1.69 lb.
Sulfur dioxide	2.97 lb.
High-level nuclear waste	0.0037 lb.
Low-level nuclear waste	0.0003 cu. ft.

Figure 1

Typical profile of municipal electricity supply, a mix of sources with varying waste and pollution consequences. The gaseous wastes are associated with the fossil fuels, the sulfur primarily with the coal. Note the relative quantities of wastes from fossil vs. nuclear fuel. Bear in mind, too, that amounts of coal ash generated are not tabulated; it is not a waste product monitored and regulated in the same way as the gaseous air pollutants or radioactive waste.

households, as we acquire more and more appliances that run on electricity. By 2009, 99.9% of U.S. households had a refrigerator, and 23% had two or more; 93% had a conventional oven, the majority of these electric; 79% reported having a clothes dryer and 59% a dishwasher. From 1978 to 2009, the proportion of households having a microwave oven went from 8% to 96%. Thirty-four percent of U.S. households reported using electricity as their primary source of heat, 83% had some sort of air conditioning, and 61% had central air conditioning. By 2009, 99% of U.S. households had a color television, 78% had two or more, and 8% had five or more (!). From 1990 to 2009, U.S. households with personal computers increased from 16% to 76%, and this percentage continues to rise. In 2009, 90% of households reported having rechargeable electronic devices, and 7% had nine or more.

A hidden issue with electricity is efficiency, or lack of it. Only *one-third* of the energy consumed in generating electricity is delivered to the end user as power (figure 2A), so as electricity consumption grows, the corresponding energy consumption grows three times as fast (figure 2B). The other two-thirds is lost, most as waste heat during the process of conversion from the energy of the power sources (for example, the chemical energy of fossil fuels) to electricity, and the rest in transmission and distribution. This would be something to keep in mind in choosing among appliances, heating systems, and so on, powered by different forms of energy, such as electricity versus natural gas.

Part of the rise in residential electricity consumption is a result of what have been called "vampire appliances," electric devices that for various reasons (such as inclusion of a clock, the need to be operated by remote control, or consumers' desire for "instant on," very fast startup) draw current all the time, not just when actively in use. Avoiding such appliances would help restrain demand. So would conservation through the use of more-efficient devices—EnergyStar appliances and compact fluorescent bulbs in place of incandescent lightbulbs. Sometimes, though, addressing one problem can exacerbate another; for example, compact fluorescents contain small amounts of the toxic metal mercury, so they require more careful disposal than do incandescent bulbs.

The inefficiency of the ways we now generate our electricity also has a bearing, indirectly, on the utility of hydrogen fuel cells. In a hydrogen fuel cell, which operates somewhat like a battery, hydrogen and oxygen are combined to generate electricity, typically to power a vehicle, and this process is actually quite efficient. The only "waste" product is water, so fuel-cell-powered cars are very clean, and the technology already exists. So far, so good. But if we are to run millions of vehicles on fuel cells, where will the hydrogen come from? The obvious answer is water, of which we have a plentiful global supply. However, the usual method for separating water into hydrogen and oxygen rapidly and in quantity uses electricity. Thus, before we can embrace fuel cells on a large scale, we need either to figure out how to generate still *more* electricity cleanly and sustainably, or to develop alternate ways to produce the necessary large quantities of hydrogen.

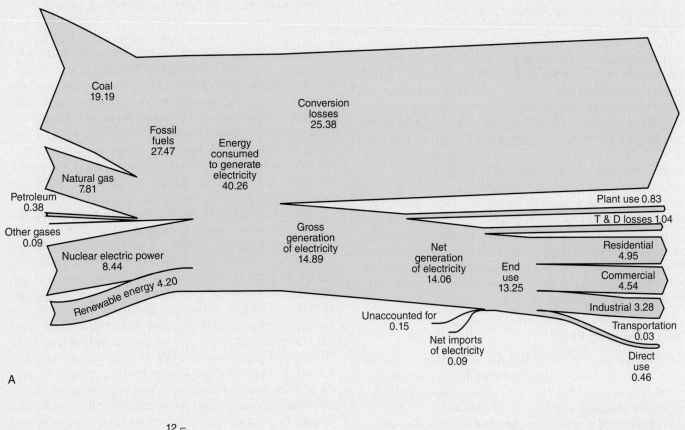

A

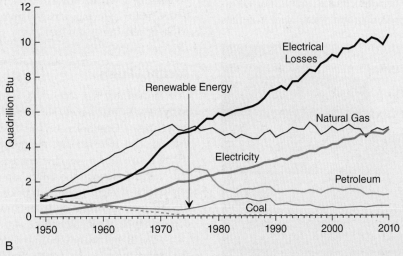

B

Figure 2

(A) Electricity flow: Energy sources on the left, outputs on the right. Units are quadrillion Btu. Note the huge "conversion losses" resulting during conversion of other forms of energy to electricity. "T & D losses" are additional losses during electricity transmission and distribution. (B) Residential and commercial energy consumption in the United States is increasingly in the form of electricity, with correspondingly larger losses. This graph is for residential energy consumption.

Source: Annual Energy Review 2010, *U.S. Energy Information Administration.*

Biofuels

The term *biomass* technically refers to the total mass of all the organisms living on earth. In an energy context, the term **biofuels** has become a catchall for various ways of deriving energy from biomass, from organisms or from their remains. Indirectly, biomass-derived energy is ultimately solar energy, since most biofuels come from plant materials, and plants need sunlight to grow. Biofuels could also be thought of as "unfossilized fuels" because they represent fuels derived from living or recent organisms rather than ancient ones and hence have not been modified extensively by geologic processes acting over long periods of time.

The possibilities for biofuels are many. Most of these fuels fall into three broad categories: wood, waste, and alcohol fuels. Some are used alone, others "cofired" (burned in combination with conventional fuels); see figure 15.31. All biofuels are burned to release their energy, so they share the carbon-dioxide-pollution problems of fossil fuels. Some, like wood, also contribute particulate air pollutants. However, unlike the fossil fuels, biofuels are renewable. We can, in principle, produce and replenish them on the same time scale as that on which we consume them. This is their appeal, together with the fact that we can produce them domestically, thus reducing dependence on energy imports.

Waste-Derived Fuels

In some sense, much of the wood burned as biomass fuel is also waste, especially from logging, lumber-milling, and similar operations. However, most of the truly waste-related biomass fuels involve agricultural or other wastes that would historically have been burned in a field or dumped in a landfill. For example, there is increasing interest in burning waste plant materials

Figure 15.31

Biomass (wood chips) is cofired with coal at the Northern Indiana Public Service Company electricity generating station in Bailey, IN; the biomass contributes about 5% of the energy. Other plants run entirely on wood chips or other combustible waste.

Photograph by Kevin Craig, courtesy National Renewable Energy Laboratory.

after a crop is harvested. Some such crop wastes may require processing beforehand; sugar-cane waste is an example. The combustible portion of urban refuse can be burned to provide heat for electric generating plants. More-extensive use of incineration as a means of solid-waste disposal, both in the United States and abroad, would further increase the use of biomass for energy worldwide.

Some waste-derived fuels are liquids. Research is ongoing on ways to derive inoffensive liquid fuels from animal manures, which are rich in organic matter. Experimental vehicles have been designed to run on used vegetable oil from food-frying operations. Some diesel-powered vehicles can run on *biodiesel,* a general term for fuels derived from vegetable oil or animal fats (including wastes), or a blend of such oil with petroleum diesel fuel. The alcohol fuels discussed below are now mainly produced from corn, but processes to derive alcohol from plant wastes rather than potential food are being developed.

Another waste-derived biomass fuel growing in use, *biogas,* could be called "gas from garbage." When broken down in the absence of oxygen, organic wastes yield a variety of gaseous products. Some of these are useless, or smelly, or toxic, but among them is methane (CH_4), the same compound that predominates in natural gas. Sanitary-landfill operations, described in chapter 16, are suitable sites for methane production as organic wastes in the refuse decay. Straight landfill gas is too full of impurities to use alone, but landfill-derived gas can be blended with purer, pipelined natural gas (another example of cofiring) to extend the gas supply. Methane can also be produced from decaying manures. Where large quantities of animal waste are available, as for example on feedlots, it may also be possible to create biogas by microbial digestion in enclosed tanks, again reducing waste-disposal and pollution problems while generating useful fuel.

Alcohol Fuels

One biofuel that has received special attention is alcohol. Initially, it was extensively developed mainly for incorporation into *gasohol.* Gasohol as originally created was a blend of 90% gasoline and 10% alcohol, the proportions reflecting a mix on which conventional gasoline engines could run. Gasohol came into vogue following the OPEC oil embargo of the 1970s. Its popularity waned when oil prices subsequently declined, though most gasoline still contains some ethanol (the most common type of fuel alcohol).

The higher the proportion of alcohol in the mix, of course, the further the gasoline can be stretched. There is, in principle, no reason why engines cannot be designed to run on 20%, 50%, or 100% alcohol, and, in fact, vehicles have been made to run solely on alcohol. Recently, increasing numbers of vehicles have been designed to run on *E85,* a blend of 85% ethanol, 15% gasoline. Such vehicles, most common in the Midwest and South, are now the majority of alcohol-fueled vehicles. Some can run either on E85 or on regular gasoline, though they must be specifically designed to do so.

Government policy can encourage expanded use of alcohol and alcohol/gasoline blends. The Clean Air Act Amendments of 1990 included provisions for reduced vehicle emissions, and alcohol is somewhat cleaner-burning. The Energy Policy Act of 1992 required gradual replacement of some government, utility-company, and other "fleet" vehicles with "alternative-fuel" vehicles, which would include vehicles powered by straight alcohol or E85. (This is more immediately feasible for fleet vehicles because they can be refueled at central locations stocking their special fuels.) The same act extends an excise-tax break on ethanol-blend fuels. Large gains in ethanol consumption remain to be realized, however. More than 500,000 vehicles now run on E85, but this number is small compared to the 136 million passenger cars and over 100 million vans, pickups, and SUVs on the road.

Substantial concerns about ethanol fuels are also being raised as world production rises sharply (figure 15.32). One important consideration is that to grow, harvest, and derive ethanol from corn or other starting plant material takes energy. Depending on the process, indeed, it can take *more* energy than is released burning the ethanol. Thus the alcohol fuels as currently developed are not so much a major new source of energy as a way to reduce dependence on imported liquid petroleum—if the energy to make the ethanol comes from something else. Diversion of potential food crops such as corn for producing fuel has driven up food prices and sparked some protests around the world. It has been calculated that the corn equivalent of the energy represented by a fuel tank of ethanol for a large SUV could almost feed one person for a year, raising questions about whether biofuel is a suitable use of food crops. In the United States, the push to grow more corn for ethanol production has created pressure to grow crops on marginal cropland previously set aside for erosion-control purposes under the Conservation Reserve Program discussed in chapter 11. And studies have suggested that, to the extent that more forests and grasslands are cleared to provide additional cropland for biofuel production, that land-use change will increase net greenhouse-gas emissions by decreasing carbon "sinks" (see chapter 18). Research is underway to develop effi-

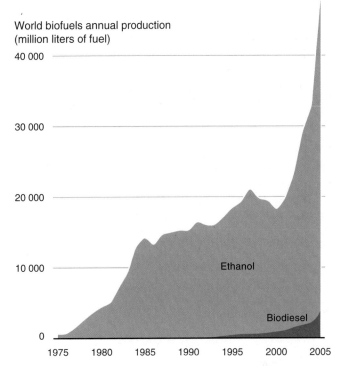

Figure 15.32

Rising oil prices have spurred a sharp rise in biofuel production, primarily ethanol. The two principal producers currently are the U.S. and Brazil.

Reprinted with permission UNEP. UNEP/GRID-Arendal Maps and Graphics Library 2009. Edited by Joel Benoit. http://maps.grida.no/library/files/the-production-of-biodiesel-and-ethanol-has-increased-substantially-in-recent-years

cient means to produce ethanol from nonfood crops such as switchgrass, but any process involving growing crops specifically for biofuel production will involve some of the foregoing issues as well. Over the long term, learning to produce ethanol from waste organic matter may be the best option.

Summary

As we contemplate the limits on supplies of petroleum and conventional natural gas and the environmental impacts of coal, we find an almost bewildering variety of alternatives available. None is as versatile as liquid and gaseous petroleum fuels, or as immediately and abundantly available as coal. Some of the already-viable, clean, renewable alternatives are "placebound" and, in any case, have limited ultimate potential (e.g., hydropower, geothermal power). Nuclear fission produces minimal emissions but entails waste-disposal problems, and concerns about reactor safety; the fuel reprocessing necessary if fission-power use is to be greatly expanded raises security concerns. Solar and wind energy, though free, renewable, and clean to use, are so diffuse, and so variable over time and space, that without substantial technological advances they are impractical for energy-intensive applications. Biofuels, like fossil fuels, yield carbon dioxide (and perhaps other pollutants also), and may require substantial expansion of cropland,

potentially increasing soil-erosion problems. This incomplete sampling of alternatives and their pros and cons illustrates some of the complexity about the future of our energy sources.

The energy-use picture in the future probably will not be dominated by a single source as has been the case in the fossil-fuel era. Rather, a blend of sources with different strengths, weaknesses, and specialized applications is likely. Different nations, too, will make different choices, for reasons of geology or geography, economics, or differing environmental priorities (figure 15.33).

In the meantime, as the alternatives are explored and developed, vigorous efforts to conserve energy would yield much-needed time to make a smooth transition from present to future energy sources. However, considerable pressures toward increasing energy consumption exist in countries at all levels of technological development.

Figure 15.33

(A) Shares of fossil and non-fossil fuels in energy consumption for selected countries and the world, 2009. Except for Iceland, total of geothermal plus solar energy used in countries shown is less than 1% of energy used. (B) In part, national choices in renewable energy are dictated by geography and geology. Total energy needs are also a factor: Iceland can rely so heavily on renewables, as shown in (A), partly because its consumption is so modest—about ¼ of 1% of the energy used by the United States.

(A) Source: International Energy Agency. (B) Map courtesy United Nations Environment Programme/GRIDA. Reprinted with permission UNEP.

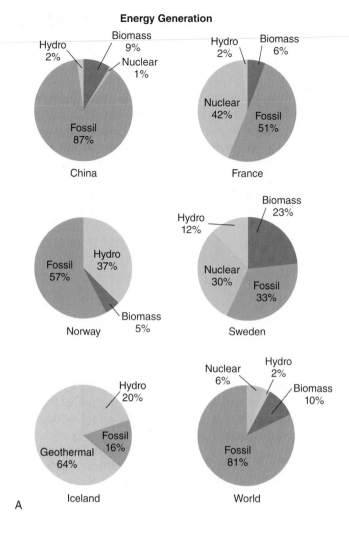

Energy Generation

A

Key Terms and Concepts

biofuels 360	core meltdown 338	fusion 336	hot-dry-rock 351
breeder reactor 337	decommissioning 339	geothermal energy 349	ocean thermal energy conversion 354
chain reaction 336	fission 336	geothermal gradient 351	photovoltaic cells 346

Exercises

Questions for Review

1. Briefly describe the nature of the fission chain reaction used to generate power in commercial nuclear power plants. How is the energy released utilized?

2. If the nuclear power option is pursued, breeder reactors and fuel reprocessing will be necessary. Why? What additional safety and security concerns are then involved?

3. What is "decommissioning" in a nuclear-power context?

4. Describe the fusion process, and evaluate its advantages and present limitations.

5. In what areas might solar energy potentially make the greatest contributions toward our energy needs? Explain.

6. What technological limitations do solar and wind energy currently share?

7. Explain the nature of geothermal energy and how it is extracted.

8. What factors restrict the use of geothermal energy in time and in space? How do hot-dry-rock geothermal areas expand its potential?

9. Assess the potential of (a) conventional hydropower and (b) tidal power to help solve impending energy shortages.

10. What is the basis of ocean thermal energy conversion; in what areas is it most viable and why?

11. Describe two issues, other than technological ones, that limit the locations in which wave energy might be harnessed.

World Potential Renewable Energy

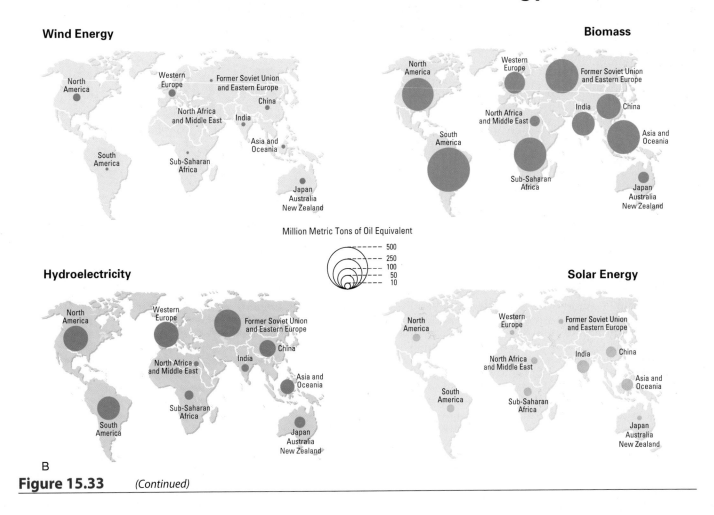

Figure 15.33 *(Continued)*

12. What are biofuels? Describe two examples.

13. Describe any three issues of concern with rapid growth in the production of fuel ethanol from corn.

14. Choose any two energy sources from this chapter and compare/contrast them in terms of the negative environmental impacts associated with each.

Exploring Further

1. Investigate the history of a completed commercial nuclear-power plant project. How long a history does the project have? Did the plant suffer regulatory or other construction delays? How much electricity does it generate? How long has it been operating, and when is it projected for decommissioning?

2. While most alternative energy sources must be developed on a large scale, solar space heating can be installed building by building. For your region, explore the feasibility and costs of conversion to solar heating.

3. A commonly used utility-scale wind turbine has an electricity-generating capacity of 1500 kilowatts. If the average U.S. home used 11,500 kilowatt-hours of electricity in 2010, how many homes' electricity could be supplied by one such turbine, operating at maximum capacity? Why is this number really an upper limit?

4. Look up data on the generating capacity of a commercial wind farm or solar-electric facility, preferably close to your region. From the area occupied by the facility, calculate the area of the same generators that would be required to replace a 1000-megawatt fossil or nuclear power plant. What assumptions are you making in calculating your result?

5. Consider having a home energy audit (often available free or at nominal cost through a local utility company) to identify ways to conserve energy.

6. Choose any nonfossil energy source and investigate its development over recent decades, in the United States, in another nation, or globally. How significant a factor is it in worldwide energy production currently? How do its economics compare with fossil fuels? What are its future prospects?

Waste Disposal

High-consumption technological societies tend to generate copious quantities of wastes. In the United States alone, in 2010 each person generated, on average, about 4.43 pounds (more than 2 kilograms) of what is broadly called "garbage" every day. That represents an increase of more than 70% over 1960 per-capita waste production. Every five years, each average American generates a mass of waste greater than the mass of the Statue of Liberty! That, plus industrial, agricultural, and mineral wastes, together amount to an estimated total of 4 billion tons of solid waste produced in the United States alone each year. Most of the 40 billion gallons of water withdrawn daily by public water departments end up as sewage-tainted wastewater, and more-concentrated liquid wastes are generated by industry. Each day, the question of where to put the growing accumulations of radioactive waste materials becomes more pressing. Proper, secure disposal of all these varied wastes is critical to minimizing environmental pollution. According to data from the Recycling Council of Ontario, North America has 8% of the world's population, consumes one-third of the world's resources—and produces almost half the world's nonorganic trash. In this chapter, we will survey various waste-disposal strategies and examine their pros and cons.

Spoils (grey and white) from the Escondida copper mine in Chile's Atacama Desert. The spoils are pumped in solution (green) into the basin in which they then dry out. For scale, note the retaining dam, the straight line at lower left, which is 1 km (about 0.6 miles) long.

Image courtesy of Image Science and Analysis Laboratory, NASA Johnson Space Center

Solid Wastes—General

The sources of solid waste in the United States are many. More than half of these wastes are linked to agricultural activities, with the dominant component being waste from livestock. Most of this waste is not highly toxic except when contaminated with agricultural chemicals, nor is it collected for systematic disposal. The volume of such waste, therefore, is seldom realized.

The other major waste source is the mineral industry, which generates immense quantities of spoils, tailings, slag, and other rock and mineral wastes. Materials such as tailings and spoils are generally handled onsite—as, for example, when surface mines are reclaimed or as shown in the chapter-opening photograph. The amount of waste involved makes long-distance transportation or sophisticated treatment of the wastes uneconomical. The weathering of mining wastes can be a significant water-pollution hazard, depending on the nature of the rocks, with metals and sulfuric acid among the principal pollutants. Shielding the pulverized rocks from rapid weathering with a soil cover is a common control/disposal strategy. In addition, certain chemicals used to extract metals during processing are toxic and require special handling in disposal like other industrial wastes.

Much of the attention in solid-waste disposal is devoted to the comparatively small amount of municipal waste, an esti-mated 2–5% of the solid-waste stream. It is concentrated in cities, is highly visible, and must be collected, transported, and disposed of at some cost. The following section explores historical and present strategies for dealing with municipal wastes.

Nonmining industrial wastes likewise command a relatively large share of attention, despite their relatively small volume, because many industrial wastes are highly toxic. Most of the highly publicized unsafe hazardous-waste disposal sites involve improper disposal of industrial chemical wastes. Some of the principal industrial solid-waste sources are shown in figure 16.1.

While industrial wastes may supply the largest amounts of toxic materials, municipal waste is far from harmless. Aside from the organic materials like food waste and paper, a wide variety of poisons is used in every household: corrosive cleaning agents, disinfectants, solvents such as paint thinner and dry-cleaning fluids, insecticides and insect repellents, and so on. These toxic chemicals together represent a substantial, if more dilute, potential source of pollution if carelessly handled.

Municipal Waste Disposal

A great variety of materials collectively make up the solid-waste disposal problem that costs municipalities several billion dollars each year (figure 16.2). The complexity of the waste-disposal problem is thus compounded by the mix of different materials to be dealt with. The best disposal method for one kind of waste may not be appropriate for another.

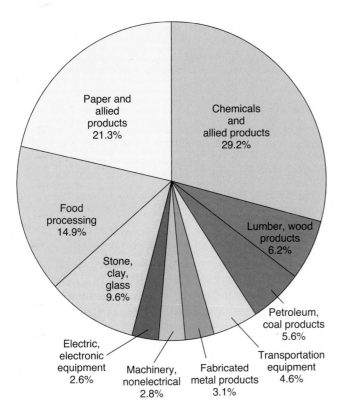

Figure 16.1

Principal industrial solid-waste source industries. All these wastes together amount to only a few percent of the quantities of mineral and agricultural wastes.

Data from U.S. Environmental Protection Agency.

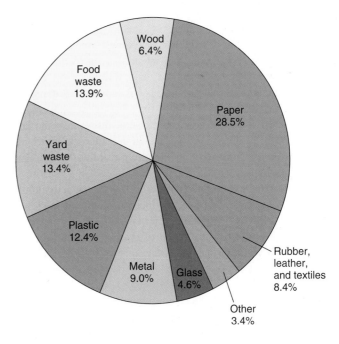

Total: 250 million tons

Figure 16.2

Typical composition of municipal solid waste in 2010, before recycling. Individual municipalities' waste composition may vary considerably.

Data from U.S. Environmental Protection Agency.

A long-established method for solid-waste disposal that demands a minimum of effort and expense has been the open dump site. Drawbacks to such facilities are fairly obvious, especially to those having the misfortune to live nearby. Open dumps are unsightly, unsanitary, and generally smelly; they attract rats, insects, and other pests; they are fire hazards. Surface water percolating through the trash can dissolve out, or leach, harmful chemicals that are then carried away from the dump site in surface runoff or through percolation into ground water. Trash may be scattered by wind or water. Some of the gases rising from the dump may be toxic.

All in all, open dumps are an unsatisfactory means of solid-waste disposal. Yet even now, after decades of concerted efforts to close all open-dump sites, the Environmental Protection Agency has estimated that there may still be several hundred thousand illicit open dumps in the United States.

Sanitary Landfills

The major share of municipal solid waste in the United States ends up in **sanitary landfills** (figure 16.3). The method has been in use since the early twentieth century. In a basic sanitary-landfill operation, a layer of compacted trash is covered with a layer of earth at least once a day. The earth cover keeps out vermin and helps to confine the refuse. Landfills have generally been sited in low places—natural valleys, old abandoned gravel pits, or surface mines. When the site is full, a thicker layer of earth is placed on top, and the land can be used for other purposes, provided that the nature of the wastes and the design of the landfill are such that leakage of noxious gases or toxins is minimal (figure 16.3B). The most suitable uses include parks, pastureland, parking lots, and other facilities not requiring much excavation. The city of Evanston, Illinois, built a landfill up into a hill, and the now-complete "Mount Trashmore" is a ski area. Golf courses built over old landfill sites are increasingly common. Attempts to construct buildings on old landfill sites may be complicated by the limited excavation possible because of refuse near the land surface, by the settling of trash as it later decomposes (which can put stress on the structures), and by the possibility of pollutants escaping from the landfill site.

Pollutants can escape from improperly designed landfills in a variety of ways. Gases are produced in decomposing refuse in a landfill just as in an open dump site, though the particular gases differ somewhat as a result of the exclusion of air from landfill trash. Initially, decomposition in a landfill, as in an open dump, proceeds aerobically, consuming oxygen and producing such products as carbon dioxide (CO_2) and sulfur dioxide (SO_2). When oxygen in the covered landfill is used up, anaerobic decomposition yields such gases as methane (CH_4) and hydrogen sulfide (H_2S). If the surface soil is permeable, these gases may escape through it. Sealing the landfill to prevent the free escape of gases can serve a twofold purpose: reduction of pollution, and retention of useful methane as described in chapter 15. The potential buildup of excess gas pressure requires some venting of gas, whether deliberate or otherwise. Where the quantity of usable methane is insufficient to be recoverable

Compacted waste filling trench   Original land surface

Daily 6-inch earth cover

A

B

Figure 16.3

Sanitary landfills. (A) Basic principle of a sanitary landfill (partial cutaway view). The "sanitary" part is the covering of the wastes with soil each day. New landfills must meet still more-stringent standards. (B) New Lyme (Ohio) Landfill site after restoration.

(B) Photograph by Harry Weddington, courtesy U.S. Army Corps of Engineers.

economically, the vented gases may be burned directly at the vent over the landfill site to break down noxious compounds.

If soil above or below a landfill is permeable, **leachate** (infiltrating water containing dissolved chemicals from the refuse) can escape to contaminate surface or ground waters (figure 16.4). This is a particular problem with landfills so poorly sited that the regional water table reaches the base of the landfill during part or all of the year. It is also a problem with older landfills that were constructed without impermeable liners beneath. Increasing awareness particularly of the danger of groundwater pollution has led to improvements in the location and design of sanitary landfills. They are now ideally placed over rock or soil of limited permeability (commonly clay-rich soils, sediments, or sedimentary rocks, or unfractured bedrock) well above the water table. Where the soils are too permeable, liners of plastic or other waterproof material may be used to contain percolating

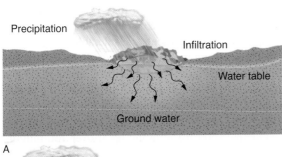

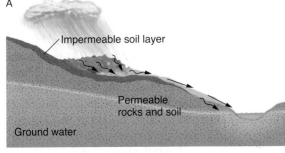

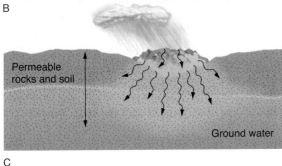

Figure 16.4

Leachate can escape from a poorly designed or poorly sited landfill to contaminate surface or ground waters. (A) Direct contamination of ground water; water table intersects landfill. (B) Leachate runs off over sloping land to pollute lake or stream below, despite impermeable material directly under landfill site. (C) Lack of impermeable liner below allows leachate to infiltrate to ground water.

leachate, or thick layers of low-permeability clay (several meters or more in thickness) may be placed beneath the site before infilling begins.

Unfortunately, there is another potential problem with landfills sealed below with low-permeability materials, if the layers above are fairly permeable. Infiltrating water from the surface will accumulate in the landfill as in a giant bathtub, and the leachate may eventually spill out and pollute the surroundings (figure 16.5A). Use of low-permeability materials above as well as below the refuse minimizes this possibility (figure16.5B). Leachate problems are lessened in arid climates where precipitation is light, but unfortunately, most large U.S. cities are not located in such places. Thus, modern landfills are carefully monitored to detect developing leachate-buildup problems, and in some cases, leachate is pumped out to prevent leakage (though this does, in turn, create a liquid-waste-disposal problem).

A subtle pathway for the escape of toxic chemicals may be provided by plants growing on a finished landfill site that is

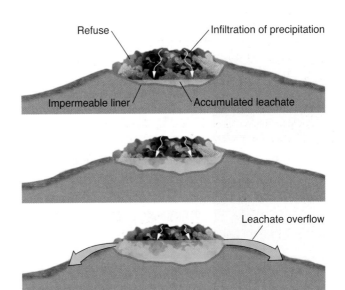

Figure 16.5

(A) The "bathtub effect" caused by the accumulation of infiltrated leachate above impermeable liner in unmonitored landfill. Overflow may not occur for months or years. (B) Installation of impermeable plastic membrane and gas vents, K. I. Sawyer Air Force Base, Michigan.

(B) Photograph by Harry Weddington, courtesy U.S. Army Corps of Engineers.

not covered by an impermeable layer. As plant roots take up water, they also take up chemicals dissolved in the water, some of which, depending on the nature of the refuse, may be toxic. This possibility indicates the need for caution in using an old landfill site for cropland or pastureland.

Thus, a well-designed modern municipal landfill is a complex creation, not just a dump topped with dirt (figure 16.6). Still, however carefully crafted, a landfill involves a commitment of land. A rule of thumb for a sanitary landfill for municipal

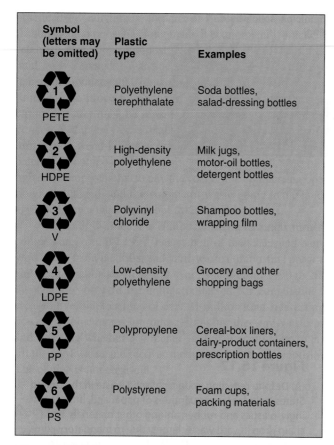

Symbol (letters may be omitted)	Plastic type	Examples
1 PETE	Polyethylene terephthalate	Soda bottles, salad-dressing bottles
2 HDPE	High-density polyethylene	Milk jugs, motor-oil bottles, detergent bottles
3 V	Polyvinyl chloride	Shampoo bottles, wrapping film
4 LDPE	Low-density polyethylene	Grocery and other shopping bags
5 PP	Polypropylene	Cereal-box liners, dairy-product containers, prescription bottles
6 PS	Polystyrene	Foam cups, packing materials

Figure 16.13

Use of standard symbols facilitates separation of distinct types of plastics for recycling. Sometimes the same symbol with a numeral "7" is used for "other." Certain products (e.g., milk jugs and soda bottles) are consistently made with one type of plastic; others (e.g., containers for detergents or dairy products) may be made of different types depending upon the desired properties—clear container versus opaque, flexible versus rigid, and so on.

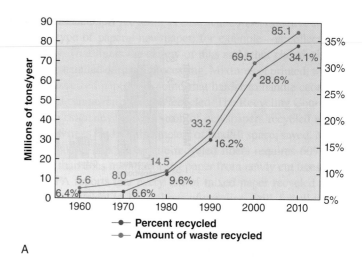

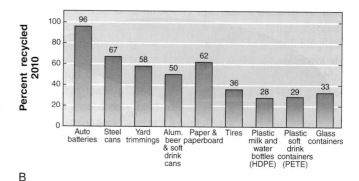

Figure 16.14

(A) Both the amount of municipal solid waste recycled, and the percentage, have risen sharply in the last two decades. (B) The recycling rate varies greatly by material. Note that although the two types of plastic containers highlighted here are recycled at rates close to 30%, only 8% of plastics overall are recycled.

Source: U.S. Environmental Protection Agency.

bricks salvaged for use as paving for campus paths, and so on. The United States and other countries are developing "waste exchanges," whereby one industry's discarded waste becomes another's raw materials. Ideally, this both reduces waste disposal and saves the waste generator money: For example, an aluminum smelting firm in Maryland that had been paying $70 a ton in landfill fees to dispose of its leftover powdered fluorite found a taker in Pittsburgh who would pay $20 a ton; one company's waste isopropyl alcohol became another's cleaning solvent. The Canadian Waste Materials Exchange has dealt in such diverse materials as glove-leather scrap, oat hulls, and fish-processing waste.

There are also international exchanges of materials for recycling, though precise data are not always available. One recognized area of growing concern is "e-cycling," recycling of electronics waste. The concern relates particularly to toxic elements in electronics: lead in cathode-ray tubes and circuit boards; cadmium in semiconductors; mercury in switches, circuit boards, lamps, and batteries; and more. Much international

e-cycling occurs in south and east Asia (figure 16.15) where environmental regulations and laws protecting workers may be limited or nonexistent. Efforts are underway to improve the conditions under which this e-cycling occurs so that such recycling will genuinely be "green."

Whether or not source reduction in the volume of municipal solid wastes has been successful (which is hard to assess), there has been a clear shift in the handling of municipal wastes collected in the United States, toward recovery (for composting or recycling) and away from land disposal (figure 16.16). Still, some materials are too difficult to reuse efficiently, for reasons already outlined, and these will continue to require ultimate disposal. Beyond municipal wastes are toxic by-products of industrial processes that are not themselves useful, or are too toxic for safe handling during extensive reprocessing. These require more-specialized and careful disposal. Also, many of the highly toxic industrial wastes are liquids rather than solids, which may require somewhat different handling from solid wastes.

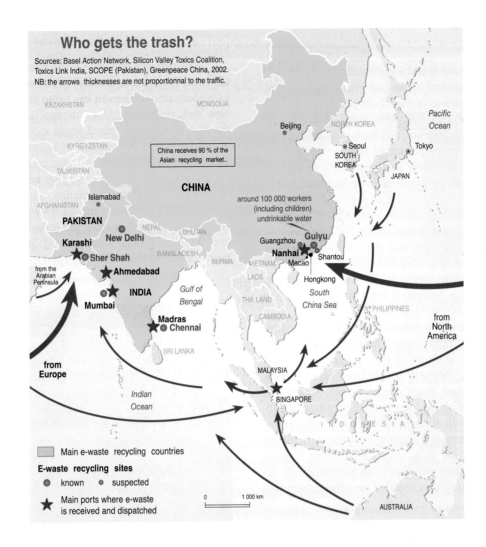

Who gets the trash?

Sources: Basel Action Network, Silicon Valley Toxics Coalition,
Toxics Link India, SCOPE (Pakistan), Greenpeace China, 2002.
NB: the arrows thicknesses are not proportionnal to the traffic.

China receives 90 % of the
Asian recycling market..

around 100 000 workers
(including children)
undrinkable water

Main e-waste recycling countries

E-waste recycling sites
- known   - suspected

★ Main ports where e-waste
is received and dispatched

0   1 000 km

Figure 16.15

Low labor costs and limited environmental regulations drive a flow of electronic waste from industrialized nations—especially in Europe and North America—to China, India, and Pakistan for e-cycling.

Graphic by Philippe Rekacewicz, UNEP/GRID-Arendal Maps and Graphics Library. Reprinted with permission UNEP. http://maps.grida.no/go/graphic/who-gets-the-trash.

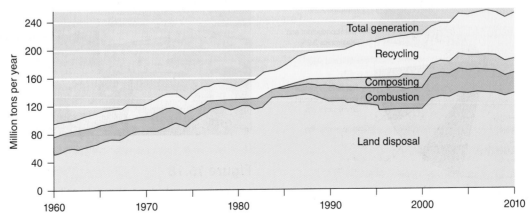

Figure 16.16

The trend in municipal waste disposal is away from traditional "disposal," but there are limits to the feasibility of the alternatives.

Source: U.S. Environmental Protection Agency.

Secure Landfills

For some years, waste-disposal specialists have believed that, in principle, it is possible to design a **secure landfill** site for toxic solid and liquid wastes. An example of a recommended design is shown in figure 16.19. The wastes are put in sealed drums before disposal. Beneath the drums are layers of plastic and/or compacted clay to contain any unexpected leaks. Wells and piping are installed so that the ground water below and around the site can be checked periodically for any sign of leakage of the waste chemicals. Excess accumulating leachate can be pumped out before it leaks out. Such a system provides multiple safeguards against accidental environmental contamination, and the monitoring wells allow prompt detection of any leaks. The design shares many features with modern municipal landfills, but with still more provisions for waste containment and site monitoring.

Unfortunately, a growing body of evidence indicates that no site is truly secure, even if conscientiously designed. Carefully compacted clay may be very low in permeability but is probably never completely impermeable, especially over long time intervals. Chemical and biological reactions in the wastes and leachate can rupture or decompose plastic, and the stress caused by the weight of wastes and cover can fracture a clay liner. Even relatively innocuous municipal waste can prove hard to contain. When built, the "Mount Trashmore" landfill in Evanston, Illinois, was hailed for its state-of-the-art design. But monitoring wells later revealed detectable leakage of at least a dozen volatile organic compounds deemed high-priority toxic pollutants by the Environmental Protection Agency, including benzene, toluene, vinyl chloride, and chloroform. Leakage from "secure" toxic-waste dumps, in which hundreds or thousands of barrels of concentrated toxic liquid chemicals are stored, has far more potential for harm. The historical response to detection of leakage from one toxic-waste dump has been to dig up as much of the hazardous material and contaminated soil as possible and transfer it to a more secure landfill. There is now some question as to whether a wholly different disposal method might be preferable.

Deep-Well Disposal

Another alternative for disposal of liquid industrial waste is injection into deep wells (figure 16.20). This method has been practiced since World War II. The rock unit selected to receive the wastes must be relatively porous and permeable (commonly, sandstone or fractured limestone), and it must be isolated by low-permeability layers (for example, shale) above and below. The subsurface geology must be known in sufficient detail that there is reasonable confidence that the disposal stratum remains

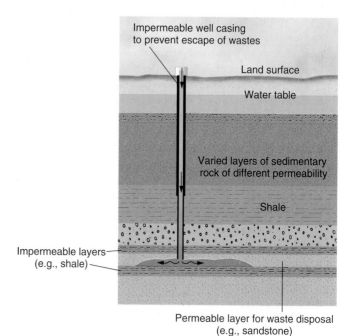

A

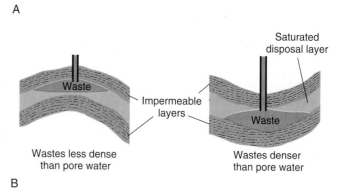

B

Figure 16.20

Deep-well disposal for liquid wastes. (A) Basic design: Wastes are placed in a deep permeable layer that is geographically and geologically isolated by low-permeability strata. (B) Containment of wastes assisted by geologic structures, much as petroleum is trapped.

Figure 16.19

A secure landfill design for toxic-waste disposal, including provisions for leachate containment and for monitoring the chemistry of subsurface water nearby.

isolated for some distance from the well site in all directions. Information about that geology may be derived from many sources: direct drilling to obtain core samples that provide a vertical section of the rock units present; geophysical studies that provide data on depths to and thicknesses of different rock layers, and on the distribution of ground water; geologic mapping on the basis of cores, surface outcrops, and geophysical data to interpolate between points sampled directly.

These disposal wells are hundreds to thousands of meters deep, far removed from the surface, and below the regional water table. The pore water in the disposal stratum should be brackish or saline water not suitable for a water supply. Where the well intersects any shallower aquifers that are or might be used for water supply, it must be snugly lined (cased) to prevent leakage of the wastes into those aquifers. Local well water is monitored to detect any accidental leaks promptly.

Movement of deep ground water is generally slow, and the assumption is that by the time the toxic chemicals have migrated far enough laterally to reach a usable aquifer or another body of water, they will have become sufficiently diluted not to pose a threat. This presumes knowledge of the toxicity of the chemicals in low concentrations. When the wastes are more or less dense than the ground water they displace and not miscible with it, folds or other geologic structures may help to contain them and slow their spread, as oil traps contain petroleum (figure 16.20B). The behavior of chemicals that dissolve in the pore water is much less well understood. They can diffuse through the water more rapidly than the water itself moves, so that even if deep ground-water transport is slow, contaminant migration may not be.

Costs for deep-well disposal are comparable to or somewhat less than those associated with "secure" landfill sites. The rate of waste disposal in a deep well is limited by the permeability of the rocks of the disposal stratum, while landfills have no equivalent limitation. The region's geology must be such that suitable strata exist for disposal by injection, while landfills can be constructed in a much greater variety of settings. Like landfills, deep injection wells may leak. Finally, as noted in chapter 4, deep-well waste injection may trigger earthquakes in faulted rocks. A flurry of small earthquakes in the central United States during 2011 has been associated with injection of waste water from natural-gas production (not from the fracking itself). While none of these events was damaging—the largest were about magnitude 5—they do reinforce the need for care in employing this disposal method. The various geologic constraints on possible deep-well disposal sites are thus more restrictive than those for landfill sites, and deep-well disposal is correspondingly less common.

Other Strategies

There are now furnaces designed to incinerate fine streams of liquid at very high temperatures. This permits the destruction of toxic liquid organic chemicals, often with no worse by-product than carbon dioxide, and may ultimately prove the best way to deal with them. Some liquid wastes can be neutralized or broken down by chemical treatment, which may avoid the necessity for ultrasecure disposal altogether. As with solid wastes, it may even

be possible to use certain liquid "wastes" via waste exchanges. The nitric acid used in quantity by the electronics industry to etch silicon wafers can be neutralized to produce calcium nitrate and then incorporated in high-grade fertilizers. Spent acid used in the steel industry is rich in dissolved iron and can be used at geothermal power plants to control hydrogen sulfide gas emissions, which react with the iron in solution to precipitate iron sulfides.

These and other means of handling toxic liquid industrial wastes will continue to be developed. What these methods generally have in common is that they are designed to deal either with specialized types of waste or with limited quantities. Volumetrically, at least, a far larger liquid-waste disposal problem is posed by that very commonplace material, sewage.

Sewage Treatment

Problems arising from organic matter in water include oxygen depletion and "algal bloom," as described in chapter 17. These problems, and concern about the spread of disease through biological contamination of drinking-water supplies by *pathogenic* (disease-causing) organisms, provide the motivation for effective sewage disposal. Much of the approximately 40 billion gallons of water withdrawn for public water supplies in the United States each day winds up mixed with sewage. So does urban surface runoff water collected in storm drains, along with a portion of the water used in rural areas. Appropriate treatment strategy varies with population density and local geology.

Septic Systems

On an individual-user level, modern sewage treatment typically involves a septic system of some kind (figure 16.21). Wastes are first transferred to a settling tank in which solids settle out, to be broken down slowly through bacterial action. The remaining liquid carries a load of dissolved organic matter and of microorganisms—some pathogenic—whose metabolism requires little or no oxygen. The dissolved organic matter represents food for those microorganisms. The liquid is allowed to seep out through porous pipes into the soil of the **leaching field** or **adsorption field.** There, oxygen is available, in the pore spaces, and aerobic soil microorganisms that can use that oxygen in metabolizing the organic matter compete for the nutrients with the microorganisms in the sewage, breaking down the organic matter and destroying some pathogens. Passage through the soil, especially if it is fine-grained, also filters the liquid, removing remaining fine suspended solids and even the larger pathogenic organisms. Inorganic reactions in the soil can also break down undesirable compounds in the sewage. Ideally, by the time any of the liquid reaches either the surface or groundwater supplies, it has been purified of biological and many chemical contaminants. Some compounds do remain in solution, however; nitrate is ordinarily the most significant potential pollutant among them. (Should the settling tank require pumping out, disposal of the septage can be a further problem, for few sites will accept it, and municipal sewage-treatment facilities are reluctant to take on the added load.)

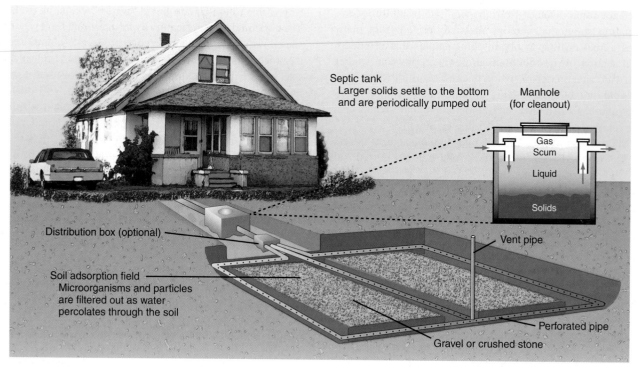

Figure 16.21

Basics of septic tank system: A settling tank for solids and the slow release of liquids into the soil of the leaching field for natural decomposition.

There are several geologic requirements for a properly functioning septic system. The soil must be sufficiently permeable that the fluids will flow through it rather than merely backing up in the septic tank, but not so permeable that the flow into water supplies or out at the surface occurs before the wastes have been sufficiently purified. Ordinarily, the water table should be well below the level of the septic system: first, to avoid immediate groundwater contamination by raw sewage; second, because oxygen levels in saturated soil are often too low to permit rapid aerobic breakdown of the organic matter. There must be sufficient soil depth so that the wastewater will be adequately filtered by the time it reaches either surface or bedrock; this filtration usually requires at least 60 centimeters of soil above the pipes and 150 centimeters below. The leaching field should not extend to within about 15 meters of any body of surface water, for similar reasons. (The exact dimensions required are controlled by soil characteristics.)

If a well is used to obtain drinking water on the same site, it must be far enough removed from the septic system that partially decomposed sewage does not reach the well, or else it should tap an aquifer that is isolated from the septic system by low-permeability material. Where many houses in a single area rely on septic systems, they must be spaced far enough apart so that they do not collectively saturate the soil with raw sewage and overwhelm the natural capacity of the soil and the microorganisms within it to handle the waste. The necessary spacing depends, in part, on the sizes of leaching fields to be accommodated. The required size of each leaching field is controlled, in turn, by soil permeability (table 16.2) and the number of persons to be served. Considerations such as these and assessments of the potential impact of the nitrate released from these systems lead local authorities to stipulate minimum lot sizes where septic tanks are to be used. Typical lot sizes might be one-half acre to one acre per dwelling, but again, appropriate limits are controlled largely by local geology.

As is evident from the foregoing discussion, the sewage treatment associated with a septic system is entirely natural, and its thoroughness thus highly variable from site to site. One

Table 16.2	Relationship between Soil Character and Required Extent of Leaching Field

Soil Type*	Adsorption Area Needed (square meters per bedroom)
gravel or coarse sand	6.5
fine sand	8.3
sandy loam	10.6
clay loam	13.9
sandy clay	16.2
clay with minor sand or gravel	23.1
heavy clay	unsuitable

*Arranged broadly in order of decreasing permeability from top to bottom of table; all else being equal, the more permeable the soil, the smaller the area needed.

Source: Data from Water in Environmental Planning, by T. Dunne and L. B. Leopold, p. 187. Copyright © 1978 W. H. Freeman and Company.

can enhance the suitability of the site—for example, by embedding the plumbing of the leaching field in "imported" sand if the soil is naturally too clay-rich to be suitable—but the natural soil chemistry, biology, and physical properties largely control the effectiveness of the system. Also, if toxic household wastes are dumped in the system, many will be untouched by the chemical processes and microbial activity that attack and decompose the organic wastes (and some might even kill those helpful microorganisms).

Just as natural wetlands can serve a water-cleaning role, so artificial wetlands and ponds can enhance a septic system and increase its capacity, with the aid of plants. The conventional subsurface leaching field can be replaced by one or more gravel-filled basins planted with wetland plants. While the effluent stays below the surface, the plants' roots reach down into the wet gravel. The plants consume some of the nutrients in the water, and their roots host quantities of the bacteria that decompose the sewage. The cleaned water may then be channeled into a pond supporting additional plant life and waterfowl. Most such schemes at present are small, experimental operations, but their aesthetic appeal can make them an attractive option where the wastewater load is not too great.

Municipal Sewage Treatment

In urbanized areas, population density is far too high to permit effective sewage treatment by septic systems. More than 75% of the U.S. population is now served instead by sewer systems. In fewer than 5% of the cases is completely untreated sewage released from such systems, but the completeness of municipal sewage treatment does vary. The basic steps involved are summarized in figure 16.22A.

Primary treatment is usually physical processing only, involving removal of solids, including inorganic trash, fine sediment, and so on, that has been picked up by the wastewater. The screens remove large objects like floating trash. Sand and gravel then settle in the grit chamber. The flow is slowed, and suspended organic and inorganic matter settles into the bottom of the sedimentation tank. Oil or grease may be skimmed off the surface. Where industrial wastes are also present, some preliminary chemical treatment, such as neutralizing excess acidity or alkalinity, may be necessary. Otherwise, the dissolved organic matter and other dissolved chemicals remain in solution after this stage.

Secondary treatment of the remaining liquid is mainly biological. The effluent from primary treatment is aerated, and bacteria and fungi act on the dissolved and suspended organic matter to break it down. The treated sewage passes through another settling tank to remove solids, which now contain the useful microorganisms, too. This sludge can be cycled back to mix with fresh input water and air to continue the breakdown process. Sometimes, the water is also chlorinated to disinfect it when biological activity is completed. Primary plus secondary treatment together can reduce suspended solids and oxygen demand (see chapter 17) by about 90%, nitrogen by one-half, and phosphorus by one-third. Other dissolved chemicals remain.

A weakness in many municipal treatment systems, even those that practice secondary treatment, is that large amounts of rain or meltwater runoff in a short time can overload the capacity of treatment plants. At such times, some raw sewage may be released untreated. Also, as noted, even secondary treatment has little impact on dissolved minerals and potentially toxic chemicals, which remain in the water. Recently, the safety of chlorination has been questioned because, in some communities where hydrocarbon impurities were already present in the water supply, chlorination was found to lead to the formation of chlorinated hydrocarbons, such as chloroform. Some of these compounds may be carcinogenic. However, the chlorination procedure often can be adjusted to minimize the production of chlorinated hydrocarbons. The value of chlorination in the destruction of pathogenic organisms probably far outweighs the danger from this largely controllable problem.

After secondary treatment, wastewater can also be subjected to various kinds of tertiary, or advanced, treatment and purified sufficiently to be recycled as drinking water. Fine filtration, passage through activated charcoal, distillation, further chlorination, and various chemical treatments to remove dissolved minerals and other chemicals are some possible forms of tertiary treatment. Such thorough tertiary wastewater treatment may cost up to five times what primary and secondary treatment together do, which is why tertiary treatment is uncommon. Only a few percent of the U.S. population is served by systems undertaking tertiary sewage treatment. It is now practiced mainly in a few communities where water shortages compel water recycling. On the other hand, tertiary treatment of all U.S. municipal wastewater would cost only a few tens of dollars per person per year. As a means of extending dwindling drinking-water supplies, tertiary treatment is likely to find increasing use in the future. (On the other hand, it is becoming ever more challenging to make waste water safe to drink, as more diverse chemicals find their way into municipal sewage; for example, in recent years there has been growing concern about accumulating medications that people flush down the drain and into the sewer system.)

Wastewater treatment can yield other benefits besides drinking water. The city of Arcata, California, uses sewage that has passed through primary treatment to supply water to marshes and ponds, where natural microbial action and plant growth purify the water further and remove nutrients from it. Ultimately, the water purity is such that the output water can be used at a fish hatchery. The Tapia Water Reclamation Facility in northwestern Los Angeles County takes 10 million gallons of sewage per day and transforms it into water of high enough quality to use for irrigation. And Calpine, the company that currently operates The Geysers geothermal-power complex, is transporting close to 20 million gallons a day by pipelines from Santa Rosa and Lake County, California, to recharge the geothermal reservoir and increase power production.

A by-product of sewage treatment is quantities of sludge, which can present a solid-waste disposal problem. Chicago's municipal sewage treatment facilities process 1.4 billion gallons of sewage and generate approximately 600 tons of sludge (on a dry weight basis) *every day*. What can be done with such wastes? One option, because sludge is rich

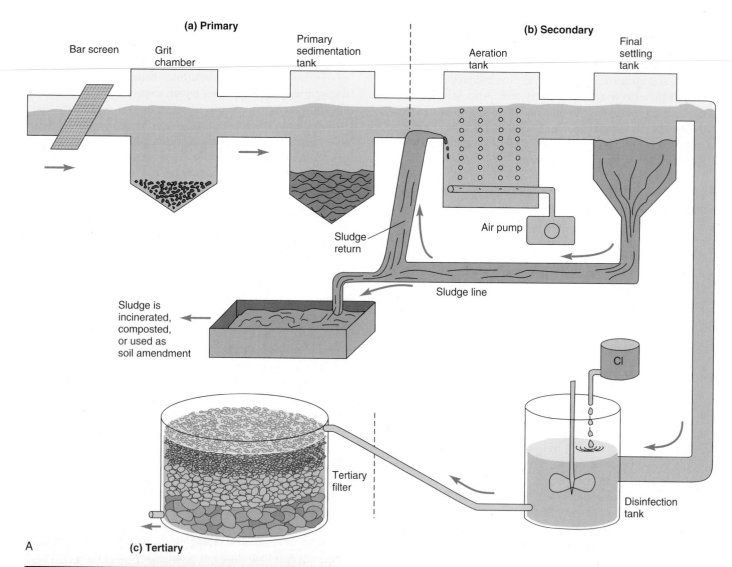

(a) Primary

Bar screen
Grit chamber
Primary sedimentation tank

(b) Secondary

Aeration tank
Final settling tank

Sludge return

Air pump

Sludge line

Sludge is incinerated, composed, or used as soil amendment

Cl

Tertiary filter

Disinfection tank

A   (c) Tertiary

B

Figure 16.22

(A) Schematic diagram of primary, secondary, and tertiary stages of municipal sewage treatment. See text for fuller description of the role of each stage. (B) Dallas Central Wastewater Treatment Plant.

(A) Modified after U.S. Environmental Protection Agency;
(B) Photograph by Bill Empson, U.S. Army Corps of Engineers.

in organic matter and nutrients, is to use it as fertilizer. In the early 1970s, for example, Chicago developed a plan for using some of its sludge to fertilize 11,000 acres of formerly strip-mined land during reclamation—thereby neatly solving several environmental problems at once. Austin, Texas, composts and transforms sludge into "Dillo Dirt" with which parks and athletic fields are fertilized; it is also sold to gardeners. (The name comes from the city's mascot, the armadillo.) The city of Milwaukee has been selling the sludge as the organic fertilizer "milorganite" for decades. However, much urban sludge contains unacceptably high concentrations of heavy metals or other toxic industrial chemicals, which have to be removed with the organic solids, and which are too dispersed through the sludge for efficient extraction. This may limit the use of such sludge fertilizer to land not to be used for food crops or pastureland (parks, golf courses, and so on) and may even necessitate its disposal in landfill sites like other solid wastes.

In speaking of sewage treatment, it is worth keeping in mind the disparities between the industrialized and other nations. According to the World Resources Institute, it has been estimated that over 90% of the sewage in developing countries is discharged into surface waters with no treatment whatever. A 2008 compilation for the U.N. Environment Programme indicated that in many parts of the world, over 80% of the wastewater reaching fresh or coastal waters is untreated; areas where this is true include the Caribbean, West and Central Africa, and South and East Asia. It should also be noted that while the focus here has been on U.S.-style sewage transport and treatment, which involve a great deal of water, treatment of human wastes does not necessarily require those water-intensive approaches. In this country, composting toilets are perhaps the most familiar example of a non-aqueous system. Other nations, where water is in shorter supply, have made larger-scale use of such approaches, and they become increasingly appealing as water shortages become more acute.

Radioactive Wastes

Radioactive wastes differ somewhat in character from chemical wastes. Thus, proposals for the disposal of solid and liquid radioactive wastes usually differ somewhat from the methods used for other wastes, although some radioactive materials are chemical toxins as well.

Radioactive Decay

Some isotopes, some of the possible combinations of protons and neutrons in atomic nuclei, are basically unstable. Sooner or later, unstable nuclei decay. In doing so, they release radiation: alpha particles (nuclei of helium-4, with two protons, two neutrons, and a +2 charge), beta particles (electrons, with a −1 charge), or gamma rays (electromagnetic radiation, similar to X rays or microwaves but shorter in wavelength and more penetrating). A given decay may emit more than one type of radiation.

Radioactive decay is a statistical phenomenon. It is impossible to predict the instant at which an atom of a radioactive element will decay. However, over a period of time, a fixed percentage of a group of atoms of a given radioactive isotope (radioisotope) will decay. The idea is somewhat like flipping a coin: One cannot know beforehand whether a given flip will come up heads or tails, but over a great many flips, one can expect that half will come up heads, half tails.

Each different radioisotope has its own characteristic rate of decay, often described in terms of a parameter called **half-life.** The half-life of a radioisotope is the length of time required for half the atoms of that isotope initially present in a system to decay. The concept is illustrated in figure 16.23. Suppose, for example, that the half-life of a particular radioactive isotope is ten years and that, initially, there are 2000 atoms of it present. After ten years, there will be about 1000 atoms left; after ten more years, 500 atoms; after ten more years, 250; and so on. After five half-lives, only a few percent of the original amount is left; after ten half-lives or more, the fraction left is vanishingly small.

Half-lives of different radioactive elements vary from fractions of a second to billions of years, but for any specific radioactive isotope, the half-life is constant. Uranium-238 has a

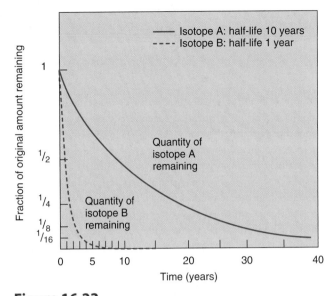

Figure 16.23

The phenomenon of radioactive decay: Each radioisotope decays at its own characteristic rate, defined by its half-life.

half-life of about 4.5 billion years; carbon-14, 5730 years; radon-222, 3.82 days. The half-life of a given isotope is constant regardless of the physical or chemical state in which it exists, whether it is in a mineral or dissolved in water, in the atmosphere or deep in the crust at high temperature and pressure. Spontaneous radioactive decay cannot be accelerated or slowed down by chemical or physical means. Because this is so, naturally occurring radioisotopes are powerful tools for studying the earth's history, as described in appendix A. Unfortunately, it also means that radioactive wastes will continue to be radioactive, and each isotope will decay at its own rate, regardless of how the wastes are handled. The wastes cannot be made to decay faster to get rid of the radioactive material more quickly, and they cannot be treated to make them nonradioactive. This is a key difference between radioactive wastes and many chemical toxins; many of the latter can, through proper treatment, be broken down or neutralized, which increases the disposal options available for dealing with them.

Effects of Radiation

Alpha and beta particles and gamma rays all are types of ionizing radiation. That is, they can strip off electrons from atoms or split molecules into pieces. Depending on which particular atoms or molecules in an organism are affected and the intensity of the radiation dose, the results could include genetic mutations, cancer, tissue burns—or nothing significant at all. Alpha particles are more massive but not very energetic. They can be very damaging but cannot travel far through matter; they can be stopped by a sheet of paper. Alpha-emitting radioisotopes are thus dangerous only when inhaled or ingested. Beta particles are lighter and less damaging but travel farther; they can be stopped by wood. Gamma rays are the most energetic and penetrating, but can be stopped by concrete.

The Ghost of Toxins Past: Superfund

Disposal of identifiably toxic wastes in the United States is carefully controlled today. It has not always been so. In some cases, disposal companies hired by other firms to dispose of toxic wastes have done so illicitly and inappropriately. In other instances, the waste-generating companies dumped or buried their own wastes onsite. Decades later, when the firm responsible is long gone, pollutants may ooze from the land surface or may suddenly be detected in ground water. Sometimes, the waste disposer can be found but denies that the problem is serious and/or declines to spend the money to clean up the site. Sometimes, the responsible company is no longer in business.

In response to these and other similar situations, in late 1980, Congress passed the Comprehensive Environmental Response, Compensation, and Liability Act, which, among other provisions, created what is popularly known as Superfund. The original $1.6 billion trust fund was generated primarily through taxes on oil and on producers of forty-two specific hazardous chemicals. Superfund monies are intended to pay for immediate emergency cleanup of abandoned toxic-waste sites and of sites whose owners refuse to clean them up. This represents a major step toward rectification of past waste-disposal abuses following recognition of the problem. Superfund was reauthorized several times, its trust fund increasing to $13.6 billion before its taxing authority expired in 1995. Since that time, its operations have been supported by federal appropriations. The American Recovery and Reinvestment Act of 2009 provided an infusion of $600 million to Superfund to help speed up cleanup activities.

Meanwhile, there has been increasing realization that the size of the problem is far greater than originally believed. At the time

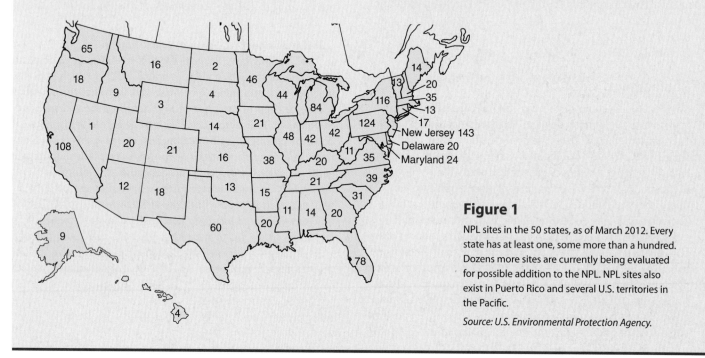

Figure 1

NPL sites in the 50 states, as of March 2012. Every state has at least one, some more than a hundred. Dozens more sites are currently being evaluated for possible addition to the NPL. NPL sites also exist in Puerto Rico and several U.S. territories in the Pacific.

Source: U.S. Environmental Protection Agency.

How much damage is done by a given dose of radiation? How small a dose is harmless? Scientists do not know precisely. Most controlled radiation studies have been done on animals, not humans. The most definitive data on radiation effects on people come from accidental exposures of scientists or technicians to sizable doses of radiation or from observing the after-effects of the explosion of atomic weapons in Japan during World War II. In those cases, the doses received by the victims are generally very high and very imprecisely known, and their relevance to concerns about exposure to low levels of radiation is not clear in any case. Data from Chernobyl and Fukushima may, in time, provide some clarification.

The effects of low doses of radiation on humans are particularly hard to quantify for several reasons. One is that some of the consequences, such as slowly developing cancers, do not appear for many years after exposure. By that time, it is much harder to link cause and effect clearly than it is, for example, in the case of severe radiation burns resulting immediately from a massive dose. A related problem is that many of the results of radiation exposure—like cancer or mutations—can have other causes, too. There are hosts of known or suspected *chemical* carcinogens and mutagens; how does one uniquely identify the cause of each particular cancer or mutation, years after exposure to whatever caused it?

Superfund was created, Congress envisioned cleanup of about 400 toxic-waste sites. By 1987, over 950 "priority" sites had been identified, and the number continued to rise. As of March 2012, 1661 sites had been placed on the National Priorities List (NPL) (figure 1), and 62 more listings were pending. Many other sites posing less-acute risks have also been identified. Since 1980, the Environmental Protection Agency has worked with state agencies and tribal councils to assess or clean up over 47,000 other sites. Overall, EPA estimates that 1 in 4 Americans lives within 3 miles of one or more Superfund sites.

Some Superfund NPL sites have been well publicized and notorious. One famous example is Love Canal, New York, where Hooker Chemical dumped 21,000 tons of waste chemicals from 1942 through 1952. The wastes were subsequently covered up, and homes and a school built nearby. Unusually high rates of cancer and other health problems among residents eventually led to recognition of the pollution, and Love Canal was placed on the NPL in 1983. Other sites may be smaller or less well-known, but all NPL sites are considered to pose major hazards. Not all involve what might be considered typical toxic wastes, either; the Libby vermiculite mine described in chapter 2 is another Superfund site.

Cleanup of NPL sites is meticulously planned and executed. It can therefore be both expensive and slow. Even after construction is completed, it may be several more years before treatment of the site is sufficient for EPA to delete it from the NPL. About ⅔ of NPL sites are at the "construction completed" stage, but only 359 have been deleted from NPL yet (figure 2). Love Canal was not deleted from the NPL until 2004.

More problem areas continue to be identified. The average cost of cleaning up a Superfund site is over $30 million, not counting any associated litigation, and some projections now put eventual total cleanup costs of Superfund sites at around *$1 trillion*. Where responsible parties can be identified, they do contribute to the cleanup costs—over $24 billion through fiscal year 2005—but massive cleanup bills remain. It has been suggested that additional funding might be provided by a tax on *disposers* of toxic waste, rather than the generators of it. There is concern, however, that a tax added at the

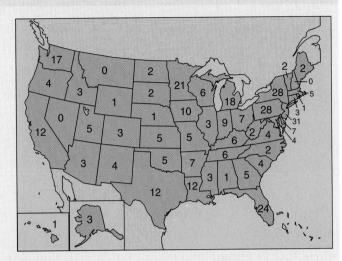

Figure 2

Superfund sites deleted from NPL, by state. Subtracting these numbers from those in figure 1 gives a sense of the scope of problem sites still being addressed.

Data from U.S. Environmental Protection Agency.

disposal end might tend to encourage more illicit dumping of toxic waste, ultimately compounding an already large problem.

In the meantime, every bit of hazardous waste cleaned up and removed from one spot still has to be disposed of in another. . . . Thousands of non-NPL abandoned toxic-waste sites remain to be addressed, though work progresses on hundreds of these each year. And Superfund deals only with abandoned or inoperative sites in need of cleanup. The EPA has also identified an estimated 2500 *active* disposal sites that may require corrective action, too. Waste minimization becomes increasingly appealing as these problematic sites multiply.

Curious about the location and status of current NPL sites in your state? You can start with the interactive map at www.epa.gov/superfund/sites/query/queryhtm/nplfin.htm.

Another uncertainty arises from the question of linearity in cause and effect (figure 16.24). If a population is exposed to a certain dose of a certain type of radiation and one hundred cases of cancer result, does that mean that one-tenth the exposure would induce ten cases of cancer, and so on? Or is there a minimum harmful level of radiation exposure, below which we are not at risk?

We are, after all, surrounded by radiation. Cosmic rays bombard us from space; naturally occurring radioactive elements decay in the soil beneath our feet and in the wood and brick and rock and concrete in our homes. (Your granite countertop is a—very minor—natural source of radiation; even your

potassium-rich banana contains some radioactive potassium-40!) Perhaps, in consequence, humans have developed a tolerance of radiation doses that are not significantly above natural background levels, as with line B in figure 16.24. Nobody can avoid some natural exposure to radiation (table 16.3). We increase our exposure with every medical X ray, every jet airplane ride, every journey to a high mountaintop, and every time we wear a watch with a luminous dial. Indeed, each of us is radioactive, a consequence of naturally occurring carbon-14 and other radioisotopes in the body. Therefore, any additional exposure from radioactive waste (or uranium in mine tailings, or other nonroutine sources) is just that, an incremental increase over our inevitable, constant

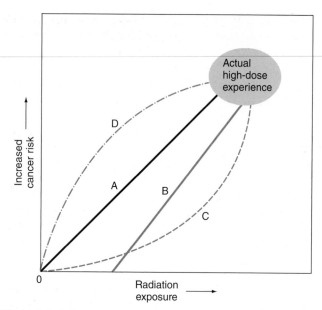

Figure 16.24

Current U.S. radiation-exposure limits are based on line A, "linear—no threshold": If a massive dose of radiation causes a certain increase in cancer, we can extrapolate back on a straight line to zero dose, zero increased risk. Line B is "linear with threshold": it assumes we can tolerate small doses with no ill effects but that risk increases in direct proportion to radiation exposure. Or maybe low doses add some risk but large doses are proportionately much more hazardous (line C); or maybe low doses increase risk a lot, and beyond that, higher doses don't increase risk much more (line D). None of us has zero exposure, and cancer has many causes, so we simply can't pin down the shape of the curve, for lack of clear data.

exposure to radiation. Does such a small exposure to low-level radiation represent a significantly increased risk? How small is small? Frustratingly, perhaps, there are no simple, direct answers to questions such as these. Individuals' background exposure to radiation is universal, yet quite variable; for this and other reasons discussed above, it is essentially impossible to determine definitively the effects of additional low radiation doses. That uncertainty, together with a lack of public understanding of natural radiation exposure, contributes to the controversy concerning radioactive-waste disposal.

Nature of Radioactive Wastes

As noted in chapter 15, nuclear fission reactor wastes include many radioisotopes. The isotopes of most concern from the standpoint of radiation hazards are not those of very short half-life (if an isotope has a half-life of 10 seconds, it is virtually all gone in a few minutes) or very long half-life (if an isotope's half-life is 10 billion years, very little of it will be decaying over a human lifetime). It is the elements with half-lives of the order of years to hundreds of years that pose the greatest problems: They are radioactive enough to be significant hazards, yet will persist in the environment for some time.

Some of the waste radioactive isotopes are also toxic chemical poisons, so they are dangerous independently of their radioactivity, even if their half-lives are long. Plutonium-239, with a half-life of 24,000 years, is one example of such an isotope. Other isotopes pose special hazards because the corresponding elements are biologically concentrated, often in one particular organ, leading to the possibility of a concentrated radiation source in the body. Included in this category are such products as iodine-131

Table 16.3	Typical Natural Radiation Exposure
Source	**Dosage, mrem/yr***
radon	200
external (cosmic rays, rocks, soil, masonry building materials)	55
medical (mainly X rays)	53
internal (in the body)—includes doses from food, water, air	39
consumer products	10
other, including normal operation of nuclear power plants	3
Where and how you live can affect your exposure	
worldwide radioactive fallout	4 mrem/yr
chest X ray	10–39 mrem
lower-gastrointestinal-tract X ray	500 mrem
jet travel—for each 2500 miles, add	1 mrem
TV viewing, 3 hours/day	1 mrem/yr
cooking with natural gas, add	6 mrem/yr

*Radiation dosages are expressed in units of *rems* or, for very small doses, *millirems* (mrem). A *rem* ("radiation equivalent man") is a unit that takes into account the type of radiation involved, the energy of that radiation, the amount of it to which one has been exposed, and the effect of that type of radiation on the body. The U.S. average annual dosage is 360 mrem.

Source: U.S. Department of Energy.

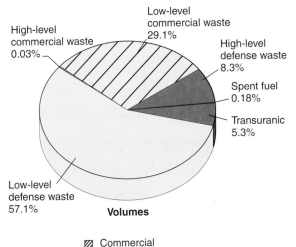

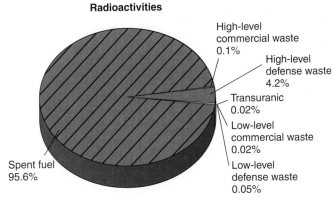

Figure 16.25

By volume, high-level waste is a small fraction of the radioactive-waste-disposal problem—but it accounts for nearly 100% of the radioactivity, and thus the radiation hazard.

Data from U.S. Department of Energy report DOE/RW-0006.

(half-life, 8 days), which would be concentrated, like any iodine, in the human thyroid gland; iron-59 (45 days), which would be concentrated in the iron-rich hemoglobin in blood; cesium-137 (30 years), which is concentrated in the nervous system; and strontium-90 (29 years) which, like all strontium, tends to be concentrated with calcium—in bones, teeth, and milk.

Radioactive wastes are often classified collectively as **low-level** or **high-level.** The division is an imprecise one. The low-level wastes are relatively low-radioactivity wastes, not requiring extraordinary disposal precautions. Volumetrically, they account for over 90% of radioactive wastes generated (figure 16.25). Some low-level wastes are believed to be harmless even if released directly into the environment. In this category, for example, fall the routine small releases of radioactive gas from operating fission reactors. Liquid low-level wastes—from laundering of protective clothing, from decontamination processes, from floor drains—are often released also, diluted as necessary so that the concentration of radioactivity is suitably low. Solid low-level wastes, such as filters, protective

clothing, and laboratory materials from medical and research laboratories, are commonly disposed of in landfills. Some solid low-level wastes from commercial reactor operations are held on-site until their radioactivity has decreased essentially to background levels, at which point they are treated like ordinary trash. Other low-level wastes are collected onsite until enough has accumulated to warrant shipment for disposal.

Federal legislation required states to take the responsibility for their own low-level wastes in the early 1990s (though they may enter into multi-state compacts, and several of these exist). States that benefit most from nuclear power and research and other facilities using nuclear materials thus bear more of the burden of the disposal of the corresponding low-level wastes. There is often concern over local disposal of low-level wastes (the NIMBY phenomenon again). As a result, no states have yet developed the mandated disposal sites to the point of actual disposal, and in some cases, planning is not far advanced. However, there are three sites currently licensed by the Nuclear Regulatory Commission to receive low-level wastes for disposal, one each in South Carolina, Utah, and Washington state, and states may contract with these facilities for disposal of their low-level nuclear wastes.

Spent reactor fuel rods and by-products from their fabrication and reprocessing are examples of high-level wastes. Given that they represent much more concentrated and intense sources of radiation, there is correspondingly greater concern about their proper disposal. (Reprocessing more spent fuel rods to recover plutonium–239 or re-enrich the uranium for fuel would not solve the problem. Aside from the cost and security issues that currently limit interest in reprocessing, spent fuel contains dozens of different radioactive isotopes among the fission products that have no practical use and would still constitute a significant quantity of high-level waste.)

Over the years, various more-or-less elaborate disposal schemes have been proposed. Currently, these materials sit in temporary storage—often on-site at nuclear reactors and other nuclear facilities—awaiting "permanent" disposal (figure 16.26). The basic disposal problem with the chemically varied high-level wastes is how to isolate them from the biosphere with some confidence that they will stay isolated for thousands of years or longer. In 1985, the Environmental Protection Agency specified that high-level wastes be isolated in such a way as to cause fewer than 1000 deaths in 10,000 years (the expected fatality rate associated with unmined uranium ore); in 2008, they increased the required time period of waste isolation to 1 million years. The sections that follow survey some of the many strategies proposed for dealing with these high-level and other concentrated radioactive wastes.

Historical Suggestions: Space, Ice, and Plate Tectonics

Some of the more exotic schemes proposed in the past for disposal of high-level radioactive wastes involved putting wastes in space, perhaps rocketing them into the sun. Certainly, that would get rid of the wastes permanently—if the rocket launch were successful. However, many launches are not successful (as the shuttle *Challenger* explosion dramatically demonstrated) or

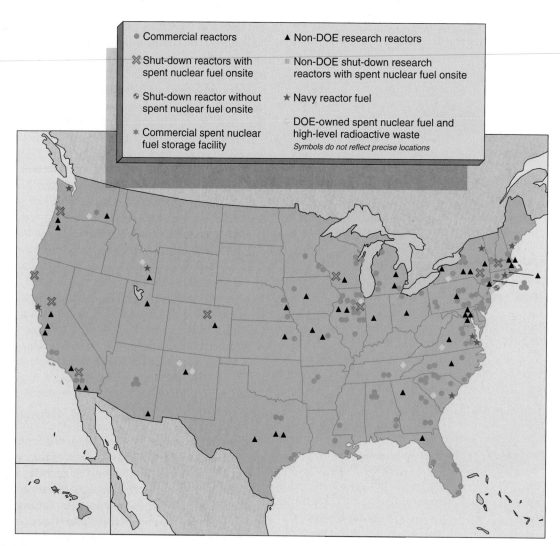

Figure 16.26

Locations (approximate) of spent nuclear fuel elements and high-level radioactive wastes in the United States awaiting disposal. Currently, over 60,000 metric tons (over 130 million pounds) of such wastes have accumulated, and the amount continues to increase.

Source: From U.S. Department of Energy, Office of Civilian Radioactive Waste Management.

are aborted, and the payloads return to earth, sometimes in an uncontrolled fashion. Furthermore, since many spacecraft would be required, the costs would be extremely high, even if successful launches could somehow be guaranteed.

Others have proposed putting encapsulated wastes under thick ice sheets, with Antarctica a suggested site. High-level radioactive wastes generate heat. They could melt their way down through the ice sheet, to be well isolated eventually by the overlying blanket of ice, which would refreeze above. But glacier ice flows, and it flows at different rates at the top and bottom of a glacier. How would one keep track of waste movement through time? What if future climatic changes were to melt the ice sheet? For various reasons, disposal of radioactive wastes in Antarctica is presently prohibited by international treaty, and no other, thinner ice sheets are seriously under consideration.

Subduction zones also have been considered as possible repositories for high-level wastes. The seafloor trenches above

subduction zones are often sites of fairly rapid accumulation of sediments eroded from a continent. In theory, these sediments would cover the wastes initially, and the wastes would, in time, be carried down with the subducting plate into the mantle, well out of the way of surface-dwelling life. However, subduction is a slow process; the wastes would sit on the sea floor for some time before disappearing into the mantle. Even at sedimentation rates of several tens of centimeters per year, large waste containers would take years to cover. Seawater, especially seawater warmed by decaying wastes, is highly corrosive. Might the wastes not begin to leak out into the ocean before burial or subduction was complete? How would the wastes be attached to the down-going plate to ensure that they would actually be subducted? Currently, the behavior of sediments in subduction zones is not fully understood. Geologic evidence suggests that some sediments are indeed subducted with down-going oceanic lithosphere, but some, on the other hand, are "scraped off" by the overriding plate and not subducted. The uncertainties and weaknesses certainly appear to be far too great for this scheme to be satisfactory.

Seabed Disposal

The deep-sea floor is a relatively isolated place, one with which humans are unlikely to make accidental contact. Much of this area is covered by layers of clay-rich sediment that may reach hundreds of meters in thickness, and studies have shown that the clays have remained stable for millions of years. Interest is growing in radioactive-waste disposal schemes that involve sinking waste canisters into these deep, stable clays (in areas well removed from both active plate boundaries and any seafloor resources, notably manganese nodules). The low permeability of clays means restricted water circulation in the sediments. Many of the elements in the wastes, even if they did escape from the containers, would be readily adsorbed by the fine clay particles. Any dissolved elements escaping into the ocean would be entering the dense, cold (1 to 4°C, or 34 to 39°F), saline, deep waters that mix only very slowly with near-surface water. By the time any leaked material reached inhabited waters, it would be greatly diluted, and many of the shorter-half-life isotopes would have decayed. The long-term geologic stability of the deep-sea floor is appealing, especially to nations having radioactive wastes to dispose of but not located in particularly stable areas of the continents, or having such high population densities that they lack remote areas for disposal sites. An international Seabed Working Group currently has this option under active study. It has already concluded that seabed disposal is technically feasible but that more research on the long-term safety of the method is still needed.

Bedrock Caverns for Liquid Waste

Some high-level radioactive wastes, such as by-products of fuel reprocessing, are presently held in liquid form rather than solid, in part to keep the heat-producing radioactive elements dilute enough that they do not begin to melt their storage containers. These liquid wastes are currently stored in cooled underground tanks, holding up to 1 million gallons apiece, which from time to time have leaked spectacularly: At the federal radioactive-waste repository in Hanford, Washington, more than 300,000 gallons of high-level liquid wastes leaked from their huge storage tanks in the late twentieth century. (Currently, remediation efforts are in process, including bioremediation through which microorganisms break down organic compounds and convert toxic metals to less-mobile compounds; but these methods can do nothing to reduce the radioactivity. Secure initial disposal is clearly preferable to later remediation.)

Caverns hollowed out of low-permeability, unfractured rocks—igneous rocks such as basalt or granite, for example—have been suggested as more secure, permanent disposal sites for high-level liquid wastes. The rocks' physical properties would have to be thoroughly investigated beforehand to ensure

that the caverns could indeed contain the liquid effectively. The wells leading down to the caverns, through which wastes would be pumped, would have to be tightly cased to prevent shallower leaks. The long-term concern, however, is the geologic stability of the bedrock unit chosen. Faulting or fracturing of the rocks would create conduits through which the highly radioactive liquid could escape rapidly, perhaps to the surface or into nearby aquifers. While it is possible to identify with a high degree of probability the areas that are likely to remain stable for thousands or millions of years, stability cannot be absolutely guaranteed, and liquid wastes clearly are potentially very mobile.

Serious consideration has been given to solidifying liquid high-level wastes prior to disposal because of the much lower mobility of solids. Certain minerals, ceramics, and glasses have proven to be relatively resistant to leaching over centuries or even millennia in the natural environment. Conversion of the wastes into a relatively stable solid form would be a first step in many high-level waste disposal schemes. The waste would then be sealed in canisters and the canisters placed in some kind of bedrock cavern, old mine, or similar space. Solidification would not eliminate the uncertainty regarding bedrock stability, but it would render the wastes less susceptible to rapid escape if fracturing breached the storage volume.

Bedrock Disposal of Solid High-Level Wastes

A variety of bedrock types are being investigated worldwide as host rocks for solid high-level-waste disposal. The general design with any bedrock disposal site or repository would involve the **multiple barrier** (or *multibarrier*) **concept:** surrounding solid waste with several different types of materials to create multiple obstructions to waste leakage or invasion by ground water. A major variable is the nature of the host rock. Each of the rock types highlighted below has some particular characteristics that make it promising as a potential host rock, as well as some potential drawbacks.

Granite is a plentiful constituent of the continental crust. Granites are strong and very stable structurally following excavation. They have low porosity; the common minerals of granite (quartz, feldspars) are quite insoluble in a temperate climate. A concern with respect to granite is the possible presence of fractures (natural or created during excavation) that would allow solutions to permeate and wastes, perhaps, to escape. One might expect that, below a depth of about 1 kilometer, the pressures would close any fractures. However, field measurements suggest that this is not necessarily the case. Half a dozen nations worldwide are investigating granitic rock as a potential host rock for radioactive-waste disposal. The United States and Sweden have a joint testing program in progress on granite at Stripa, Sweden; Finland is evaluating a repository in granitic gneiss.

Thick *basalts,* such as the Columbia River plateau basalts described in chapter 5, are another option. Fresh, unfractured basalt is a strong rock. Basalt consists of high-temperature minerals and sometimes glass, so it can withstand the elevated temperatures of high-level wastes, and it has fairly high thermal conductivity, to allow that heat to be dispersed. However, basalts commonly contain porous zones of gas bubbles; they may also weather easily. Both phenomena contribute to zones of weakness. Fractures in the basalt and even the planes between successive flows may provide water conduits.

Massive deposits of *tuff*, pyroclastic materials deposited during past large-scale explosive volcanic eruptions, are another possible host rock for solid high-level wastes. *Welded tuffs* are those that were so hot at the time of deposition that they fused; these deposits are not unlike basalts in many of their physical properties. But they are very brittle and easily fractured, which raises questions about their integrity with respect to water flow. Other tuffs have been extensively altered, producing abundant *zeolites*. These hydrous silicate minerals—the same group used in water softeners—are potentially valuable in a waste-disposal site for their ion-exchange capacity. They should tend to capture and hold wastes that might begin to migrate from a repository. However, zeolitic tuffs are physically weaker, more porous, and more permeable than welded tuffs; furthermore, zeolites begin to break down and dehydrate at relatively low temperatures (100 to 200°C, or about 200 to 400°F), which reduces their ion-exchange potential and also releases water that might facilitate waste migration.

Shale and other clay-rich sedimentary rocks are another option. As noted earlier in the seabed-disposal discussion, clays may adsorb migrating elements, providing a measure of containment for the waste. Shale is also low in permeability, and somewhat plastic under stress, though less so than salt (see below). On the other hand, shaly rocks are weak, fracture readily, and may be interlayered with much more permeable sedimentary units. At elevated temperatures, clays, like zeolites, may decompose and dehydrate. Some clays are stable to higher temperatures than are most zeolites, but those that would most actively adsorb waste components tend to lose their adsorptive efficiency at those higher temperatures. So, overall, clays may not have clear advantages over zeolites in disposal sites. Belgium and Italy have investigated shaly rocks as disposal units.

In the United States and Germany, much consideration has been given to high-level waste disposal in thickly bedded *salt* deposits or salt domes. Salt has several special properties that may make it a particularly suitable repository. It has a high melting point, higher in fact than those of most rocks. Consequently, salt can withstand considerable heating by the radioactive wastes without melting. Although soluble under wet conditions, salt typically has low porosity and permeability. Under dry conditions, then, it can provide very "tight" storage with minimal leakage potential. Salt's low porosity and permeability result, in part, from its ability to flow plastically under pressure while remaining solid. It is therefore somewhat self-sealing. If it were fractured by seismic activity or by accumulated stress in the rocks, flow in the salt could seal the cracks. Most other rock types are more brittle under crustal conditions and do not so readily self-seal. Finally, evaporites are not particularly scarce resources. Using an old salt mine for radioactive-waste disposal would not be a hardship in that context.

Waste Isolation Pilot Plant (WIPP): A Model?

While commercial high-level wastes await permanent disposal, the U.S. government *has* inaugurated a disposal site in southeastern New Mexico for some of its own high-level wastes, **transuranic wastes** (wastes containing radioactive elements heavier than uranium—including plutonium) created during the production of nuclear weapons. The Waste Isolation Pilot Plant, or WIPP, officially became the first U.S. underground repository for such wastes on 26 March 1999, when it received its first waste shipment. This event was the culmination of twenty years of site evaluation, planning, and construction following congressional authorization of the site in 1979.

The disposal unit at WIPP is bedded salt, underlain by more evaporites and overlain by mudstones. The disposal layer is 2150 feet below the surface. The region is now dry, so the water table is low, and examination of the geology indicates that it has been stable for over 200 million years, offering reassurance with respect to future stability and waste containment. The facility is shown in figure 16.27.

In its first ten years of operation, WIPP received more than 2 million cubic feet of transuranic wastes (about 1/3 of its total capacity), from 14 sites around the country. To address public concerns about the transport of these highly radioactive materials, they are shipped in extremely sturdy containers (figure 16.28), and all the WIPP trucks are tracked by satellite. Collectively they have traveled over 16 million miles to and from WIPP collecting the wastes for disposal, without incident.

WIPP is ultimately expected to receive transuranic wastes from some two dozen sites across the nation. However, WIPP's authorization extends only to defense-related transuranic waste, not to commercial or other high-level radioactive wastes. For these, site analysis and development are still in progress.

The Long Road to Yucca Mountain: A Dead End?

Much of the delay in the selection of a site for disposal of high-level radioactive waste from commercial nuclear reactors has resulted from the need for thorough investigation of the geology of any proposed site. However, in the United States at least, there is also a political component to the delay. Simply put, nobody wants to host a high-level-waste disposal site. (Many in New Mexico were unenthusiastic about WIPP before it was built, too.)

The Nuclear Waste Policy Act of 1982 provided for the establishment of two high-level-waste disposal sites for commercial reactor waste, one in the western United States, one in the East, both to be identified by the Department of Energy (DOE). The same Act created the Office of Civilian Radioactive Waste Management and established the Nuclear Waste Fund (NWF), supported by spent-fuel fee charges on waste generators and a fee of 0.1 cent per kilowatt-hour on nuclear electricity. This sounds small, but to date, over $31 billion has been paid into NWF. The NWF was to finance disposal-site selection and investigation; over $10 billion has been spent so far.

In 1986, DOE announced three candidates for the western site: Hanford, Washington (where disposal would be in basalt);

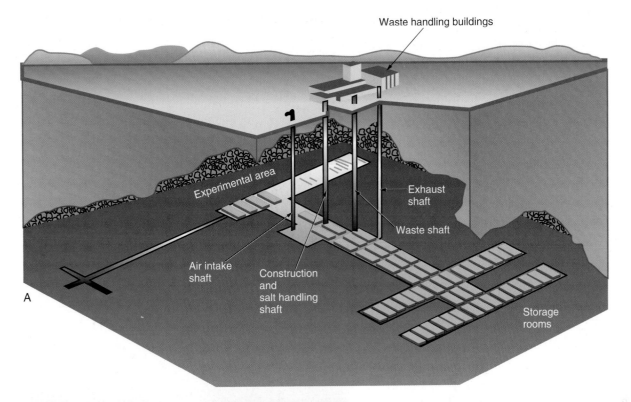

Waste handling buildings

Experimental area

Exhaust shaft

Waste shaft

Air intake shaft

Construction and salt handling shaft

Storage rooms

A

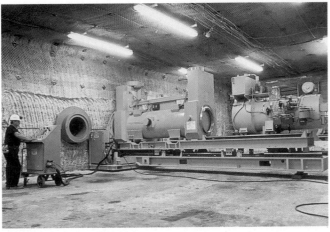

B

Figure 16.27

Waste Isolation Pilot Plant (WIPP) in New Mexico. (A) Plan of storage facility. (B) Waste canisters are emplaced, by remotely controlled machinery, into holes bored into salt walls.

Sources: (A) After Office of Environmental Restoration and Waste Management, Environmental Restoration and Waste Management: An Introduction, *U.S. Department of Energy. (B) Photograph courtesy of U.S. Department of Energy.*

Yucca Mountain, Nevada (ash-flow tuff); and Deaf Smith County, Texas (salt). All three states protested. So did Tennessee, where—at Oak Ridge—the waste-packaging site was to be built to prepare the wastes for disposal. Original plans would next have called for a billion-dollar testing program to be undertaken at each of the three proposed disposal sites, to be followed by selection of one to be *the* western site. Congress subsequently called a year-long halt to further action.

Then, in late 1987, Congress suddenly moved decisively to end the uncertainties—at least those of choice of site. It was decided that the Yucca Mountain site (figure 16.29) would be developed, and investigation of the other sites halted, at a cost savings estimated at $4 billion. The state of Nevada would receive $20 million per year and "special consideration" of certain research grants by way of compensation.

Properties of the site that contribute to its attractiveness include (1) the tuff host rock; (2) generally arid climate, with no perennial streams and limited subsurface water percolation in the area; (3) low population density; (4) low regional water table, so the repository would be 200 to 400 meters above the water table; and (5) apparent geologic stability. There are no known useful mineral or energy resources in the vicinity. The site is also located on lands already owned by the federal government, and overlaps the Nevada Test Site previously used for nuclear-weapons testing.

There are several geologic issues to be considered. Seismicity is an obvious concern: there are a number of faults in the area, some related to the collapse of ancient volcanic calderas, some related to the formation of the mountains, some related to plate motions. However, current seismicity is low, and evidence suggests limited local fault displacement for the last 10 million years. Still, repository design would provide for earthquakes up to magnitude 6.8 or so, the largest plausible on the known faults. The most recent volcanism in the region is a cinder cone that may be as young as 20,000 years old, and basaltic volcanism is possible within the next 10,000 years or so. There are also questions about how likely the region is to stay as dry as it currently is, particularly in light of

Trupact-II

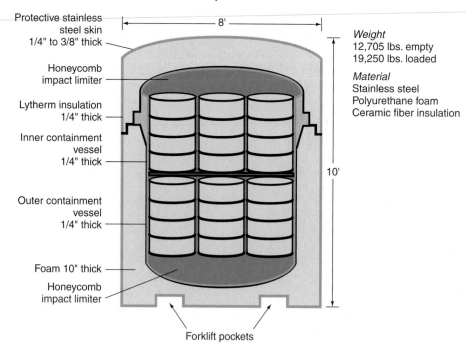

Protective stainless steel skin 1/4" to 3/8" thick

Honeycomb impact limiter

Lytherm insulation 1/4" thick

Inner containment vessel 1/4" thick

Outer containment vessel 1/4" thick

Foam 10" thick

Honeycomb impact limiter

Forklift pockets

8'

10'

Weight
12,705 lbs. empty
19,250 lbs. loaded

Material
Stainless steel
Polyurethane foam
Ceramic fiber insulation

Figure 16.28

Wastes destined for WIPP are contained in specially designed vessels such as this even before transport to the site. Similar vessels would be used for waste headed for Yucca Mountain or an alternate repository.

Source: U.S. Department of Energy.

A

Figure 16.29

(A) View to the south of Yucca Mountain; coring activities are visible along the ridge. The desert site is 90 miles from Las Vegas, the closest sizeable population center. (B) Cross section of proposed waste-disposal site at Yucca Mountain, Nevada. Generalized east-west section, showing potential water and water-vapor flow paths; the "welded" and "unwelded" units are tuffs. Repository level highlighted (yellow).

(A) Photograph courtesy Office of Civilian Radioactive Waste Management/DOE (B) After Site Characterization Plan Overview, *U.S. Department of Energy, 1988.*

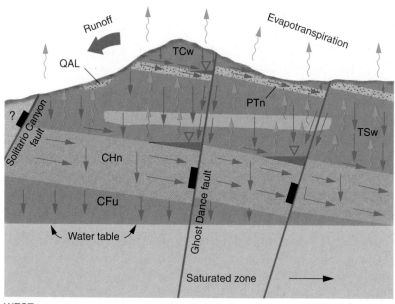

WEST   EAST

QAL   Alluvium (stream deposits)
TCw   Tiva Canyon welded unit
PTn   Paintbrush nonwelded unit
TSw   Topopah Spring welded unit
CHn   Calico Hills nonwelded unit
CFu   Crater Flat (undifferentiated) unit

↓   Liquid-water flow

〰   Water-vapor flow

▰   Normal fault

▽   Possible perched-water zone

B

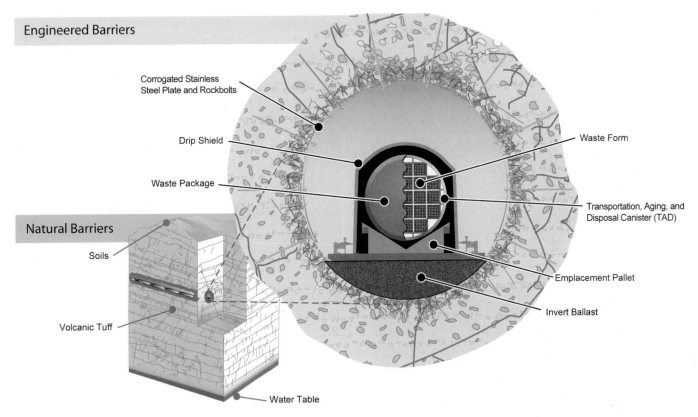

Engineered Barriers

Corrogated Stainless
Steel Plate and Rockbolts

Drip Shield

Waste Package

Waste Form

Transportation, Aging, and
Disposal Canister (TAD)

Emplacement Pallet

Invert Ballast

Natural Barriers

Soils

Volcanic Tuff

Water Table

Figure 16.30

Both natural and engineered barriers contribute to secure containment of wastes in the Yucca Mountain design.

Diagram courtesy Office of Civilian Radioactive Waste Management/DOE.

anticipated global climate change, so the rocks' response to increased water is relevant.

However positive the geologic factors, waste containment would not rely solely on geology (figure 16.30). In addition to the natural barriers are such engineered barriers as a stable waste form, durable container, drip shield over the waste to deflect downward-percolating water, steel-plate liner for the tunnels, and more. This illustrates the multibarrier concept previously mentioned.

Site-investigation activities have included very detailed structural, hydrologic, and geochemical studies, and more detailed analysis of geologic-stability issues (figure 16.31). The evaluation has been comprehensive, with laboratory studies supplementing research in the field. The site characterization *plan* was a nine-volume, 6000$^+$-page document, the draft and final environmental impact statements much longer. Meanwhile, the high-level wastes remain on hold, in temporary storage, as the Yucca Mountain timetable slips. Site characterization began in 1986. At one time, the projection was for the start of construction in 1998 and first waste disposal in 2003. In fact, the final site analysis was not completed until late in fiscal year 2001, the final environmental impact statement was issued by the Department of Energy (DOE) in February 2002, and the site was officially designated by Congress and the president for a high-level waste repository in July 2003. (Objections are still being raised by the state of Nevada and others.) Nuclear Regulatory Commission (NRC) review of DOE's 2008

license application was expected to take three to four years; officials hoped for construction to begin in time for waste disposal to start in 2017, their estimate of the "best achievable" timetable.

In 2009, President Obama suspended all funding for the Yucca Mountain project, after over 20 years and $9 billion spent studying Yucca Mountain and planning for the waste–disposal site. This did not halt NRC review of the license application, which continued until late 2011, when the matter became mired in legal action, where it remains as this is being written. The future of the project now is highly uncertain. An additional point to bear in mind is that the Yucca Mountain repository design was to accommodate 70,000 metric tons of high-level waste—and we have accumulated almost that much already, as noted in figure 16.26. Indeed, even before Yucca Mountain funding was suspended, the secretary of energy made a report to Congress citing the need for a second repository if Congress did not lift the statutory limit of 70,000 metric tons of waste at Yucca Mountain.

No High-Level Radioactive Waste Disposal Yet

At least two dozen nations are making plans to dispose of high-level nuclear wastes; but so far, *no* high-level radioactive wastes from power-reactor operation have been consigned to permanent disposal sites elsewhere in the world by any nation. All this waste—some solid, some liquid—is in temporary storage awaiting selection and construction of disposal sites. Geologists

A

B

C

Figure 16.31

Tunnels have been bored into Yucca Mountain using alpine mining equipment (A) to provide access to study the rocks in place, including their response to heat (B) and their hydrologic properties (C).

Photographs courtesy Office of Civilian Radioactive Waste Management/DOE.

Radioactive-waste disposal thus remains an issue even if the expanded nuclear-fission power option is not pursued. Currently, France, Russia, Japan, India, and China reprocess most spent reactor fuel elements, which, as noted earlier, reduces the volume of high-level waste but does not eliminate the problem. Like the U.S., Canada, Finland, and Sweden do not reprocess, but plan simple disposal of spent fuel. Worldwide, spent fuel is accumulating at over 10,000 metric tons a year. In half a century of nuclear-power generation, close to 300,000 metric tons of spent fuel have been produced, about a third of which have been reprocessed. A growing number of nations have accumulated high-level wastes, not only from past power generation, but also from nuclear weapons production and radiochemical research, including medical applications. All of that waste must ultimately be put somewhere in permanent disposal. (The only alternative is to keep the wastes at the surface, closely monitored and guarded to protect against accident or sabotage, assuming that long-term political stability is more likely than geologic stability—a debatable assumption, surely.) But it remains an open question where the wastes will be put, and how soon.

can identify sites that have a very high probability of remaining geologically quiet and secure for many generations to come, but absolute guarantees are not possible. Uncertainties about the long-term integrity and geologic stability of any given site, coupled with increasing public resistance in regions with sites under consideration, have held up the disposal efforts. Sweden has narrowed its choice of a geological disposal site to two possibilities, but as with Yucca Mountain, an actual repository remains unbuilt. Finland had begun construction on a bedrock disposal site (in granitic gneiss), but actual waste disposal is years away even if the Finnish government ultimately approves. WIPP is the only geological repository in operation.

Summary

Methods of solid-waste disposal include open dumping (largely phased out for health reasons), use of landfills, incineration on land or at sea, and disposal in the oceans. Incineration contributes to air pollution but can also provide energy. Landfills require a commitment of space, although the land may be used for other purposes when filled, and useful methane produced by waste breakdown can be recovered from some landfills. Composting may be the best option for yard waste. "Secure" landfills for toxic solid or liquid industrial wastes should isolate the hazardous wastes with low-permeability materials and include provisions for monitoring groundwater chemistry to detect any waste leakage. Increasing evidence suggests that construction of wholly secure landfill sites is not possible. Deep-well disposal and incineration are alternative methods for handling toxic liquid wastes. Reduction in initial waste volume, and recycling where feasible, both help to minimize the quantity of waste requiring disposal.

The principal liquid-waste problem in terms of volume is sewage. Low-population areas can rely on the natural sewage treatment associated with septic tanks, provided that soil and site characteristics are suitable. Municipal sewage treatment plants in densely populated areas can be constructed to purify water so well that it can even be recycled as drinking water, although most municipalities do not go to such expensive lengths.

A complicating feature of radioactive wastes is that they cannot be made nonradioactive; each radioisotope decays at its own characteristic rate regardless of how the waste is handled. Given the long half-lives of many isotopes in high-level wastes and the fact that some are chemical poisons as well as radiation hazards, long-term waste isolation is required. Disposal in bedrock, tuff, salt deposits, or on the deep-sea floor are among the methods under particularly serious consideration worldwide. In the United States, the Waste Isolation Pilot Plant has begun to receive the government's transuranic waste; Yucca Mountain, Nevada, had been designated for development of a permanent disposal site for commercial high-level wastes, but actual disposal is still years away even if a decision to proceed here is ultimately made.

Key Terms and Concepts

adsorption field 379	leachate 366	multiple barrier concept 389	source separation 373
half-life 383	leaching field 379	sanitary landfill 366	transuranic wastes 390
high-level waste 387	low-level waste 387	secure landfill 378	

Exercises

Questions for Review

1. What two kinds of activities generate the most solid wastes?

2. What advantages does a sanitary landfill have over an open dump? Describe three pathways through which pollutants may escape from an improperly designed landfill site.

3. Describe any three features of a modern municipal landfill that reduce the risk of pollution, and explain their function.

4. Landfills and incinerators have in common the fact that both can serve as energy sources. Explain.

5. Use of in-sink garbage disposal units in the home is sometimes described as "on-site disposal." Is this phrase accurate? Why or why not?

6. Compare the relative ease of recycling of (a) glass bottles, (b) paper, (c) plastics, (d) copper, and (e) steel, noting what factors make the practice more or less feasible in each case.

7. Compare the dilute-and-disperse and concentrate-and-contain philosophies of liquid-waste disposal.

8. Outline the relative merits and drawbacks of deep-well disposal and of incineration as disposal strategies for toxic liquid wastes.

9. What kinds of limitations restrict the use of septic tanks?

10. Municipal sewage treatment produces large volumes of sludge as a by-product. Suggest possible uses for this material, and note any factors that might restrict its use.

11. Describe two reasons that it is difficult to assess the hazards of low-level exposure to radiation.

12. Evaluate the advantages, disadvantages, and possible concerns relating to disposal of high-level radioactive wastes in (a) subduction zones, (b) sediments on the deep-sea floor, (c) tuff, and (d) bedded salt.

13. What characteristics of the Yucca Mountain site make it attractive as a possible disposal site for high-level wastes? Outline three geologic concerns relative to the long-term security of waste containment there.

Exploring Further

1. Where does your garbage go? How is it disposed of? Is any portion of it utilized in some useful way? If your community has a recycling program, find out what is recycled, in what quantities, and what changes have occurred over the last five or ten years in materials recycled and levels of participation.

2. For a period of one or several weeks, weigh all the trash you discard. If every one of the 300 million people in the United States or the 30 million in Canada discarded the same amount of trash, how much would be generated in a year? Suppose that each of the more than 7 billion people on earth did the same. How much garbage would that make? Is this likely to be a realistic estimate of actual world trash generation? Why or why not?

3. Investigate your community's sewage treatment. What level is it? What is the quality of the outflow? How does the storm-sewer system connect with the sewage-treatment system?

4. Choose a single Superfund NPL site, or several sites within a state or region, and check the history and current status of cleanup activities there. A website including all the NPL (National Priorities List) sites by region, which would be a place to start, is at http://www.epa.gov/superfund/sites/npl/index.htm.

5. Check on the current status of the Yucca Mountain license application at www.nrc.gov/ waste/ hlw–disposal/ yucca–lic–app. html, or read the secretary of energy's report on the need for a second repository, at www. ymp.gov/sites/prod/files/edg/media/Second_Repository_Rpt_120908.pdf.

CHAPTER 17

Water Pollution

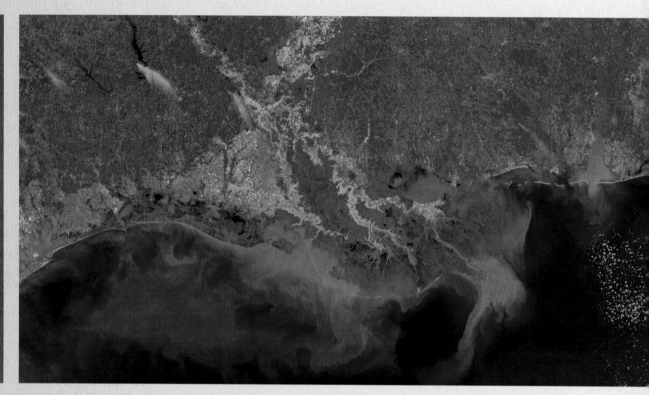

As noted in chapter 11, the more-or-less fresh ground water and surface water on the continents constitutes less than 1% of the water in the hydrosphere. Pollution has tainted much of the most accessible of this reserve: lakes, streams, water in shallow aquifers near major population centers. The number and variety of potentially harmful water pollutants are staggering.

A *pollutant* is sometimes defined simply as "a chemical out of place," a substance found in a high enough concentration in some setting that it creates a nuisance or hazard. (Dust and ash can also be pollutants, as we will see in the next chapter.) In this chapter, we review some important categories of water pollutants and the nature of the problems each presents. We also look at some possible solutions to existing water-pollution problems. While it is true that substantial quantities of surface water and ground water are unpolluted by human activities, much of this water is far removed from population centers that require it, and the quality of much of the most accessible water has indeed been degraded.

In the spring, the Atchafalaya River delivers a heavy load of suspended sediment to the Gulf of Mexico, clouding the water. What is not visible is that the water also carries dissolved nutrients, especially fertilizer that has run off farmland along with soil.

NASA image courtesy Norman Kuring, Ocean Color Team.

General Principles

Any natural water—rainwater, surface water, or ground water—contains dissolved chemicals. Some of the substances that find their way naturally into water are unhealthy to us or to other life forms, as, unfortunately, are some of the materials produced by modern industry, agriculture, and just people themselves.

Geochemical Cycles

All of the chemicals in the environment participate in geochemical cycles of some kind, similar to the rock cycle. Not all environmental cycles are equally well understood.

For example, a very simplified natural cycle for calcium is shown in figure 17.1. Calcium is a major constituent of most common rock types. When weathered out of rocks, usually by solution of calcium carbonate minerals or by the breakdown of calcium-bearing silicates, calcium goes into solution in surface water or ground water. From there it may be carried into the oceans. Some is taken up in the calcite of marine organisms' shells, in which form it may later be deposited on the sea floor after the organisms die. Some is precipitated directly out of solution into limestone deposits, especially in warm, shallow water; other calcium-bearing sediments may also be deposited. In time, the sediments are lithified, then metamorphosed, or subducted and mixed with magma from the mantle, or otherwise reincorporated into the continental crust as the cycle continues.

Subcycles also exist. For instance, some mantle-derived calcium is leached directly into the oceans by the warmed seawater or other hydrothermal fluids interacting with the fresh seafloor basalts at spreading ridges; after precipitation, it may be reintroduced into the mantle without cycling through the crust. Some dissolved weathering products are carried into continental waters—such as lakes—from which they may be directly redeposited in continental sediments, and so on.

Residence Time

Within the calcium cycle outlined, there are several important calcium reservoirs, including the mantle, continental crustal rocks, submarine limestone deposits, and the ocean, in which large quantities of calcium are dissolved. A measure of how rapidly the calcium cycles through each of these reservoirs is a parameter known as **residence time.** This can be defined in a variety of ways. For a substance that is at saturation level in a reservoir, so that the reservoir can hold no more of it, residence time is given by the relationship:

$$\text{Residence time (of a substance in a reservoir)} = \frac{\text{Capacity (of the reservoir to hold that substance)}}{\text{Rate of influx (of the substance into the reservoir)}}$$

This can be illustrated by a nongeologic example. Consider a hotel with 100 single rooms. Some lodgers stay only one night, others stay for weeks. On the average, though, ten people

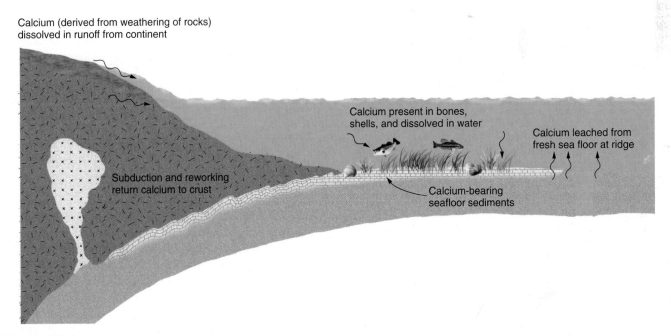

Figure 17.1

Simplified calcium cycle: Calcium weathered out of rocks is washed into bodies of water. Some remains in solution, some is taken up by organisms, and some is deposited in sediments. The deposited calcium is, in time, reworked through the rock cycle into new rock.

Calcium (derived from weathering of rocks) dissolved in runoff from continent

Subduction and reworking return calcium to crust

Calcium present in bones, shells, and dissolved in water

Calcium leached from fresh sea floor at ridge

Calcium-bearing seafloor sediments

check out each day, so (assuming this is a popular hotel that can always fill its rooms) ten new people check in each day. The residence time of lodgers in the hotel, then, is

$$\text{Residence time} = \frac{100 \text{ persons}}{10 \text{ persons/day}} = 10 \text{ days}$$

In other words, people stay an average of ten days each at this hotel.

There are many other situations in which the quantity of a substance in a reservoir stays approximately constant (a *steady-state* situation) though the reservoir is not at capacity. In such a case, residence time could be defined as

$$\frac{\text{Amount of the substance in the reservoir}}{\text{Rate of input (or outflow) of the substance}}$$

To pursue our hotel analogy, consider a competing hotel in the next block that has similar turnover of clientele, but a much lower occupancy rate—only 50 of its 100 rooms are typically occupied. Then if 10 people check in and out each day, the residence time of patrons in the second hotel is only (50 persons)/(10 persons/day) = 5 days; each patron stays only half as long, on average, in the second hotel. Or, assuming for the moment that world population is a constant 7 billion, then if 130 million people are born each year and 130 million die (the numbers would have to be equal in order for the population to stay constant), the residence time of people on earth would be (7 billion) ÷ (130 million) = 54 years. That would not mean that everyone would live on earth for 54 years, but that would be the *average* life span. (In fact, the death rate is lower, so world population is growing and life span is longer than this.)

Which brings us to a more qualitative way of viewing residence time: as the average length of time a substance remains in a system (or reservoir). This may be a more useful working definition for substances whose capacity is not limited in a given reservoir (the continental crust, for instance) or for substances that do not achieve saturation but are removed from the reservoir by decomposition after a time, as is the case with many synthetic chemicals that cycle through water or air.

Of various reservoirs in figure 17.1, the one in which the residence time of calcium can be calculated most directly is the ocean. The capacity of the ocean to hold calcium is basically controlled by the solubility of calcite. (If too much calcium and carbonate have gone into solution, some calcite is removed by precipitation.) The rate of influx can be estimated by measuring the amount of dissolved calcium in all major rivers draining into the ocean. The resulting calculated residence time of calcium in the oceans works out to about 1 million years. In other words, the average calcium atom normally floats around in the ocean for 1 million years from the time it is flushed or leached into the ocean until it is removed by precipitation in sediments or in shells.

Oceanic residence times for different elements vary widely. Sodium, for instance, is put into the oceans less rapidly (in smaller quantity) than calcium, and the capacity of the ocean for sodium is much larger. (Sodium chloride, the main dis-solved salt in the ocean, is very soluble; what sodium is removed is taken up by clay-rich sediments.) The residence time of sodium in the oceans, then, is longer than that of calcium: 100 million years. By contrast, iron compounds, as noted earlier, are not very soluble in the modern ocean. The residence time of iron in the ocean is only 200 years; iron precipitates out rather quickly once it is introduced into the sea. Residence times will differ for any given element in different reservoirs with different capacities and rates of influx.

Human activities may alter the natural figures somewhat, chiefly by altering the rate of influx. In the case of calcium, for instance, extensive mining of limestone for use in agriculture (to neutralize acid soil) and construction (for building stone and the manufacture of cement) increases the amount of calcium-rich material exposed to weathering; calcium salts are also used for salting roads, and these salts gradually dissolve away into runoff water. These additions are probably insufficient to alter appreciably the residence times of materials in very large reservoirs like the ocean. Locally, however, as in the case of a single lake or stream, they may change the rate of cycling and/or increase a given material's concentration throughout a cycle.

Residence Time and Pollution

In principle, one can speak of residence times of more complex chemicals, too, including synthetic ones, in natural systems. Here, there is the added complication that not enough is really known about the behavior of many synthetic chemicals in the environment. There is no long natural history to use for reference when estimating the capacities of different reservoirs for recently created compounds. Some compounds' residence times are limited by rapid breakdown into other chemicals, but breakdown rates for many are slow or unknown. So, too, is the ultimate fate of many synthetic chemicals. Uncertainties of this sort make evaluation of the seriousness of industrial pollution very difficult.

Where residence times of pollutants are known in particular systems, they indicate the length of time over which those pollutants will remain a problem in those reservoirs, and thus the rapidity with which the pollutants will be removed from that system. Suppose, for example, that a factory has been discharging a particular toxic compound into a lake. Some of the compound is drained out of the lake by an outflowing stream. The residence time of the water, and thus of the toxic chemical in the lake, has been determined to be 20 years. If that is the average length of time the water and synthetic chemical remain in the lake, and if input of the toxin ceases but flushing of water through the lake continues, then after one residence time (20 years), half of the chemical should have been removed from the lake; after a second 20 years, another half; and so on. Mathematically, the decrease in concentration follows the same pattern as radioactive decay, illustrated in the last chapter. One can then project, for instance, how long it would take for a safely low concentration in the water to be reached. The calculations are more complex if additional processes are involved (if, for instance, some of the toxin decomposes, and some is removed into lake-bottom sediment), but the principles remain the same.

However, substances removed from one system, unless broken down, will only be moved into another, where they may continue to pose a threat. If our hypothetical toxic chemical is at saturation concentration in a lake and is removed only by precipitation into or adsorption onto the lake-bottom sediments, then if the rate of input into the lake doubles, so will the rate of output, and the concentration in the sediments will be twice as high. And just because a toxin is removed into sediments, it is not really "gone"; if the sediments are disturbed or the water chemistry changes, the toxin may go back into solution again.

Trace Elements, Pollution, and Health

Much of the concern about pollution reflects concern about possible negative effects on health, whether of humans or other organisms. Centuries ago, scientists began to observe geographic patterns in the occurrence of certain diseases. For example, until the twentieth century, the northern half of the United States was known as the "goiter belt" for the high frequency of the iodine-deficiency disease goiter. Subsequent analysis revealed that the soils in that area were relatively low in iodine; consequently, the crops grown and the animals grazing on those lands constituted an iodine-deficient diet for people living there. When their diets were supplemented with iodine, the occurrence of goiter decreased.

Such observations provided early evidence of the importance to health of *trace elements,* those elements that occur in very low concentrations in a system (rock, soil, plant, animal, etc.). Trace-elements' concentrations are typically a few hundred ppm or less. Their effects may be beneficial or toxic, depending on the element and the organism. To complicate matters further, the same element may be both beneficial and harmful, depending upon the dosage; fluorine is an example. Teeth are composed of a calcium phosphate mineral—apatite. Incorporation of some fluoride into its crystal structure makes the apatite harder, more resistant to decay. This, too, was recognized initially on the basis of regional differences in the occurrence of tooth decay. Persons whose drinking water contains fluoride

concentrations even as low as 1 ppm show a reduction in tooth-decay incidence. This corresponds to a fluoride intake of 1 milligram per liter of water consumed; a milligram is one thousandth of a gram, and a gram is about the mass of one raisin. Fluoride may also be important in reducing the incidence of osteoporosis, a degenerative loss of bone mass commonly associated with old age. Normal total daily fluoride intake from food and water ranges from 0.3 to 5 milligrams.

Fluoride is not an unmixed blessing. Where natural water supplies are unusually high in fluoride, so that two to eight times the normal dose of fluoride is consumed, teeth may become mottled with dark spots. The teeth are still quite decay-resistant, however; the spots are only a cosmetic problem. Of more concern is the effect of very high fluoride consumption (twenty to forty times the normal dose), which may trigger abnormal, excess bone development (bone sclerosis) and calcification of ligaments. A **dose-response curve** is a graph illustrating the relative benefit or harm of a trace element or other substance as a function of the dosage. Fluoride in humans would be characterized by the dose-response curve shown in figure 17.2.

The prospect of reduced tooth decay is the principal motive behind fluoridation of public water supplies where natural fluoride concentrations are low. Because extremely high doses of fluoride might be toxic to humans (fluorine is an ingredient of some rat poisons), there has been some localized opposition to fluoridation programs. However, the concern is a misplaced one, for the difference between a beneficial and toxic dose is enormous. The usual concentration achieved through purposeful fluoridation is 1 ppm (1 milligram per liter). A lethal dose of fluoride would be about 4 grams (4000 milligrams). To take in that much fluoride through fluoridated drinking water, one would not only have to consume 4000 liters (about 1000 gallons) of water, but would have to do so within a day or so, because fluoride is not accumulative in the body; it is readily excreted.

As a practical matter, there are limits on the extent to which detailed dose-response curves can be constructed for humans. The limits are related to the difficulties in isolating the effects of individual trace elements or compounds and in having

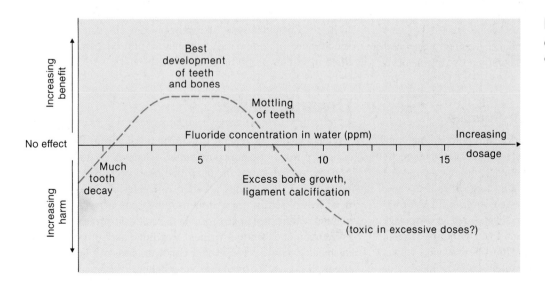

Figure 17.2

Generalized dose-response curve for fluoride in humans.

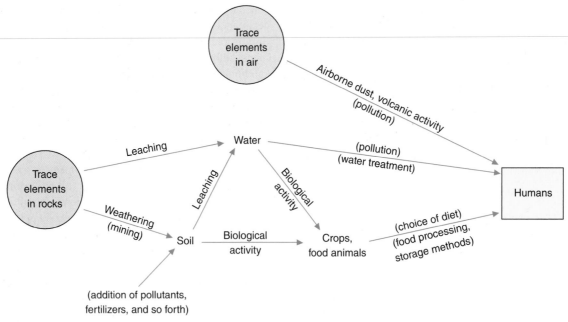

Figure 17.3

Pathways through which trace elements enter the body, including processes through which trace-element concentrations may be modified and other substances added to our intake. Human influences are in parentheses.

representative human populations that have been exposed to the full range of doses of interest. Controlled experiments on plant or animal populations or tissue cultures are possible, but there is then the difficulty of trying to extrapolate the results to humans with any confidence. It is also true that different individuals, and persons of different ages, may respond somewhat differently to any chemical, so any dose-response curve must either be specific to a particular population, or it must be only an approximation. Such considerations make it difficult to establish safe exposure limits for hazardous materials and permissible levels of pollutant emissions.

The challenge of isolating the effects on health of any one substance is also illustrated, in part, by the complex paths by which elements move from rocks or air into the body (figure 17.3). A variety of natural processes and human activities modify their concentrations along the way. In addition, we are exposed to other materials that can affect our health, and a huge and growing variety of synthetic chemicals as well as naturally occurring elements and chemical compounds.

Point and Nonpoint Pollution Sources

Sources of pollution may be subdivided into point sources and nonpoint sources (figure 17.4). **Point sources,** as their name suggests, are sources from which pollutants are released at one readily identifiable spot: a sewer outlet, a steel mill, a septic tank, and so forth. **Nonpoint sources** are more diffuse; examples would include fertilizer runoff from farmland, acid drainage from an abandoned strip mine, or runoff of sodium or calcium chloride from road salts, either directly off roadways or via storm drains. The point sources are often easier to identify as

potential pollution problems. They are also easier to monitor systematically. It is a comparatively straightforward task to evaluate the quality of water from a single sewer outfall or industrial waste pipe, and also to control and treat it, if necessary. Sampling runoff from a field in a representative way is more difficult, and it may also be more difficult to assess potential negative effects on human health. So, both for practical reasons and because its historic focus has been particularly on health concerns, the EPA's National Pollutant Discharge Elimination System has concentrated on controlling point sources of water pollution.

Identifying the sources of water pollutants once a problem is recognized can likewise be a challenge. Sometimes, the source can be identified by the overall chemical "fingerprint" of the pollution, for rarely is a single substance involved. Different pollutant sources—municipal sewage, a factory's wastewater, a farmer's particular type of fertilizer—are characterized by a specific mix of compounds, which may, in some cases, be identified in similar proportions in polluted water.

Organic Matter

Organic matter in general (as distinguished from the smaller subset of toxic organic compounds, discussed in the next section) is the substance of living and dead organisms and their by-products. It includes a variety of materials, ranging from dead leaves settling in a stream to algae on a pond. Its most abundant and problematic form in the context of water pollution is human and animal wastes. Feedlots and other animal-husbandry activities create large concentrations of animal wastes (figure 17.5), and the trend in U.S. agriculture is toward ever-denser

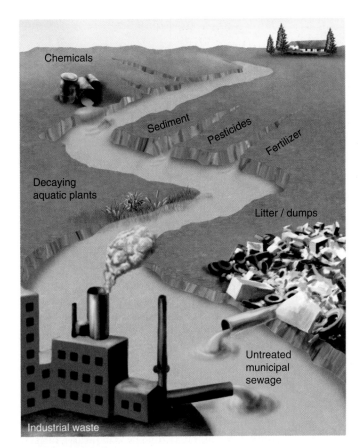

Figure 17.4

Point sources (such as industrial plants and storm-sewer outflow) and nonpoint sources (for example, fields and dispersed litter) of water pollution in a stream. Downward infiltration of contaminant-laden water from the nonpoint sources eventually also pollutes ground water.

Source: After USDA Soil Conservation Service.

Figure 17.5

Uncontrolled runoff from this livestock yard may be a major contributor of organic waste to local streams or shallow ground water.

Photo by Tim McCabe, courtesy USDA Natural Resources Conservation Service.

concentrations of cattle, or pigs, or poultry in such operations. Food-processing plants are other sources of large quantities of organic matter discharged in wastewater.

Sewage becomes mixed with a large quantity of wastewater not originally contaminated with it: Domestic wastewater is typically all channeled into one outlet, and in municipalities, domestic wastes, industrial wastes, and frequently even storm runoff collected in street drains all ultimately come together in the municipal sewer system. Moreover, sewage-treatment plants are commonly located beside a body of surface water, into which their treated water is discharged. To the extent that the wastewater is incompletely treated, the pollution of the surface water is increased.

Aside from its potential for spreading disease, organic matter creates another kind of water-pollution problem. In time, organic matter in water is broken down by microorganisms, especially bacteria, just as in a landfill. If there is ample oxygen in the water, **aerobic decomposition** occurs, consuming oxygen, facilitated by aerobic organisms. This depletes the dissolved oxygen supply in the water. Eventually, so much of the oxygen may be used up that further breakdown must proceed by **anaerobic decomposition** (without oxygen). Anaerobic decay produces a variety of noxious gases, including hydrogen sulfide (H_2S), the toxic gas that smells like rotten eggs, and methane (CH_4), which is unhealthful and certainly of no practical use in a body of water. More importantly, anaerobic decay signals oxygen depletion. Animal life, including fish, requires oxygen to survive—no dissolved oxygen, no fish. When the dissolved oxygen is used up, the water is sometimes described as "dead," though anaerobic bacteria and some plant life may continue to live and even to thrive.

Biochemical Oxygen Demand

The organic-matter load in a body of water is described by a parameter known as **biochemical oxygen demand,** or **BOD** for short. The BOD of a system is a measure of the amount of oxygen required to break down the organic matter aerobically; the more organic matter, the higher the BOD. The BOD may equal or exceed the actual amount of dissolved oxygen in the water. As oxygen is depleted, more tends to dissolve in the water to restore chemical equilibrium, but, initially, this gradual reoxygenation lags behind oxygen consumption. Measurements of water chemistry along a stream below a sizable source of organic waste matter characteristically define an **oxygen sag curve** (figure 17.6), a graph of dissolved-oxygen content as a function of distance from the waste source that shows sharp depletion near the source, recovering downstream. Persistent oxygen depletion is more likely to occur in a body of standing water, such as a lake or reservoir, than in a fast-flowing stream. Flowing water is better mixed and circulated, and more of it is regularly brought into contact with the air, from which it can take up more oxygen. Climate also plays a role, because oxygen dissolves more readily in cold water than in warm (table 17.1). Removal of organic matter to reduce BOD is a key function of municipal sewage-treatment plants.

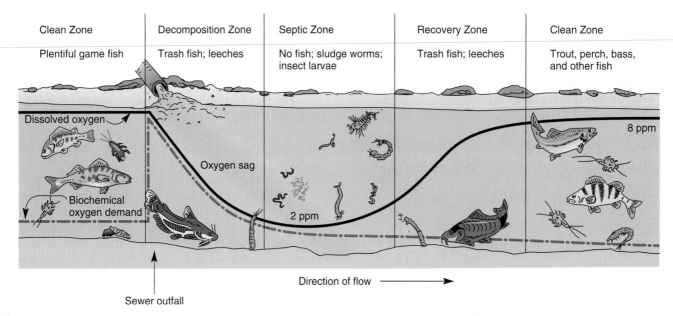

Clean Zone — Plentiful game fish
Decomposition Zone — Trash fish; leeches
Septic Zone — No fish; sludge worms; insect larvae
Recovery Zone — Trash fish; leeches
Clean Zone — Trout, perch, bass, and other fish

Dissolved oxygen
Oxygen sag
8 ppm
Biochemical oxygen demand
2 ppm

Direction of flow

Sewer outfall

Figure 17.6

Typical oxygen sag curve downstream from an organic-matter input, and its effect on both dissolved oxygen and the fauna in the stream. The dissolved-oxygen content of this stream is normally 8 ppm; it drops to 2 ppm below the sewer outfall where BOD spikes abruptly.

Table 17.1	Dissolved Oxygen as a Function of Temperature	
Temperature, °C (°F)		**Dissolved oxygen, mg/L**
0	(32)	14.6
5	(41)	12.8
10	(50)	11.3
15	(59)	10.2
20	(68)	9.2
25	(77)	8.6

As temperature goes up, dissolved oxygen concentrations decline. (1 mg/L is approximately 1 ppm.)

Source: U.S. Geological Survey Water Resources Division.

Eutrophication

The breakdown of excess organic matter not only consumes oxygen, but it also releases a variety of compounds into the water, among them nitrates, phosphates, and sulfates. The nitrates and phosphates, in particular, are critical plant nutrients, and an abundance of them in the water strongly encourages the growth of plants, including algae. Excess fertilizer runoff from farmland can also put excessive nutrients into water. This development is known as **eutrophication** of the water; the water body itself is then described as *eutrophic*. The exuberant algal growth that often accompanies or results from eutrophication has also been called "algal bloom" and may appear as slimy green scum on the water (figure 17.7). Once such a condition develops, it acts to continue to worsen the water quality (figure 17.8). Algal growth proceeds vigorously in the photic (well-lighted) zone near the water surface until the plants are killed by cold or crowded out by other plants. The dead plants sink to the bottom, where they, in turn, become part of the organic-matter load on the water, increasing the BOD and, as they decay, re-releasing nutrients into the water. Again, the consequences are most acute in very still water, where bottom waters do not readily circulate to the surface to take up oxygen. It should be noted that eutrophication can be a normal process in the development of a natural lake. Pollution, however, may accelerate or enhance it to the point that it becomes problematic. Aside from the aesthetic impact of algal bloom, oxygen depletion in eutrophic waters can kill aquatic organisms such as fish that require oxygen to survive.

Thorough waste treatment, as described in chapter 16, reduces the impact of organic matter, at least in terms of oxygen demand. However, as already noted, breakdown of the complex organic molecules releases into the water simpler compounds that may serve as nutrients. Human or animal wastes are not the only agents in sewage that contribute to eutrophication, either. Fertilizer runoff from farmland is a factor, as noted later in the chapter. Phosphates from detergents are potential nutrients, too. The phosphates are added to detergents to enhance their cleaning ability, in part by softening the water. Realization of the harmful environmental effects of phosphates in sewage has led to restrictions on the amounts used in commercial detergents, but some phosphate component is still allowed in most places. And most sewage treatment does not remove all of the dissolved nutrient (nitrate and phosphate) load.

A

B

Figure 17.7

(A) Typical algal bloom on a pond in summer. In this case, the source of the nutrients is waterfowl. The seemingly benign action of feeding flocks of ducks and geese around a suburban pond like this one not only alters the birds' behavior and teaches them to depend on handouts; it also leads to an increased pollutant load of organic fertilizer in the pond, and resultant degradation in water quality. (B) In extreme cases, algae may cover the whole water surface, as on this Wisconsin lake.

Industrial Pollution

Hundreds of new chemicals are created by industrial scientists each year. The rate at which new chemicals are developed makes it impossible to demonstrate the safety of the new chemicals as fast as they are invented. To "prove" a chemical safe, it would be necessary to test a wide range of doses on every major category of organism—including people—at several stages in the life cycle. That simply is not possible, in terms of time,

money, or laboratory space (not to mention the ethics of testing potential toxins on humans). Tests are therefore usually made on a few representative populations of laboratory animals to which the chemical is believed most likely to be harmful, or perhaps on tissue cultures. Whether other organisms will turn out to be unexpectedly sensitive to the chemical will not be known until after they have been exposed to it in the environment. The completely safe alternative would be to stop developing new chemicals, of course, but that would deprive society of many important medicinal drugs, as well as materials to make life safer, more comfortable, or more pleasant. The magnitude of the mystery about the safety of industrial chemicals is emphasized by a National Research Council report. A committee compiled a list of nearly 66,000 drugs, pesticides, and other industrial chemicals and discovered that *no* toxicity data at all were available for 70% of them; a complete health hazard evaluation was possible for *only 2%*. In 1990, the American Chemical Society recorded the *ten millionth* entry in its registry of new chemicals created or identified since 1957. Complete toxicity assessment for that many substances would obviously be a formidable task.

Inorganic Pollutants—Metals

Many of the inorganic industrial pollutants of particular concern are potentially toxic metals. Manufacturing, mining, and mineral-processing activities can all increase the influxes of these naturally occurring substances into the environment and locally increase concentrations from harmless to toxic levels. Examples of metal pollutants commonly released in wastewaters from various industries are shown in table 17.2.

Table 17.2	Principal Trace Metals in Industrial Wastewaters
Industry	**Metals**
mining and ore processing	arsenic, beryllium, cadmium, lead, mercury, manganese, uranium, zinc
metallurgy, alloys	arsenic, beryllium, bismuth, cadmium, chromium, copper, lead, mercury, nickel, vanadium, zinc
chemical industry	arsenic, barium, cadmium, chromium, copper, lead, mercury, tin, uranium, vanadium, zinc
glass	arsenic, barium, lead, nickel
pulp and paper mills	chromium, copper, lead, mercury, nickel
textiles	arsenic, barium, cadmium, copper, lead, mercury, nickel
fertilizers	arsenic, cadmium, chromium, copper, lead, mercury, manganese, nickel, zinc
petroleum refining	arsenic, cadmium, chromium, copper, lead, nickel, vanadium, zinc

Source: Data from J. E. Fergusson, Inorganic Chemistry and the Earth. Copyright © 1982 Pergamon Press, Oxford, England.

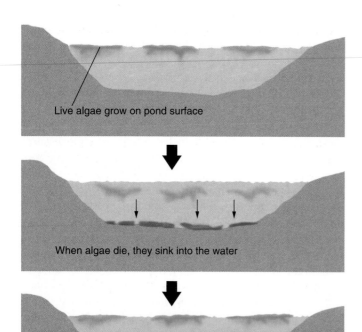

A

Figure 17.8

(A) Schematic diagram showing effects of algal bloom on water quality: (top) Abundant growth of algae in sunlit shallow water when nutrients are abundant. (middle) In colder weather, algae die and sink to the lake bottom. (bottom) The next growing season, more algae thrive at the surface while older material decays at the lake bottom, increasing BOD and releasing more nutrients to fuel new growth. Nutrients supplied to the Gulf of Mexico by the Mississippi River (B), primarily through runoff from agricultural land, support abundant growth of plankton (microscopic plants) near the water surface in the Gulf. When the algae die and settle toward the bottom, they increase BOD, deplete oxygen, and create a "dead zone" covering thousands of square miles (C). High temperatures in the Gulf area exacerbate the problem; recall table 17.1.

(B) Maps courtesy U.S. Geological Survey National Water Quality Assessment Program. (C) Map courtesy U.S. Environmental Protection Agency Gulf of Mexico program.

Labels within diagram A:
- Live algae grow on pond surface
- When algae die, they sink into the water
- Decomposition of dead algae re-releases nutrients to feed next generation of algae growing above

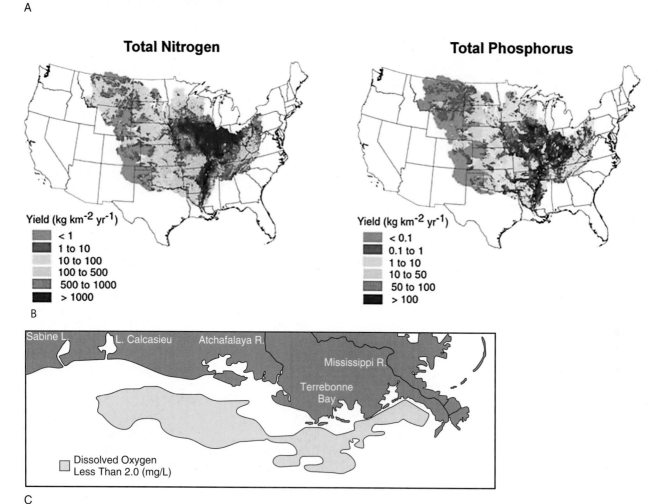

Total Nitrogen

Yield (kg km^{-2} yr^{-1})
- < 1
- 1 to 10
- 10 to 100
- 100 to 500
- 500 to 1000
- > 1000

Total Phosphorus

Yield (kg km^{-2} yr^{-1})
- < 0.1
- 0.1 to 1
- 1 to 10
- 10 to 50
- 50 to 100
- > 100

B

Sabine L. L. Calcasieu Atchafalaya R. Mississippi R. Terrebonne Bay

Dissolved Oxygen Less Than 2.0 (mg/L)

C

The element mercury is a case in point. In nature, the small amounts of mercury (1 to 2 ppb) found in rocks and soil are weathered out into waters (where mercury is typically found in concentrations of a few ppb) or into the air (less than 1 ppb), from which it is quickly removed with precipitation. Mercury in water may be taken up by plants and animals; eventually it is removed into sediments. Its residence time in the oceans is about 80,000 years.

Mercury is one of the **heavy metals,** a group that includes lead, cadmium, plutonium, and others. A feature that the heavy metals have in common is that they tend to accumulate in the bodies of organisms that ingest them. Therefore, their concentrations increase up a food chain (figure 17.9), a phenomenon known as **biomagnification.** Some marine algae may contain heavy metals at concentrations of up to 100 times that of the water in which they are living, though concentrations 4 to 6 times that of the host water are more typical. Small fish eating the algae develop higher concentrations of heavy metals in their flesh; larger fish who eat the smaller fish concentrate the metals still further. The flesh of predator fish high up a multistage food chain may show mercury concentrations 100,000 to 1 million times that of the water in which they swim. Birds and mammals (including people) who eat those fish may concentrate it further, though the diversity of most people's diets is greater than that of most animals and will moderate the effects of consuming mercury-rich fish.

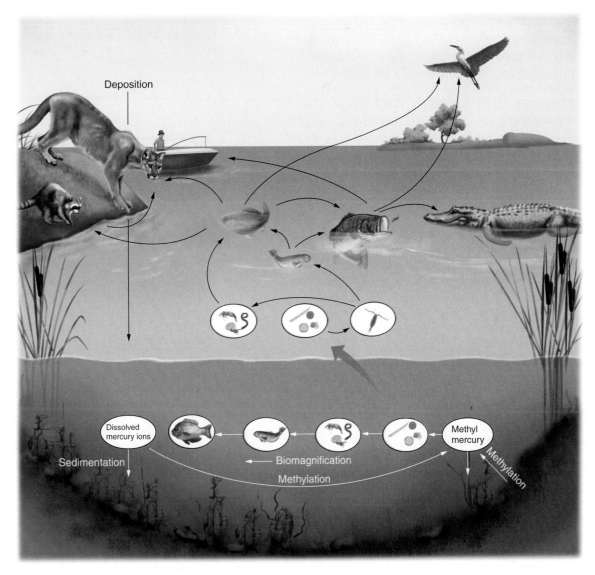

Figure 17.9

Simplified outline of mercury behavior in a body of water: Mercury deposited in the water may be converted to methylmercury, in which form it is concentrated by biomagnification. Microorganisms containing methylmercury are consumed by small fish, the larger fish eat the smaller fish, and so on up to birds and mammals higher up the chain. Accumulative toxins like the heavy metals may become very highly concentrated in creatures at the top of the chain. From a few parts per billion in the water, mercury concentrations in fish can easily reach several parts per million in fish, and this is enough to pose a health threat.

Modified after USGS Fact Sheet 063–00.

For those who are aware of the mercury in the silver fillings in teeth, a word of reassurance may first be in order. The "silver" that dentists put in teeth is really an amalgam of mercury and silver, but elemental mercury is not the highly toxic form, for it is not readily absorbed by the body; silver fillings pose no known health threat. The severely toxic form is *methylmercury*, an organic compound of mercury produced by microorganisms in water and taken up by plants and small organisms at the bottom of the food chain. It is as methylmercury that mercury accumulates up the food chain.

Even so, in most natural settings, heavy-metal accumulations in organisms are not very serious because the natural concentrations of these metals are low in waters and soils to begin with. The problems develop when human activities locally upset the natural cycle. Mining and processing heavy metals in particular can increase the rate at which the heavy metals weather out of rock into the environment, and these industries have also discharged concentrated doses of heavy metals into some bodies of water.

Methylmercury acts on the central nervous system and the brain in particular. It can cause loss of sight, feeling, and hearing, as well as nervousness, shakiness, and death. By the time the symptoms become apparent, the damage is irreversible. The effects of mercury poisoning may, in fact, have been recognized long before their cause. In earlier times, one common use of mercury was in the making of felt. Some scholars have suggested that the idiosyncrasies of the Mad Hatter of *Alice in Wonderland* may actually have been typical of hat makers of the period, whose frequent handling of felt led to chronic mercury poisoning, commonly known as "hatter's shakes." Mercury has also been used as a fungicide. In Iraq, in 1971–1972, severe famine drove peasants to eat mercury-treated seed grain instead of saving it to plant. Five hundred died, and an estimated 7000 people were affected. See also Case Study 17.1.

Mercury is not the only toxic heavy metal. Cadmium poisoning was most dramatically demonstrated in Japan when cadmium-rich mine wastes were dumped into the upper Zintsu River. The people used the water for irrigation of the rice fields and for domestic use. Many developed *itai-itai* (literally, "ouch-ouch") disease, cadmium poisoning characterized by abdominal pain, vomiting, and other unpleasant symptoms. Fear of the toxic-metal effects of plutonium is one concern of those worried about radioactive-waste disposal.

For some industrial applications, substitutes for toxic metals can be found. Mercury is no longer used in the manufacture of felt, for instance. However, high concentrations of toxic metals may exist in natural geological materials. A volatile metal, mercury may be emitted during explosive volcanic eruptions, but these are short-lived, and the metal is quickly removed from the atmosphere by rain or snow. Human activities such as ore processing and manufacturing have had a more lasting impact (figure 17.10). Coal can contain up to several parts per million of mercury; the burning of coal is now the source of more than 85% of mercury pollution in the United States, according to the EPA.

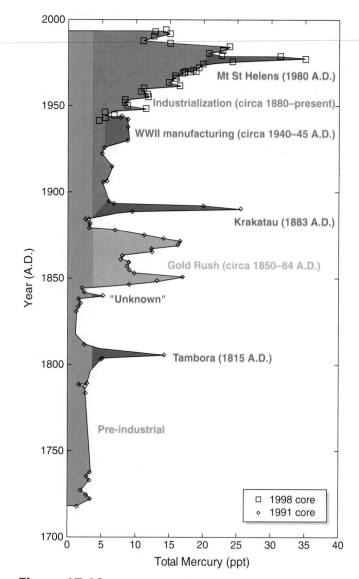

Figure 17.10

A USGS study of nearly 100 ice cores from a Wyoming alpine glacier shows clear evidence of both natural and human influence on mercury. There has clearly been a recent decline from peak mercury levels, likely due to pollution-control regulations on incinerators. But mercury emissions from coal-fired power plants are not yet regulated, in part because of legal challenges to EPA rulemaking. (ppt = parts per trillion)

After USGS Fact Sheet 051-02.

Likewise, potentially toxic metals can occur naturally in lakes, streams, and ground water at concentrations above U.S. drinking-water standards. Arsenic is an example.

The toxic effects of large doses of arsenic are well documented. It can cause skin, bladder, and other cancers; in large doses, it is lethal. The effects of low doses are much less clear. In some cultures, arsenic is deliberately consumed (in moderation) to promote a clear complexion and glossy hair. It is accumulative in the body, and over time, some tolerance to its toxicity may be developed. Arsenic is not known to be an essential nutrient—there is no recommended minimum daily dose, as there is for iron, or zinc, or copper, or several other metals—but whether it is harmful even in the smallest quantities is not known. Drinking-water limits for such elements are established by extrapolation from high doses known to be toxic to a dose low enough to correspond to a perceived acceptable level of risk (e.g., one death per million persons, or serious illness per 100,000, or whatever) coupled with statistical studies of the incidence of toxic effects in populations naturally exposed to various doses of the substance, often with an additional safety margin thrown in.

For many years, the drinking-water standard for arsenic was set at 50 ppb (or about 50 millionths of a gram per liter of water). More-recent research led the National Research Council, in 1999, to call for a new, lower limit for arsenic in drinking water. In January 2001, the EPA published a new limit of 10 ppb, to take effect in January 2006 (to give municipalities time to adjust their processing of water accordingly). Objections were immediately raised, in part because natural arsenic levels in ground water in much of the country already exceed that level (figure 17.11), and in part because the evidence justifying the lower standard was not definitive. In March 2001, the administration withdrew the new standard, but it was later reimposed, and did indeed become effective in January 2006.

Nor is arsenic in ground water a problem unique to the United States. The World Health Organization estimates that over 100 million people in South and Southeast Asia drink ground water containing harmful concentrations of arsenic. The problem is particularly severe in Bangladesh. Ironically, the issue has developed only since the 1970s, when the people were encouraged to switch from using water from lakes and streams (often contaminated with human or animal wastes or disease-causing organisms) to using well water. Now, 90% of the population drinks well water, despite the fact that potentially dangerous arsenic levels in the ground water have been recognized since the 1990s, and by now many cases of arsenic poisoning have been seen there. Solutions—treating the water, drilling deeper wells to cleaner aquifers—are being developed, but the problem will not be solved quickly.

Other Inorganic Pollutants

Some nonmetallic elements commonly used in industry are also potentially toxic to aquatic life, if not to humans. For example, chlorine is widely used to kill bacteria in municipal water and

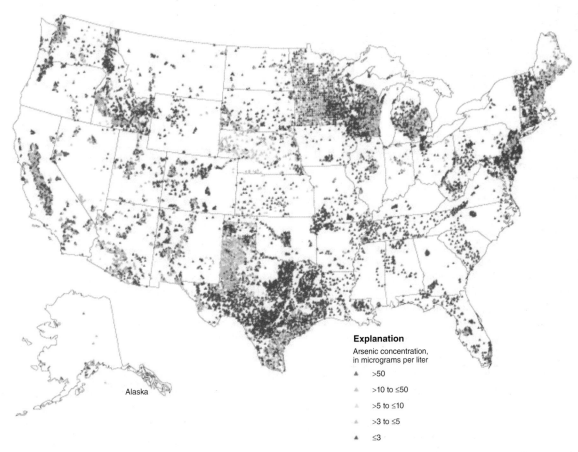

Explanation
Arsenic concentration, in micrograms per liter

▲ >50
▲ >10 to ≤50
▲ >5 to ≤10
▲ >3 to ≤5
▲ ≤3

Alaska

Figure 17.11
Sampling of over 31,000 wells indicates that arsenic in ground water in many parts of the United States naturally exceeds the EPA drinking-water standard of a maximum of 10 micrograms per liter (μg/L), and new data suggest significant toxicity at still lower concentrations.

Map courtesy U.S. Geological Survey.

Lessons from Minamata

In the early twentieth century, Minamata was a small coastal town in western Kyushu, Japan, with a long history as a village of farmers and fishermen. Industrial development was in its early stages. At the time, release of industrial wastewater into the ocean was common in many places worldwide, and so it was here. The major local industrial firm,

Chisso Corporation, paid the local fishermen to compensate them for any possible damage to their livelihood as a result.

In 1932, Chisso began to manufacture acetaldehyde, a chemical used in the production of plastics. At that time, the production of acetaldehyde, in turn, required mercury. In the early 1950s, after the

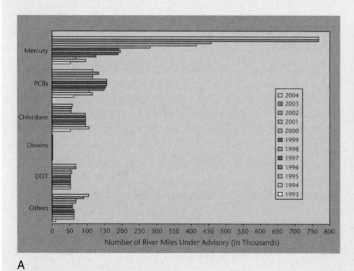

A

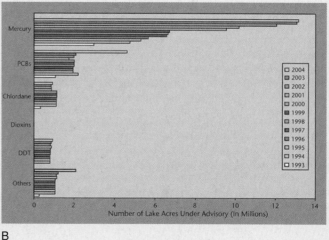

B

Figure 1

Number of river miles (A) and lake acres (B) under fish consumption advisories for various pollutants, 1993–2004. Note that the only fast-growing problem over this period is mercury. Note also the sharp rises since 1999, especially in the river miles under advisory, and compare with the ice-core data of Figure 17.10; recent expansion of coal-fired power plants may be reversing the drop shown in that figure.

Graphs after U.S. Environmental Protection Agency.

sewage treatment plants and to destroy various microorganisms that might otherwise foul the plumbing in power stations. Released in wastewater, it can also kill algae and harm fish populations.

Acids from industrial operations used to be a considerably greater pollution problem before stringent controls on their release were imposed. Acid mine drainage, however, remains a serious source of surface- and groundwater pollution, especially in coal- and sulfide-mining areas, as described in chapters 13 and 14. The acids, in turn, may leach additional toxic metals from rocks, tailings, or soil.

The toxic effects of certain asbestos minerals were not manifested or well defined until long after their initial release into the environment by human activities. Asbestos minerals have been prized for decades for their fire-retardant properties and have been used in ceiling tiles and other building materials for years. Wastes from asbestos mining and processing were dumped into many bodies of water, including the Great Lakes—they were, after all,

believed to be just inert, harmless mineral materials. By the time the carcinogenic effects of some asbestos minerals were realized, asbestos workers had been exposed and segments of the public had been drinking asbestos-bearing waters for twenty years or more. The full impact of this sustained exposure is still being assessed. While the potential risk from inhalation is best documented, it is worth noting that ingestion of asbestos minerals has been implicated as a cause of cancer, at least in some cases: In Japan, many prefer rice coated with powder, and the powder is mainly pulverized talc that occurs geologically in association with some types of asbestos. It has been suggested that this may account for higher incidence of stomach cancer.

Unfortunately, the harmful effects of such inorganic pollutants are often slow to develop, which means that decades of testing might be required to reveal the dangers of such materials. The costs of such long-term testing are extremely high, generally prohibitively so.

end of World War II, demand for plastics—and thus acetaldehyde—soared, as did production at Minamata.

It was the cats who soon provided the first clue to the developing tragedy, although its significance initially went unrecognized. The cats of Minamata began to behave oddly, "dancing," falling into the ocean and drowning. Then, strange symptoms appeared in people. They began to stumble, to slur their speech, to suffer convulsions and fits of trembling. Babies were born deformed. Some victims became paralyzed. Some died. The collection of symptoms was dubbed "Minamata disease," and researchers scrambled to identify its cause.

We know now that it was acute mercury poisoning. Mercury in Chisso's wastewater was flushed into Minamata Bay, the water of which only slowly mixes with the open ocean. As methylmercury, it was taken up in algae, then fish and shellfish, then the cats and the people who ate the fish, its concentration biomagnified along the way. Among the human population, the fishermen, who typically ate the most fish, were most acutely affected.

Fishing has long been prohibited in Minamata Bay, where even now the sediments are laden with mercury. A memorial garden has been planted to honor and commemorate the victims. The total number of victims may never be known precisely; milder cases of mercury poisoning have symptoms that cannot be tied uniquely to mercury. The toll of those acutely poisoned is grim enough: more than 1400 have died so far, and thousands more suffer lingering effects. A 2001 study suggested that tens of thousands of persons may have been affected altogether, including many living outside Minamata Bay.

As the toxicity and the accumulative quality of methylmercury were realized, testing of fish for mercury was initiated in many nations.

The results were sufficiently alarming that during the 1960s and 1970s, people were advised to limit their intake of tuna and certain other fish with particularly high mercury contents. (Over 95% of the mercury in fish is in the form of methylmercury.) After some years of enhanced pollution control, particularly of point sources such as industrial wastewater, the mercury levels, and the concerns, abated.

However, that is far from the end of the story, for it turns out that wastewater is not the only contributor to mercury pollution problems.

Mercury is a very volatile (easily vaporized) metal. When mercury vapor is put into the atmosphere—whether by coal burning, waste incineration, smelting of mercury-bearing ores, battery manufacturing, or whatever—it disperses widely before being deposited on land and water. In recent decades, human activity has more than doubled the concentration of mercury in the atmosphere, correspondingly increasing that deposition. What begins as a potential air-pollution issue becomes a water-pollution problem. (In fact, because mercury in air is typically elemental mercury, it becomes a health threat only after it enters water bodies where it can be methylated.)

The increasing use of coal for energy, together with expanded sampling, have led to sharply increasing numbers of mercury-related fish consumption advisories on lakes and rivers (figure 1). By 2004, some 21 states had 100% of their lakes and rivers under advisory for various pollutants, mercury chief among them. The EPA has estimated that in some watersheds, the airborne mercury load to lakes and streams would have to be reduced by more than 75% to bring methylmercury concentrations in fish down to acceptable levels.

Organic Compounds

The majority of new chemical compounds created each year are organic (carbon-containing) compounds. Many thousands of these compounds, naturally occurring and synthetic, are widely used as herbicides and pesticides, as well as being used in a variety of industrial processes. Examples include DDT and dioxin. Their negative effects in organisms vary with the particular type of compound: Some are carcinogenic, some are directly toxic to humans or other organisms, and others make water unpalatable. Some also accumulate in organisms as the heavy metals do.

Oil spills, discussed in chapter 14, are another kind of organic-compound pollution. At least as much additional oil pollution occurs each year from the careless disposal of used crankcase oil, dumping of bilge from ships, and the runoff of oil from city streets during rainstorms. Underground tanks and pipelines may also leak, and drilling muds and waste brines (saline pore fluids) discarded in oil fields may be contaminated with petroleum. Oil spills into U.S. waters average over 10 million gallons per year.

Another type of organic-compound pollution is the result of the U.S. plastics industry's demand for production of nearly 7 billion pounds of vinyl chloride each year. Vinyl chloride vapors are carcinogenic, and it is not known how harmful traces of vinyl chloride in water may be.

Polychlorinated biphenyls (PCBs) were used for nearly twenty years as insulating fluid in electrical equipment and as plasticizers (compounds that help preserve flexibility in plastics). Laboratory tests revealed that PCBs in animals cause impaired reproduction, stomach and liver ailments, and other problems. PCB production in the United States was banned in 1977, but approximately 900 million pounds of PCBs had already been produced and some portion of that quantity released into the environment, where it remains.

A concern that arose in the 1990s is MTBE (methyl tert-butyl ether), a petroleum derivative added to gasoline. MTBE is used to promote more-complete burning of gasoline, to reduce carbon monoxide pollution in urban areas. However, MTBE has begun to appear in analyses of ground water in areas where it is widely used. Its health effects are not known precisely but it is regarded by the U.S. EPA as a potential carcinogen; it also has an objectionable taste and odor. Studies to assess the incidence of MTBE contamination in ground water and the need for regulation are in progress, and the EPA has added MTBE to the list of substances for which it may, in the future, establish drinking-water standards.

Still more recently, the EPA has added another group of compounds to its category of "contaminants of emerging concern": pharmaceuticals, both prescription and over-the-counter. At one time the disposal method recommended to individuals with old or unwanted pharmaceuticals was to flush the medications down a sink or toilet (rather than put them in the household trash, where children or pets might get at them and be harmed). Unfortunately, this meant that those compounds that did not readily decompose by themselves became part of the wastewater stream—and municipal sewage-treatment plants do not generally have procedures for removing pharmaceuticals. So, many such compounds were added to lakes and streams in treated wastewater, where some, it appears, cause adverse health effects in aquatic animals. Traces of common medications have been found in some drinking-water supplies as well. Potential negative effects on humans, and the need for limits on these compounds in drinking water, are still being evaluated. Meanwhile, disposal of pharmaceuticals by the health-care industry is now regulated, with these compounds classified as "hazardous waste." For individuals, the EPA recommends taking advantage of pharmaceutical takeback or household hazardous-waste collection programs, if available; otherwise, they suggest mixing the medicine with something unpalatable (such as kitty litter), then sealing it up tightly and putting it in the trash.

Problems of Control

Cost and efficiency are factors in pollution control. Those substances not to be released with effluent water (or air) must be retained. This becomes progressively more difficult as complete cleaning of the effluent is approached. First, there is no wholly effective way to remove pollutants from wastewater before discharge. No chemical process is 100% efficient. Second, as cleaner and cleaner output is approached, costs skyrocket. If it costs $1 million to remove 90% of some pollutant from industrial wastewater, then removal of the next 9% (90% of the remaining 10% of the pollutant) costs an additional million dollars. To get the water 99.9% clean costs a total of $3 million, and so on. Some toxin always remains, and the costs escalate rapidly as "pure" water is approached. At what point is it no longer cost-effective to keep cleaning up? That depends on the toxicity of the pollutant and on the importance (financial or otherwise) of the process of which it is the product. Furthermore, the toxins retained as a result of the cleaning process do

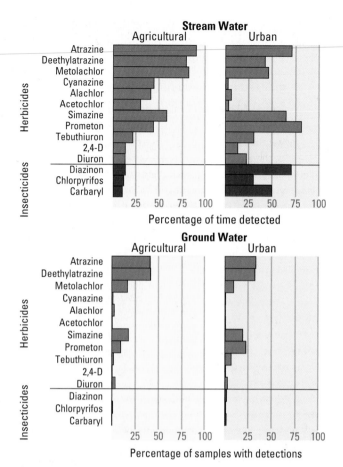

Figure 17.12

Herbicides and insecticides are everywhere in our water and, perhaps contrary to expectations, as common in urban as agricultural areas. Pesticides were detectable in nearly all stream samples in a national study.

Source: USGS Circular 1291, as revised online 2007.

not just disappear. They become part of the growing mass of industrial toxic waste requiring careful disposal.

The very nature of the ways in which hazardous compounds are used is an obvious problem. Herbicides and pesticides are commonly spread over broad areas, where it is typically impossible to confine them. The result, not surprisingly, is that they become nearly ubiquitous in the environment (figure 17.12).

Even when the threat posed by a toxic agent is recognized and its production actually ceases, it may prove long-lived in the environment. Once released and dispersed, it may be impossible to clean up. The only course is to wait for its natural destruction, which may take unexpectedly long. DDT is a case in point; see Case Study 17.2.

DDT was undoubtedly a uniquely powerful tool for fighting insect pests transmitting serious diseases. It was neither the first nor the last new compound to be found—long after it came into general use—to have significantly toxic effects. This may be inevitable given the rate of synthesis of new compounds and the complexity of testing their safety, but one unfortunate result has been the long-term persistence of toxins in water, sediments, and organisms.

The Long Shadow of DDT

The compound DDT (dichlorodiphenyl trichloroethane) was discovered to act as an insecticide during the 1930s. Its first wide use during World War II, killing lice, ticks, and malaria-bearing mosquitoes, unquestionably prevented a great deal of suffering and many deaths. In fact, the scientist who first recognized DDT's insecticidal effects, Paul Muller, was subsequently awarded the Nobel Prize in medicine for his work. Farmers undertook wholesale sprayings with DDT to control pests in food crops. The chemical was regarded as a panacea for insect problems; cheery advertisements in the popular press promoted its use in the home.

Then the complications arose. One was that whole insect populations began to develop some resistance to DDT. Each time DDT was used, those individual insects with more natural resistance to its effects would survive in greater proportions than the population as a whole. The next generation would contain a higher proportion of insects with some inherited resistance, who would, in turn, survive the next spraying in greater numbers, and so on. This development of resistance resulted in the need for stronger and stronger applications of DDT, which ultimately favored the development of ever-tougher insect populations that could withstand these higher doses. Insects' short breeding cycles made it possible for all of this to occur within a very few years.

Early on, DDT was also found to be quite toxic to tropical fish, which was not considered a big problem. Moreover, it was believed that DDT would break down fairly quickly in the environment. In fact, it did not. It is a persistent chemical in the natural environment. Also, like the heavy metals, DDT is an accumulative chemical. It is fat-soluble and builds up in the fatty tissues of humans and animals. Fish not killed by DDT nevertheless accumulated concentrated doses of it, which were passed on to fish-eating birds (figure 1). Then another deadly effect of DDT was realized: It impairs calcium metabolism. In birds, this effect was manifested in the laying of eggs with very thin and fragile shells. Whole colonies of birds were wiped out, not because the adult birds died, but because not a single egg survived long enough to hatch. Robins picked up the toxin from DDT-bearing worms. Whole species were put at risk. All of this prompted Rachel Carson to write *Silent Spring* in 1962, warning of the insecticide's long-term threat.

Finally, the volume of data demonstrating the toxicity of DDT to fish, birds, and valuable insects such as bees became so large that, in 1972, the U.S. Environmental Protection Agency banned its use, except on a few minor crops and in medical emergencies (to fight infestations of disease-carrying insects). So persistent is it in the environment that thirty years later, fish often had detectable (though much lower) levels of DDT in their tissues. Encouragingly, many wildlife populations have recovered since DDT use in the United States was sharply curtailed. Still, DDT continues to be applied extensively in other countries, with correspondingly continued unfortunate side effects.

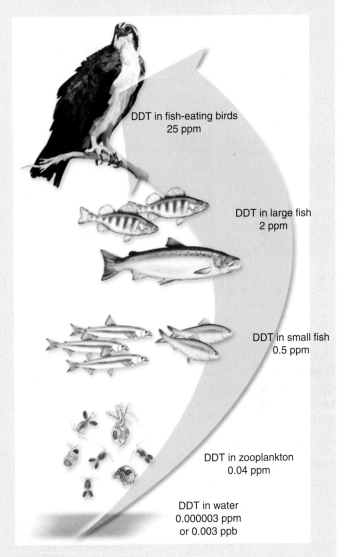

DDT in fish-eating birds
25 ppm

DDT in large fish
2 ppm

DDT in small fish
0.5 ppm

DDT in zooplankton
0.04 ppm

DDT in water
0.000003 ppm
or 0.003 ppb

Figure 1

Biomagnification can increase DDT concentration a million times or more from water to predators high up the food chain.

It should be noted that not all of the synthetic organic chemicals last long in the environment. Some (e.g., vinyl chloride) evaporate rapidly, even if released into water; some (e.g., benzene) evaporate or break down in a matter of days. Even among the pesticides, behavior varies widely. One can speak of the "half-life" of such a compound in the environment, as a measure of rate of breakdown for those that decompose in soil or water. While some, such as the insecticide malathion, have

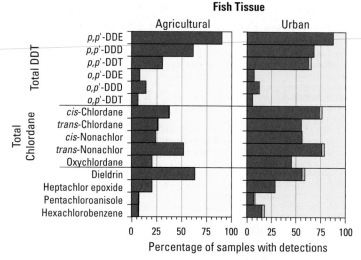

Fish Tissue

Percentage of samples with detections

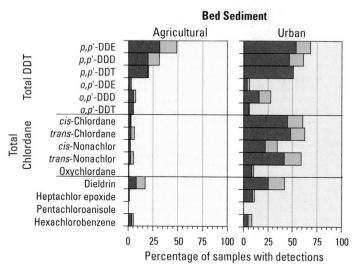

Bed Sediment

Percentage of samples with detections

Figure 17.13

A national study published in 2005 showed continued presence of pesticides in sediments and fish, even decades after the pesticides' last use: DDT was banned for most applications in 1972, and dieldrin was banned in 1974, while chlordane use continued into the late 1980s. Darker portion of each bar represents higher detection levels, above 5 ppb for fish tissue and 2 ppb in dry sediment.

Source: U.S. Geological Survey Circular 1291, as revised online 2007.

half-lives of a few days, the majority of common pesticides have half-lives on the order of a year. Among the longest-lasting, however, are dieldrin (half-life about 10 years), chlordane (over 30 years), and DDT (15 to over 120 years depending on the exact compound; see figure 17.13).

Thermal Pollution

Thermal pollution, the release of excess or waste heat, is a by-product of the generation of power. The waste heat from auto-mobile exhaust or heating systems contributes to thermal pollution of the atmosphere, but its magnitude is generally be-lieved to be insignificant. Potentially more serious is the local thermal pollution of water by electric generating plants and other industries using cooling water.

Only part of the heat absorbed by cooling water can be extracted effectively and used constructively. The still-warm cooling water is returned to its source—most commonly a stream—and replaced by a fresh supply of cooler water. The resulting temperature increase usually exceeds normal seasonal fluctuations in the stream and results in a consistently higher temperature regime near the effluent source.

Fish and other cold-blooded organisms, including a variety of microorganisms, can survive only within certain temperature ranges. Excessively high temperatures may result in the wholesale destruction of organisms. More-moderate increases can change the balance of organisms present. For example, green algae grow best at 30 to 35°C (86 to 95°F); blue-green algae, at 35 to 40°C (95 to 104°F). Higher temperatures thus favor the blue-green algae, which are a poorer food source for fish and may be toxic to some. Many fish spawn in waters some-what cooler than they can live in; a temperature rise of a few degrees might have no effect on existing adult fish populations but could seriously impair breeding and thus decrease future numbers. Furthermore, changes in water temperature change the rates of chemical reactions and critical chemical properties of the water, such as concentrations of dissolved gases, including the oxygen that is vital to fish; recall table 17.1.

Thermal pollution can be greatly reduced by holding wa-ter in cooling towers before release. (In practice, however, even where cooling towers are used, the released water is still at a somewhat elevated temperature.) The extent of thermal pollu-tion is also restricted in both space and time. It adds no sub-stances to the water that persist for long periods or that can be transported over long distances, and reduction in the output of excess waste heat results in an immediate reduction in the mag-nitude of the continuing pollution problems.

Agricultural Pollution

The same kinds of water-pollution problems that result from agricultural activities can certainly occur in urban and suburban areas, too. Fertilizers, herbicides, and insecticides are applied to parks and gardens just as they are to farmland, and their residues correspondingly appear in water; recall figure 17.12. Indeed, just as we noted that soil erosion during urbanization can be much more intense than erosion during cultivation of crops, so pesti-cide accumulations may be more notable in urban areas, as seen in figure 17.13. However, also as with soil erosion, agriculture warrants special focus in the context of water pollution because the much greater land area involved means much larger quanti-ties of potential pollutants applied (figure 17.14) and greater impact of agricultural activities on pollution extent overall.

Fertilizers and Organic Waste

The three principal constituents of commercial fertilizers are nitrates, phosphates, and potash. When applied to or incorporated in the soil,

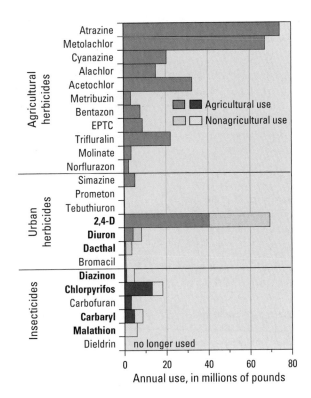

Figure 17.14

The quantities of pesticides used in agriculture dwarf the quantities used primarily in urban areas.

Source: U.S. Geological Survey Circular 1291, as revised online 2007.

they are not immediately taken up by plants. The compounds must be soluble in order for plants to use them, but that means that they can also dissolve in surface-water and groundwater bodies. These plant foods then contribute to eutrophication problems. There is typically a clear correlation with applications of fertilizer (figure 17.15), and thus the concerns are greatest in such areas.

Reduction in fertilizer applications to the minimum needed, perhaps coupled with the use of slow-release fertilizers, would minimize the harmful effects. Nitrates could alternatively be supplied by the periodic planting of legumes (a plant group that includes peas, beans, and clovers), on the roots of which grow bacteria that fix nitrogen in the soil. This practice reduces the need to apply soluble synthetic fertilizers.

Unfortunately, the use of other natural fertilizers, like animal manures, fails to eliminate the water-quality problems associated with fertilizer runoff. Manures are, after all, organic wastes. Thus, they contribute to the BOD of runoff waters in addition to adding nutrients that lead to eutrophication.

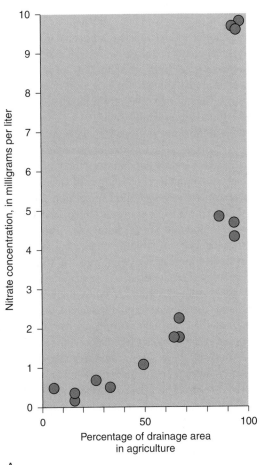

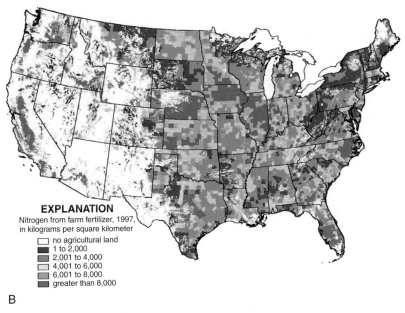

B

Figure 17.15

(A) Study in the Willamette Basin shows typical strong correlation of nitrate in spring stream runoff with proportion of drainage basin in agricultural use. Other studies show higher nitrate in runoff during the growing season, when fertilizers have been applied. (B) Potential nonpoint-source pollution from nitrogen in commercial fertilizer application, 1997. Fertilizer sales and applications have continued to rise since then.

Sources: (A) Data from U.S. Geological Survey Circular 1225, (B) From USGS Scientific Investigations Report 2006-5012.

A

Figure 17.16

Anaerobic digester for pig manure generates biogas while containing waste.

Photograph courtesy California Polytechnic State University and National Renewable Energy Lab.

Likewise, runoff of wastes from domestic animals, especially from commercial feedlots, is a potential problem. One concern that arises in connection with some such operations is not unlike the problem of municipal sewage-treatment plants overwhelmed by the added volume of runoff water from major storms. It is increasingly common for the wastes to be diverted into shallow basins where they may be partially decomposed and/or dried and concentrated for use as fertilizer. If poorly designed, the surrounding walls (often made of earth) may fail, releasing the contents. Even if the structure holds, in an intense storm, the added water volume may cause overflowing of the wastes. One approach that has shown promise is to build broad channels downslope of the waste source, planted with grass or other vegetation, to act as filter strips for the runoff. The filter strip catches suspended solids and slows fluid runoff, allowing time for decomposition of some dissolved constituents, and nutrient uptake by the vegetation (depending on season), plus increased infiltration to reduce the volume of runoff leaving the site. Studies have shown reductions of up to 80% in dissolved phosphorous, nitrogen, and other constituents in the surface runoff water leaving the filter strip, so the practice clearly helps to protect surface-water quality. The primary potential disadvantage is that, by increasing infiltration, the filter strip may increase the dissolved-nutrient load in ground water below the site.

An alternative, productive solution to the excess-manure problem would be more extensive use of these wastes for the production of methane for fuel. This has already been found to be economically practical where large numbers of animals are concentrated in one place (figure 17.16).

Sediment Pollution

Some cases of high suspended-sediment load in water occur naturally (figure 17.17). However, these are commonly localized situations, and may be temporary as well. In many agricultural

Figure 17.17

Sediment clouds stream waters. The stream at right is murky with suspended sediment, probably because of bank collapse. Note the contrast with the clear tributary at left, and the fact that the waters of the two streams do not immediately mix where they join. Junction of Soda Butte Creek and the Lamar River, Yellowstone National Park.

areas, sediment pollution of lakes and streams is the most serious water-quality problem and is ongoing. Farmland and forest land together are believed to account for about 75% of the 3 billion tons of sediment supplied annually to U.S. waterways. Of this total, the bulk of sediment is derived from farmland; sediment yields from forest land are typically low.

Sediment pollution not only causes water to be murky and unpleasant to look at, swim in, or drink, it reduces the light available to underwater plants and blankets food supplies and the nests of fish, thus reducing fish and shellfish populations. Channels and reservoirs may be filled in, as noted in earlier chapters. The sediment also clogs water filters and damages power-generating equipment. Some sediment transport is a perfectly natural consequence of stream erosion, but agricultural development typically increases erosion rates by four to nine times unless agricultural practices are chosen carefully.

Chapter 12 examined some strategies for limiting soil loss from farmland. From a water-quality standpoint, an additional approach would be to use settling ponds below fields subject to serious erosion by surface runoff (figure 17.18). These ponds do not stop the erosion from the fields, but by impounding the water, they cause the suspended soil to be dropped before the water is released to a lake or stream. (This may reduce sediment pollution at a cost of increased erosion along a channel downstream, but the sediment pollution is often the more serious concern.) Settling ponds can also be helpful below large construction projects, logging operations, or anywhere ground disturbance is accelerating erosion.

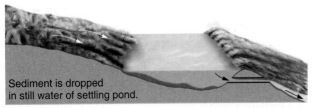

Surface runoff carries
a suspended sediment load.

Sediment is dropped
in still water of settling pond.

Outflow water is clear
of suspended sediment.

A

B

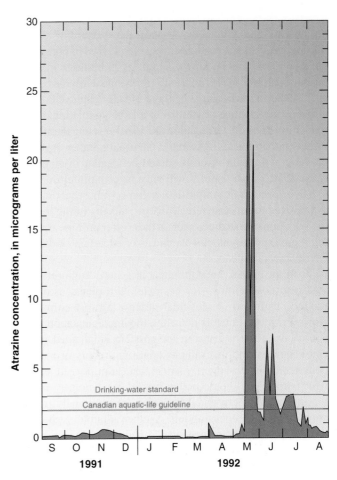

Figure 17.18

(A) Settling ponds enhance water quality and reduce soil loss, reducing sediment pollution and thereby also trapping pollutants adsorbed onto the sediment particles. Removal of sediment from the water, however, may increase erosion below the settling pond. (B) Heavy surface runoff may contribute to both flooding and soil erosion, muddying floodwaters with sediment pollution. This "watershed dam" in eastern Iowa traps the sediment-laden waters, reducing flooding downstream and containing the sediment, which will settle out of the basin's still water.

(B) Photograph by Lynn Betts, courtesy USDA Natural Resources Conservation Service.

Figure 17.19

A study in the central Nebraska corn belt showed a sharp rise in pesticide runoff (here illustrated by atrazine) following early-spring application to fields. Spring and summer rains continued to wash off significant, though declining, amounts. While the average annual concentration in runoff did not exceed drinking-water standards, peak concentrations were up to ten times that standard.

Source: After USGS Circular 1225.

Pesticides

Herbicides (chemical weed-killers) and insecticides (chemicals used to kill insect pests) are a significant source of pollution in agriculture. Most such compounds now in use are complex organic compounds of the kinds described in the "Industrial Pollution" section earlier in the chapter and are in some measure toxic to humans or other life-forms. U.S. farmers use more than a billion pounds of pesticides each year, of which about 90% are synthetic organic compounds. The use of insecticides in agriculture doubled, and the use of herbicides more than quadrupled, from 1965 to 1980, though the use of both has leveled off since then.

The agricultural chemicals themselves are not the sole problem, either. Many of the chemicals used in producing the end-use products are themselves toxic and persistent (dioxin, for instance); so are some of the early products of herbicide and pesticide decomposition, when and if they do begin to break down.

Commonly, spring applications of agricultural chemicals are followed by concentration spikes in runoff water (figure 17.19); clearly, a great deal of what is applied is wasted. The potential risks could be reduced by eliminating routine applications of such chemicals as a sort of preventive medicine. Applying a remedy only when a problem is demonstrated to exist may initially reduce synthetic chemical use at the cost of crop yield. However, over the longer term, the results may be more economical, in part because the bugs or weeds will not as readily develop resistance to the substance being used.

A growing variety of nonchemical insect-reduction strategies are also finding increasing acceptance. Some are specific to a single pest. For instance, the larvae of a small wasp that is harmless to people and animals prey parasitically on tomato hornworms and kill them. A particular bacterium *(Bacillus thuringiensis)* attacks and destroys several types of borer insects but again produces no known ill effects in humans, mammals, birds, or fish, and leaves no persistent chemical residues on food.

Other remedies may be more broadly applicable. An example is the practice of sterilizing a large population of a particular insect pest by irradiation and then releasing the sterilized insects into the fields. Assuming the usual number of matings, the next generation of pests should be smaller in number because some of the matings will have involved infertile insects. Other growers bait simple insect traps with scents carefully synthesized to simulate female insects, thereby luring the males to their doom. Procedures such as these offer promise for reducing the need for synthetic chemicals of uncertain or damaging environmental impact.

In the 1990s, breakthroughs in genetic engineering created new possibilities—for example, corn plants modified to produce their own insecticidal defense against corn borers. However, several factors constrain the implementation of such approaches. One is concern for possible unintended negative consequences (in this example, apparent toxicity of the corn's pollen to monarch-butterfly larvae that eat milkweed leaves on which stray wind-borne pollen has fallen). Another is regulatory-agency concern about the safety of genetically modified crops for human consumption. Another is public resistance to the general concept of genetically engineered foods, independent of any specific health or safety concerns.

Reversing the Damage— Surface Water

All of the foregoing pollutants together have significantly affected surface waters over broad areas of the United States, as might be expected from such figures as 12.15, 17.8B, and Case Study 17.1 figure 1. Reduction in pollutant input is one strategy for addressing such problems, but it requires time to succeed, especially with relatively stagnant waters such as lakes. A variety of more-aggressive treatment methods are available where indicated for specific circumstances.

Many water pollutants, including phosphates, toxic organics, and heavy metals, can become attached to the surfaces of fine sediment particles on a lake bottom. After a long period of pollutant accumulation, the bottom sediments may contain a large reserve of undesirable or toxic chemicals, which could, in principle, continue to be re-released into the water long after input of new pollutants ceased. Wholesale removal of the contaminated sediments by dredging takes the pollutants permanently out of the system. However, they still remain to be disposed of, commonly in landfills.

Dredging operations must be done carefully to minimize the amount of very fine material churned back into suspension in the water. Fine resuspended sediment, with its high surface-to-volume ratio, tends to contain the highest concentrations of pollutants adsorbed onto grain surfaces. Also, if fine resuspended sediments remain suspended for some time, the increased water turbidity may be harmful to life in the lake. The process, moreover, can be expensive.

Dredging projects in the United States so far have tended to be small, usually involving less than 1 million cubic meters of sediment, and most are sufficiently recent that the long-term impacts are not known. Dredging is used most widely in the United States for emergency cleanup of toxic wastes. When 250 gallons of PCBs spilled into the Dunwamish Waterway near Seattle in 1974, the EPA chose to dredge nearly 4 meters' thickness of contaminated sediment from the waterway and thereby succeeded in recovering 220 to 240 gallons of the spilled chemicals.

Another way to reduce the escape of contaminants from bottom sediments is to leave the sediments in place but to isolate them wholly or partially with a physical barrier. Over limited areas, like lagoons and small reservoirs, impermeable plastic liners have been placed on top of the sediments and held in place by an overlying layer of sand. The permanence of the treatment is questionable, and it has not been used on a large scale. Compacted clay layers of low permeability might, in principle, serve a similar purpose, but this treatment has not been attempted on any significant scale.

The addition of salts of aluminum, calcium, and iron to sediment changes the sediment chemistry, fixing phosphorus in the sediment and thereby reducing eutrophication. The technique has been used in small lakes in Wisconsin, Ohio, Minnesota, Washington, Oregon, and elsewhere. While not always successful, the method has often reduced phosphate levels in the water and corresponding algal growth. The cost compares favorably with the use of synthetic algicides to destroy the algae. However, careless overtreatment may prove toxic to fish, and, in any event, the treatment must be repeated every two to three years. An alternative, biological means of addressing eutrophication in ponds and lakes that has been developed recently involves introducing algae-eating (nonnative) fish that have been sterilized to prevent their overwhelming native species. (It can also be noted that algal blooms may not, in fact, be uniformly undesirable: Recent studies of a large, severely polluted lake in China have indicated that algae may be scavenging mercury and arsenic from the water, suggesting that heavy-metal concentrations might be reduced by encouraging algal blooms, then collecting and disposing of the algae!)

Active decontamination is most often used in response to toxic-waste spills. The specific treatment methods depend on the toxin involved. In 1974, a toxic organic-chemical herbicide washed into Clarksburg Pond in New Jersey from an adjacent parking lot. Many fish were killed, and local wildlife dependent on the pond for water were endangered. The contamination also threatened to spread to ground water by infiltration and to the Delaware River to which the pond water flows overland. The EPA set up an emergency water-cleaning operation. The herbicide was removed principally by activated charcoal through which the water was filtered. Subsequent tests showed the ground water to be uncontaminated, and within two years, fish were again plentiful in the pond.

Artificial aeration addresses oxygen depletion in a lake. There are several possible procedures, including bubbling air or oxygen up through the waters, thereby providing more oxygen to oxygen-starved deep waters, or simply circulating the water mechanically, cycling up deep waters to the surface where they can dissolve oxygen directly from the atmosphere. Many small artificial ponds are now designed with fountains whose function is not purely decorative: spraying water into the air enhances aeration through the droplets' considerable surface area. When successful, aeration can transform water from an anaerobic to an aerobic condition, to the great benefit of fish populations (a common reason for the action). However, when air rather than pure oxygen is bubbled through the water under pressure at depth, an excess of dissolved nitrogen may develop in the water, which is toxic to some fish.

From a public-health perspective, the principal concern with respect to quality of surface water relates to its use for drinking water. Here the news is generally good, as the most common contaminants in most areas, especially agricultural regions, are excess nutrients—and in those areas, most public supplies are drawn from ground water, not surface water. However, the ground water in these areas is often contaminated too.

Groundwater Pollution

Groundwater pollution, whether from point or nonpoint sources, is especially insidious because it is not visible and often goes undetected for some time. Municipalities using well water must routinely test its quality. Homeowners relying on wells may be less anxious to go to the trouble or expense of testing, particularly if they are unaware of any potential danger. Yet a recent comprehensive analysis of water quality in domestic (private) wells has suggested potential health concerns in a significant fraction of cases (figure 17.20). Also, in most instances, the passage of pollutants from their source into an aquifer used for drinking water is slow because that passage occurs by percolation of water through rock and soil, not by overland flow. There may therefore be a significant time lapse between the introduction of a pollutant into the system in one spot and its appearance in ground or surface water elsewhere. Conversely, groundwater pollution in karst areas, with their rapid drainage, may spread unexpectedly swiftly. The "out of sight, out of mind" aspect of ground water contributes to the problem. So does imperfect understanding of the interrelationships between surface and ground water in specific cases.

The Surface–Ground Water Connection Explored

Given the nature of groundwater recharge, the process can readily introduce soluble pollutants along with recharge water. For example, agricultural pesticides are applied at the surface. But those pesticides neither used up nor broken down may remain dissolved in infiltrating water as well as runoff, resulting in pesticide contamination of ground water, as previously de-

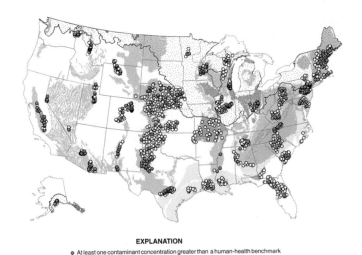

EXPLANATION
● At least one contaminant concentration greater than a human-health benchmark
○ No contaminant concentration greater than a human-health benchmark

Figure 17.20

Sampling of nearly 1400 wells spanning 45 states and 30 major aquifers revealed at least one contaminant presenting a health concern in 23% of cases. (Shaded areas in different colors indicate different aquifer systems.)

From USGS Circular 1332, 2009.

scribed. Acid mine drainage, storm-sewer runoff, and other surface-water pollutants are likewise potential groundwater pollutants. The susceptibility of the ground water to pollution from such sources is a function of both inputs and local geology (figure 17.21).

Acid rain (discussed in chapter 18) can be a particular problem because many toxic metals and other substances are more soluble in more acidic water. How severely groundwater chemistry is affected is related, in part, to the residence time of the ground water in its aquifer before it is extracted through a well or spring. That is, if water residence time is short, as with the shallow aquifers in figure 17.22, there is less opportunity for chemical reactions in the aquifer to buffer or moderate the impacts of acid rain. Longer residence times may allow more moderation of the chemistry of the acid waters as well as the concentrations of various dissolved constituents, providing more opportunity for breakdown of potentially toxic compounds.

Altogether, substantial groundwater-quality problems exist in the United States. Some are natural; for example, high dissolved mineral content, arsenic (figure 17.11), or radium (Case Study 11 figure 2). Some are due to human activities (pollution), as seen in figure 17.23. The nature of the problems and their causes can vary greatly from place to place. Not surprisingly, pesticides (or their breakdown products) are common in ground water beneath agricultural areas. Nitrate concentrations, too, tend to be higher in ground water below agricultural areas than in ground water under undeveloped areas. On the other hand, as noted, fertilizers, insecticides, and herbicides are not unique to agricultural regions; they may be used abundantly in urban/residential areas, too. And where improvements in municipal sewage treatment

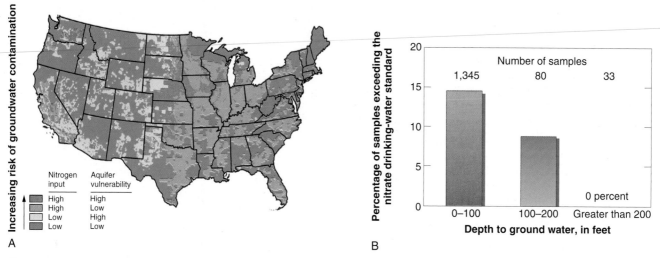

Figure 17.21

(A) Susceptibility of ground water to nitrate pollution is related not only to nitrogen applied in fertilizer (as shown in figure 17.15B) but to speed of drainage; "well-drained" soils are permeable soils through which water infiltrates readily. (B) In a national study of major aquifers, nitrate concentrations in ground water decreased as depth to aquifer increased; most samples exceeding the drinking-water standard of 10 milligrams per liter were from shallow aquifers. This may reflect more mixing of contaminated surface water with uncontaminated water in deep aquifers, or protection of deeper aquifers by thicker cover of low-permeability material.

Source: U.S. Geological Survey National Water Quality Assessment Program.

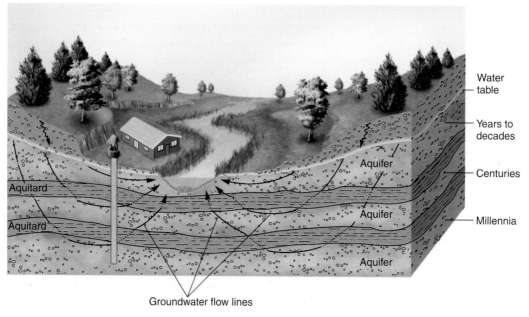

Groundwater flow lines

Figure 17.22

Groundwater flow rates may vary widely; in deep aquifers beneath multiple aquitards, centuries or more may elapse between infiltration in the recharge area and appearance in surface water, depending upon the rate of passage through aquitards. Some deep confined aquifers may be entirely isolated from surface waters on a human timescale.

have been made to reduce ammonia concentrations, this has often been accomplished by converting the nitrogen in the ammonia to nitrate—thereby increasing nitrate concentrations in the water system. In virtually all cases involving human activity, a groundwater-pollution problem has evolved from surface-water pollution.

Groundwater pollution has sometimes appeared decades after the industry or activity responsible for it has ceased to operate and has disappeared from sight and memory. Chemicals dumped or spilled into the soil long ago might not reach an aquifer for years. Even after the source has been realized, so large an area may have been contaminated that cleanup is

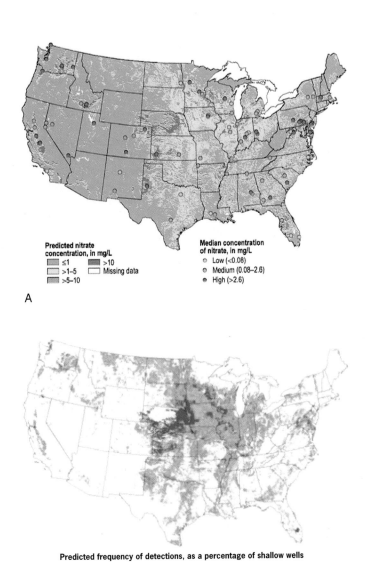

A

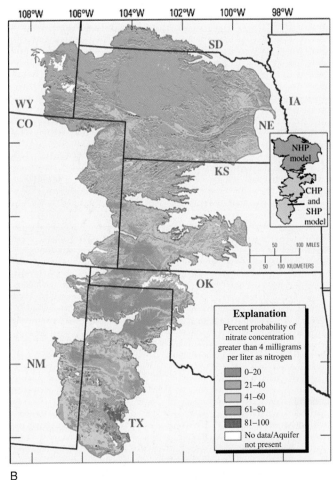

B

Figure 17.23

(A) Circles represent sampled shallow groundwater wells. Using these data together with information on local geology, hydrology, and fertilizer application allows modeling to predict groundwater nitrate levels across unsampled areas. (B) Detailed modeling of this type for the High Plains Aquifer suggests high dissolved nitrate concentrations over much of the aquifer. (C) Similar modeling for the herbicide atrazine again suggests that widespread groundwater contamination is to be expected.

From (A) USGS Circular 1350; (B) USGS Circular 1337; (C) USGS Circular 1291.

C

impractical and/or prohibitively expensive. This is a problem with many old, abandoned toxic-waste dump sites such as Superfund sites discussed in chapter 16. Groundwater pollution from nonpoint sources, like farmland, may also be so widespread that cleanup is not feasible.

Some strategies for addressing groundwater contamination are explored later in the chapter. With ground water, however, it may be especially important to avoid or limit pollution initially, because it is typically more difficult to clean up afterward than is polluted surface water. So, for example, there is pressure to replace underground fuel storage tanks of metal (which may corrode, and leak) with tanks of more corrosion-resistant materials, such as plastic or fiberglass (which, however, sometimes crack and leak themselves). Pollution avoided is pollution that does not have to be treated.

Bearing in mind the potential of groundwater pollution from surface-water infiltration, one can also assess the vulnerability of a groundwater system to such pollution in terms of how well protected it is. Such an assessment may, in turn, suggest how closely the groundwater system should be monitored. For the sources of surface-water pollution are many. In addition to those already cited, there are a variety of casual pollution sources related to improper disposal: crankcase oil dumped into a storm sewer by a home mechanic, solvents and cleaning fluids poured down the drain by homeowners or dry cleaners or painters or furniture refinishers, oil that was deposited on roads by traffic flushed off and running into storm sewers in a heavy rain, even chemicals from a college science laboratory thoughtlessly flushed down a laboratory drain—all these, many simply reflecting individual carelessness, may pollute ground water.

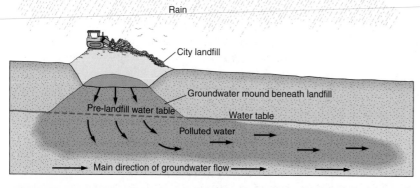

Rain

City landfill

Groundwater mound beneath landfill

Pre-landfill water table

Water table

Polluted water

Main direction of groundwater flow

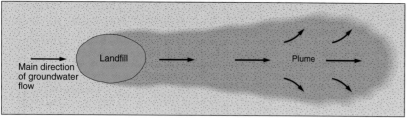

Main direction of groundwater flow

Landfill

Plume

A

Figure 17.24

Groundwater contaminant plumes.
(A) Schematic view, in cross section (top) and map view (bottom). Note that the plume tends to spread as it flows. When it reaches a well or body of surface water, serious pollution may appear very suddenly. (B) A groundwater contaminant plume from the Norman (Oklahoma) Landfill Environmental Research Site. Note that concentration distributions differ for different substances. (mg/L = milligrams per liter, approximately equal to ppm)

(B) After U.S. Geological Survey Fact Sheet 040-03.

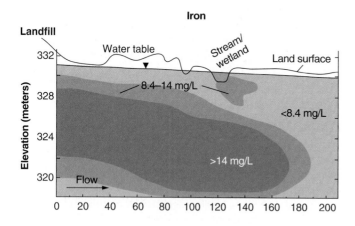

Iron

Landfill

Water table

Stream/wetland

Land surface

8.4–14 mg/L

<8.4 mg/L

>14 mg/L

Flow

Elevation (meters)

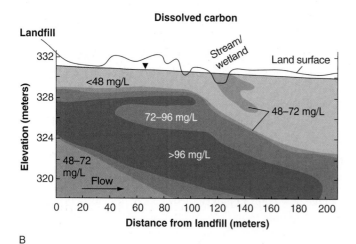

Dissolved carbon

Landfill

Stream/wetland

Land surface

<48 mg/L

72–96 mg/L

48–72 mg/L

48–72 mg/L

>96 mg/L

Flow

Elevation (meters)

Distance from landfill (meters)

B

Tracing Pollution's Path

Groundwater pollutants migrate from a point source in two ways. As ground water flows, it carries pollutants along. Dissolved constituents can also diffuse through the water, even when it is not moving, from water in which their concentrations are high toward areas of lower concentration. (If you drop a drop of dye or food coloring into a glass of water, the color will slowly disperse throughout the glass.) Depending on the nature of the pollutants, they may gradually break down as they move away from their source, as by reaction with aquifer rocks or through the action of microorganisms.

Where pollutant migration is primarily by groundwater flow, a **contaminant plume** develops down-gradient from the point source. This is a tongue of polluted ground water that may extend for hundreds of meters from the source (figure 17.24). Leachate plumes seep from leaky landfills, ill-designed toxic-waste sites, and other pollutant sources.

Their direction will be constrained by the prevailing direction of groundwater flow, but the extent of each plume and concentration of each contaminant must be determined by sampling via wells. Some pollutants float or sink if they are less or more dense than water and do not mix with it just as is true of some liquid wastes injected in wells (recall figure 16.20B). Some dissolve and diffuse very rapidly. Understanding the behavior of the plumes and the fate of the various pollutants is important to containing and cleaning up the contaminants; studies are underway at a number of current and former waste sites, such as the Norman Landfill from which the data of figure 17.24B come.

Reversing the Damage— Ground Water

Earlier sections of this chapter noted various means of reducing the flow of pollutants into water. The approach of reducing or stopping the input of further pollutants, then waiting for natural processes to remove or destroy the pollutants already in the system ("monitored natural attenuation"), is sometimes the only approach technically or economically possible in the case of contaminated ground water, given its relative inaccessibility. Systematic monitoring of ground water can be difficult without an extensive (and usually expensive) network of wells because, by the time the pollution problems are recognized, the contaminants have typically spread widely in the aquifer system. Even after the original source of the pollutants is removed or contained, further groundwater contamination may occur intermittently for some time, too. If pollutants have adsorbed onto soil in the unsaturated zone, some may be dissolved or dislodged each time precipitation percolates through that soil, adding a new pulse of pollution to the groundwater system. In other cases, pollutants may have diffused into fine sediments or become associated with organic matter in an aquifer, and may seep back out into the ground water long after introduction of new contaminants into the system has been stopped. The slow migration and limited mixing of most ground waters complicates *in situ* treatment of contamination; bear in mind the way ground water occurs, in cracks and pores dispersed throughout a large volume of rock. Often then, polluted ground water is only treated after it is extracted for use.

Decontamination After Extraction

This approach may be called the **pump-and-treat** method. It is most often used to limit the spread of contamination, as described in the Rocky Mountain Arsenal situation later in this chapter, but may also be employed when impure ground water must be used for some purpose such as a municipal water supply. The type of treatment depends on the intended water use and the particular pollutants involved.

Inorganic compounds can be removed from extracted ground water by adjusting its acidity (pH). Addition of alkalies may result in the precipitation of many heavy or toxic metals as hydroxides. Many of the same metals may be precipitated as sulfides, which are relatively insoluble, or as carbonates. All of these strategies yield a solid sludge (the precipitate) that contains the toxic metals and must still be disposed of.

Just as microorganisms can be used to decompose organic matter in sewage, microbial activity can break down a variety of organic compounds in ground water after it is extracted. The organisms can be mixed in bulk with the water; after treatment, the biomass must then be settled or filtered out. Alternatively, the organisms can be fixed on a solid substrate, and the contaminated water passed over them.

Air stripping encompasses a set of methods by which volatile organic pollutants are transferred from water into air and thus removed from the water. The mechanical details of the process vary, but each case involves some form of aeration of the water, followed by separation of the gas. The pollutants are still present in the gas phase and must be removed for alternate disposal.

Activated charcoal (activated carbon), used in a variety of filters, adsorbs organic compounds dissolved in ground water. Again, the compounds themselves remain intact and require disposal.

In short, treatment methods available *after* ground water is extracted are many and span the same broad range as methods of treating surface-water pollution or purifying a municipal water supply. But as long as the polluted ground water remains in the ground, unless it is contained (which is not common, because of costs), the contaminants can go right on spreading, carried with the water flow or diffusing through the fluid.

In Situ Decontamination

Treatment of contaminated ground water in place is possible only when the extent and nature of the pollution are well defined; the methods used are very site- and pollutant-specific. For inorganic pollutants, such as heavy metals, immobilization is one strategy. Injection wells placed within the contaminated zone are used to add chemicals that cause the toxic substances to precipitate and thus to become stationary. The uncontaminated water can then be extracted for use.

Biological decomposition *(bioremediation)* is effective on a broad range of organic compounds, many of which can be broken down by microorganisms. Organisms that will "eat" certain types of contaminants are injected into the contaminated ground water. The process can be stimulated, in individual cases, by the addition of oxygen or nutrients to accelerate growth of the microorganisms; additional microorganisms (indigenous or foreign) can be introduced to attack the particular compound(s). These approaches may be used to accelerate cleanup in cases where monitored natural attenuation is the primary strategy, too. Not all organics can be eliminated in this way, and occasionally, objectionable residues affect water taste and odor. However, most of the twenty-four "priority organic pollutants" identified by the Environmental Protection Agency can be successfully attacked by biological means. Genetic engineering may also allow the development of strains of microorganisms targeted to specific pollutants, increasing the efficiency of bioremediation efforts.

With petroleum products that do not mix well with water, but tend to float on it (as with natural oil and gas deposits), wells that reach barely to water-table level may selectively pump out contaminant-rich fluid, leaving much purer water behind.

A more recently developed approach is the *permeable reactive barrier,* in which pollutants are broken down as the ground water flows through the barrier. Such an approach has successfully dealt with toxic organic compounds that previously escaped from the site of the Denver Federal Center (figure 17.25).

Damage Control by Containment—The Rocky Mountain Arsenal

One human activity can have multiple environmental impacts. We noted in chapter 4 that deep-well disposal of liquid waste at the Rocky Mountain Arsenal, near Denver, Colorado, was identified

A

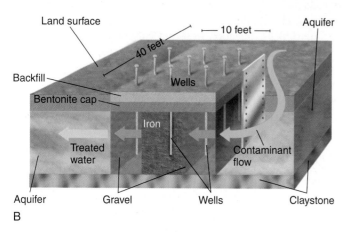

B

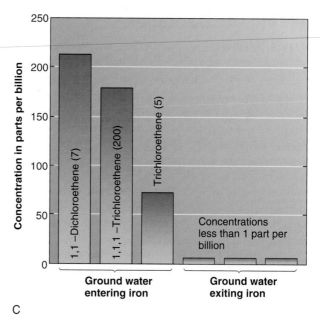

C

Figure 17.25

(A) A trench is dug across the groundwater flow path, and the reactive material installed. (B) Ground water flows through gravel into permeable barrier containing metallic iron, which reacts with and destroys the organic solvents, leaving clean water to flow on through the aquifer (C).

(A) Photograph courtesy David Naftz/U.S. Geological Survery;
(B–C) After U.S. Geological Survey, National Wetlands Research Center.

as the cause of a series of small earthquakes in the area. Careless surface disposal of toxic wastes at the same site also resulted in contamination of local ground water. Beginning in 1943, wastewater containing a variety of organic and inorganic chemicals was discharged into unlined surface ponds. The principal contaminants were the organics diisomethylphosphonate and dichloropentadiene.

The water table in the shallow aquifer in the area is locally only 2 to 5 meters below the ground surface. Water from this aquifer is widely used for irrigation and for watering livestock. Incidents of severe crop damage were reported in the early 1950s. To limit further contamination, an asphalt-lined disposal pond was constructed in 1956, but, in time, the liner leaked. Contaminants already in the groundwater system continued to spread. When new complaints of damage to crops and livestock were made in the early 1970s, the Colorado Department of Health investigated. They detected a variety of toxic organic compounds in water and soil on the arsenal property; some of these compounds were present, though in much lower concentrations, in water drawn from municipal supply wells off the site. The Department of Health ordered a cessation of the waste discharges and cleanup of the existing pollution, with provision to prevent further discharge of pollutants from the arsenal property.

Because groundwater contamination was already so widespread, the only feasible approach to containing the further spread of pollutants was to halt them at the arsenal boundaries. This effort was combined with cleanup activity (figure 17.26). A physical barrier of clay-rich material was placed along the arsenal side of the boundary at several locations on the north and west sides, where the natural groundwater flow was outward from the arsenal. On the arsenal side of the barrier, wells extract the contaminated ground water. It is treated by a variety of processes, including filtration, chemical oxidation, air stripping, and passage through activated charcoal. The cleaned water is then reintroduced into the aquifer on the outward side of the barrier. The net groundwater flow regime outside of the arsenal has been only minimally disrupted, and pollutants in nearby water supplies have been greatly reduced. This, then, is a version of pump-and-treat remediation: the water is not immediately used after treatment, but is returned, cleaned, to the aquifer system.

The efforts have not been inexpensive. Over $25 million has been spent on the control and decontamination activities, and the boundary barrier system will have to continue in operation indefinitely as the slowly migrating contaminants in the ground water continue to move toward the limits of the site. These activities illustrate both the decontamination and control success that can be achieved in selected cases and the value (practical and economic) of appropriate planning to prevent contamination in the first place.

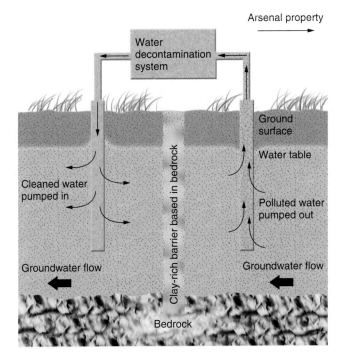

Figure 17.26

Sketch of boundary control system for contaminated ground water at Rocky Mountain Arsenal. Again, the barrier—in this case, an impermeable one—is placed across the path of groundwater flow.

Figure 17.27

Obviously polluted runoff water from mine dump, Leadville, Colorado.

Photograph © David Hiser/Stone/Getty Images.

New Technology Meets Problems from the Past: California Gulch Superfund Site, Leadville, Colorado

Mining near Leadville began in 1859; the sulfide-rich ores were mined for gold, silver, lead, and zinc. In the process, tailings piles were scattered over about 30 square kilometers (about 12 square miles) in and around Leadville. Sulfide-mineral residues in the tailings weather to produce sulfuric-acid-rich drainage. The acid drainage, in turn, leaches a number of metals, some of them toxic, including lead, cadmium, and arsenic (figure 17.27). The hazard is obvious where polluted surface drainage is visible. Where it is not, however, visual inspection does not help much in assessing degrees of risk from acid drainage. Recently, the use of high-resolution airborne spectrometers has provided a powerful tool for assessing such sites.

Investigators selected a single tailings pile and made an aerial traverse across it (figure 17.28A). Comparing spectral data to analysis of surface samples, they could identify zones of different mineralogy across the pile by the distinctive spectrographic signatures of the minerals: different minerals reflect different wavelengths of radiation, just as they reflect different wavelengths of visible light, resulting in different colors. These mineral zones corresponded to varying degrees of acidity of drainage (figure 17.28B). In general, increasing acidity (lower pH) correlated with higher solubility/leachability of metals. In short, the mineralogy—which could be determined by overflight—was an indicator of drainage geochemistry. This offers a much quicker way to map relative degrees of hazard from dissolved toxic metals over a broad area than does surface sampling and conventional chemical analysis (figure 17.29).

There are close to 50,000 sites in the United States alone, now inactive, where metals were once mined. Each of these represents a potential acid-drainage hazard. Having efficient ways to identify quickly those sites, or parts of sites, posing the greatest risk allows prioritization of cleanup efforts.

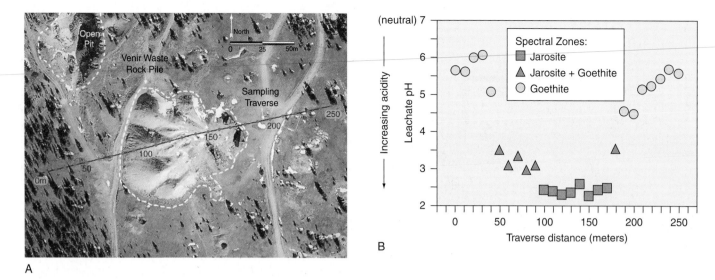

A

B

Figure 17.28

(A) Traverse track across a tailings pile at California Gulch. Red dots are sampling sites. (B) Different minerals found across the tailings pile correspond to different acidity (pH) of leachate; goethite is an iron hydroxide, jarosite a hydrous iron sulfate formed by weathering of sulfides in acid conditions. Many toxic metals at this site are much more leachable in acid waters.

Source: U.S. Geological Survey Spectroscopy Lab.

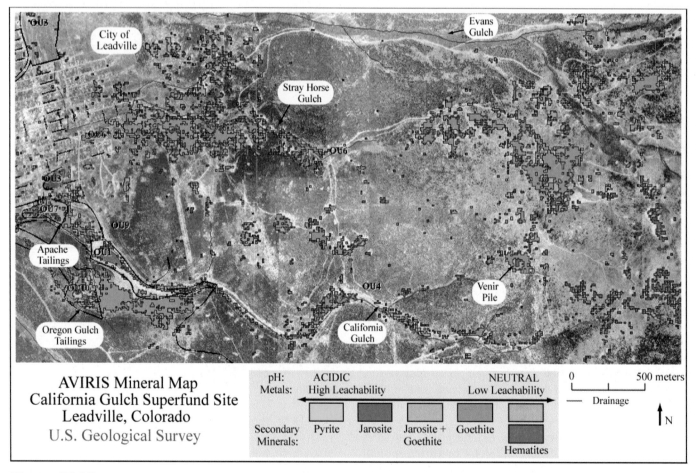

Figure 17.29

Airborne-spectroscopic scan shows distribution of minerals, which correspond to varying degrees of acidity of water and thus of leachability of metals; combined with drainage information, the image helps target areas of greatest water-pollution danger.

Source: U.S. Geological Survey Spectroscopy Lab.

Summary

Much of the concern about pollution relates to the effects of trace elements or other chemicals on health. A dose-response curve describes the degree of benefit or harm from a particular substance in varying doses. The same substance may be beneficial or harmful depending upon the dose, and different organisms may respond quite differently to the same chemical.

A variety of substances, some naturally occurring and some added by human activities, cycles through the hydrosphere. How long each substance spends in a particular reservoir is described by the residence time of that substance in the reservoir. The materials of greatest concern in the context of pollution are usually toxic ones with long residence times in the environment. Little is known about many synthetic chemicals—their movements, longevity in natural systems, method of breakdown, and often even the extent of their toxicity.

In addition to adding many new and sometimes-toxic chemicals to the environment, industrial activities can unbalance the natural cycles of other harmful substances, such as the heavy metals, by increasing the rate at which they are introduced into the environment. These elements share with some other chemicals the tendency to accumulate in organisms and increase in concentration up the food chain, becoming a greater hazard higher up in the chain. Power plants may present a thermal, rather than a chemical, pollution threat.

Sediment pollution is a problem particularly in agricultural areas, where better erosion control could greatly reduce it. Organic matter, whether contributed by domestic sewage, agriculture, or industry, increases the biochemical oxygen demand (BOD) of the water, making it inhospitable to oxygen-breathing organisms. Where the water is rich in plant nutrients from organic wastes, fertilizer runoff, or other sources, such as phosphate from detergents, a eutrophic condition develops that encourages undesirably vigorous growth of algae and other plant life. The toxic organic compounds in herbicides and insecticides are widely used in both agricultural and urban areas, and their residues are often persistent in both ground and surface waters. Some of these compounds also show biomagnification in organisms.

Reduction in pollutant release into aqueous systems is the most common strategy for reducing pollution problems. As a general rule, pollutants from point sources are more easily controlled or confined than are pollutants from nonpoint sources. More-aggressive physical and/or chemical treatments are sometimes used in water-quality restoration efforts, particularly in the case of small, still water bodies, or to combat toxic-chemical spills. However, even if harmful materials are contained rather than released into the environment or are removed from water or underlying sediment, they still pose a waste-disposal problem. Groundwater pollution is a particular challenge, not only because it is often less readily detected, but because the nature of ground water makes it harder to treat *in situ*. Bioremediation and the use of permeable reactive barriers are among techniques that can successfully address groundwater pollution without disrupting flow patterns. New technological tools can assist in focusing efforts to remediate water-pollution problems on areas of most acute need.

Key Terms and Concepts

aerobic decomposition 401
anaerobic decomposition 401
biochemical oxygen demand (BOD) 401

biomagnification 405
contaminant plume 420
dose-response curve 399
eutrophication 402

heavy metals 405
nonpoint source 400
oxygen sag curve 401
point source 400

pump-and-treat 421
residence time 397

Exercises

Questions for Review

1. What is a dose-response curve? Sketch an example for a substance that is essential in small doses but harmful at high doses.

2. Explain the concept of residence time; illustrate it with respect to some dissolved constituent in seawater or another water reservoir.

3. In what way do human activities most commonly alter the cycles of naturally occurring elements?

4. What is biomagnification, and why is it a particular concern of humans? Give an example of a chemical that is subject to biomagnification.

5. Why do potentially harmful health effects of organic compounds constitute a major area of concern?

6. What is BOD? How is it related to the oxygen sag curve often noted in streams below sources of organic-waste matter?

7. What is thermal pollution? From what activity does it principally originate? In what sense is it a less-worrisome kind of pollution than most types of chemical pollution?

8. Cite at least three possible sources of the nutrients that contribute to eutrophication of water; explain the concept. Why are eutrophic conditions generally considered undesirable?

9. What kinds of pollution can be reduced by the use of settling ponds? Explain.

10. Briefly describe two after-the-fact approaches to reducing water pollution, and note their limitations.

Exploring Further

1. Investigate your own water quality. Ask your local water department for information—or, if your water comes from a private source such as a well, contact the health department for information on how to have your water quality tested. If you use a municipal water supply, look into its water quality at water.epa.gov/drink/local/. What, if any, constituents exceed the EPA drinking-water standards?

2. As the largest freshwater lakes in the United States, the Great Lakes receive considerable attention and study; in 2004 the president declared them a "national treasure" by executive order. Investigate the current status of pollution and remediation efforts, in any of the Great Lakes, at epa.gov/greatlakes/.

3. Learn more about fish-consumption advisories at water.epa.gov/scitech/swguidance/fishshellfish/fishadvisories/. Follow the link from that page to check the status of any such advisories in your area.

4. A lake contains 200 million m^3 of water. It is fed and drained by streams; the average inflow (and outflow) is 15 m^3/sec.
   (a) What is the residence time of water in the lake, in days?
   (b) A pollutant in the lake has reached a concentration of 75 ppm before its input is stopped. Thereafter, it is removed in solution in the outflow stream. Given the meaning of residence time, half the lake water is replaced in a period equal to the residence time, so in this case the quantity of the pollutant is reduced by half in that time. Approximately how long will it take for the concentration of the pollutant to reach 10 ppm or less?

Air Pollution

CHAPTER 18

Donora, Pennsylvania, 1948: "The fog closed over Donora on the morning of Tuesday, October 26th. . . . By Thursday, it had stiffened . . . into a motionless clot of smoke. That afternoon, it was just possible to see across the street . . . the air began to have a sickening smell, almost a taste . . ." (Roueché, p. 175). By Friday, residents with asthma and other lung disorders began to find it difficult to breathe. Then more of the residents took sick, becoming nauseous, coughing and choking, suffering from headaches and abdominal pains. The fog persisted for five days. Altogether, nearly 6000 people were stricken by the polluted fog; twenty of them died.

Donora was a mill town with a steel plant, a wire plant, and a zinc and sulfuric acid plant among its industries. The litany of toxic chemicals in its air identified during later investigations included fluoride, chloride, hydrogen sulfide, sulfur dioxide, and

cadmium oxide, along with soot and ash. These materials were all routinely released into the air, but not generally with such devastating effects.

The events of late October 1948 in Donora might be called an acute air-pollution episode—sudden, obvious, and dramatic. Other still more serious single episodes are known. For four days in early December 1952, weather conditions trapped high concentrations of smoke and sulfur gases from coal burning in homes and factories over London, England; an estimated 3500 to 4000 people died in consequence, with many more made ill. Acute pollution events of this kind are, fortunately, rare. The health impact of moderately elevated levels of pollutants found widely over the earth, especially near urban areas, is more difficult to assess.

Air pollution is costly, and not only in terms of health. The Council on Environmental Quality has estimated *direct* costs at

---

Air pollution spreads across political boundaries even more readily than does water pollution. Here, easterly winds carry smoke from West Africa across the Atlantic. The carbon monoxide is produced mainly by fires used in slash-and-burn agriculture. Red dots indicate fire locations identified by satellite.

NASA image courtesy Jacques Delcloitres, MODIS Land Rapid Response Team at NASA GSFC.

$16 billion per year in the United States alone, including over $1 billion just for cleaning soiled items and $500 million in damage to crops and livestock. Worldwide, estimated costs of forest-product harvest reduction due to air pollution exceed $40 billion per year.

In addition, there are the sizable costs in illness, medical expenses, absenteeism, and loss of production, which are somewhat harder to quantify. One estimate suggests that reduction of air-pollution levels by 50% in major urban areas would save more than $2 billion per year in health costs. Implicit in the decreased health costs is an increase in longevity, decrease in illness, and improvement in quality of life among those now adversely affected by air pollution. The financial and human considerations together are powerful incentives for trying to limit air pollution.

Atmospheric Chemistry—Cycles and Residence Times

The atmosphere consists of three principal elements. On average, nitrogen (mostly as N_2) comprises nearly 77% of the total by weight; oxygen (O_2), about 23%; the inert gas argon, close to 1%. Locally, water vapor can be significant, up to 3% of the total. Everything else in the atmosphere together makes up much less than 1% of it.

Materials cycle through the atmosphere as they do through other natural reservoirs. One can speak of the residence times of gases or particles in the atmosphere just as one can discuss residence times of chemicals in the ocean. As with dissolved substances, residence times of gases are influenced by the amounts present in a given reservoir such as the atmosphere, and the rates of addition and removal. Oxygen, for example, is added to the atmosphere during photosynthesis by plants; it is removed by oxygen-breathing organisms, by solution in the oceans, by reaction with rocks during weathering, and by combustion. Its residence time in the atmosphere is estimated at 7 million years.

Carbon dioxide (CO_2) has an even more complex cycle (figure 18.1). It is added to the atmosphere by volcanic eruptions and as a product of respiration and combustion, and it is

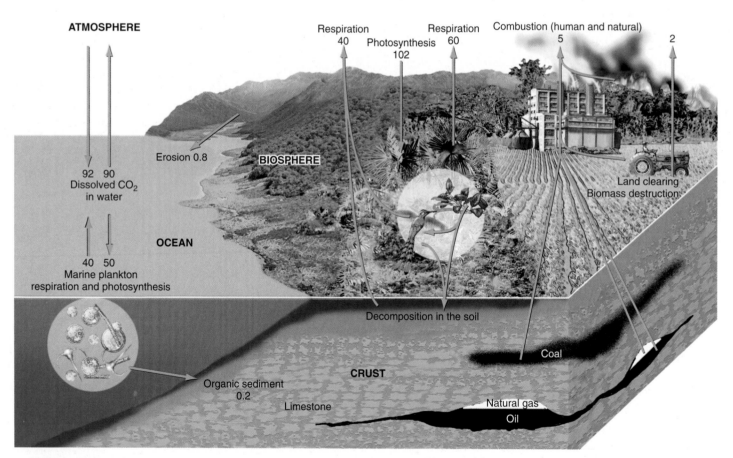

Figure 18.1

Reservoirs and fluxes in the global carbon cycle. Reservoirs are indicated in capital letters, fluxes from one reservoir to another by arrows. Numbers are billions of tons of carbon. While flux changes due to human activities seem relatively small, they can have a perceptible effect on atmospheric CO_2 concentrations.

Source: Intergovernmental Panel on Climate Change (IPCC), 2002.

removed during photosynthesis and by solution in the oceans. It is further removed from the oceans by precipitation in carbonate sediments. The concentration of carbon dioxide in the atmosphere is about 390 ppm, far less than that of oxygen, and carbon dioxide has a correspondingly much shorter residence time, estimated at less than 100 years. Some of the consequences of increased CO_2 content in the atmosphere were explored in chapter 10. One of the factors complicating projections of the greenhouse effect relates to imprecise knowledge of the carbon fluxes between pairs of reservoirs shown in figure 18.1. For example, consider the difficulty of estimating precisely the amount of CO_2 consumed by plants worldwide during photosynthesis. What might seem a far simpler problem, estimating net CO_2 flux between atmosphere and oceans, is not so simple after all, for the extent to which CO_2 dissolves in the oceans varies with temperature, and air and water temperatures, in turn, vary both regionally and seasonally. Still, the broad outline of the global carbon cycle, in terms of reservoirs and processes, is reasonably well understood, as shown. (However, as noted earlier, CO_2 is not the only greenhouse gas, and climate feedbacks can be many and varied.)

The chemically inert gases, including argon and trace amounts of helium and neon, have virtually infinite residence times: they do not, by definition, react chemically with other substances, so they are not readily removed by natural processes. The geochemical cycle of nitrogen is so complex that its overall residence time in the atmosphere is not known, although residence times of individual compounds have been estimated. They vary considerably, from about 44 million years for N_2 to a matter of weeks for the trace amounts of nitric acid (HNO_3). As with many synthetic water pollutants, the fate and residence times of anthropogenic air pollutants, those added by human activity, often are also poorly known. The chlorofluorocarbon compounds, discussed later in this chapter, are just one case in point.

Types and Sources of Air Pollution

Most air pollutants are either gases or particulates (fine, solid particles). The principal gaseous pollutants are oxides of carbon, nitrogen, and sulfur. They share some common sources but create distinctly different kinds of problems. The principal sources for each of the major pollutants and emissions are indicated for the United States in figure 18.2.

Particulates

The **particulates** include soot, smoke, and ash from fuel (mainly coal) combustion, dust released during industrial processes, and other solids from accidental and deliberate burning of vegetation. In other words, particulate air pollutants generated by human activity are largely derived from point sources. Estimates of the magnitude of global anthropogenic contributions vary widely, from about 35 million tons/year (two-thirds from combustion) to 180 million tons/year (mostly industrial). Additional particulates are added by many natural processes,

including volcanic eruptions, natural forest fires, erosion of dust by wind, and blowing salt spray off the sea surface.

Typically neither natural nor anthropogenic particulates have long residence times in the atmosphere. Generally, they are quickly removed by precipitation, usually within days or weeks. In rare cases, such as fine volcanic ash shot high into the atmosphere by violent eruptions, the finest, lightest material may stay in the air for as long as several years. Traditionally, particulate pollution has been considered a local problem, most severe close to its source and for a short time only.

The nature of the problem(s) posed by particulate pollution depends somewhat on the nature of the particulates. Certainly, dense smoke is unsightly, whatever its makeup. In areas showered with ash from a power station or industrial plant, cleaning expenses increase. Fine rock and mineral dust of many kinds has been shown to be carcinogenic when inhaled, and growing evidence suggests that the finest particulates may be especially unhealthy. Many particulates are also chemically toxic. Coal ash may contain both heavy metals and uranium stuck to the particles, for example. The dust was a significant concern after the 2001 destruction of the World Trade Center. This dust was a complex mixture of materials, including glass fibers, concrete fragments, gypsum wallboard, (chrysotile) asbestos insulation, paper, and metal-rich particles. Care had to be taken during cleanup because some of this dust could be harmful if inhaled. Particulates may also have different health and environmental effects depending on their size, so beginning in 1999, the EPA began tracking emissions of both particulates with particle diameters less than 10 microns (about 0.0004 inches) and "fine particulates," defined as those less than 2.5 microns across.

In the United States, particulate emissions from many industrial and commercial sources have been sharply reduced in recent decades (figure 18.3); dust raised by such activities as agriculture, construction, and driving on unpaved roads, all remaining significant sources, is harder to control. Globally, the control of particulate pollution can be a matter of health, aesthetics, and even of climate. Satellites are increasingly being used to track particulates and other pollutants, and even to distinguish between pollutants from natural processes and those resulting from human activities (figure 18.4). Worldwide, particulate pollution in the form of soot is most intense in the air above China, which burns a great deal of coal, and central Africa, where fires are commonly used in agriculture (as in slash-and-burn farming). But that soot drifts widely through the atmosphere, as shown in the chapter-opening image also, even reaching the Arctic (figure 18.5). There, as it settles out of the air to darken the surface, it can affect temperatures and weather patterns, and accelerate melting of Arctic ice. Soot settling in the Himalayas likewise appears to be accelerating the melting of Tibetan glaciers.

Carbon Gases

The principal anthropogenic carbon gases are carbon monoxide and carbon dioxide. Carbon dioxide is not generally regarded as a pollutant *per se*. Some is naturally present in the atmosphere.

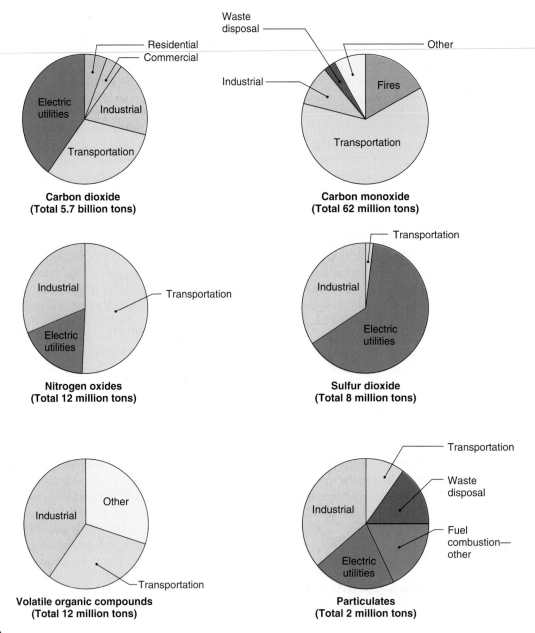

Figure 18.2

Principal sources of U.S. air pollutants. "Other" may include natural sources; particulates total excludes natural dust.

Source: CO₂, 2010 data from Environmental Protection Agency (EPA) Inventory of U.S. Greenhouse Gas Emissions and Sinks, 1990–2010; *other data for 2011, from EPA* National Emissions Inventory 2012.

It is essential to the life cycles of plants, and it is not unhealthful to humans, especially in the moderate concentrations in which it occurs. It is a natural end product of the complete combustion of carbon-bearing fuels:

$$C \ + \ O_2 \ = \ CO_2$$

carbon   oxygen carbon dioxide

As such, carbon dioxide is continually added to the atmosphere through fossil-fuel burning, as well as by natural processes, including respiration by all oxygen-breathing organisms.

At one time, it was believed that natural geochemical processes would keep the atmospheric carbon dioxide level constant, that the oceans would serve as a "sink" in which any temporary excess would dissolve to make, ultimately, more carbonate sediments. It is now apparent that this is not true. As noted in chapter 10, atmospheric carbon dioxide levels have increased about 40% in the last century and a half because of heavy fossil-fuel use, and concentrations continue to climb, spurring concerns about resultant climate change. An estimated 32 billion tons of anthropogenic carbon dioxide are added to the atmosphere each year.

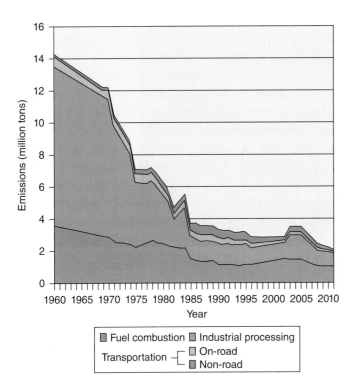

Figure 18.3

U.S. particulate emissions were among the first to be controlled, declining even before the various gases were addressed, but declining more sharply after the Clean Air Acts and establishment of the EPA to monitor compliance.

Source: U.S. Environmental Protection Agency National Emissions Inventory 2012.

Carbon monoxide (CO), while volumetrically far less important than carbon dioxide, is more immediately deadly. It is produced during the incomplete combustion of carbon-bearing materials, when less oxygen is available:

$$2C \quad + \quad O_2 \quad = \quad 2CO$$
carbon   oxygen   carbon monoxide

Anthropogenic carbon monoxide, nearly all of it from fossil-fuel burning, does not persist in the atmosphere for very long. Within a few months, carbon monoxide reacts with oxygen in the air to create more excess carbon dioxide. Actually, human activities add far less carbon monoxide to the atmosphere than do natural sources on a global basis. Nevertheless, it may constitute a serious *local* health hazard, and occasionally a regional one as well (figure 18.6).

Carbon monoxide is an invisible, odorless, colorless, tasteless gas. Its toxicity to animals arises from the fact that it replaces oxygen in the hemoglobin in blood. The vital function of hemoglobin is to transport oxygen through the bloodstream. Carbon monoxide molecules can attach themselves to the hemoglobin molecule in the site that oxygen would normally occupy; moreover, carbon monoxide bonds there more strongly. The oxygen-carrying capacity of the blood is thereby reduced. As carbon monoxide builds up in the bloodstream,

cells (especially brain cells) begin to die from the lack of oxygen, and, eventually, enough cells may fail to cause the death of the whole organism. The difficulty of detecting carbon monoxide, together with the sleepiness that is an early consequence of a reduced supply of oxygen to the brain, may explain the number of accidental deaths from carbon monoxide poisoning in inadequately ventilated spaces.

Carbon monoxide does not remain in the blood indefinitely. If someone in the early stages of carbon monoxide poisoning is removed to fresh air or, if possible, given concentrated medical oxygen, the carbon monoxide is gradually released (though some irrevocable brain damage may have occurred). The problem is to recognize in time what is taking place. Carbon monoxide can build up to dangerous levels wherever combustion is concentrated and air circulation is poor.

In the U.S., as in most industrialized countries, the single largest anthropogenic source of carbon monoxide in the atmosphere, by far, is the automobile (recall figure 18.2). Nonfatal cases of carbon monoxide poisoning have been widely reported in congested urban areas with high traffic density. (Traffic police officers in Tokyo have had to take periodic "oxygen breaks" to counteract the accumulation of carbon monoxide in their blood.) This is one argument in favor of cleaning up auto emissions and making engines burn fuel as efficiently as possible. Another is simply that, if an engine is producing carbon monoxide, it is wasting energy. Burning carbon completely to make carbon dioxide releases three times the energy that is produced from incomplete combustion to carbon monoxide. When fuel is in finite supply, we should try to extract as much energy as possible from our fuels. Some progress in reducing CO emissions has been made. Even if anthropogenic production of carbon monoxide could be eliminated, however, the problems associated with carbon dioxide remain.

Sulfur Gases

The principal sulfur gas produced through human activities is sulfur dioxide, SO_2. More than 50 million tons are emitted worldwide each year. About two-thirds of this amount is released from coal combustion in factories, power-generating plants, and, in some places, home heating units. Most of the rest is released during the refining and burning of petroleum. Within a few days of its release into the atmosphere, sulfur dioxide reacts with water vapor and oxygen in the atmosphere to form sulfuric acid (H_2SO_4), which is a strong and highly corrosive acid:

$$SO_2 \quad + \quad H_2O \quad + \quad \tfrac{1}{2}O_2 =$$
sulfur dioxide   water vapor   oxygen

$$H_2SO_4$$
sulfuric acid

Much of this is scavenged out of the atmosphere in the form of acid rain (discussed later in the chapter) to contribute to acid runoff. This, then, is another example of an air-pollution problem that becomes a water-pollution problem. As long as it

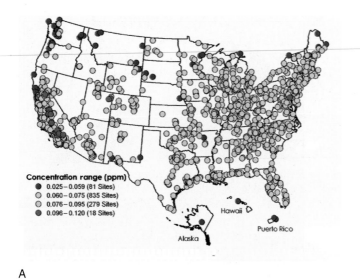

A

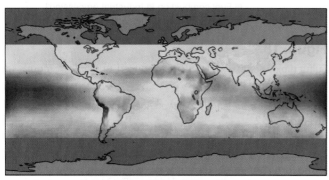

Tropospheric ozone (Dobson units)

12   55

B

Figure 18.8

(A) Snapshot of maximum ozone levels in 2010: for each site, the value is the fourth-highest 8-hour ozone concentration. The EPA's air-quality standard for 8-hour ozone concentration is 0.075 ppm. (B) The higher concentration of cities and industry in the Northern Hemisphere means higher near-surface ozone levels there; once generated, the ozone is swept around the globe by westerly winds. This image is based on satellite data from 1979–2000.

(A) U.S. Environmental Protection Agency, Our Nation's Air–Status and Trends Through 2010. *(B) Image based on data from Jack Fishman, NASA Longley Research Center.*

enriched in ozone, the so-called **ozone layer** (figure 18.9). That ozone can absorb further ultraviolet radiation, shielding the earth's surface from it. **Ultraviolet** (UV) **radiation** can cause skin cancer (it is what makes excessive tanning and sun exposure unhealthy). The presence of the ozone layer decreases that risk. Ultraviolet radiation spans a range of wavelengths shorter than those of visible light (figure 18.10). UVA, the longest-wavelength UV, is the least damaging. UVC, the shortest, would be highly damaging, but it is largely absorbed by oxygen and water vapor in the atmosphere. It is the intermediate-wavelength UVB, also potentially damaging, that

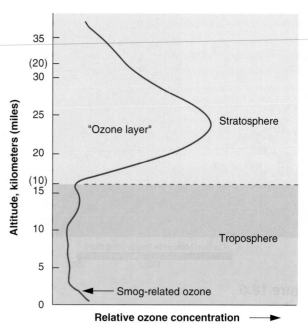

Figure 18.9

About 10% of atmospheric ozone is in the troposphere (the region that accounts for most weather). The other 90% is in the ozone-enriched region of the stratosphere known as the "ozone layer." Even there, ozone concentration is less than half a part per million, at most.

After National Oceanic and Atmospheric Administration.

is particularly absorbed by ozone. Measurements worldwide document that the amount of damaging UV reaching the surface increases as the ozone in the ozone layer decreases (figure 18.11). Scientists of the U.N. Environment Programme estimate that each 1% reduction in stratospheric ozone could result in a 3% increase in nonmelanoma skin cancers in light-skinned people, in addition to increased occurrence of melanoma, blindness related to development of cataracts, gene mutations, and immune-system damage.

Normally, oxygen (O_2) molecules in the stratosphere absorb UVC radiation, which splits them into two oxygen atoms, each of which can attach to another O_2 molecule to make O_3. The O_3 molecule, in turn, absorbs UVB radiation, which splits it into O_2 plus a free oxygen atom, and the latter, colliding with another O_3, causes regrouping back into two O_2 molecules. Over time, a balance between ozone creation and destruction develops.

The concentration of ozone in the ozone layer varies seasonally and with latitude (figure 18.12A). Ozone production rates are most rapid near the equator, as a consequence of the strong sunlight there, but generally, the total ozone in a vertical column of air increases with latitude (that is, increases toward the poles) as a result of the balance between natural ozone production and destruction rates, and atmospheric circulation patterns. Thus, UVB exposure is typically higher at low latitudes and lower near the poles, as seen in figure 18.12B.

It has been recognized for decades that the concentration of ozone over Antarctica decreases in the Antarctic winter. However, in the 1980s, it became apparent that the extent of

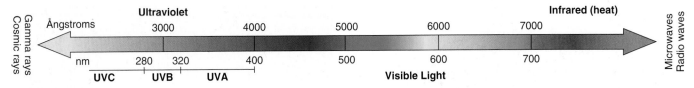

Figure 18.10

Ultraviolet radiation is to the shorter-wavelength, higher-energy side of the visible-light spectrum. 1 nm (nanometer) = 10^{-9} meters = one-billionth of a meter = 10 Å (Ångstroms)

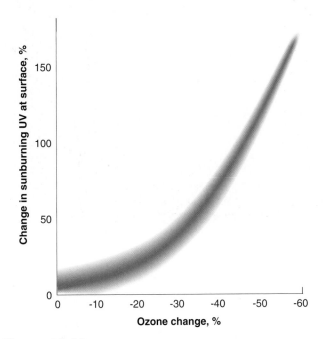

Figure 18.11

As ozone above thins, more damaging UV radiation reaches the surface.

After World Meteorological Organization, UNEP, and NOAA.

the depletion was becoming more pronounced each year (figure 18.13). This extreme thinning of the protective shield of stratospheric ozone came to be described (somewhat misleadingly) as an **ozone hole.** It is really just a roughly circular region over which stratospheric ozone is relatively depleted, or sparser than over the surrounding area. (Think of a foggy area with a brighter patch in which the fog has dissipated somewhat.)

Measurements showed unexpectedly high concentrations of reactive chlorine compounds in the air of that region, compounds that could have been derived from anthropogenic chlorofluorocarbons (CFCs). A clear negative correlation was found between atmospheric concentrations of ClO and of ozone (O_3) in the Arctic. While correlation does not demonstrate cause-and-effect, it was highly suggestive, and no more convincing explanation was forthcoming. Further studies clarified the role of the CFCs.

The chemistry of ozone destruction is complex. It appears that CFCs remain in the atmosphere for up to a century, slowly migrating upward and breaking down over time. Chlorine atoms are freed by this CFC decomposition, which seems to occur especially readily in polar clouds during the winter, when ice crystals provide a surface on which the key chemical reactions can take place. The chlorine reacts with ozone to form chlorine monoxide (ClO). Moreover, a single chlorine atom can destroy many ozone molecules, as seen in figure 18.14.

Reduction in ozone has also been detected over the Northern Hemisphere (figure 18.15), and in March 2011 a pronounced "ozone hole" was detected over the North Pole. It too was correlated with ClO concentrations in the stratosphere (figure 18.16), supporting the CFC link. The development of this hole was attributed to prolonged, unusually cold temperatures in the stratosphere that winter. If the stratosphere continues to cool, as discussed in Case Study 10, we may see more such episodes.

Recognizing the evident threat of ozone destruction by CFCs, the global community banded together to support the Montreal Protocol on Substances that Deplete the Ozone Layer. First adopted in 1987, strengthened in 1990, and signed by 175 nations, the Montreal Protocol stipulated phaseout of production of CFCs and other ozone-depleting compounds by the year 2000. One remaining concern is that some populous developing countries have been reluctant to sign, so the long-term effectiveness of the Montreal Protocol remains to be seen. Another concern is that even in developed nations, there has been some delay in the CFC phaseout; and the compounds being substituted for CFCs themselves can contribute to ozone depletion, though not as strongly. Also, the residence times of CFCs and their breakdown products are years to decades, so their effect on the atmosphere continues long after their production ceases. And old air conditioners and refrigerators continue to leak CFCs. In 2011 some researchers suggested that they could see evidence that recovery of the ozone in the stratosphere is beginning. However, recent projections suggest that the ozone layer may not recover fully until about 2065 in the mid-latitudes, later in Antarctica.

Lead

One air pollutant that has been greatly reduced is lead. Most of the lead once released into the atmosphere was emitted by automobile exhaust systems. While lead is not naturally a significant component of petroleum, one particular lead compound, tetraethyl lead, had been a gasoline additive since the 1940s as an antiknock agent to improve engine performance.

Lead is one of the heavy metals that accumulate in the body. It has a variety of harmful effects, including brain damage in high concentrations. Mild lead poisoning in the nervous system can cause depression, nervousness, apathy, and other psychological disorders, as well as learning difficulties. Acute lead

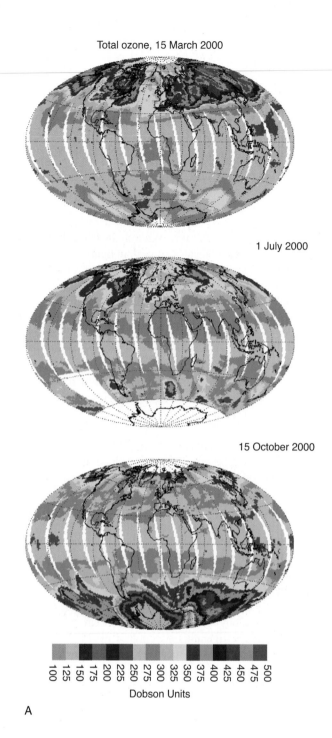

Total ozone, 15 March 2000

1 July 2000

15 October 2000

100 125 150 175 200 225 250 275 300 325 350 375 400 425 450 475 500
Dobson Units

A

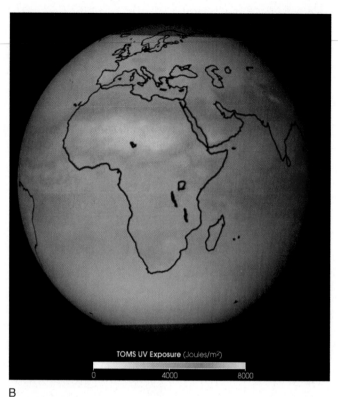

TOMS UV Exposure (Joules/m²)
0   4000   8000

B

Figure 18.12

(A) Typical seasonal variations in global ozone distribution as measured by NASA's Total Ozone Mapping Spectrometer (TOMS). 1 Dobson unit = 0.001 cm thickness of ozone (at standard temperature and pressure) distributed over the whole vertical column of atmosphere. White areas indicate lack of data. (B) December 2000 estimated UVB exposure based on a month of ozone measurements by TOMS.

(A) Images courtesy NASA. (B) Data courtesy TOMS science team, NASA Goddard Space Flight Center.

engine parts and decrease their useful life.) Also, illegal use of leaded gasoline in cars equipped with catalytic converters destroys the converters' effectiveness and results in the emission of more unburned hydrocarbons, carbon monoxide, and other pollutants. Beginning in the early 1970s, the Environmental Protection Agency began to mandate reductions in the lead content of gasoline, with lead phased out almost completely from automotive fuel by 1987. Lead consumed in gasoline dropped 70% from 1975 to 1982. Concentrations of atmospheric lead decreased correspondingly, by about 65% over the same period, and this, in turn, was reflected in greatly reduced lead levels in blood (figure 18.17A). For a time, a limited quantity of leaded fuel remained available, primarily for the benefit of old engines in farm equipment that for technical reasons need some lead to protect their valves. By 1996, a complete ban on leaded gasoline took effect, and lead emissions have been virtually eliminated (figure 18.17B).

Unfortunately, this is not yet quite the end of the lead-pollution story. Studies have shown that lead from gasoline emitted in vehicle exhaust was commonly deposited in and on

poisoning from exhaust fumes alone has not been documented. However, children in urban areas who breathed lead-laden air and also consumed chips of old lead-based paint indeed could develop high and harmful lead levels in their blood. It has been estimated that 5 to 10% of inner-city children may have suffered from lead poisoning.

Lead levels allowed in most paints have been greatly reduced, and lead can be left out of gasoline. Equally good engine performance can be obtained from higher-octane gasolines at the cost of a few more cents a gallon. (In the long run, in fact, leaded gasoline could be the more costly because some lead compounds produced when leaded gasoline is burned coat

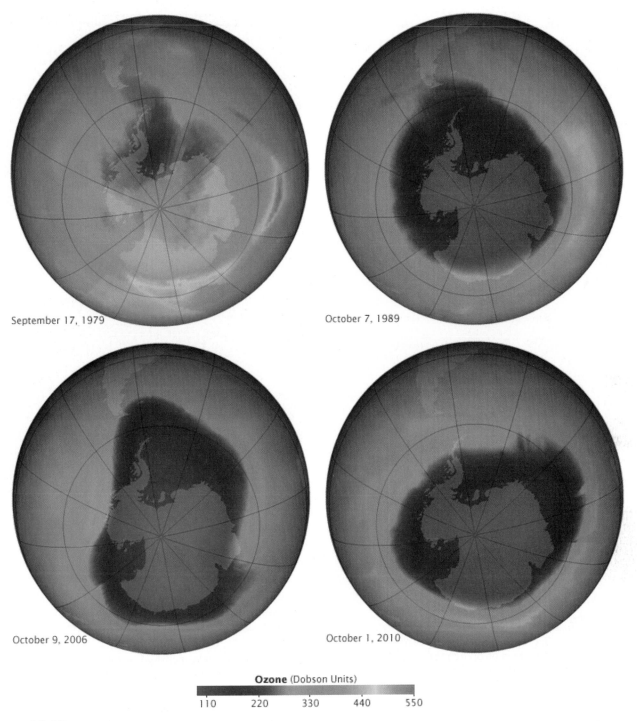

Figure 18.13

Measurements of ozone over the Southern Hemisphere showed a deepening "hole" beginning in 1979. The seasonal ozone depletion continued to intensify for years after the banning of CFCs; 2006 showed the deepest hole yet. By now, measurements suggest, the situation has stabilized, but ozone recovery is yet to come.

NASA images courtesy NASA Ozone Hole Watch.

soils adjacent to highways, reaching concentrations of hundreds of parts per million along busy urban streets. From there it may be leached into surface or ground water, transported with the soil in runoff, or simply stirred up with fine dust, to be redistributed or even inhaled. Even now, significant numbers of urban children still develop unhealthy levels of lead in their blood, with values higher in summer when they are more exposed to soil outdoors. So lead may no longer be a notable

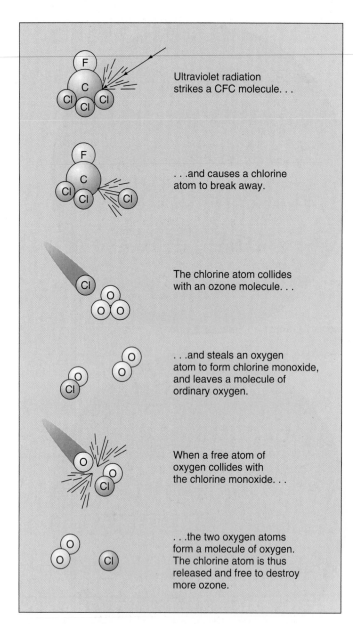

Figure 18.14

Schematic of the role CFCs are believed to play in ozone destruction. This would involve destruction of ozone beyond what normally occurs by interaction with ultraviolet light.

Images courtesy NASA.

Ultraviolet radiation strikes a CFC molecule. . .

. . .and causes a chlorine atom to break away.

The chlorine atom collides with an ozone molecule. . .

. . .and steals an oxygen atom to form chlorine monoxide, and leaves a molecule of ordinary oxygen.

When a free atom of oxygen collides with the chlorine monoxide. . .

. . .the two oxygen atoms form a molecule of oxygen. The chlorine atom is thus released and free to destroy more ozone.

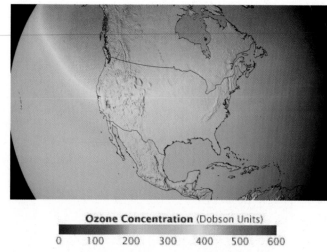

Ozone Concentration (Dobson Units)

0   100   200   300   400   500   600

A

B

Figure 18.15

Northern Hemisphere ozone in spring 1974 (A) and 1994 (B) shows clear decline, though concentrations are still far higher than over Antarctica.

NASA images courtesy Goddard Space Flight Center Scientific Visualization Studio.

air pollutant, but clearly, it will be some years yet before the residue of the tetraethyl lead from gasoline ceases to be a pollution concern.

Other Pollutants

Volatile, easily vaporized organic compounds are a major component of air pollution. Their principal sources are (1) transportation, especially automobiles emitting unburned gasoline, and (2) industrial processes, which release a variety of organics including unburned hydrocarbons and volatile organic solvents. Unburned hydrocarbons are not themselves highly toxic, but they play a role in the formation of photochemical smog and may react to form other compounds that irritate eyes and lungs.

Other air pollutants are of more localized significance. Heavy metals other than lead, for example, can be a severe problem close to mineral-smelting operations. Such elements as mercury, lead, cadmium, zinc, and arsenic may be emitted either as vapors or attached to particulate emissions from smelters. While they are quickly removed from the atmosphere by precipitation, the metals may then accumulate in local soils, from which they are concentrated in organic matter, including growing plants. They can constitute a serious health hazard if accumulated in food or forage crops.

Finally, there are air pollutants of particular concern only indoors; see Case Study 18.

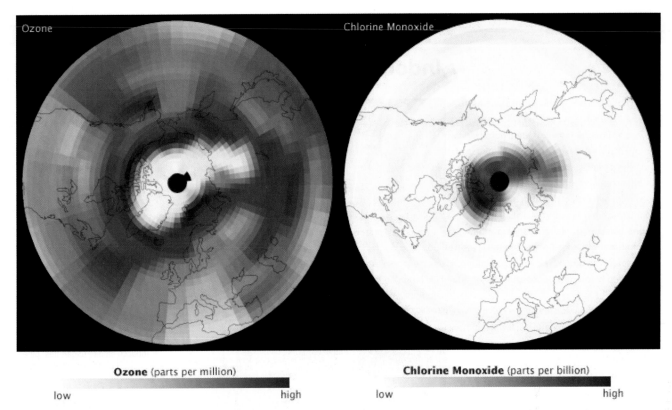

Ozone (parts per million)

low   high

Chlorine Monoxide (parts per billion)

low   high

Figure 18.16

March 2011 "ozone hole" over the Arctic, a product of ClO action in an extended period of severe cold in the stratosphere. Clearly, high ClO concentrations correspond to areas of most profound ozone loss.

NASA image by Eric Nash and Robert Simmon using data from the Aura Microwave Limb Sounder Team.

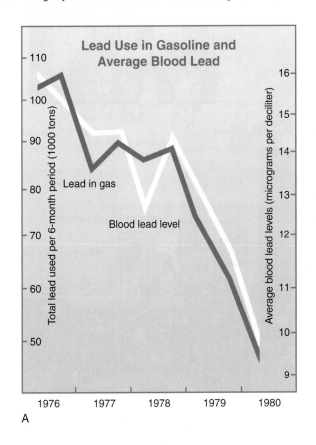

A

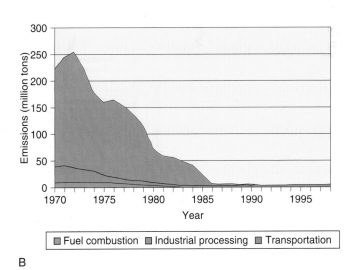

B

Figure 18.17

(A) Declines in lead used in gasoline were directly reflected in lower lead levels in blood. (B) By now, lead emissions are negligible.

(A) After L. Whiteman, "Trends to Remember: The Lead Phase Down," EPA Journal, *May/June 1992, p. 38. (B) From U.S. Environmental Protection Agency* National Emissions Trends 1990–1998.

Acid Rain

Acidity is reported on the **pH scale.** The pH of a solution is inversely proportional to the hydrogen-ion (H^+) concentration in the solution; the pH scale, like the Richter magnitude scale, is a logarithmic one (figure 18.18). Neutral liquids, such as pure water, have a pH of 7. Acid substances have pH values less than 7; the lower the number, the more acidic the solution. Typical pH values of common household acids like vinegar and lemon juice are in the range of pH 2 to 3. Alkaline solutions, like solutions of ammonia, have pH values greater than 7. All natural precipitation is actually somewhat acidic as a result of the solution of gases (for instance, CO_2) that make acids (such as carbonic acid, H_2CO_3). However, carbonic acid is what is termed a weak acid, meaning that a relatively small proportion of H_2CO_3 molecules dissociate to release the H^+ ions that contribute to acidity. **Acid rain** is rain that is more acidic than normal. While many gases in the air contribute to the acidity of rain, discussions of acid rain focus on the sulfur gases that react to form atmospheric sulfuric acid, a strong acid that dissociates extensively.

As noted in chapter 14, acid rain can contaminate water supplies. It can damage structures and outdoor sculptures through its corrosive effects (figure 18.19). Plants may suffer defoliation, or their growth may be stunted as changes in soil-water chemistry reduce their ability to take up key nutrients such as calcium. Fish may be killed in acidified streams; or, the adult fish may survive but their eggs may not hatch in the acidified waters. Acidic water more readily leaches potentially toxic metals from soils, transferring them to surface- or groundwater supplies, and reduces the capacity of soils to neutralize further additions of acid. Acid water leaching nutrients from lake-bottom sediments may contribute to algal bloom. Impacts of acid rain on ground water were also explored in chapter 17. Studies have found a correlation between rainfall acidity and potentially toxic levels of mercury in lakes in northern North America. Other metals that may be linked

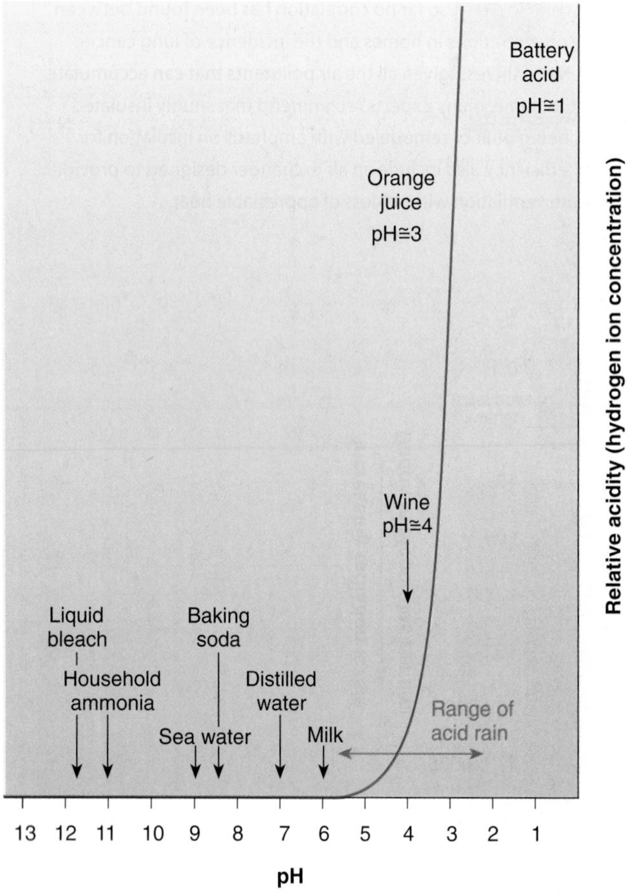

A

B

Figure 18.19

(A) Closeup of a grotesque decorating a bell tower shows erosion and pitting of the limestone. (Neck of figure has also been patched.) (B) Detail of church doorway in Bordeaux, France, shows extreme acid-rain damage to limestone (left and center); column at right has been restored/replaced, but it is impossible to recover all the detail of the original, given the extent of damage.

Figure 18.18

The pH scale is a logarithmic one; small differences in pH value translate into large differences in acidity. Range of acid precipitation is compared with typical pH values of common household substances.

to acid rain include lead, zinc, selenium, copper, and aluminum; recall that acid mine drainage is correlated with a higher concentration of leached metals. The accumulative properties of the heavy metals, in turn, raise concerns about toxic levels in fish and consequent risks to humans and other fish-eaters.

Regional Variations in Rainfall Acidity and Impacts

Rainfall is demonstrably more acidic in regions downwind from industrial areas releasing significant amounts of sulfur among their emissions. These observations form the basis for decisions to control sulfur pollution. What is less clear is the precise role that anthropogenic sulfur plays in the sulfur cycle.

For one thing, there is little information about natural background levels of sulfur dioxide (SO_2) and related sulfur species. Historically, pure rainwater has been assumed to have a pH of about 5.6 in areas away from industrial sulfur sources. This is the pH to be expected as a consequence of the formation of carbonic acid due to natural carbon dioxide in the air. Recent research suggests that natural background pH may be somewhat lower as a result of other chemical reactions in the atmosphere and may vary from place to place as a consequence of localized conditions. Natural rainfall acidity may correspond to a pH as low as 5. However, air over all the globe is now polluted to some extent by human activities. Therefore, the chemistry of precipitation in wholly unpolluted air cannot be determined precisely enough to assess anthropogenic impacts fully. It does appear that, overall, more sulfate is added to the air by sea spray containing dissolved sulfate

minerals than by human activities. Perhaps anthropogenic sulfate only locally aggravates an existing situation.

It may, in fact, be the case that an increase in rainfall acidity due to the formation of nitric acid (another strong acid) from nitrogen oxides formed in internal-combustion engines is a much more significant problem in and near urban areas. The focus on sulfuric acid in rain partially reflects the fact that sulfur gas pollution from stationary sources is more readily controllable and partially recognizes the more long-range and global impact of anthropogenic sulfuric acid, including the political aspects of the issue (see below).

The nature of the local geology on which acid rain falls can strongly influence the severity of the acid's impact. Chemical reactions between rain and rock are complex. Farmers and gardeners have known for years that certain kinds of rock and soil react to form acidic pore waters, while others yield alkaline water. For instance, because limestone reacts to make water more alkaline, it serves, to some extent, to neutralize acid waters. The effects of acid rain falling on carbonate-rich rocks and soils are therefore moderated. Conversely, granitic rocks and the soils derived from them are commonly more acidic in character. They cannot buffer or moderate the effects of acid rain; the acid rain just makes the surface and subsurface waters more acidic in such regions. Some operators of coal-fired plants in the Midwest have argued that the predominantly granitic soils of the Northeast are principally at fault for the existence of very acid lakes and streams in the latter region. Certainly, the geology does not *reduce* the problem. However, the rainfall in the Northeast, with an average pH around 4.5, is clearly acidified (figure 18.20).

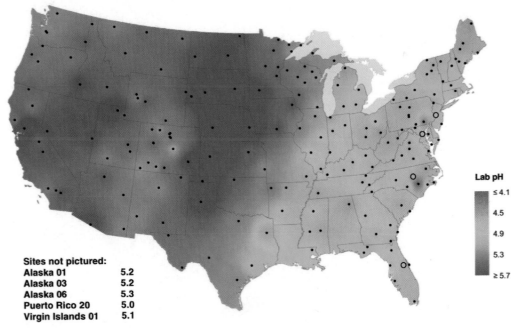

Sites not pictured:

Alaska 01	5.2
Alaska 03	5.2
Alaska 06	5.3
Puerto Rico 20	5.0
Virgin Islands 01	5.1

Lab pH

≤ 4.1
4.5
4.9
5.3
≥ 5.7

Figure 18.20

Rainfall is acidic everywhere, but in the United States, it is especially so in the northeast (2010 data; pH as measured in the lab). Dots are sampling locations.

Map from National Atmospheric Deposition Program (NRSP-3) (2012). NADP Program Office, Illinois State Water Survey, 2204 Griffith Dr., Champaign, IL 61820.

Nor is the northeastern United States the only area concerned about sulfuric acid generated in this country. The Canadian government has expressed great concern over the effects of acid rain caused by pollutants generated in the United States—especially the industrialized upper Midwest and Northeast—that subsequently drift northeast over eastern Canada. Much of that area, too, is underlain by rocks and soils highly sensitive to acid precipitation and is already subject to acid precipitation associated with Canadian sulfate sources. In Europe, acid rain produced by industrial activity in central Europe and the United Kingdom falls on Scandinavia. Acid rain, then, has become an international political concern.

More information is needed about natural water chemistry in various (unpolluted) geologic settings before the role of acid rain can be evaluated fully. The science of pinpointing sulfur sources is also becoming more sophisticated as means are developed to "fingerprint" those sources by the assemblage of gases present and their composition. This will allow cause-and-effect relationships to be determined more accurately.

Negative impacts of acid rain are certainly well documented. Increasingly, damage from related phenomena—acid snow and even acid fog—are being recognized. In snowy regions with acid precipitation, rapid spring runoff of acid snowmelt can create an "acid shock" to surface waters and their associated communities of organisms. Where sulfur gas release has been reduced, some recovery of water quality in lakes and streams has occurred. Control of sulfur emissions is plainly desirable. Some strategies are discussed later in this chapter.

Air Pollution and Weather

Thermal Inversion

Air-pollution problems are naturally more severe when air is stagnant and pollutants are confined. Particular atmospheric conditions can contribute to acute air-pollution episodes of the kinds mentioned at the beginning of this chapter. A frequent culprit is the condition known as a **thermal inversion** (figure 18.21). Within the lower atmosphere, air temperature normally decreases

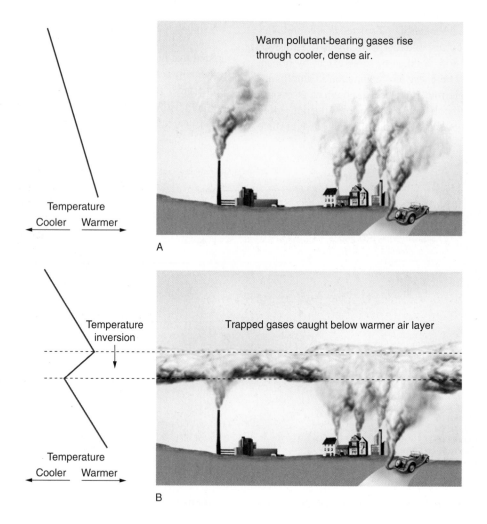

Figure 18.21

Effect of thermal inversion in trapping air pollution. Lines at left indicate relative temperature at various altitudes. (A) Normal conditions permit warm pollutants to rise and disperse through cooler air above. (B) Inversion traps warm pollutant gases under or in a stable warm-air layer.

as altitude increases. This is why mountaintops are cold places, even though direct sunshine is stronger there. In a thermal inversion, there is a zone of relatively warmer air at some distance above the ground. That is, going upward from the earth's surface, temperatures decrease for a time, then increase in the warmer layer (inversion of the normal pattern), then ultimately continue to decrease at still higher altitudes. Inversions may become established in a variety of ways. Warmer air moving in over an area at high altitude may move over colder air close to the ground, thus creating an inversion. Rapid cooling of near-surface air on a clear, calm night may also lead to a thermal inversion.

Most air pollutants, as they are released, are warmer than the surrounding air—exhaust fumes from automobiles, industrial smokestack gases, and so on. Warm air is less dense than colder air, so ordinarily, warm pollutant gases rise and keep rising and dispersing through progressively cooler air above. When a thermal inversion exists, a warm-air layer becomes settled over a cooler layer. The warm pollutant gases rise only until they reach the warm-air layer. At that point, the pollutants are no longer less dense than the air above, so they stop rising. They are effectively trapped close to the ground and simply build up in the near-surface air. Sometimes, the cold/warm air boundary is so sharp that it is visible as a planar surface above the polluted zone (figure 18.22). The Donora, Pennsylvania, episode mentioned at the beginning of the chapter was triggered by an inversion. Its effect on a freight train's smoke was described by physician R. W. Koehler: "They were firing up for the grade and the smoke was belching out, but it didn't rise.... It just spilled out over the lip of the stack like a black liquid, like ink or oil, and rolled down to the ground and lay there" (Roueché, p. 175).

Every one of the half-dozen major acute air-pollution episodes of the twentieth century has been associated with a thermal inversion, as have many milder ones. Certainly, air pollution can be a health hazard in the absence of a thermal inversion. The inversion, however, concentrates the pollutants more strongly. Moreover, inversions can persist for a week or longer, once established, because the denser, cooler air near the ground does not tend to rise through the lighter, warmer air above.

Certain geographic or topographic settings are particularly prone to the formation of thermal inversions (figure 18.23).

Figure 18.22

Sunset view looking southwest across upstate New York clearly shows a layer of smog trapped below clouds (white) by a thermal inversion.

Image courtesy of Earth Sciences and Image Analysis Laboratory, NASA Johnson Space Center

Los Angeles has the misfortune to be located in such a spot. Cool air blowing across the Pacific Ocean moves over the land and is trapped by the mountains to the east beneath a layer of warmed air over the continent. Donora, Pennsylvania, lies in a valley. When a warm front moves across the hilly terrain, some pockets of cold air may remain in the valleys, establishing a thermal inversion. Germany's coal-rich Ruhr Valley suffered so from pollution trapped by a prolonged inversion during the winter of 1984–1985 that schools and many factories had to be closed and the use of private automobiles banned until the air

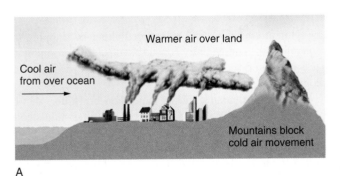

A

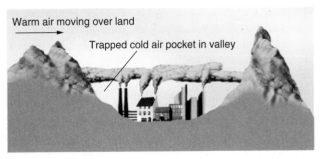

B

Figure 18.23

Influence of topography on establishing and maintaining thermal inversions. (A) Cool sea breezes blow over Los Angeles and are trapped by the mountains to the east. (B) Cold air, once settled into a valley, may be hard to dislodge.

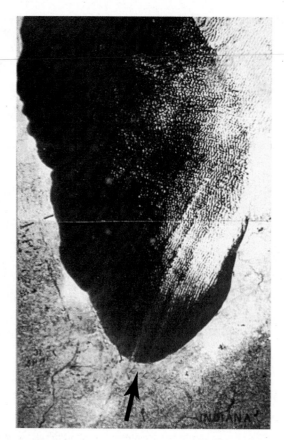

Figure 18.24

Air-pollution plumes from the Chicago, Illinois, and Gary/Hammond, Indiana, area bring clouds and snow to the area to the northeast. Black area is Lake Michigan; arrow points to source of pollution and initial trend.

Photograph courtesy NASA.

cleared. As with most weather conditions, nothing can be done about a thermal inversion except wait it out.

Impact on Weather

While weather conditions such as thermal inversions can affect the degree of air pollution, air pollution, in turn, can affect the weather in ways other than reducing visibility, modifying air temperature, and making rain more acidic. This is particularly true when particulates are part of the pollution. Water vapor in the air condenses most readily when it has something to condense on. This is the principle behind cloud seeding: Fine, solid crystals are spread through wet air, and water droplets form on and around these seed crystals. Particulate pollutants can perform a similar function.

This is illustrated in figure 18.24. Southwest of Lake Michigan lies the industrialized area of Chicago, Illinois, and Gary and Hammond, Indiana. The Gary/Hammond area, especially, is a major steel milling region. At the time this satellite photograph was taken, waste gases from the mills were being blown northeast across the lake (note faint smoke plumes at bottom end of lake, just left of center). At first, they remained almost invisible, but as water vapor condensed, clouds appeared and, in this scene,

snow crystals eventually formed. A large area of snow in Michigan is on a direct line with the pollution plumes from the southwest. The particulate pollutants in the exhaust gases facilitate the condensation and precipitation.

Recent studies have suggested that the role of aerosols in weather is complex, and their effects can vary depending on their type and size. Dark particulates can absorb sunlight and warm an area, while light-colored particulates and droplets may promote cooling by reflecting sunlight back into space. These temperature effects can influence air-circulation patterns, water evaporation, and cloud formation, and thus precipitation patterns. Studies in China have implicated sooty particulate pollution in more-frequent flooding in south China and droughts in north China. NASA researchers have also proposed that dust from Saharan dust storms, carried across the Atlantic, was partly responsible for a relatively quiet hurricane season in 2006: The dust may have reduced the warming of surface water that supplies energy to storm systems, and suppressed convection of water vapor upward in the atmosphere to precipitate as rain. So, in some special circumstances, particulate pollution can be a good thing.

Toward Air-Pollution Control

Air-Quality Standards

Growing dissatisfaction with air quality in the United States led, in 1970, to the Clean Air Act Amendments, which empowered the Environmental Protection Agency to establish and enforce air-quality standards (see also chapter 19). The standards developed are designed to protect human health, clear the visible pollution from the air, and prevent crop and structural damage and other adverse effects. When meteorologists and others report on local air quality, they are referring to the Air Quality Index, developed by EPA to translate pollutant concentrations into descriptors that the general public can readily understand. The AQI is illustrated in figure 18.25. On any given day, the AQI may be different for different pollutants. Overall air quality is assigned the descriptors and color applicable to the pollutant with the highest (worst) AQI that day. The AQI provides a quick way to assess air quality and improvements over time (figure 18.26).

Control Methods

Air-pollutant emissions can be reduced either by trapping the pollutants at the source or by converting dangerous compounds to less harmful ones prior to effluent release. A variety of technologies for cleaning air exist.

Where particulates are the major concern, filters can be used to clear the air, with the fineness of the filters adjusted to the size range of particles being produced. Filters may be 99.9% efficient or better, but they are commonly made of paper, fiber, or other combustible material. These cannot be used with very high-temperature gases.

An alternate means of removing particulates is electrostatic precipitators or "scrubbers." In these, high voltages charge the particulates, which are then attracted to and caught on

Air Quality Index (AQI) Value	Descriptor	Meaning
0–50	Good	Air quality is satisfactory and air pollution poses little or no risk.
51–100	Moderate	Air quality is acceptable, but a few persons unusually sensitive to the pollutant may experience negative health effects.
101–150	Unhealthy for sensitive groups	While the general public is not likely to be affected, members of sensitive groups are more likely to experience health effects. (For example, ozone at this level may be problematic for those with lung disease; those with lung or heart disease are at risk from particulates.)
151–200	Unhealthy	The population at large may experience negative health effects, and impact on sensitive groups will be more severe.
201–300	Very Unhealthy	Everyone may experience more-serious health effects.
> 300	Hazardous	This AQI level triggers health warnings of emergency conditions. The whole population is more likely to be affected.

Figure 18.25

For each of six major air pollutants regulated by the Clean Air Act Amendments (ground-level ozone, particulates, carbon monoxide, sulfur dioxide, nitrogen dioxide, and lead), an air-quality standard (concentration or exposure limit) has been set. An AQI = 100 corresponds to the standard level for the particular pollutant. Various ranges of AQI are assigned different descriptors based on likely health effects, and color codes that allow the general public to understand easily the corresponding degree of health risk.

Adapted from U.S. Environmental Protection Agency.

charged plates. Efficiencies are usually between 98 and 99.5%. These systems are sufficiently costly that they are practical only for larger operations. They are widely used at present in coal-fired electricity-generating plants.

Wet scrubbers are another possibility. In these scrubbers, the gas is passed through a stream or mist of clean water, which removes particulates and also dissolves out some gases. Wet chemical scrubbing, in which the gas is passed through a water/chemical slurry rather than pure water, is used particularly to clean sulfur gases out of coal-fired plant emissions. Currently, the most widely used designs use a slurry of either lime (calcium oxide, CaO) or limestone (CaCO$_3$), which reacts with and thus removes the sulfur gases. Wet scrubbing produces contaminated fluids, which present equipment-cleaning and/or waste-disposal problems.

Both carbon monoxide and the nitrogen oxides are best controlled by modifying the combustion process. Lower combustion temperatures, for example, minimize the production of nitrogen oxides and moderate the production of carbon monoxide. The use of afterburners to complete combustion is another way to reduce carbon monoxide emission, by converting it to carbon dioxide.

Combustion is an effective way to destroy organic compounds, including hydrocarbons, producing carbon dioxide and water. More-complex scrubbing or collection procedures for organics also exist but are not discussed in detail here. Such measures are necessary for low-temperature municipal incinerators burning materials that might include toxic organic chemicals that can only be thoroughly decomposed in the high-temperature incinerators designed for toxic-waste disposal.

Altogether, the United States spends over $100 billion a year on air-pollution abatement. We have already seen some of the results in reduced emissions. Improvements have been seen in

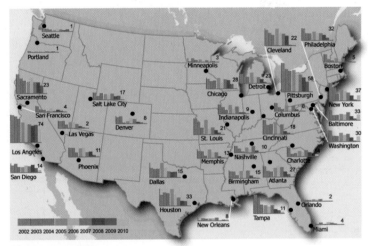

Figure 18.26

Number of days per year on which overall AQI was over 100, for selected cities. For most, air quality continues to improve as standards are tightened and new pollution-control techniques are implemented.

U.S. Environmental Protection Agency, Our Nation's Air—Status and Trends Through 2010.

deposition as well. Phase I of the 1990 reauthorization of the Clean Air Act Amendments targeted a number of sources of sulfur pollution, with a view to reducing acid rain. Sulfate and acidity (hydrogen ion concentration) both showed substantial reductions downwind of Phase I targets. Reductions in acid precipitation, in turn, have resulted in reduced acidity of surface waters in the northeastern United States. Although concentrations of both nitrate and sulfate in precipitation remain somewhat elevated locally, both have shown notable declines across the U.S. (figure 18.27).

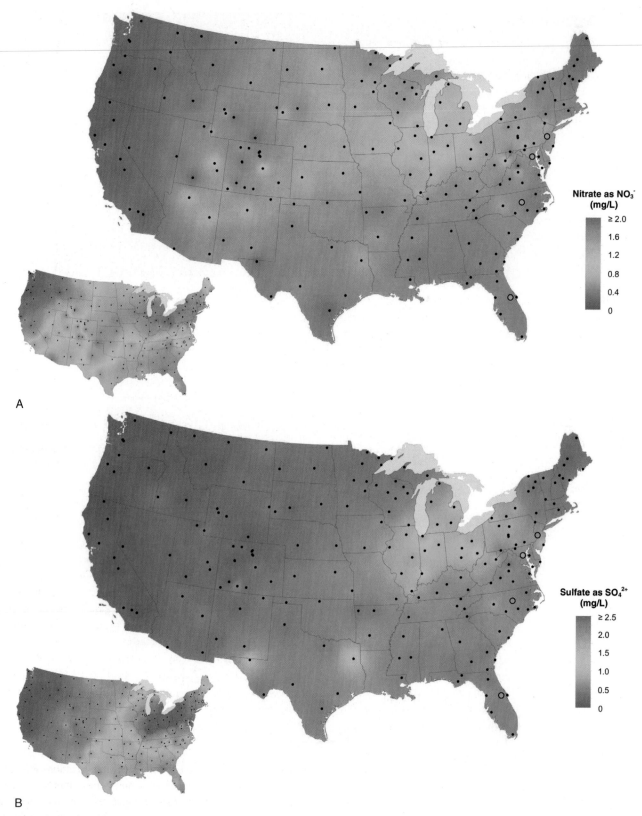

Figure 18.27

(A) Nitrate (NO_3^-) concentration in precipitation (mg/L) reflects particularly nitrogen-oxide emissions from urban and industrial areas upwind; (B) sulfate (SO_4^{2-}) forms from sulfur-dioxide emissions. Large maps show 2010 data; inset maps, 1994. Dots represent sampling locations.

Maps from National Atmospheric Deposition Program (NRSP-3) (2012). NADP Program Office, Illinois State Water Survey, 2204 Griffith Dr., Champaign, IL 61820.

Automobile Emissions

As part of the efforts to reduce air pollution, the Environmental Protection Agency has imposed limits on permissible levels of carbon monoxide and hydrocarbon emissions from automobiles. Many car manufacturers use catalytic converters to meet the standard. In a catalytic converter, a catalyst (commonly platinum or another platinum-group metal) enhances the oxidation, or combustion, of the hydrocarbons and carbon monoxide to water and carbon dioxide. In this process, catalytic converters are quite effective. Unfortunately, they also promote the production of sulfuric acid from the small quantities of sulfur in gasoline, as well as the formation of nitrogen oxides. Emissions of these compounds may be *increased* by the use of catalytic converters. While the increased sulfuric acid emission by vehicles using catalytic converters is minor compared to the output from coal-fired power plants, it nevertheless makes automobile exhaust more irritating to breathe where exhaust fumes are confined. The environmental impacts of the nitrogen oxides may be more negative than previously assumed, which may dictate a need to control these more carefully as well.

Improvements in fuel economy reduce all emissions simultaneously, of course, as less fuel is burned for a given distance traveled. The National Highway Transportation Safety Administration has been setting fuel economy standards for new cars since 1978, and for new light trucks (pickups, vans, and SUVs) since 1979—all motivated by the Arab oil embargo of the 1970s. Raising fuel efficiency of new vehicles gradually raises the average fuel efficiency of all vehicles on the roads, but the latter always lags behind the new-vehicle standards. For model year 1979, the new-vehicle standards were 19 mpg for cars and 17.2 mpg for light trucks, but the averages for all vehicles on the road that same year were 14.6 and 11.6 mpg, respectively. From 1990 to 2010, the standard for new passenger cars was 27.5 mpg, but the average fuel economy for all cars in 1990 was only 20.2 mpg, rising to 22.6 mpg by 2008. Furthermore, the depressed fuel prices of the mid-1980s contributed to a rebound in the popularity of larger, less fuel-efficient cars and the rising popularity of vans and sport utility vehicles. The new-vehicle standard for these (classified as light trucks) was only 20.0 mpg in 1990 and 22.5 mpg in 2008, while average fuel economy of all light trucks on the road was 16.1 mpg in 1990 and just 18.1 mpg in 2008. Still, the stricter fuel-economy requirements are slowly improving overall average vehicle mileage, and additional limits on emissions (independent of mileage ratings) for both cars and trucks have helped bring total air pollutant emissions down.

Carbon Sequestration

Historically, CO_2 has not been a pollutant monitored by EPA for public-health reasons (though recent legal decisions may change this in future). Clearly, however, it is a concern from a climatological point of view. Recently, there has been increased interest in various methods of **carbon sequestration,** also known as *carbon capture and storage:* storing carbon in some

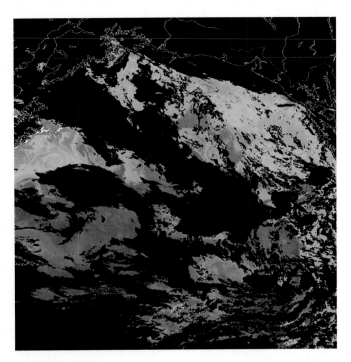

Figure 18.28

In this NASA satellite image of part of the Gulf of Alaska, brighter colors indicate higher chlorophyll levels, in turn reflecting abundance of phytoplankton (algae). Small bright patch at bottom center is algal bloom resulting from a 2002 iron-fertilization experiment.

Image courtesy Jim Gower, Bill Crawford, and Frank Whitney of Institute of Ocean Sciences, Sidney BC; the IOS SERIES team; and NASA.

reservoir to remove it from atmospheric circulation, using physical or biological means.

Enhancement of natural carbon sinks is one approach. Reforestation is one example; a lush rain forest's vegetation contains much more carbon than a typical grassland or field of cultivated crops. A more novel strategy is ocean fertilization to promote growth of algae, the idea being that as the algae die and sink deep into the ocean, they will take their carbon with them. Small experiments have shown that indeed, fertilizing sea surface water with iron, a key algal nutrient often in short supply, can promote the desired algal bloom (figure 18.28). Remaining uncertainties include the potential impacts on broader ocean ecosystems, and the long-term fate of the carbon submerged with the algae.

To address the voluminous CO_2 output of industrial processes, a variety of physical methods are being considered for carbon dioxide capture and storage. Carbon dioxide could be pumped into the oceans, to dissolve in the water (at depths below about 1 km), or deposited by pipeline on the sea floor (below 3 km), where it would be expected to form a denser-than-water "lake" of liquid CO_2. Either approach raises questions about the long-term stability of the CO_2 storage, and in the case of dissolved CO_2, the effect on seawater chemistry. A number of geological reservoirs have also been considered (figure 18.29A). One might pump CO_2 into coal seams too thin or low-quality to

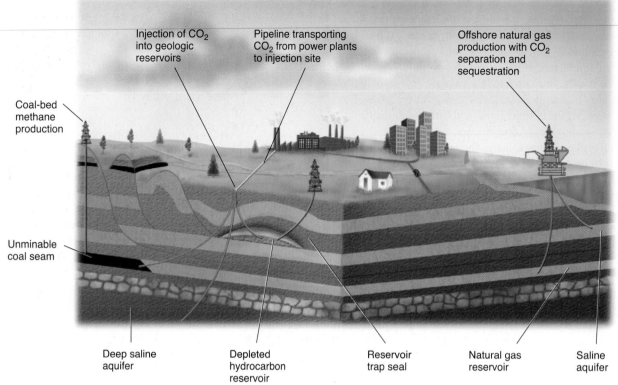

Coal-bed methane production

Injection of CO$_2$ into geologic reservoirs

Pipeline transporting CO$_2$ from power plants to injection site

Offshore natural gas production with CO$_2$ separation and sequestration

Unminable coal seam

Deep saline aquifer

Depleted hydrocarbon reservoir

Reservoir trap seal

Natural gas reservoir

Saline aquifer

A

B

Figure 18.29

(A) Examples of geological methods of carbon sequestration. (B) Natural-gas extraction and carbon sequestration occur together at the Sleipner field off the coast of Norway.

(A) After USGS Fact Sheet FS-026–03; (B) Photograph © Øyvind Hagen/ Statoil

mine, perhaps extracting useable coal-bed methane in the process. The CO$_2$ could be pumped into depleted petroleum reservoirs (perhaps to enhance oil recovery), or into ground water too salty, too high in dissolved minerals, to be of interest as a water source. Experiments and theoretical calculations have been done on more-exotic possibilities, such as reacting CO$_2$ with mafic minerals in ultramafic rocks to produce carbonate minerals that would be geologically stable and thus keep the carbon "locked up" for a very long time.

Some commercial carbon-sequestration facilities are already in operation (figure 18.29B). For example, natural gas from the Sleipner field offshore from Norway is high in associated CO$_2$. But Norway has a high carbon-emission tax. So it is economically advantageous for Statoil, the company operating the field, to separate the CO$_2$ and inject it back into saline pore fluid in a permeable sandstone below the sea floor, rather than releasing it into the atmosphere. Such sequestration activities will become more common if the economics are right.

Summary

The principal air pollutants produced by human activities are particulates, gaseous oxides of carbon, sulfur, or nitrogen, and volatile organic compounds. The largest source of this pollution is combustion, and especially transportation. For some pollutants, anthropogenic contributions are far less than natural ones overall, but human inputs tend to be spatially concentrated. Control of air pollutants is complicated because so many of them are gaseous substances generated in immense volumes that are difficult to contain; once released, they may disperse widely and rapidly in three dimensions in the atmosphere, thwarting efforts to recover or remove them. Climatic conditions can influence the severity of air pollution: Strong sunlight in urban areas produces photochemical smog, and temperature inversions in the atmosphere trap and concentrate pollutants. Air pollutants, in turn, may influence weather conditions by contributing to acid rain, by providing particles on which water or ice can condense, and by blocking or absorbing solar radiation. Ozone in photochemical smog can be a serious health threat, especially to those with respiratory problems. Yet in the "ozone layer" in the stratosphere, it provides important protection against damaging UV radiation. Imposition of air-quality and emissions standards by the Environmental Protection Agency for pollutants with potential health consequences has led to measurable improvements in air quality over the last four decades, particularly with respect to pollutants for which anthropogenic sources are relatively important, such as lead. However, even when pollutants are controlled, the environment may be slow to recover. For example, despite international agreements curtailing the release of the chlorofluorocarbons that contribute to the destruction of stratospheric ozone, CFC levels remain high as a consequence of the compounds' decades-long residence times, so ozone concentrations have not yet recovered proportionately. A relatively new approach to air-pollution abatement, carbon sequestration, may help to reduce CO_2 pollution, which, while not directly a health issue, is certainly an ongoing environmental concern.

Key Terms and Concepts

acid rain 442	ozone hole 435	particulates 429	thermal inversion 444
carbon sequestration 449	ozone layer 434	pH scale 442	ultraviolet radiation 434

Exercises

Questions for Review

1. Describe the principal sources and sinks for atmospheric carbon dioxide.

2. Carbon monoxide is a pollutant of local, rather than global, concern. Explain.

3. What is the origin of the various nitrogen oxides that contribute to photochemical smog?

4. What is photochemical smog, and under what circumstances is this problem most severe?

5. Ozone is an excellent example of the chemical-out-of-place definition of a pollutant. Explain, contrasting the effects of ozone in the ozone layer with ozone at ground level.

6. What evidence links CFC production with ozone-layer destruction? What is the role of UV radiation in forming and destroying ozone, thus protecting us from UV rays from the sun?

7. Lead additives in gasoline were effectively eliminated decades ago in the United States. So why is lead from vehicle exhaust still a pollution concern in urban areas?

8. What radiation hazard is associated with indoor air pollution, and why has it only recently become a subject of concern?

9. What pollutant species is believed to be primarily responsible for acid rain? What is the principal source of this pollutant?

10. Explain briefly why the seriousness of the problems posed by acid rain may vary with local geology.

11. Describe the phenomenon of a thermal inversion. Give an example of a geographic setting that might be especially conducive to the development of an inversion. Outline the role of thermal inversion in intensifying air-pollution episodes.

12. Federal emissions-control regulations for automobiles have led to improvements in air quality with respect to some but not all pollutants in auto exhaust systems. Explain briefly.

13. What is "carbon sequestration"? Describe two geological approaches and one biological approach to carbon sequestration.

Exploring Further

1. EPA monitors air quality and issues air-pollution hazard warnings. The daily air quality forecast for your area may be available on **www.weather.com/activities/health/airquality/**. If so, keep track, over a period of weeks, of the weather conditions on days when warnings are issued, and see what patterns emerge.

2. Examine air-quality trends for your local area at **www.epa.gov/air/airtrends/where.html**. What pollutants have declined, and by how much? Have any increased? If so, consider why.

3. Check the latest information on stratospheric ozone depletion. A good place to start is at NASA's Global Change Master Directory, specifically **gcmd.gsfc.nasa.gov/Resources/FAQs/ozone.html**. Investigate the validity of claims, made by those opposing CFC restrictions, that natural chlorine sources (sea spray, volcanic gases) can account for ozone depletion. (See NetNotes for information sources.)

4. Use the data in table 15.1 to calculate the average fuel economy (mpg) of the U.S. light-duty vehicle fleet (cars plus pickups/vans/SUVs). If burning 1 gallon of gasoline produces 19.6 pounds of CO_2, how much CO_2 would be produced annually by a vehicle achieving that fuel economy that is driven 11,000 miles in a year?

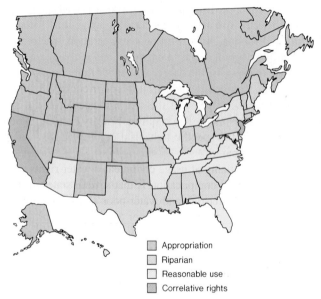

- Exclusively appropriation
- Exclusively riparian
- Both doctrines recognized

Figure 19.1

Distribution of underlying principles of surface-water law applied in the United States.

Source: Data from New Mexico Bureau of Mines Circular 95.

- Appropriation
- Riparian
- Reasonable use
- Correlative rights

Figure 19.2

Distribution of principles of groundwater law applied in the United States. "Riparian" here refers to the English Rule or Rule of Capture described in the text, water rights being related to owning land overlying an aquifer rather than bordering a stream or lake.

Source: Data from New Mexico Bureau of Mines Circular 95.

English Rule, or "Rule of Capture," gives property owners the right to all the ground water they can extract from under their own land. This is sometimes considered analogous to riparian surface-water rights. However, it ignores the reality of lateral movement of ground water. Ground water does not recognize property lines, of course, so heavy use by one landowner may deprive adjacent property owners of water as ground water flows toward active well sites. The American Rule, or "Rule of Reasonable Use," limits a property owner's groundwater use to beneficial use in connection with the land above, and it includes the idea that one person's water use should not be so great that it deprives neighboring property owners of water. Unfortunately, as with surface-water laws, there is no consistent definition of what constitutes "beneficial use." These rules have then been variously combined with appropriation or riparian philosophies of water rights.

Some states, notably California, have specifically addressed the problem of several landowners whose properties overlie the same body of ground water (aquifer system), specifying that each is entitled to a share of the water that is proportional to his or her share of the overlying land *(correlative rights)*. Many other states have introduced a prior-appropriation principle into the determination of groundwater rights, as indicated in figure 19.2.

One factor that complicates the allocation of ground water is its comparative invisibility. The water in a lake or stream can be seen and measured. The extent, distribution, movement, and quantity of water in a given aquifer system may be so imperfectly known that it is difficult to recognize at what level one individual's water use may deprive others who draw on the same water source.

As an aside, note that similar problems arise with other reservoirs of fluids underground—specifically, oil and gas. Rights to petroleum from an oil field underlying the Iraq-Kuwait border were a point of contention between those nations prior to the 1990 invasion of Kuwait by Iraq.

Resource Law: Minerals and Fuels

Mineral Rights

For more than a century, U.S. federal laws have included provisions for the development of mineral and fuel resources. The fundamental underlying legislation, even today, is still the 1872 Mining Law. Its intent was to encourage exploitation of mineral resources by granting mineral rights and often land title to anyone who located a mineral deposit on federal land and invested some time and money in developing that deposit. Federal land could be converted to private ownership for $2.50 to $5 an acre—and as those prices were set in the legislation, they would remain in effect for any minerals still subject to the 1872 law, even decades later when inflation had raised land values. The individual (or company) did not necessarily even have to notify the government of the exact location of the deposit, nor was the developer required to pay royalties on the materials mined. The style of mining was not regulated, and there was no provision for reclamation or restoration of the land when mining operations were completed. What this amounted to was a colossal giveaway of minerals on public lands (and land too) in the interest of resource development.

Subsequent laws, particularly the Mineral Leasing Act of 1920 and the Mineral Leasing Act for Acquired Lands of 1947, began to restrict this giveaway. Certain materials, such as oil, gas, coal, potash, and phosphate deposits, were singled out for different treatment. Identification of such deposits on federally controlled land would not carry with it unlimited rights of exploitation. Instead, the extraction rights would be leased from the government for a finite period, and royalties had to be paid, compensating the public in some measure for the mineral or fuel wealth extracted from the public lands. Still, no requirements concerning the mining or any land reclamation were imposed, and the so-called hard-rock minerals (including iron, copper, gold, silver, and others) were exempted. The Outer Continental Shelf Lands Act of 1953 extended similar leasing provisions to offshore regions under U.S. jurisdiction.

Sparked by a 1989 General Accounting Office report recommending reform of some provisions of the 1872 Mining Law, intense debate has been taking place in Congress about possible reform of that law. Under reform proposals in a 1995 bill passed by the House of Representatives, rights to ore deposits still governed by the 1872 law would not be so freely granted, but would be treated more like the rights to oil, coal, and other materials already explicitly removed from the jurisdiction of the 1872 law. Fairer (nearer market) prices for land would have to be paid, as would royalties, and strict mine-reclamation standards would have to be met. Again, an obvious motive is to assure the public, in whose behalf federal lands are presumably held, fair value for mineral rights granted, a return on minerals extracted, and environmental protection. Industry advocated lower royalty rates, deduction of certain mineral-extraction expenses from the amount on which royalties would be paid, and other differences from the House-passed bill, backing a less radical reform passed by the Senate in 1996. Both bills stalled in committee. Subsequent attempts to revive the reform efforts failed. From time to time during the late 1990s, Congress did pass moratoria on the issuing of new mineral rights. Until a revision is passed, however, the 1872 Mining Law, for all its perceived flaws, remains in effect.

Mine Reclamation

Growing dissatisfaction with the condition of many mined lands eventually led to the imposition of some restrictions on mining and post-mining activities in the form of the Surface Mining Control and Reclamation Act of 1977. This act applies only to coal, whether surface-mined or deep-mined, and attempts to minimize the long-term impact of coal mining on the land surface. Where farmland is disturbed by mining, for example, the land's productivity should be restored afterward. In principle, the approximate surface topography should also be restored. However, in practice, where a thick coal bed has been removed over a large area, there may be no nearby source of sufficient fill material to make this possible, so this provision may be disputed in specific areas. (This is one of the fiercely contested aspects of mountaintop-removal coal mining, discussed in chapter 14: Shearing off the top of a mountain and putting the tailings in a valley profoundly alters the topography.) Similarly, the act stipulates that groundwater recharge capacity be restored, but particularly if an aquifer has been

removed during mining, perfect restoration may not be possible. Still, even when a complete return to pre-mining conditions is not possible, this legislation and other laws patterned after it go a long way toward preserving the long-term quality of public lands and preventing them from becoming barren wastes. Unfortunately, it does not require reclamation of lands mined before it was enacted, which may be continuing sources of acid runoff. The National Environmental Policy Act (discussed later in the chapter) has provided a vehicle for controlling mining practices for other kinds of mineral and fuel deposits; and Superfund helps deal with "grandfathered" unreclaimed sites that present acute pollution problems.

Many states impose laws similar to or stricter than the federal resource and/or environmental laws on lands under state control. However, the impact of the federal laws alone should not be underestimated. Nearly one-third of the land in the United States, most of it in the west, is under federal control (figure 19.3). Indeed, part of the pressure for reforming the 1872 Mining Law comes from state land-management agents in the western states, where the great disparities between mineral-rights and mineral-development laws on federal and on state land create some tensions. For example, it may be economically advantageous under current laws for a company to seek mineral rights on, and mine, federal rather than adjacent state-controlled lands; but the states (like the federal government) receive no returns from mineral extraction on federal lands within their borders, while they do have to deal with problems such as air and water pollution associated with the mining and mineral processing.

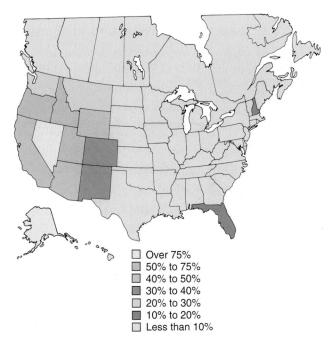

- ☐ Over 75%
- ☐ 50% to 75%
- ☐ 40% to 50%
- ☐ 30% to 40%
- ☐ 20% to 30%
- ☐ 10% to 20%
- ☐ Less than 10%

Figure 19.3

Federal land ownership in the United States. Of the federal lands, about 40% is under the control of the Bureau of Land Management, 30% is the national forest system, and about 15% each, the national park system and the national wildlife refuge system.

Source: Data from 2007 National Resources Inventory, *2009.*

International Resource Disputes

As noted in chapter 13, additional legal problems develop when it is unclear in whose jurisdiction valuable resources fall. International disputes over marine mineral rights have intensified as land-based reserves have been depleted and as improved technology has made it possible to contemplate developing underwater mineral deposits.

Traditionally, nations bordering the sea claimed territorial limits of 3 miles (5 kilometers) outward from their coastlines. Realization that valuable minerals might be found on the continental shelves led some individual nations to assert their rights to the whole extent of the continental shelf. That turns out to be a highly variable extent, even for individual countries. Off the east coast of the United States, for example, the continental shelf may extend beyond 150 kilometers (about 100 miles) offshore. The continental shelf off the west coast is less than one-tenth as wide. Nations bordered by narrow shelves quite naturally found this approach unfair. Some consideration was then given to equalizing claims by extending territorial limits out to 200 miles offshore. That would generally include all of the continental shelves and a considerable area of sea floor as well. The biggest beneficiaries of such a scheme would be countries that were well separated from other nations and had long expanses of coastline. Closely packed island nations, at the other extreme, might not realize any effective increase in jurisdiction over the old 3-mile limit. Landlocked nations, of course, would continue to have no seabed territorial claims at all.

Law of the Sea and Exclusive Economic Zones

In 1982, after eight years of intermittent negotiations collectively described as the "Third U.N. Conference on the Law of the Sea," an elaborate convention on the Law of the Sea emerged that attempted to bring some order out of the chaos. Resources are only one area addressed by the treaty. Territorial waters are extended to a 12-mile limit, and various navigational, fishing, and other rights are defined. **Exclusive Economic Zones** (EEZs) extending up to 200 nautical miles from the shorelines of coastal nations are established, within which those nations have exclusive rights to mineral-resource exploitation. These areas are not restricted to continental shelves only. (Indeed, part of the idea behind the 200-mile limit was to help "geologically disadvantaged" nations lacking broad continental shelves on which to claim resource rights.) Some deep-sea manganese nodules fall within Mexico's Exclusive Economic Zone; Saudi Arabia and the Sudan have the rights to the metal-rich muds on the floor of the Red Sea. Included in the EEZ off the west coast of the United States is a part of the East Pacific Rise spreading-ridge system, the Juan de Fuca rise, along which sulfide mineral deposits are actively forming. Areas of the ocean basins outside EEZs are under the jurisdiction of an International Seabed Authority. Exploitation of mineral resources in areas under the Authority's control requires payment of royalties to it, and the treaty includes some provision for financial and technological assistance to developing countries wishing to share in these resources.

The United States was one of only four nations (out of 141) to vote against the convention, at the time objecting principally to some of the provisions concerning deep-seabed mining. However, the United States has tended to abide by most of the treaty's provisions. In March 1983, President Reagan unilaterally proclaimed a U.S. Exclusive Economic Zone that extends to the 200-mile limit offshore. This move expanded the undersea territory under U.S. jurisdiction to some 3.9 billion acres, more than 1½ times the land area of the United States and its territories (figure 19.4).

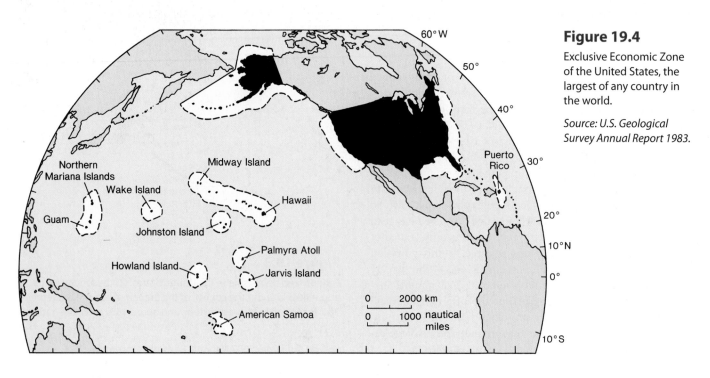

Figure 19.4

Exclusive Economic Zone of the United States, the largest of any country in the world.

Source: U.S. Geological Survey Annual Report 1983.

So far, detailed exploration of this new domain has been minimal. However, certain kinds of mineral and fuel potential are already recognized. For example, off the Atlantic coast, there is potential for finding oil and gas deposits in the thick sediment deposited as the Atlantic Ocean formed—though there is also considerable concern about environmental impacts of oil and gas exploration on the continental shelf. "Black smoker" hydrothermal vents and related hydrothermal activity on the Pacific continental margin have produced sulfide and other metal deposits. Sulfides may also be associated with the Pacific island margins (consider the islands' volcanic origins), and on the flanks of those islands, in many places, mineral crusts rich in cobalt and manganese have been formed. The western and northern margins of Alaska are, like the Atlantic coast, thickly blanketed with sediment hosting known and potential petroleum deposits. Altogether, quite a variety of resources may be recoverable from the EEZ of the United States (figure 19.5).

A provision of the Convention on the Law of the Sea that has just begun to take effect is prompting a flurry of seafloor mapping and seems likely to lead to some conflicts. The 200-mile limit does encompass the whole width of the continental margin in many cases, but not all. The provision in question allows a nation to claim rights beyond the 200-mile limit, extending its EEZ, if it can prove that the area in question represents a "natural prolongation" of its landmass. Approximately 30 nations, including Australia, Canada, India, and Mexico, may be able to extend their EEZs under this provision. If a nation's continental margin extends beyond 200 miles offshore, it may claim jurisdiction over resources out to 350 miles offshore, or as much as 100 miles out from the point on the continental slope corresponding to a water depth of 2500 meters (the conventional base of the continental slope beyond the shelf), depending on the thickness of sediments and other criteria. Also, with global warming shrinking Arctic ice cover, Arctic resources become more accessible. Thus, there is growing interest among many

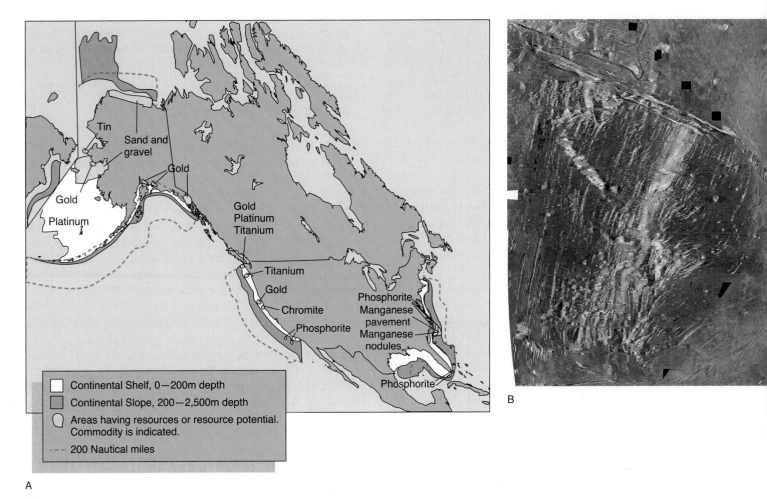

A

Figure 19.5

(A) Possible mineral and rock resources in the North American portion of the EEZ of the United States. (B) Detailed sonar map of a portion of the Juan de Fuca ridge off the northwestern United States. The ridge system runs from upper right to lower left; a transform fault slices across the top of the image at the upper end of the ridge zone.

(A) Source: Adapted from B. A. McGregor and T. W. Oldfield, The Exclusive Economic Zone: An Exciting New Frontier, *U.S. Geological Survey, p. 7.*
(B) Mosaic of images from USGS Open-File Report 2010–1332.

nations in staking their maximum possible jurisdictional claims in the Arctic. While the United States is not party to the Convention and thus not directly affected by the latest provisions, some indication of the possible magnitude of such claims can be seen in figure 19.5: Note how far north of Alaska the continental slope extends in some areas, and realize that the Convention provisions could extend jurisdiction up to 100 miles beyond that.

The Convention creates a Commission on the Limits of the Continental Shelf that will review such claims, which must be supported by detailed mapping of seafloor topography (such as that shown in figure 19.5 B) and sediment thickness. The first applications were due to the panel in 2009. As with the 200-mile limit, there is plenty of potential for overlapping or conflicting claims by different nations bordering the same shelf. It will be interesting to see how these conflicts are resolved.

Antarctica

Antarctica has presented a jurisdictional problem not unlike that of the deep-sea floor. Prior to 1960, seven countries—Argentina, Australia, Chile, France, New Zealand, Norway, and the United Kingdom—had laid claim to portions of Antarctica, either on the grounds of their exploration efforts there or on the basis of geographic proximity (see figure 19.6). (Note the British claim based in part on the location of the Falkland Islands, site of a brief "war" between the U.K. and Argentina.) Several of these claims overlapped. None was recognized by most other nations, including five (Belgium, Japan, South Africa, the United States, and the former Soviet Union) that had themselves conducted considerable scientific research in Antarctica.

After a history of disagreement punctuated by occasional violence, these twelve nations signed a treaty in 1961 that set aside all territorial claims to Antarctica. Various additional points were agreed upon: (1) the whole continent should remain open; (2) military activities, weapons testing, and nuclear-waste disposal should be banned there; and (3) every effort should be made to preserve the distinctive Antarctic flora and fauna. Other nations subsequently signed also; there are now twenty nations that are "consultative parties" to the Antarctic treaty.

The treaty did not address the question of mineral resources, however, in part because the extent to which minerals and petroleum might occur there was not realized, and in part because their exploitation in Antarctica was then economically quite impractical anyway. Not until 1972 was the question of minerals even raised. The issue continued to be a source of contention for some years. Finally, in 1988, after six years of negotiations, the consultative parties (including the United States) produced a convention to regulate mineral-resource development in Antarctica: the Convention on the Regulation of Antarctic Mineral Resource Activities (CRAMRA). Under this convention, no such activity—prospecting, exploration, or development—could occur without prior environmental-impact assessment; significant adverse impacts would not be permitted. Actual mineral-resource development would require the unanimous consent of a supervisory commission composed mainly of representatives of the consultative parties.

In the years after CRAMRA's formulation, debate has continued. Sixteen of the twenty consultative parties must approve the treaty in order for it to take effect. Part of the reason for resistance to ratification was a concern that CRAMRA, by establishing guidelines for mineral-resource development, was effectively opening up the continent to such development (and possible resultant negative environmental impacts, safeguards notwithstanding). In the United States, this negative reaction to CRAMRA led to the passage of the Antarctic Protection Act of 1990. Briefly, the Act prohibits Antarctic mineral prospecting and development by U.S. citizens and companies and urges other nations to join in an indefinite ban on such activities and to consider banning them permanently. The 1991 Protocol on Environmental Protection to the Antarctic Treaty indeed involved an indefinite prohibition on activities related to development of geologic resources. In 2001–2002, the U.S. EPA issued a rule on environmental impact statements (described in a later section) for other nongovernmental activities—including tourism and scientific study—to be undertaken in Antarctica. Unfortunately, there is growing pressure for resource development just as scientists and others are realizing the importance of studies in Antarctica—a setting relatively undisturbed by human activities—to understanding global climate change.

Pollution and Its Control

At the outset of a discussion of laws regulating pollution, one might legitimately ask what legal basis or justification exists in the United States for antipollution legislation. Have we, in fact, any legal right to a clean environment? The primary Constitutional justification depends on an interpretation of the Ninth Amendment: "The enumeration in the Constitution of certain rights shall not be construed to deny or disparage others retained by the people." The argument made is that the right to a clean, healthful environment is one of those individual rights so basic as not to have required explicit protection under the Constitution.

This interpretation generally has not been upheld by the courts. However, those harmed or threatened by pollution have often sought relief through lawsuits alleging nuisance or negligence on the part of the polluter. Nuisance and negligence are varieties of *torts* (literally, "wrongs" in Latin) that are violations of individual personal or property rights, punishable under civil law, and distinct from violations of basic public rights. The government has also exercised its authority to promote the general welfare by enacting legislation to restrict the spread of potentially toxic or harmful substances (including pollutants).

Water Pollution

Water pollution ought not to have been a significant problem in the United States since the start of the twentieth century, when the Refuse Act of 1899 prohibited dumping or discharging refuse into any body of navigable water. Because most streams drain into larger streams, which ultimately become navigable, or into large lakes, this act presumably banned the pollution of

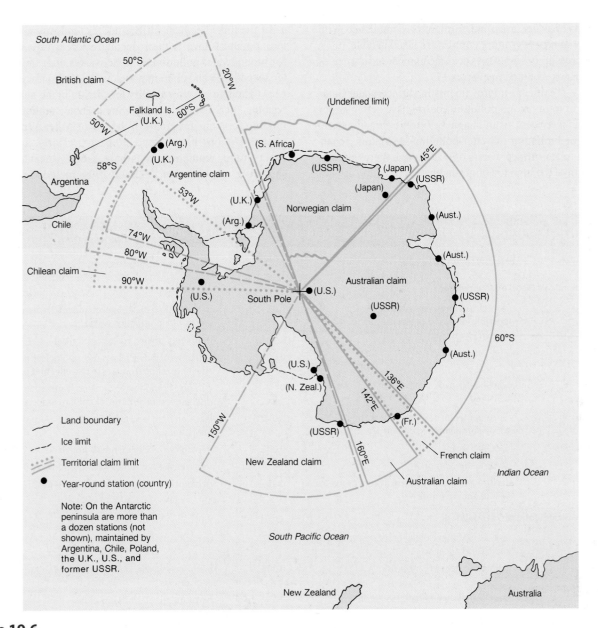

Figure 19.6

Conflicting land claims in Antarctica prior to the 1961 treaty. Note especially the overlapping claims.

Source: Data from C. Craddock, et al., Geological Maps of Antarctica, Antarctic Map Folio Series, Folio 12, copyright 1970 by the American Geographical Society; and J. H. Zumberge, "Mineral Resources and Geopolitics in Antarctica," American Scientist 67, pp. 68–79, 1979.

most lakes and streams. Unfortunately, the enforcement of the law left much to be desired.

In succeeding decades, stricter and more specific anti-water-pollution laws have been enacted. The Federal Water Pollution Control Act of 1956 focused particularly on municipal sewage-treatment facilities. The much broader Water Quality Improvement Act of 1970 and the subsequent amendments in the Clean Water Act of 1977 also address oil spills and various chemical pollutants, from both point and nonpoint sources. One of the objectives was to eliminate the discharge of pollutants into navigable waters of the United States by 1985—something that was, in theory, mandated eighty-six years earlier. The Clean

Water Act and amendments nominally required all municipal sewage-treatment facilities to undertake secondary treatment by 1983, although practice lagged somewhat behind mandate; only about two-thirds of municipalities actually met that deadline. Another provision was that "best available technologies" to control pollutant releases by twenty-nine categories of industries were to be in place by July 1984. (This concept of "best available" treatment is discussed further later in the chapter.)

The fundamental underlying objective is "to restore and maintain the chemical, physical, and biological integrity of the Nation's waters." To this end, the Environmental Protection Agency (EPA) is directed to establish water-quality criteria for various types of water uses and to monitor compliance with water-quality standards (some examples are listed in table 19.1). This is an increasingly complex task as the number of substances of concern grows, as noted in chapter 17.

Locally, significant improvements in surface-water quality resulting from improved pollution control have certainly occurred, but as is evident from chapter 17, more remains to be done. A major limitation of water-pollution legislation generally is that it treats surface waters only, although there is growing realization of the urgency of groundwater protection as well

as of the links between surface and ground water. Recently, in fact, polluters have been prosecuted successfully, if indirectly, for groundwater pollution by toxic wastes after the ground water seeped into and contaminated associated surface waters.

The fate of water-quality legislation in the near future has become increasingly cloudy. Since 1991, authorization and funding for many programs of the Clean Water Act and its later reauthorizations have lapsed. Repeated efforts since 1992 to enact further reauthorizations have failed. Recent versions of such bills, introduced in 1995 and 1996, included controversial

Table 19.1	EPA's Primary Safe Drinking-Water Standards for Selected Chemical Components		
Contaminants	Health Effects	MCL* (ppm)	Significant Sources
Inorganic Chemicals (Total of 16 regulated)			
arsenic	skin, circulatory-system damage; possible increased cancer risk	0.01	geological; pesticide runoff from orchards; glass and electronics production waste
barium	increased blood pressure	2	petroleum drilling waste; geological
cadmium	kidney damage	0.005	corrosion of galvanized pipes; geological; mining and smelting; runoff from waste batteries and paints
fluoride	skeletal damage	4	geological; additive to drinking water; toothpaste; discharge from fertilizer, aluminum processing
lead	developmental delays in infants and children; kidney problems, high blood pressure in adults	0.015**	leaching from lead pipes and lead-based solder pipe joints; geological
mercury (inorganic)	kidney damage	0.002	geological; refineries; runoff from landfills, croplands
nitrate	"blue-baby syndrome"	10	fertilizer, feedlot runoff; sewage; geological
selenium	gastrointestinal effects, circulatory problems, hair or fingernail loss	0.05	geological; mining; petroleum refineries
Organic Chemicals (Total of 53 regulated)			
alachlor	eye, liver, kidney, spleen problems; anemia; increased cancer risk	0.002	herbicide runoff from row crops
benzene	anemia; increased cancer risk	0.005	factory discharge; leaching from gas storage tanks and landfills
2,4-D	liver, kidney, or adrenal-gland problems	0.07	herbicide runoff from row crops
dichloromethane	liver problems; increased cancer risk	0.005	discharge from drug, chemical factories
dioxin	reproductive difficulties; increased cancer risk	0.00000003	emission from waste incineration, combustion; discharge from chemical factories
diquat	cataracts	0.02	herbicide runoff
ethylbenzene	liver or kidney problems	0.7	discharge from petroleum refineries
methoxychlor	reproductive difficulties	0.04	runoff/leaching from insecticide use on fruits, vegetables, alfalfa, livestock
polychlorinated biphenyls (PCBs)	skin changes; thymus-gland problems; immune deficiencies; reproductive or nervous-system difficulties; increased cancer risk	0.0005	runoff from landfills; discharge of waste chemicals
styrene	liver, kidney, circulatory-system problems	0.1	discharge from rubber or plastic factories; leaching from landfills
1, 2, 4-trichlorobenzene	changes in adrenal glands	0.07	discharge from textile-finishing factories
trichloroethylene	liver problems; increased cancer risk	0.005	discharge from metal degreasing sites, factories
vinyl chloride	increased cancer risk	0.002	leaching from PVC pipes; discharge from plastic factories

*MCL (Maximum Contaminant Level): Maximum concentration allowed in drinking water. These levels are set based on the public-health risks of particular contaminants (note how different the values are for different substances). They are designed to minimize the projected risks while taking into account available water-treatment methods and costs.

**Not an MCL, but an "action level" that requires municipalities to control the corrosiveness of their water; see lead sources for the reason.

Source: U.S. Environmental Protection Agency.

wetlands-protection provisions that were opposed by some private-property owners. Reauthorization continues to be problematic—and enforcement of environmental-quality standards requires funding.

Air Pollution

In the United States, air pollution was not addressed at all on the federal level until 1955, when Congress allocated $5 million to study the problem. The Clean Air Act of 1963 first empowered agencies to undertake air-pollution-control efforts. It was amended in 1965 to allow national regulation of motor-vehicle emissions. The Air Quality Act of 1965 required the establishment of air-quality standards to be based on the known harmful effects of various air pollutants. An overriding objective of this act and a series of 1970 amendments was to protect and improve air quality in the U.S., particularly from the perspective of public health. Substantial progress has been made with respect to many pollutants, as noted in chapter 18.

The Clean Air Act was reauthorized and amended in 1990. Various provisions of these Clean Air Act Amendments (CAAA) deal with many areas of air quality. The CAAA goes beyond the Montreal Protocol (chapter 18) in reducing production of chlorofluorocarbons and other compounds believed harmful to the ozone layer: It provides for reductions in industrial emissions of nearly 200 airborne pollutants; requires further reduction of vehicle emissions; sets standards for improved air quality with respect to ozone, carbon monoxide, and particulates in urban areas; and provides for reduction of SO_2 emissions from power plants, which are to be held at 50% of 1980 levels.

The SO_2/acid-rain provisions have an interesting economic component. Beginning in 1995, electricity-generating utility plants have been assigned SO_2 allowances (one allowance = 1 ton SO_2 emission per year) based on power output, and allowances can be traded between plants or utility companies for cash. Utilities that reduce their emissions below their allowed output can sell their excess SO_2 allowances, which is an economic incentive to reduce pollution. In the early years, millions of allowances were traded, at prices that occasionally exceeded $1500 each. This "cap-and-trade" approach has proven successful in achieving the desired emissions reductions, and could be applied to other pollutants.

The EPA also holds an annual "acid rain auction" of a portion of the allowances. Each year, the 8.95 million allowances (a limit fixed in 2010) are reassigned among the various existing SO_2 generators, but some are held back for the auction, in part to assure that new electricity-generating plants can acquire the allowances necessary to operate. Interestingly, however, anyone can bid on the allowances, and some individuals and environmental groups have bought allowances simply to take them off the market and permanently reduce total SO_2 emissions. This has become quite affordable, as improvements in sulfate control technologies have reduced the prices utilities are willing to pay for allowances at the auction: The year-2012 allowances were auctioned at an average of about 67 cents each. And one may bid on a single allowance: Students in an environmental-economics class at a New England college bought one SO_2 allowance, for $2.50.

The impact of the regulations on sulfate and acid rain, and certain other pollutants, was noted in chapter 18. Significant air-quality improvements have clearly occurred with respect to several pollutants. As will be seen later in the chapter, however, some air-pollutant emissions continue to rise, despite international agreements in principle to reduce them. There are also trade-offs: It was efforts to meet air-quality and emissions standards that caused the addition to gasoline of the MTBE that has now become a water-pollution concern.

Waste Disposal

The Solid Waste Disposal Act of 1965 was intended mainly to help state and local governments to dispose of municipal solid wastes. Gradually, the emphasis shifted as it was recognized that the most serious problem was hazardous or toxic wastes. The Resource Conservation and Recovery Act of 1976 includes provisions for assisting state and local governments in developing solid-waste management plans, but it also regulates waste-disposal facilities and the handling of hazardous waste and establishes cooperative efforts among governments at all levels to recover valuable materials and energy from solid waste. Hazardous wastes are defined and guidelines issued for their transportation and disposal.

The Environmental Protection Agency monitors compliance with the act's provisions. Development of waste-handling standards, issuance of permits to approved disposal facilities, and inspections and prosecutions for noncompliance with regulations are ongoing activities. However, there are tens of thousands of identified producers and transporters of hazardous waste in the United States subject to RCRA. In part because of this, the Environmental Protection Agency has delegated many of its responsibilities under the Resource Conservation and Recovery Act to the states.

Amendments to the Resource Conservation and Recovery Act in 1984 considerably expanded the scope of the original legislation. Approximately 100,000 firms generating only small amounts of hazardous waste (less than 1000 kilograms per month) are now subject to regulation under the act; originally, these small generators were exempt. Before 1984, a waste producer could request "delisting" as a generator of hazardous waste by demonstrating that its wastes no longer contained whatever hazardous chemicals (as identified by the EPA) had originally caused the operation to be listed; now, delisting requires that the waste producer demonstrate essentially that its wastes contain no identified hazardous chemicals at all. A whole new set of regulations has been imposed on landfills and underground storage tanks containing hazardous substances, which in many cases require expensive retrofitting of the disposal site to meet higher standards; landfill disposal of toxic liquid waste is banned. These and many other provisions clearly reflect growing recognition of the seriousness of the toxic-waste problem.

The U.S. Environmental Protection Agency

In 1970, the U.S. Environmental Protection Agency was established in conjunction with the reorganization of many federal agencies. Among its primary responsibilities were the establishment and enforcement of air- and water-quality standards. Under the Toxic Substances Control Act of 1976, the EPA was given the authority to require toxicity testing of all chemical substances entering the environment and to regulate them as necessary. The EPA also has a responsibility to encourage research into environmental quality and to develop a body of personnel to carry out environmental monitoring and improvement. As the largest and best-funded of the pollution-control agencies, it tends to have the most clout in the areas of regulation and enforcement. However, the scope of its responsibilities is broad, and expanding. The number and variety of situations with which the Environmental Protection Agency is expected to deal, restrictions imposed by limited budgets, and the varying emphases placed on different aspects of environmental quality by different EPA administrators are all factors that have contributed to limiting the agency's effectiveness. Legislation in the late 1980s and early 1990s tended to strengthen EPA's enforcement ability by establishing clear penalties for noncompliance with regulations. The CAAA, for example, allows EPA to impose penalties of up to $200,000 for violations and even to pursue criminal sanctions against serious, knowing violations of CAAA. It is also true that efforts of the U.S. EPA are often supplemented by activities of state EPAs. But enforcement also requires staff and budget for investigations, legal actions, and so on. Sharp budget cuts for EPA in the early twenty-first century, reflecting reduced priority in Congress for environmental-protection activities in general, certainly reduced the EPA's ability to carry out its various mandates and monitoring and enforcement activities.

A Supreme Court decision in early 2007 has the potential to expand EPA's responsibilities under the CAAA still further. Since 1999, a dozen states and many cities had been urging EPA to add carbon dioxide to the list of air pollutants that it regulates, particularly in the context of automobile emissions. The EPA's authority to regulate the six air pollutants discussed in chapter 18 derives from a mandate to protect public health, and CO_2 is not toxic or considered a threat to health. On the other hand, to the extent that CO_2 causes global warming and climate change and that negative health effects of global warming are increasingly being recognized, one could argue that CO_2 indeed poses a risk to public health. Initially, the EPA declined to take on CO_2 regulation, claiming both lack of legal authority and lack of a definitive connection between greenhouse gases and global warming. The states, cities, and other concerned groups sued, and the issue made its way to the Supreme Court. In a split decision, the Court ruled that EPA *does* have statutory authority to regulate greenhouse-gas emissions such as CO_2 from automobiles—that given the harm these gases (and resultant climate change) cause, they are clearly pollutants under the CAAA, and thus subject to such regulation. The ruling did not mean that EPA must immediately impose CO_2 emissions limits, but the pressure to begin moving in that direction sharply increased, and the agency could no longer claim that regulation of greenhouse gases is, in effect, not in its job description. Some steps have been taken toward establishing greenhouse-gas emissions limits, but as of this writing, final rules have not been adopted.

It has been suggested that the EPA set up a cap-and-trade scheme for CO_2 emissions, given the success of the SO_2-allowance scheme. However, the number and variety of significant sources of CO_2 emissions are far greater, making emissions allocations correspondingly more difficult. Vehicle emissions present an additional challenge, for while they represent, collectively, a considerable volume of CO_2, those emissions come in small quantities from millions of individual vehicles. It is not at all clear how a cap-and-trade model could be adapted to these circumstances. Meanwhile, the European Union *has* established a market for carbon emissions credits to be traded among 12,000 companies in 27 nations. (In 2011, however, hackers stole an estimated 30 million euros [about 40 million U.S. dollars] worth of carbon credits, shutting down the exchange for some weeks and suggesting that there are more than such practical issues to be addressed.)

Defining Limits of Pollution

Enforcement of antipollution regulations remains a major challenge, in part, because of the terms of antipollution laws themselves. Consider, for example, an objective of "zero pollutant discharge" by a certain year. At first inspection, "zero" might seem a well-defined quantity. However, from the standpoint of practical science, it is not. Water quality must be measured by instruments, and each instrument has its own detection limits for each chemical. No commonly used chemical-analysis techniques for air or water can detect the occasional atom or molecule of a pollutant. A given instrument might, for example, be able to detect manganese down to a concentration of 1 ppm, lead down to 500 ppb (0.5 ppm), mercury to 150 ppb. Discharges of these elements below the instrument's detection limit will register as "zero" when the actual concentrations in the wastewater might be hundreds of parts per billion. How close one must actually come to zero discharge, then, depends strongly on the sensitivity of the particular analytical techniques and instruments used.

The alternative approach is to specify a (detectable) maximum permissible concentration for pollutants discharged in wastewater or exhaust gases. But pollutants vary so widely in toxicity that one may be harmless at a concentration that for another is fatal. That means a need to specify *different* permissible maxima for individual chemicals, as in table 19.1. Given the number and variety of potentially harmful chemicals, a staggering amount of research would be needed to determine scientifically the appropriate safe upper limit for each. Thus, regulations are limited to a moderate number of toxins known to pose health risks. Even so, extensive data are needed to set "safe" exposure limits. With less-comprehensive data, the Environmental Protection Agency and other agencies may set standards that later may be found to be too liberal or unnecessarily conservative. For some naturally occurring toxic elements, they have occasionally set standards stricter than the natural ambient air or water quality in an area. To meet the discharge standards, then, a company might be required to release wastewater cleaner than the water it took in! For instance, this might be the case with arsenic, as suggested by figure 17.11.

No treatment process is perfectly efficient, and emissions-control standards must necessarily recognize this fact. For instance, 1982 incinerator standards established under the Resource Conservation and Recovery Act specify 99.99% destruction and/or removal efficiency for certain specified toxic organic compounds. While this is a high level of destruction, if 1 million gallons of some such material is incinerated in a given facility in a year, the permitted 0.01% would amount to a total of 100 gallons of the toxic waste released.

Imprecision in terminology remains a concern. The CAAA, for example, includes such expressions as "lowest achievable level" (of emissions; in the context of CFC emissions) and "nonessential" applications of CFCs and other controlled ozone-threatening substances (use in such applications banned beginning in 1992; but what is an "essential" use of CFCs?). In some instances, the introduction of economic considerations further muddies the regulatory waters, both in the framing of laws and regulations and in their implementation; see "Cost-Benefit Analysis" later in the chapter.

International Initiatives

The latter half of the twentieth century saw a rapid rise in international treaties relating to the environment (figure 19.7). Some especially notable steps were taken toward international cooperation and collective action to address problems of global impact and concern during the 1990s. Some of these initiatives relate to problems that affect localized areas, widely distributed around the world—for example, the U.N. Convention to Combat Desertification, which came into force in late 1996. Others, however, recognize that all nations share some resources—the

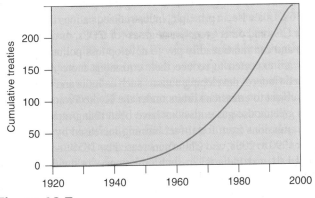

Figure 19.7

Increasing environmental awareness has led to increasing numbers of international environmental treaties.

Source: Data from U.N. Environment Programme.

atmosphere especially—and no nation individually causes or can solve problems associated with pollution of such a common global resource (though to be fair, it is certainly true that some nations have historically contributed a disproportionately large share to that pollution).

One important outgrowth of the Earth Summit in Rio de Janeiro in 1992 was the United Nations Framework Convention on Climate Change. This Convention recognizes the role of atmospheric CO_2 and other greenhouse gases in global warming and other possible climatic consequences. It calls for nations to report on their releases of greenhouse gases and efforts to reduce these and to strive for sound management and conservation of greenhouse-gas "sinks" (plants, oceans), and it calls upon developed nations to assist developing nations in such efforts. This Convention came into force, following its ratification by over 50 nations, in March 1994. Its very existence is significant, as is the promptness of ratification. The Montreal Protocol on Substances that Deplete the Ozone Layer, discussed in chapter 18, is another example of such international commitments to pollution control for the common good.

These accords reflect a relatively new concept in international law and diplomacy, the **precautionary principle.** Historically, nations' activities were not restricted or prohibited unless a direct causal link to some specific damage had been established. However, there are cases in which, by the time scientific certainty has been achieved, serious and perhaps irreversible damage may have been done. Under the precautionary principle, if there is reasonable likelihood that certain actions may result in serious harm, they may be restrained before great damage is demonstrated. So, given the known heat-trapping behavior of the greenhouse gases and documented increases in their atmospheric concentrations, nations agree to limit releases of greenhouse gases *before* substantial global warming or climate change has occurred; given the links between ozone depletion and increased UVB transmission, and between UV radiation and certain health risks such as skin cancer, they agree to reduce or stop releasing ozone-depleting compounds *before* significant negative health impacts have been documented.

This can be important to minimizing harm in cases in which there may be a significant time lag between exposure to a hazard and manifestation of the adverse health effects or those in which natural systems' response to changes in human inputs are slow. The CFCs are an example of the latter. Global emissions have been sharply reduced under the Montreal Protocol, and later amendments strengthened it and expanded its scope, so that 96 substances are now covered by it. Given the residence times of CFCs in the atmosphere, and that developing nations were given longer to phase out their production, atmospheric concentrations have only recently leveled off, and the ozone layer may not recover fully for some decades. Still, the longer nations waited to take action, the further into the future the problem would have persisted.

Of course, sometimes solving one problem can create another. Initial substitutes for the CFCs included HFCs (hydrofluorocarbons) and HCFCs (hydrochlorofluorocarbons), and their atmospheric concentrations began to rise as CFCs were phased

A conference in Cancún in 2010 yielded a framework agreement to reduce deforestation (thereby increasing CO_2 sinks) by paying nations to preserve their forests, and another to create a $100-billion-per-year "green fund" to help developing nations produce clean energy and adjust to the impacts of climate change. The next convention of the delegates in Durban, South Africa, in 2011, yielded a firm commitment to negotiate a successor treaty to the Kyoto Protocol by 2015, and to extend Kyoto while those negotiations proceed. The ultimate result remains to be seen. Meanwhile, the rising greenhouse-gas emissions of China and India, as well as those of the United States, remain unconstrained by these agreements.

Cost-Benefit Analysis

Problems of Quantification

Antipollution laws that specify using the "best available (treatment) technology economically achievable," or contain some similar clause that introduces the idea of economic feasibility, result in additional conflicts and enforcement problems. As noted in earlier chapters, costs for exhaust or wastewater treatment rise exponentially as cleaner and cleaner air and water are achieved. These costs can be defined and calculated with reasonable precision for a given mine, manufacturing plant, automobile, or whatever. Balanced against these cleanup costs are the adverse health effects or general environmental degradation caused by the pollutants. The latter are far harder to quantify and, sometimes, even to prove.

For example, the harmful effects of acid rain are recognized, and increased acidity of precipitation results, in part, from increased sulfur gas emissions. It is not possible, however, to say that, for each ton of sulfur dioxide released, x billion dollars worth of damage will be done to limestone and marble structures from acid rain, y trees will have their growth stunted, or z fish will die in an acidified lake. Even if such precise analysis were possible, what is the dollar value of a tree, a fish, or a statue? Conversely, scientists cannot say exactly how much environmental benefit would be derived from closing down a coal-fired power plant releasing large quantities of sulfur or mercury, and many people could suffer financial and practical hardship from the loss of that facility.

Similar debates arise in many areas. Allowing development on a barrier island may increase tourism and expand the local tax base, helping schools and allowing municipal improvements—but at what risk to the lives and property of those who would live on the vulnerable land, and cost to taxpayers providing disaster assistance? The dangers of increased atmospheric CO_2 are widely accepted, and fossil-fuel extraction may cause environmental disturbances of various sorts—but our energy consumption continues to grow; clearly we tend to decide, implicitly at least, that the benefits outweigh the risks, perhaps, again, because the benefits are readily perceived, while the risks are very hard to quantify (or adequate alternatives have not been found).

In the realm of pollution control, companies frequently contest strict regulation of their pollutant discharges on the grounds that meeting the standards is economically impossible and/or unreasonable. Each individual, each judge, and each jury has somewhat different ideas—of what is "reasonable"; of the value of clean air and water; of the importance of preserving wildlife, vegetation, natural topography, or geology; and of what constitutes an acceptably low health risk (assuming that that health risk can be quantified fairly precisely in the first place). Enforcement thus is very uneven and inconsistent. Virtually everyone agrees that, in principle, it is desirable to maintain a high-quality environment. But just what does that mean? What should it cost? And by whom should those costs be borne?

Cost-Benefit Analysis and the Federal Government

A month after assuming office in 1981, President Reagan issued Executive Order 12291, which decreed, in part, that a (pollution-control) regulation may be put forth (by an agency such as EPA) only if the potential benefits to society outweigh the potential costs. A Regulatory Impact Analysis must be performed on any proposed new "major" regulation, which is, in turn, defined as one that is likely to result in either (1) an impact on the economy of $100 million or more per year; (2) a major cost increase to consumers, individual industries, governmental agencies, or geographic regions; or (3) significant adverse effects on competition, employment, investment, productivity, or the ability of U.S.-based corporations to compete with foreign-based concerns.

This represented a major departure from previous policy. The Clean Water Act ordered the Environmental Protection Agency to be sensitive to costs in establishing standards, but it did not require cost-benefit analysis; the Clean Air Act prohibited considering costs in setting health standards. Now, for example, for the contaminants regulated in drinking water, the EPA defines not only the MCLs as shown in table 19.1, but also identifies "MCLGs"—Maximum Contaminant Level Goals, concentrations below which there are no known negative health effects in humans. For some of these substances (e.g., barium, cadmium, mercury, nitrate) the MCLGs are the same as the MCLs. For others (e.g., arsenic, lead, benzene, dioxin, PCBs), the MCLG is actually zero. In such cases, to quote the EPA, "MCLs are set as close to MCLGs as feasible using the best available treatment technology and taking cost into consideration." Cost-benefit analysis is still not the decisive factor in regulation, but it is at least taken into account.

Cost-benefit analysis was a key part of the debate surrounding the "Clear Skies Act," proposed in 2003 as part of President Bush's Clear Skies initiative, which would have capped emissions of sulfur and nitrogen oxides and of mercury by electricity-generating plants. The EPA, asked to do the cost-benefit analysis, projected the costs of compliance as rising to $6.3 billion per year. Set against those costs was an impressive array of benefits, including: reduction in acidification of lakes and streams in the United States from acid rain; reduced mercury in the environment; over $3 billion per year worth of benefit from improved visibility in national parks and wilderness areas; an estimated 30,000 fewer visits to hospitals and emergency rooms, 12.5 million fewer days with respiratory symptoms (meaning also, fewer days of school or work missed) per year by 2020; and at least 8400 premature deaths prevented, with $21 billion per year in health benefits. In this case, the projected benefits far outweighed the anticipated costs. However, it is clearly far easier to estimate the cost of specified effluent reductions from existing power plants using known technology than to be equally precise about, for instance, the reductions in days of respiratory symptoms Americans would suffer or improved visibility in parks, to say nothing of assigning dollar values to such benefits. This was a significant factor in ensuing years of congressional debate on this measure, which ultimately failed to get out of committee. (Since then, some limits of the kind envisioned in the Clear Skies Act have been implemented through EPA rulemaking.)

Laws Relating to Geologic Hazards

Laws or zoning ordinances restricting construction or establishing standards for construction in areas of known geologic hazards are a more modern development than antipollution legislation. Their typical objective is to protect persons from injury or loss through geologic processes, even when the individuals themselves may be unaware of the existence or magnitude of the risk. While such an objective sounds benign and practical enough, such laws may be vigorously opposed (see especially chapters 4 and 6). Opposition does not come only from real-estate investors who wish to maximize profits from land development in areas of questionable geologic safety. Individual homeowners also often oppose these laws designed to protect them, perhaps fearing the costs imposed by compliance with the laws. For example, engineering surveys may be required before construction can begin, or special building codes may have to be followed in places vulnerable to earthquakes or floods. People may also feel that it is their right to live where they like, accepting any natural risks. And some, of course, simply do not believe that whatever-it-is could possibly happen to *them*.

Occasionally, lawsuits have been filed in opposition to zoning or land-use restrictions, on the grounds that such restrictions amount to a "taking"—depriving the property owner of the right to use that property freely—without compensation. Such action is prohibited by the Fifth Amendment to the Constitution. However, the courts have generally held that such restrictions are, instead, simply the government's exercise of its police power on behalf of the public good and that compensation is thus unwarranted.

Construction Controls

Restrictive building codes can be strikingly successful in reducing damage or loss of life. After a major earthquake near Long Beach, California, in 1933, the California legislature

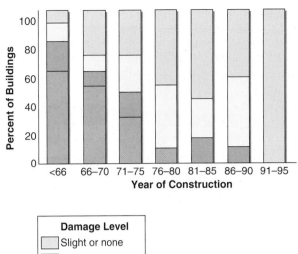

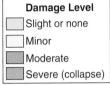

Figure 19.10

Effects of improving building codes for earthquake resistance are clearly shown in the aftermath of the 1995 Kobe earthquake: Newer buildings suffered far less damage than older ones.

passed a regulation known as the Field Act, which created improved construction standards for school buildings. Many of those standards are still being followed. Their effectiveness was demonstrated in subsequent earthquakes, notably the major one in San Fernando in 1971. It was noted afterward that damage to school buildings from the earthquake was almost wholly limited either to schools built prior to 1933 or to those later buildings whose engineers had been permitted to deviate from strict compliance with the Field Act. The 1995 Kobe earthquake further illustrated the value of earthquake-resistant design (figure 19.10). The problem of preexisting structures remains a major problem rarely addressed even by "good" building codes aimed at hazard reduction, as was illustrated yet again in the 1989 Loma Prieta earthquake and the 1994 Northridge quake. There had been talk of "retrofitting" some older buildings, freeways, and bridges to improve safety in accordance with newer design specifications, but due to cost and inertia, many modifications went unmade, and structural failures resulted. Retrofitting of privately owned buildings tends to be even rarer (figure 19.11).

Perhaps because the state is subject to so many geologic hazards, California and its municipalities have been among the leaders in passage of legislation designed to minimize the damage from geologic hazards. One successful effort in this regard has been the imposition of much-more-stringent regulations governing construction on potentially unstable, landslide-prone hillsides near Los Angeles. Over a ten-year period of rapid development (1952–1962), increasingly strict requirements for slope and soil stability studies, site grading, and construction engineering were developed. The resulting reduction in the rate

A   B

Figure 19.11

Hazard-mitigation legislation rarely requires privately owned buildings to be modified for greater resistance; the results can be unfortunate. (A) The first story of this two-story house completely collapsed in the Loma Prieta earthquake. (B) This apartment complex was severely damaged in the 1994 Northridge quake, in part because the carport on the ground level was not as strongly built as the living quarters above.

(A) Photograph by D. Schwartz, (B) photograph by E. V. Leyendecker, both courtesy of U.S. Geological Survey.

Table 19.2	Comparison of Landslide Damage With and Without Restrictive Construction Guidelines in Los Angeles, California		
	Prior to 1952	**1952–1962**	**1962–1967**
controls pertaining to landslide hazards	none	limited grading codes in place	much stricter requirements for engineering and soil studies; modern grading procedures required
sites constructed (approx.)	10,000	27,000	11,000
sites damaged (%)	1040 (10.4%)	350 (1.3%)	17 (0.2%)
total damage to residences (approx.)	$3,300,000	$2,800,000	$80,000
average damage per site built	$330	$100	$7
average cost per site damaged	$3200	$7900	$4700

Source: Data from J. F. Slosson, The Role of Engineering Geology in Urban Planning, Colorado Geological Survey Special Publication 1, 1969.

of landslide damage has been dramatic (table 19.2). Although the costs of structural damage to those homes actually damaged by landslides or soil failure has continued to be high (as the result of escalating property values), the number and percentage of individual units damaged have both dropped sharply, especially after 1963. Plainly, at least where the geologic hazards are moderate, the risk of damage from the hazards can be minimized by taking them into account and "designing around" them. This is a large part of engineering geology (see chapter 20).

Other Responses to Earthquake Hazards

Not all hazard-mitigation legislation is equally effective, as evidenced by a few California examples prompted by the 1971 San Fernando earthquake, described briefly below. Each of these bills is well-intentioned and represents a positive step toward reducing earthquake damage. Collectively, however, they also illustrate a number of weaknesses that limit their effectiveness.

The Seismic Safety Element Bill (1971) requires that communities take seismic and related hazards (ground shaking, tsunamis, and so forth) into account in planning and regulating development. This is plainly sensible in a high-seismic-risk area. But the bill does not prohibit or regulate construction explicitly; public officials are left to decide what constitutes acceptable risk in siting and building. The qualifications of the person(s) evaluating the seismic risks are not stipulated either, and there is no penalty if the community actually ignores the bill altogether.

The Dam Safety Bill (1972), despite its name, has nothing directly to do with the engineering adequacy of dams. It requires dam owners to prepare maps showing the areas that would be flooded in the event of dam failure, and it requires local authorities to prepare evacuation procedures for those areas threatened.

Building in the areas at risk is not prohibited, and those living there may not necessarily be made aware of the dangers.

The Alquist-Priolo Geologic Hazards Act (1972; amended 1975) is a more specific and detailed bill. Under its provisions, the state geologist identifies "Special Studies Zones" along active faults. Proposed construction in these zones requires review by local authorities. The principal objective is to prevent damage through fault offset. (Of course, in the case of a major fault such as the San Andreas, which can cause substantial damage tens or hundreds of kilometers from the epicenter of a large earthquake, it would be of little use to worry about whether one's house was 10 or 100 meters from the fault line, for ground shaking might well level the building even if it was not torn apart directly by the rupture of the fault zone.) Local officials are expected to consider the size of earthquakes and the extent of damage anticipated along each particular fault in deciding where or whether to allow building. Engineering and site-evaluation studies can be costly, and this prompted the 1975 amendment that *exempted* individual single-family homes. (Housing developments, apartments, and commercial buildings are still covered.) Once again, economic considerations vie with scientific ones in selecting a course of action.

Flood Hazards, Flood Insurance

When geologic catastrophes occur, federal disaster-relief funds are often part of the rescue/recovery/rebuilding operations. In other words, all taxpayers pay for the damage suffered by those who, through ignorance or by choice, have been living in areas of high geologic risks. This has struck some as inequitable and was part of the motivation behind the Flood Insurance Act (1968) and its successor, the Federal Flood Disaster Protection Act (1973). These measures provide for federally subsidized flood insurance for property owners in identified flood-hazard areas, whether in stream floodplains or in flood-prone coastal regions. Those most at risk, then, pay for insurance against their possible flood losses. The idea is that the flood insurance will replace after-the-fact disaster assistance in flood-prone areas. To "encourage" property owners to purchase the insurance, the legislation provides that it be required of those obtaining federally insured mortgages or mortgages through federally affiliated banks and lending institutions. The objective is not simply to compel homeowners to protect their own property or to make them aware that they are buying in a floodplain, but to broaden the premium base that provides funds to pay off claims when flooding does occur.

For a community to remain eligible for the insurance program, it must, in turn, enact strict floodplain-zoning regulations. Initially, all proposed new construction in the floodplain must be carefully evaluated and structures designed so as to minimize potential flood damage. Later, after the Department of Housing and Urban Development has provided detailed floodplain maps (figure 19.12), still stricter provisions are enacted that require floodproofing or elevation above the 100-year flood level of new structures in the floodplain. The community may wish to include additional provisions, such as one requiring that

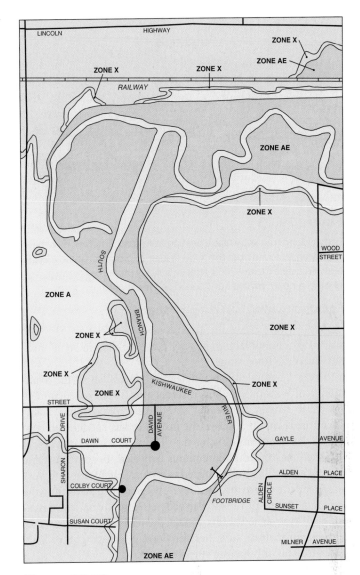

Figure 19.12

Sample section of a flood insurance rate map. Zone A is the estimated 100-year floodplain. In zone AE, height of 100-year flood is determined by detailed analysis of records. Zone X includes a variety of lower-risk situations: areas where the 100-year flood depth is estimated at less than 1 foot (narrow zone adjacent to zone A), areas protected by levees, and areas outside the 100-year but in the 500-year floodplain. Note cut-off meander, the result of earlier U.S. Army Corps of Engineers flood-hazard-reduction efforts.

any water-storage volume within the 100-year floodplain eliminated or filled in by the emplacement of new structures be compensated for by the creation of at least an equal volume of additional floodwater storage elsewhere.

The requisite flood-hazard maps can be slow in coming. As mentioned in chapter 6, detailed records collected over long periods, and careful analysis, are required to construct accurate maps indicating precisely the areas to be affected by 50-year, 100-year, or 200-year floods. The laws do not completely ban further building in the flood-prone area, and as already noted, if more buildings—"floodproof" or not—are

Figure 19.13

Barrier islands regularly take a beating from coastal storms; most have now been removed from the flood-insurance program. Grand Isle, LA after Hurricane Gustav, 2008. Note that many properties at top of photo are now permanently flooded as a result of storm erosion.

Photograph courtesy U.S. Geological Survey.

placed in a floodplain, their very presence may in some measure increase flood stages.

An unintended and unfortunate side effect of the availability of subsidized flood insurance has been to encourage people to rebuild, sometimes several times, in severely flood-prone areas and, in a sense, the insurance has encouraged continued development in such areas. The costs to the government can then far exceed the premiums collected. In some unstable coastal zones, this has happened to such an extent that Congress has voted to remove the most severely threatened areas from the program (figure 19.13; see also chapter 7). Following the devastation of Hurricane Katrina, in fact, there has been discussion of eliminating federal subsidies for flood insurance everywhere, though so far, this has not occurred.

Problems with Geologic-Hazard Mitigation Laws

A pervasive weakness in all kinds of geologic-hazard mitigation laws, including all of the examples discussed in this section, is the fact that they typically apply only to *new* structures. Only new schools in California need conform to Field Act building codes; only new hillside homes need be built using modern slope-stabilization methods; only new federally insured mortgages on flood-prone properties mandate that the property owner carry flood insurance; and so on. Where considerable development has preceded the legislation, those affected by the new law in the near term may be a very small percentage of those at risk: In the 1993 Mississippi River basin floods, many thousands of victims were uninsured, aware of neither the need for nor the availability of flood insurance. Where new construction is banned in very-high-risk areas, older structures nevertheless persist. A small move toward eliminating past poor practice would be a regulation requiring that, if an older structure in a geologically high-risk area were destroyed—whether by geologic processes or another means, such as fire—it could be rebuilt only if the new structure conformed to the newer laws for floodproofing, fault-trace

Figure 19.14

Landslide in Potrero Canyon, Pacific Palisades area, California, after 7 inches of rain in January 1956. Later building codes made newer structures safer but left old ones in areas of known risk.

Photograph by J. T. McGill, U.S. Geological Survey.

setback, slope grading, and so forth. Even modest proposals of this kind, however, commonly meet with vocal citizen opposition.

Thus, laws designed to reduce the risk of damage from geologic hazards have several problems in common. The basic scientific information on which to base sensible regulations may be lacking. Laws may be poorly written, failing to specify fully what must be done and by whom, or they may be weakened by exceptions or omissions that exempt many from their provisions. A major omission, and a very common one, is that existing structures in areas at risk may continue unaffected (figure 19.14; recall also figure 19.11).

The National Environmental Policy Act (1969)

The (U.S.) National Environmental Policy Act (NEPA) established environmental protection as an important national priority and provided for the creation of the Council on Environmental Quality in the Executive Office of the President. The **environmental impact statement** (EIS) is the most visible outgrowth of the NEPA. In this section, we will consider briefly what goes into an EIS, what the point is, and how well the legislation's objectives are being met.

The NEPA actually pertains only to federal agencies and their actions. Whenever a federal agency proposes legislation or "other major Federal action" (which can include anything from policy changes to construction projects that are wholly or partially supported by federal funds) that can significantly affect the quality of the human environment, an EIS must be prepared. Many states have adopted similar legislation related to projects involving funding or approval by state agencies, so the scope of

Figure 19.15

Number of EISs published by all federal agencies, 1973–2008.

Source: U.S. Council on Environmental Quality.

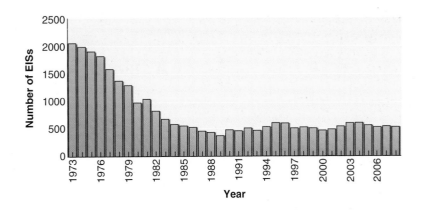

the environmental impact statement process has been considerably broadened. The preparation of environmental impact statements should ensure that, before an agency acts, it considers as fully as possible the potential environmental ramifications of its action, in order to make the best possible decisions. The discussion that follows is restricted to federally mandated EISs.

The NEPA specifies that an environmental impact statement should include

1. A description of the proposed action, its purpose, and why it is needed
2. A discussion of various alternatives (including the proposed action)
3. An indication of the environment to be affected and the environmental consequences anticipated
4. Lists of preparers of the statement and those agencies, organizations, or persons to whom copies of the statement are being sent

Additional supplementary material may also be included. The statement is expected to "rigorously explore and objectively evaluate all reasonable alternatives" (an analysis that presumably will ultimately supply appropriate justification for the particular course of action preferred by the agency). Possible environmental consequences might include not only those that are in some sense geologic (for example, altering natural processes like stream flow or runoff, causing air or water pollution, inevitable consumption of energy or other resources), but also biological ones (loss of habitat, destruction of organisms) and social ones (affecting urban quality, employment, historic or cultural resources). Cost-benefit analyses of various alternatives may be included if the agency is taking them into account.

After a draft environmental impact statement is prepared, a variety of others, ranging from other federal agencies to the general public, are invited to comment. In fact, comments should be actively solicited from all persons or organizations that are particularly likely to be interested or affected. The agency preparing the EIS is then expected to respond to those comments, which might mean modifying the proposed action or the analysis in the EIS or considering additional alternatives.

Over the first decade following passage of the NEPA, an average of more than 1000 environmental impact statements each year were prepared by federal agencies. Lawsuits were filed con-

testing about 10% of this number of projects. In many such cases, it was alleged that the EIS was inadequate, incomplete, or inaccurate. In other cases, the charge was that an EIS ought to have been prepared but was not (the responsible agency felt that no significant environmental impact was involved). Close to half the lawsuits involved citizen or environmental groups as plaintiffs. The number of environmental impact statements filed annually began to decline in the late 1970s (figure 19.15), but the number of lawsuits rose, peaking at 157 in 1982; by the mid-1980s, the number of lawsuits began to drop off sharply in turn.

In recent years, about 550 EISs have been produced each year. As might be expected, the number of suits filed against a particular agency is broadly proportional to the number of EISs that it prepares. A small number of agencies account for most of the EISs. In 2008, for example, 80% of EISs were filed by just four agencies (figure 19.16): the Department of Agriculture

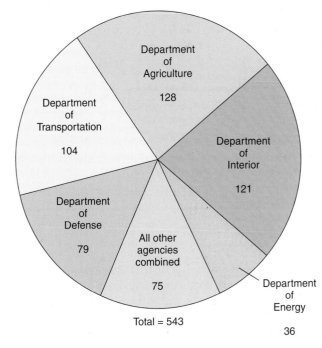

Figure 19.16

Distribution of EIS filings with the U.S. EPA in 2008, by agency.

Data from U.S. Council on Environmental Quality, 2009.

A Tale of Two Pipelines

The discovery of economically important reserves of oil along Alaska's North Slope raised the immediate question of how to transport that oil to refineries in the lower forty-eight states. It was proposed that a pipeline be constructed to carry the oil south to the port of Valdez, from which tankers could move the oil farther south. The planned route (figure 1A) crossed over 1000 kilometers of federal lands, which necessitated a permit from the government; hence the requirement of an environmental impact statement under the then newly passed NEPA.

Various possible alternatives to the proposed route included alternate pipeline routes across Alaska; transportation by some combination of pipeline, rail, and highway across both Alaska and Canada; and tanker transport directly from the North Slope (figure 1B). Transport exclusively by tanker was ultimately deemed impractical, for the climate of northern Alaska is harsh, and the harbors and surrounding seas are blocked by ice much of the year. Schemes involving pipelines across Alaska and subsequent tanker transport posed threats to both terrestrial and marine environments; routes involving only pipelines through Alaska and Canada would leave marine life intact but would disturb far more land. A pipeline through Canada would also necessarily be under the jurisdiction of a foreign government.

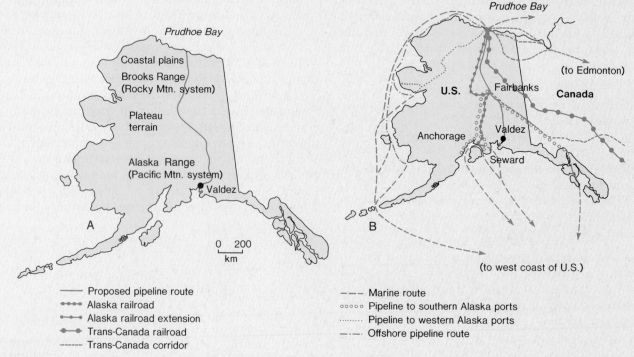

Figure 1

(A) Route of the Trans-Alaska pipeline. (B) Proposed alternative means for transport of the oil.

Source: After D. A. Brew, Environmental Impact Analysis: The Example of the Trans-Alaska Pipeline, *U.S. Geological Survey Circular 695, 1974.*

(especially in connection with forestry and range management, watershed protection/flood control, and use of pesticides and herbicides); Department of the Interior (especially relative to recreation areas, resource management, and mining and oil/gas drilling activities); Department of Transportation (mostly relating to road construction); and Department of Defense (especially in connection with U.S. Army Corps of Engineers and other projects related to navigation, dredging/filling, and watershed protection/flood control). By comparison, in that same year, 49 of 132 lawsuits filed in connection with NEPA involved the Department of Agriculture as defendant (46 of them involving the U.S. Forest Service); the Department of the Interior (31 cases), Army Corps of Engineers (15), and Department of Transportation (10) accounted for most of the rest.

The usefulness of environmental impact statements has been limited by a number of factors. We have already noted in the context of antipollution laws the difficulty of applying cost-benefit analysis to situations where not all factors have well-defined

The possible environmental impacts were many and varied. Construction and operation of overland pipelines could disturb the ground, water, vegetation, fish, and wildlife. Land must be committed to the project, which might cost wildlife habitats or inhibit the migration of land animals. Warm oil in an underground pipeline could melt frozen ground, with possible subsidence of and damage to the pipeline in the resulting mud. The pipeline would cross major fault zones; southern Alaska, in particular, is subject to relatively frequent earthquakes, and severe earthquakes also occur on the Denali Fault; pipeline rupture from any cause raised the possibility of oil spills onto land or into lakes. The use of tankers would mean possible marine oil spills also, and heavy tanker traffic could affect commercial fishing operations. On the plus side, pipeline construction would create jobs, besides providing access to major oil reserves. The influx of money and work crews and the impacts of a pipeline on the physical environment, however, would also significantly affect the lives of Alaskan natives, perhaps negatively.

On balance, the original proposed route was deemed to have the least overall adverse impact. Principal unavoidable impacts were disturbances of terrain, fish, wildlife, and some human environments; some pollution at the port of Valdez from oil discharge from treatment of tanker ballast water; and secondary effects from the influx of more people into the region. (We now know that tanker accidents can also be a significant problem.)

The analysis was a large and complex task, reflected in the six-volume final EIS. However, such detailed analysis allowed the anticipation and minimization or elimination of a variety of negative environmental effects and problems, in part through thoughtful engineering, as explored further in chapter 20. The finished pipeline (figure 2) has functioned effectively for over three decades.

Oil and gas cannot be transported interchangeably through a single pipeline. Though there is considerable natural gas associated with the North Slope oil, the market value of that gas has generally been too low to justify building a separate, costly pipeline for it. With the recent rush of development of shale gas, that situation is likely to continue.

More recently, a proposal has surfaced for a pipeline ("Keystone XL") to carry oil from the Canadian Athabasca oil sands to Oklahoma for

Figure 2

Creative engineering along the pipeline route: The pipeline crosses the Tanana River via a suspension bridge.

distribution to refineries in the Gulf Coast region. The 1700-mile-long pipeline would cross parts of seven states. Two EISs have been filed on the project. The EPA commented on both draft EISs, citing a range of concerns that it felt were not fully addressed, including air quality; pollution; impacts on water resources, wetlands, and migratory birds; possible subsidence in karst terranes; impacts on native tribes living in areas to be crossed; and adequacy of emergency-response plans. The final EISs were filed in 2008 and 2010.

Because the Keystone pipeline would cross an international boundary, a presidential permit is required. Congress, anxious for action, stipulated a two-month time limit for consideration of the permit application after filing of the second final EIS. The president declined to be rushed into approval on a project of this magnitude, and denied the permit. (Interestingly, this was also the recommendation of the State Department, lead agency for the EIS process on Keystone.) In May 2012, a new application for the presidential permit was submitted, involving a modified route ending in Nebraska. As this is being written, no action on this latest permit application has been taken. The ultimate fate of the Keystone project remains to be seen.

costs associated with them. Another potential problem is the impartiality (or lack of it) of EIS preparers or proposers. An agency preparing an EIS for a project of its own has already selected a proposed course of action, and it might, consciously or otherwise, be inclined to present that action in an unduly favorable light. Where the proposed action is to be carried out by an individual or corporation—such as a company needing a federal permit to mine on federal lands—the EIS may, in part, draw on information supplied by the proposer, which could

again be biased. There are also only general statutory requirements that the analysis in an EIS be thorough, which is no guarantee that all the appropriate kinds of specialists will be involved in the preparation of a given EIS. Audits of the accuracy with which EIS preparers predicted environmental impacts have often shown, in retrospect, rather low accuracy.

A major purpose of the National Environmental Policy Act is to allow the public to be informed about and involved in decision making related to the environment. To prevent environmental impact statements from becoming prohibitively long and verbose, the NEPA stipulates that the length should normally be 150 pages or less, or up to 300 pages for especially complex projects. However, "tiering" is allowed, whereby one EIS may be written for each level of decision making of a large or multistage project. One might then have to read through several statements of 150–300 pages each to comment intelligently on a particular project. Many people do not have the time to do that much reading, so they may not become involved at all. (For that matter, not all the relevant employees of the federal agencies involved may do their EIS homework adequately either, for the same reason.) An analysis of EISs filed in 1996 showed that the 243 draft EISs ranged in length from 55 to 1622 pages, with 20% over 300 pages of text; the 270 final EISs ranged from 12 to 1638 pages, averaging 204 pages, with 24% over 300 pages and 6% over 500 pages of text.

Nor can one always project the length of the EIS from the size of geographic area affected or the apparent complexity of the project. For example, a proposal to designate a portion of the Bering Land Bridge National Preserve as a wilderness area resulted in a 300-page final EIS. This was longer than the final EIS for a project in Wyoming that involved building a phosphate fertilizer plant, phosphate slurry pipeline, railroad spur, microwave communication system, and electrical power transmission system, and relocating a road! A project of the scale and complexity of the Trans-Alaska Pipeline, however, would be expected to involve a lengthy EIS, and it did; see Case Study 19.

A major loophole of the NEPA, in the eyes of many environmental groups, is that statements must be prepared only when the anticipated environmental impact of a proposed project is "significant," where significance is evaluated by the federal agency involved in proposing or permitting the action in question. Clearly, *significant* is an inexact term. For each project for which an EIS is filed, many more have been deemed by the agency not to warrant one, and others may dispute that conclusion. In short, as with other kinds of environmental laws, the NEPA has some very desirable objectives, but practice has often fallen short of the ideal.

Summary

Environmental laws and policies relating to geologic matters include some relating to the right to exploit certain resources, some designed to limit pollution and other kinds of environmental deterioration, and some intended to force individuals and agencies to take geologic hazards into account in development and construction. They may fall short of their goals for a variety of reasons. Terms may be inadequately defined or scientifically meaningless, making consistent enforcement difficult. Economic pressure may force pursuit of a course that may be less desirable in purely scientific terms, or a rigorous cost-benefit analysis may be impossible. There may be individual or institutional resistance or bias to combat, or a particular law may recognize so many exceptions that the majority of cases ultimately end up unregulated. These and other difficulties, many without straightforward solutions, will continue to complicate efforts to develop environmental legislation. The future may see increasing development of intergovernmental agreements on critical global environmental issues, but as those agreements commonly lack strong enforcement provisions, their effectiveness also may be limited. The National Environmental Policy Act paved the way for systematic analysis of the environmental consequences of various actions through the Environmental Impact Statement process and for citizen input, but not every project gives rise to an EIS, or to an appropriately thorough one.

Key Terms and Concepts

common but differentiated responsibility of states 464

environmental impact statement 470

Exclusive Economic Zone 456
precautionary principle 463

Prior Appropriation 453
Riparian Doctrine 453

Exercises

Questions for Review

1. Compare and contrast the basic concepts of the Riparian and Appropriation Doctrines underlying much surface-water law.

2. Why are groundwater rights inherently somewhat more difficult to define than surface-water rights?

3. What was the principal objective behind early (nineteenth-century) federal mineral-resource laws? How has the emphasis shifted over the last century?

4. What are Exclusive Economic Zones? Give two examples of types of mineral resources they might encompass.

5. Discuss some of the difficulties of defining and achieving "zero pollutant discharge."

6. Changes in federal policy have changed the degree of emphasis put on cost-benefit analysis in setting pollution-control standards. Explain briefly.

A

B

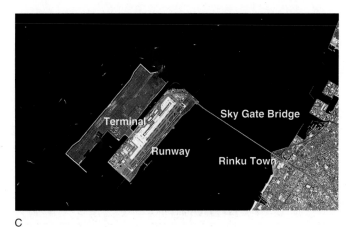

C

Figure 20.4

(A) Limestone, marble, and dozens of useful minerals were extracted from this quarry in Riverside, California; then the pits were allowed to fill, making decorative lakes, and the quarry was converted to a golf course. (B) The limestone mined here was used to build the city of St. Emilion above. The space is now an underground pottery museum; tool marks from the quarrying are still visible on walls and ceiling. Elsewhere in France, similar spaces commonly serve as "caves" in which wine is stored and aged. (C) Osaka's Kansai Airport is built on landfill. Though it does settle 2–4 cm/yr., it suffered little damage in the 1995 Kobe earthquake and essentially none in the 2011 Tohoku quake.

(C) Image courtesy NASA/GSFC/MITI/ERSDAC/JAROS and the U.S./ Japan ASTER Science team.

The alternative approach of **sequential use** involves using land for two or more different purposes, one after another. Because the different land uses need not be compatible, a greater variety of combinations is possible. Abandoned underground mines, if dry and adequately ventilated, can be used for warehouse space, manufacturing, or even office space. This has been done, for example, in the Kansas City area, in abandoned underground limestone quarries. The old quarries now include nearly 1 million cubic meters of frozen-food storage space (about one-tenth of the U.S. total), and the rock's insulating properties make the practice very energy-efficient. The great strength of rock floors makes possible the use of heavy manufacturing equipment; the controlled humidity underground is an advantage to print shops and a sailboat factory located there. Rock walls are, of course, fireproof, and they also contain noise well. On a smaller scale, abandoned mine space at Wampum, Pennsylvania, has been converted to office and laboratory space for the mining company. South Dakota is hoping to convert the nearly mile-deep Homestake gold mine to a major underground scientific research laboratory.

Not all abandoned underground mines are equally suitable for subsequent occupation. The rock structure must be sufficiently strong for safety and not prone to deterioration through slow weathering with time. Even if above the water table, the space must be protected by overlying impermeable rock from infiltration of subsurface water from above. (The deepest 2000 feet of the Homestake mine is currently flooded, since the mine's owners shut off the pumps that had been keeping it dry.) Flat-lying rock strata facilitate conversion to occupied space. There must be no likelihood of quantities of dangerous gases present; old limestone mines would be safe in this respect, while old coal mines, with possible associated methane present, probably would not be.

Alternatively, abandoned mines could be used for waste disposal (recall the proposal in chapter 16 to place high-level radioactive wastes in abandoned salt mines). Strip mines have been suggested as possible landfill sites when mining is completed; this would follow resource extraction with waste disposal, after which the land could be covered, regraded, and put to another use. Denver, Colorado, has carried out a scheme of this sort: Old gravel pits were used for sanitary landfill, and then, after they had been filled, the Denver Coliseum and its parking lots were constructed over the top. (Using such permeable materials for landfill may or may not be advisable in a given situation.)

An alternative post-landfill use might be the extraction of methane gas for energy. An abandoned quarry can be flooded to make a recreational lake. Many such sequences of activities can be imagined; see, for examples, figure 20.4. For that matter, each time a Superfund site is reclaimed and converted to a productive purpose, that, too, is sequential land use. The advantage all sequential-use arrangements share is land conservation. If one piece of land can be made to do double or triple duty, that much less land must be used for or disturbed by human activities overall. Further, it becomes easier to avoid feeling forced to use the least stable or most vulnerable areas at all.

The Federal Government and Land-Use Planning

The federal government controls approximately a third of the land in the United States. The proportion, however, varies widely among states, from less than 1% in many states of the northeast and midcontinent to over 60% of Utah, Idaho, and Alaska, and over 80% of Nevada (recall figure 19.3). Overall, the federal government owns about 30% of land in the United States. The impact of federal land-use policies, like other federal resource-related laws, is therefore felt in very different degrees in different states.

Federal land-use policies have changed through time. As with mineral-resource laws (chapter 19), federal emphasis was initially on resource development in preference to preservation. Early preservation efforts were limited. The first of the national parks, Yellowstone, was established in 1872; the first national forests were created two decades later. Still, resource development continued to be a high priority until about the middle of the twentieth century.

Federal lands can be broadly divided into two types: those intended primarily for preservation (including national parks and wilderness areas) and those on which multiple and, one hopes, compatible land uses can be allowed (for example, national forests). On the latter lands, additional uses beyond recreation or habitat preservation might include livestock grazing, mining, logging, and exploration and drilling for petroleum. Problems arise when the multiple uses allowed turn out not to be compatible after all. Livestock may outcompete wildlife for limited food; overgrazing, careless timbering, or even just too many tourists passing through may accelerate soil erosion and loss. Because the preservation function has sometimes been less successful when multiple land uses are allowed, many groups have tried to pressure the federal government to put more of its lands into the highly protected categories.

In 1978, President Carter did that with 107 million acres of Alaska. Under the Alaska Lands Act, that land—almost one-third of the state—was designated as national monuments, wilderness areas, or similarly protected land. Partially as a result of that action, mining and petroleum development now are prohibited or severely restricted on 40% of Alaska's total land. Some hailed the decision as a forward-looking move to preserve dwindling wilderness. Oil and mining interests were very unhappy at the reduction in land available for exploration and warned of possible future resource shortages. State residents who had hoped to benefit from more jobs or from taxes on those companies' profits from resource exploitation were likewise not pleased. The state, in turn, wanted more control over what was to be done with its lands. Recently, proposals to open parts of the Arctic National Wildlife Refuge to petroleum exploration have intensified the debate.

Often, controversies arising in connection with the EIS process reflect significantly different opinions about appropriate or optimal land uses for the area in question. There are also conflicting opinions about the appropriateness of resource development by private corporations on public lands, whether or not royalties are paid. So, what's a government to do? What *does* constitute managing the public lands in everyone's best interests?

Maps as a Planning Tool

Many kinds of information, geologic and otherwise, go into comprehensive land-use planning. Much of this information can most quickly and clearly be examined in map form. Any geological property or process that varies from place to place, including topography or steepness of slopes, bedrock geology, surficial materials or soil types, depth to water table, rates of cliff erosion, and so on, can be represented. Maps can also show locations of hazards past or present, such as fault zones, floodplains, and landslides. Figure 20.5 illustrates some simple examples. Nongeologic factors—vegetation, population density, or land use, for instance—may also lend themselves to representation in map form (figure 20.6).

A land-use planner can then seek sites for particular land uses based on whatever set of criteria seems most appropriate. Maps make it possible to see quickly where several different conditions are satisfied. In siting a major interstate highway, for example, a planner might seek gently sloping terrain not underlain by expansive clay soils or active fault zones, where the present population density is low. A survey of an appropriate set of maps permits the planner to find potentially suitable sites swiftly. Conversely, maps can aid in long-term land-use planning even when conversion of rural land is not immediately contemplated. If the location of a stream's 100-year floodplain or a major fault zone is known, restrictive zoning ordinances can prevent unwise development before it occurs, or special construction requirements for those areas can be imposed.

Of course, the suitability of a particular area for a specific land use cannot always be determined in a clear-cut, yes-or-no fashion. A land-use planner may instead need to consider a number of alternative sites for a project, each of which is less than fully desirable in some different way. Or there may be degrees of suitability for some purpose, such as housing developments, considering a variable like slope stability. In the latter case, the planner might stipulate different levels of intensity of

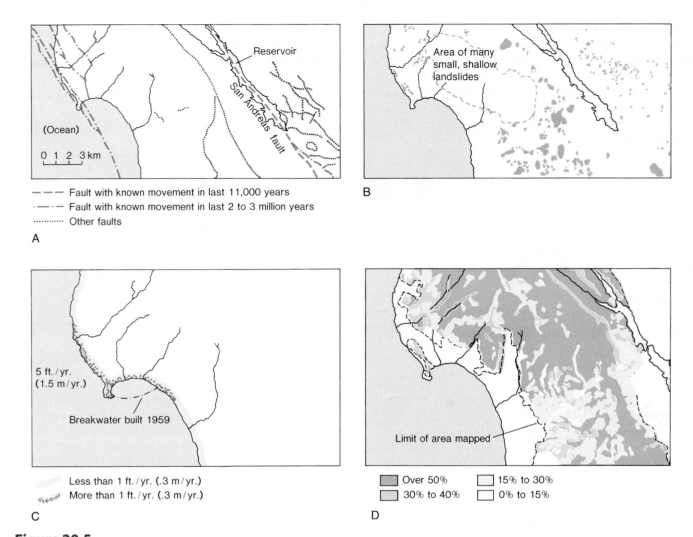

Figure 20.5

Map representation of several kinds of geologic considerations, each map focusing on a single issue: (A) Fault zones. (B) Landslides (shaded). (C) Historic coastal cliff erosion rates. (D) Slope steepness (percent grade); note that the scale of the map requires averaging over significant areas, resulting in some loss of detail.

Source: Modified from B. Atwater, "Land-Use Controls Arising from Erosion of Seacliffs, Landsliding, and Fault Movement," in U.S. Geological Survey Professional Paper 950.

the same land use in different areas. A higher density of homes might be permitted on gently sloping terrain than on steeper hillsides, for instance.

In recent decades, computers have played an increasingly important role in the land-use planning process because of their capacity to manipulate large volumes of quantitative information. A map can be broken down into a grid of numbers, each data point representing a particular property (slope, soil type, and so forth) over some area (1 square kilometer, 10 square kilometers, or whatever) (figure 20.7). The computer can then combine as many kinds of information as desired, each represented on a separate array, and produce a composite measure for each point of the grid to indicate the overall suitability of that area for the land use under consideration, which is most often urbanization/housing development (figure 20.8). The

same basic geologic or nongeologic data can be combined in different ways for different purposes by weighting various factors differently. In mapping the suitability of land for wildlife habitat or refuge, for example, abundant surface and near-surface water might be a positive factor and the presence of expansive clay soils irrelevant. A land-use planner looking at the same region with high-density housing in mind might well regard both factors as negative. The data-processing programs can be adjusted accordingly to produce summary data tailored to particular objectives.

As computers have become more powerful and their programming more sophisticated, they have increasingly been used for **geographic information systems** (GIS). A GIS is any computer-based system for storing and manipulating mappable data—in other words, data that can be associated with, or

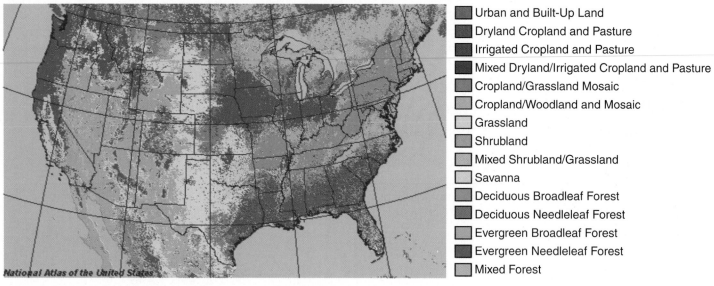

Figure 20.6

U.S. land use/cover. This map combines elements of nature of cover and urbanization. Note regional variations in dominant land use.

Map from National Atlas of the United States, based on data from U.S. Geological Survey (2011).

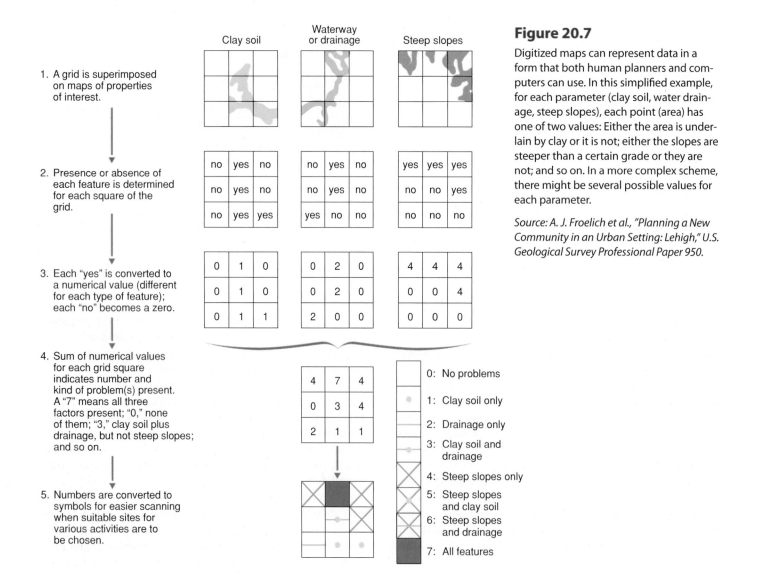

1. A grid is superimposed on maps of properties of interest.

2. Presence or absence of each feature is determined for each square of the grid.

3. Each "yes" is converted to a numerical value (different for each type of feature); each "no" becomes a zero.

4. Sum of numerical values for each grid square indicates number and kind of problem(s) present. A "7" means all three factors present; "0," none of them; "3," clay soil plus drainage, but not steep slopes; and so on.

5. Numbers are converted to symbols for easier scanning when suitable sites for various activities are to be chosen.

Figure 20.7

Digitized maps can represent data in a form that both human planners and computers can use. In this simplified example, for each parameter (clay soil, water drainage, steep slopes), each point (area) has one of two values: Either the area is underlain by clay or it is not; either the slopes are steeper than a certain grade or they are not; and so on. In a more complex scheme, there might be several possible values for each parameter.

Source: A. J. Froelich et al., "Planning a New Community in an Urban Setting: Lehigh," U.S. Geological Survey Professional Paper 950.

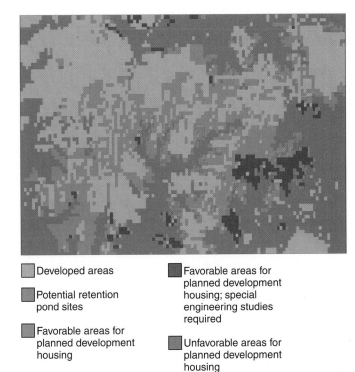

	Developed areas
	Potential retention pond sites
	Favorable areas for planned development housing

| | Favorable areas for planned development housing; special engineering studies required |
| | Unfavorable areas for planned development housing |

Figure 20.8

The computer can rapidly combine digitized data into a composite map for land-use planning, showing more- and less-suitable areas for development or any other intended purpose.

Source: A. J. Froelich et al., "Planning a New Community in an Urban Setting: Lehigh," U.S. Geological Survey Professional Paper 950.

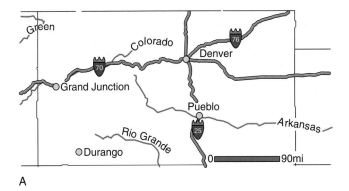

A

B

C

Figure 20.9

Sample GIS representations from USGS National Map Viewer. A roadmap-style map of this region near Denver (A) shows only major rivers, interstate highways, and cities. Adding the topography layer (B) or land cover (C) adds much more information.

Images generated by ArcIMS Viewer, courtesy U.S. Geological Survey.

plotted on, points on a map. The data are often stored and displayed in so-called *layers,* where each layer is a distinct type of information (figure 20.9). The results can be examined visually and/or analyzed by computer, which can combine data from multiple layers.

Maps as tools have limitations, too. One is the matter of scale: The information presented must be given in sufficient detail that features of interest will, in fact, show up in the data. On a map on which 1 centimeter represents 1 kilometer of distance on the earth's surface (1:100,000 scale), all kinds of relatively small features—cliffs, sinkholes, small streams, and so on—might not appear at all; yet these features could be of great concern to a prospective home owner. (The astronauts had the same problem with early lunar landings. Images of the moon made from earth, covering very large expanses of the moon's surface, were used to direct them to "fairly flat" areas. But when those proposed landing areas could be seen close up on a finer scale, many were found to be strewn with boulders several meters across, which definitely presented a hazard to smooth and safe landings!) At the same time, superimposing too much detail can obscure information (figure 20.10).

Sometimes, the difficulty is that the information itself is available on only a gross scale. Maybe the only maps of vegetation or surficial geology were made from satellite images, for instance. Sometimes, the data are simply unavailable. Producing many kinds of topographic, geologic, or other earth-science maps means compiling many observations or measurements, which takes time, personnel, and funds. Even in the United States, which is rather thoroughly mapped by world standards, not all types of information have been obtained for all areas. In these cases, land-use decisions may have to be based on incomplete data, perhaps supplemented by

How Green Is My—Golf Course?

Jan, Henri, Ravi, and Judd stood on the first tee and squinted toward the green far below. Then Henri glanced past the green. "What the heck is that?" he asked. "Looks like a construction site."

"Not exactly; it's a landfill," said Jan. "In fact, you're standing on landfill right now." (See figure 1.)

"I'm *what*? Golfing on garbage?"

"Sure. It's a way to reuse the part of the landfill they've finished and bring in some revenue for the city, too. Don't worry," she added, as he wrinkled his nose. "It doesn't get smelly when the day warms up, if that's what you're expecting. Modern landfills are a lot cleaner than you think."

The day worked out much better than the dubious Henri had expected. Except for a little muffled truck clatter when the wind blew from the active part of the operation and they were on that side of the course, and a cautionary sign or two about not fishing in the water hazards, he'd never have noticed the landfill, really. So he didn't even wince when Ravi said, at the end of the round, "Okay, next week let's go play in a floodplain!" (See figure 2.)

Historically, golf courses (of which the United States has more than 15,000) have often been roundly criticized as environmentally unfriendly land uses. By nature, they sprawl over a good deal of real estate (and as equipment has improved and even amateurs hit the ball farther, designers of new courses have tended to build them ever longer); a typical new golf course occupies about 150 acres. Natural habitat has often been cleared to make way for vast expanses of tidy close-mown grass, kept lush by lavish applications of water and fertilizer and protected by herbicides and insecticides. The water use has been especially criticized in deserts and other dry settings where a golf

Figure 1

This golf course west of Chicago, Illinois, is built on finished section of landfill; ongoing landfill operation continues on adjacent land (in distance and at right).

Figure 2

This course is unusable when the river floods, but there is minimal damage cost—effective use of floodplain land. River Heights golf course, DeKalb, IL.

course can consume a million gallons of water in a day; chemical runoff could pose pollution problems. However, things are changing for the better as a result of increased environmental sensitivity, reinforced by laws requiring environmental impact assessment of new construction.

Intensive research is producing turf grasses that require minimal use of pesticides and other chemical additives and need less water. Course designers are reducing the amount of area planted to short turf, leaving cactus-studded "waste areas" in desert courses in places the golfers shouldn't be hitting anyway, or letting much of the deep rough become very deep indeed (figure 3), perhaps even engaging in prairie restoration. Preservation of wetland or wooded habitat around the course can actually provide sanctuary for wildlife in an area otherwise being overrun by urban sprawl, while enhancing the course visually.

As our golfing friends discovered, building courses on landfills is increasingly common practice. The first such proposal was, in fact, put

Figure 3

It adds to the challenge to have long expanses of unmanicured land on each hole, but is generally kinder to the environment.

forth in 1930 in New York; now there are several dozen examples nationwide, a third of them in California, and more under development. The newest twist on this sequential-use concept is a true ugly-duckling transformation, beginning with a Superfund site near Butte, Montana, site of accumulated toxic wastes from an old copper-smelting operation in the town of Anaconda. The hazardous material has been carefully covered up, sealed with clay and plastic, and—with the help of Jack Nicklaus as designer, more than $40 million from ARCO (the site's last owner when the smelting ceased, faced with even higher potential costs with conventional site cleanup), and the collaboration and blessing of the EPA—the site has been transformed into the Old Works, an attractive, wildflower-decked golf course (figure 4). Once completed, it was deeded to the local residents, so the operation not only eliminated a hazard, but now offers a new source of revenue to the region, which suffered economically from the cessation of mining and smelting activities.

Figure 4

The Old Works golf course in Anaconda, MT, was a Superfund project founded on a sprawling abandoned smelting operation. To preserve some of the historic flavor of the site, designer Jack Nicklaus kept rows of brick smelters along one side of the course, and filled the bunkers with crushed black slag rather than sand.

A

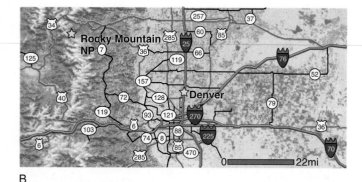

B

Figure 20.10

When topography and land cover are mapped on a finer scale, more information can be displayed; compare (A) with figure 20.9C. But adding the highways layer (B) obliterates some detail.

Images generated by ArcIMS Viewer, courtesy U.S. Geological Survey.

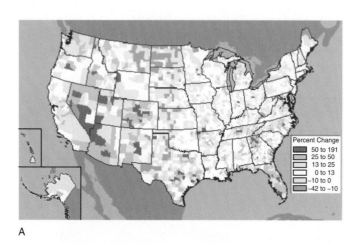

A

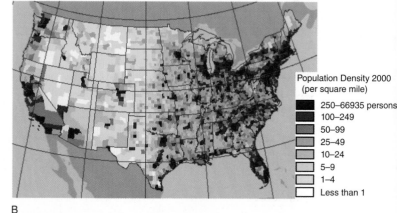

B

Figure 20.11

(A) Population change by county, 1990–2000. (B) Population density, 2000. Despite rapid population growth, much of the West remains sparsely settled, in part due to limited water resources (recall chapter 11).

Maps from National Atlas of the United States.

broad, reconnaissance-type surveys of the area under study. Sometimes, too, the problem is not so much lack of data as conflict among competing options. The pressure to develop areas where risks are high or incompletely assessed is particularly acute where development is rapid, especially if population density is already high so land available for new development is scarce (compare figure 20.11 with figure 20.6).

Satellite imagery can both assist with and supplement conventional mapping, and supply key data for GIS analysis. As we have seen in the context of resource exploration, satellites can efficiently survey broad areas with specialized sensors to distinguish different types of rocks, vegetation, etc., and resolution is improving all the time. The information on topography and geology that they provide is often directly relevant to land-use planning. Satellite images made during flood events facilitates accurate flood-hazard mapping. Current land-use patterns can be identified, urbanization monitored, and land-use changes tracked (figure 20.12). Satellite data are becoming increasingly important to both local and global land-use discussions, as the impact of land-use decisions, in turn, is recognized in new contexts, notably that of global climate change.

The interested reader is referred to the online material on maps and satellite imagery found on the text's website.

Engineering Geology— Some Considerations

Engineering geologists address a broad range of issues and problems related to the topics of previous chapters—slope stability, earthquake-resistant design, groundwater and surface-water resources, mine construction and development, coastal

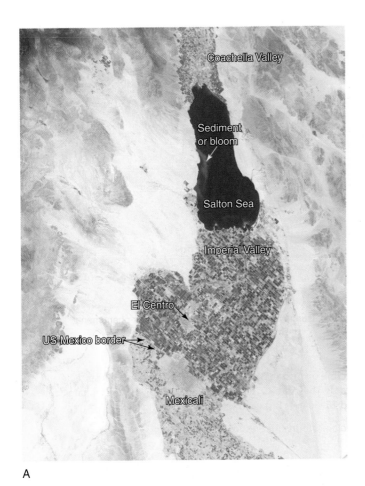

A

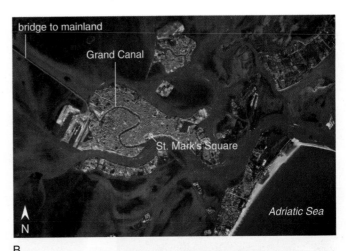

B

C

Figure 20.12

(A) In this image, the extent of agricultural development supported by water from the Salton Sea is evident; the U.S./Mexico border stands out sharply, as agricultural land use is much less developed on the drier Mexican side. (B) The vulnerability of the densely urbanized island city of Venice, Italy, to rising sea level is strikingly evident in this photograph. Satellites can monitor incursions of water into St. Mark's Square and other landmarks. (C) Land-use change in progress: Deforestation in eastern Bolivia reduces a carbon sink, as agricultural development radiates outward from small communities established by resettlement for the purpose of growing soybeans.

Images courtesy of Earth Sciences and Image Analysis Laboratory, NASA Johnson Space Center.

erosion, stability of foundations, building of highways and tunnels, and more. Obviously, not every geologic factor or process is equally important to every project, but a major construction project—for example, a housing development—might require consideration of a very broad range of geologic matters. Some are outlined below.

What rock types are present? Are they strong enough to support the proposed structure(s)? Do they fracture easily? Are there structural features within the rocks—folds, faults, bedding planes—that make the rocks' properties nonuniform and that should be taken into account? How porous and permeable are the rocks? Over the longer term, are they unusually prone to erosion or weathering?

Many of the same questions about porosity, permeability, strength, and stability might be asked about the soils. Are any slopes likely to give way to landslides? Does the construction itself have the potential to trigger slides? How cohesive is the soil? How compressible and prone to settling? How elastic? Do the soils tend to expand and contract as moisture content varies? Are they subject to hydrocompaction? This problem of expansive clay soils as a construction hazard is not a trivial one (figure 20.13). Their failures take few, if any, lives, by contrast with more obvious hazards like floods and earthquakes, but the total structural damage to homes, commercial buildings, pavements, and utilities in the United States each year has been estimated as approximately equal to the total costs of all other geologic hazards combined.

What about water? What quality and quantity of surface water or ground water is available? What are the surface runoff patterns? Does part of the site now serve as a recharge area for

much as 70 feet below the surface, could cost an estimated $85,000 *each*. The changing climatic conditions and their impact on engineering issues will also need to be taken into account if and when a natural-gas pipeline is constructed to tap North Slope gas resources.

Even where some permafrost remains at depth, warmer temperatures mean later freezing at the surface in the fall, and earlier melting in the spring. But the heavy equipment used in oil exploration and drilling cannot navigate the soft muck of the melted permafrost; that equipment can only be used in permafrost areas while the ground is solidly frozen. In areas of the Alaskan tundra, the number of days that this equipment can be deployed has shrunk dramatically over the past three decades, from 200 or more per year to only about 100. To the extent that the dwindling of the permafrost is related to warming caused by CO_2 from the burning of fossil fuels, some see real irony here.

Testing and Scale Modeling

Relatively modern additions to the arsenal of tools available to the geological engineer are the use of scale models of natural systems, and testing—actual or theoretical—of the behavior of natural and construction materials.

Many scale-model experiments take place in the laboratory. Nowadays, indeed, much "testing" is done using sophisticated computer models rather than by building physical models or working with actual samples of materials. However, it is often still useful to conduct physical experiments and observe the results. We have already noted this in the context of earthquake-resistant design, but scale models have been applied to other geologic hazards as well.

The photographs in figure 20.17 show a 1:500-scale model of the Swift Dam and Reservoir on the south flank of Mount St. Helens. The model was built by Western Canada Hydraulic Laboratories for the dam's owners, Pacific Power and Light Company of Portland, Oregon.

High-density muds like those used in drilling oil wells simulated volcanic mudflows entering the reservoir from the creeks that feed it. Figures 20.17B and 20.17C show the progress of one such flow, from the model Swift Creek, into the reservoir. With each test, maximum wave height at the model dam and overall reservoir response to mudflow input were closely observed. At the request of the State of Washington, maximum mudflow volumes corresponding to about half the reservoir volume were tested. Under these conditions, waves from single flows, and the final reservoir height, nearly reached the dam crest.

Experiments such as these can play a key role in engineering safer structures in high-risk areas.

Case Histories, Old and New

This section briefly outlines several case histories, in partial illustration of the range of problems encountered in engineering geology and some approaches to their solution.

A

B

C

Figure 20.17

Completed model reservoir being filled with water to normal depth prior to test runs (A). Mudflow scaled to correspond to 44,000 acre-feet is released from model Swift Creek into reservoir (B); as flow progresses into reservoir (C), water height increases and waves lap at the dam.

Photographs by J. E. Peterson, USGS Photo Library, Denver, CO.

The Leaning Tower of Pisa

The Leaning Tower of Pisa is not Pisa's only bell tower, nor is it the only one that leans. At least two other bell towers built during the same period have also tilted, by as much as 5 degrees from the vertical. The underlying reason is the same in all cases: soft, unstable ground. But it is the elegant, freestanding bell tower of the Pisa Cathedral (figure 20.18) that is by far the greatest tourist attraction among these towers, so it has drawn the most vigorous efforts to stabilize it.

Pisa is underlain by several layers of sediment. The upper 10 meters consist mainly of silt, with about 30 meters of soft marine clay below that. These two layers are water-rich and compressible under load, and can also flow under stress. Below them are much more rigid sand layers.

The Leaning Tower was built in several phases between A.D. 1173 and 1370. It began to tilt even before it was completed. The weight of the tower forced water out of the silt and clay layers, compacting them, and they may also have flowed somewhat. Once the tilt was well established, it seemed to be self-reinforcing—that is, as the structure began to lean, more and more pressure was concentrated on the lower side, causing more flow and compaction of the unstable clay and silt layers and more tilt. The 58-meter-high tower has also sunk about 2½ meters into the soft ground. By 1911, the top had tilted more than 5 meters from vertical, and careful measurements thereafter showed that the tilt was continuing to increase. When a similarly built bell tower at the Cathedral of Pavia collapsed in 1989, the Leaning Tower of Pisa was closed to tourists, and what would become an eleven-year stabilization effort was begun.

Initial suggestions included physical methods, such as boring and selectively removing some material from the north (high) side of the foundation to reduce the tilt, and chemical methods, such as treating the clays to stiffen them and prevent further sinking of the settled south side. In 1993, 900 tons of lead were placed on the high side of the Tower's base to anchor and counterweight it, halting further tilting. In 1999, a careful three-year program of selective soil extraction from under the north side was implemented to reduce the tilt a bit and stabilize the tower. By 2002, the Leaning Tower, though still leaning, was pronounced stabilized, and was opened to visitors again.

Figure 20.18

The Leaning Tower of the Pisa Cathedral has been leaning to some extent since the twelfth century. The leaning began during construction, and builders' attempts to compensate resulted in some curvature in the tower itself.

Photo © Goodshot/Fotosearch RF.

The Panama Canal

The idea of a canal across the Isthmus of Panama was suggested as early as 1528 by one of Cortez's men, but it was not until 1880 that a French company began excavation. The initial design, patterned after the Suez Canal, was for a sea-level cut from Atlantic to Pacific, with no locks—this being the preference not of the engineers, but of those financing and directing the project. The Suez Canal, however, was excavated mainly through sand and soil, not rock, and in a climate with much less rainfall to promote sliding. Also, because of the higher elevation of the Isthmus of Panama, a much larger volume of material would have had to be excavated for a sea-level canal across Panama.

Little or no investigation of the geology of the Panama Canal site was made before excavation began. Such investigation would have revealed layers of young volcanic rocks, lava flows, and pyroclastic deposits, interbedded with some shale and sandstone. Because the rocks dip in many places toward the canal, excavation removed the support from some of these rock layers, which then tended to slide (figure 20.19A). Sliding was facilitated by very high rainfall, which averages 215 centimeters (85 inches) a year. Elsewhere, as the canal was dug, the weight of the rocks on the side caused flow and buckling of the rocks at the bottom of the excavation, which might rise 10 meters or more, requiring redigging to open the canal and removal of material from the sides to relieve some of the pressure (figure 20.19B).

Added to the geologic obstacles were heat, humidity, and rampant diseases, such as the mosquito-spread yellow fever, which caused many deaths among the workers. After seven frustrating years, the French company was bankrupt. A five-year lull in excavation was followed by renewed efforts by a second French firm, but it, too, later abandoned the attempt.

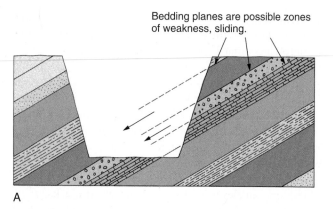
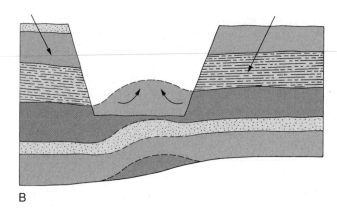

Bedding planes are possible zones of weakness, sliding.

A

B

Figure 20.19

Geologic factors complicated construction of the Panama Canal. (A) Dipping beds sliding into a canal reduce the canal's capacity, requiring more excavation. (B) Removing the weight of the rocks in the canal may allow buckling of rocks below, which are under pressure from the rocks on the canal's sides.

Figure 20.20

(A) Excavation of the Panama Canal, 1907. (B) Landslides remain a threat today.

(A) Courtesy Panama Canal Commission; photograph (B) from USGS Open-File Report OFR 01-0276.

A

The United States acquired the rights to take over the project in 1903. At that point, the whole project design was reconsidered, and instead of a sea-level canal, a more complex design was chosen. A river was dammed to create a lake in mid-isthmus, and locks were to be built on either side to carry ships, in several stages, from sea level up to lake level and back down to sea level on the other side. One technical advantage of the lock design was that it would reduce the total volume of excavation required. An ecological benefit was that the lock design would better separate the Atlantic and Pacific oceans; a concern with a sea-level canal and associated free flow of water and organisms between oceans was the possible impact on various ecosystems that had previously been separated by land.

Even so, finishing the canal wasn't easy (figure 20.20). A geologist was not invited to examine the excavation and associ-

B

ated landslides until 1910, by which time many large slides were already beyond control. Sliding and excavation, in fact, continued long past the time of the canal's nominal completion and substantially increased its costs. Early consideration of the geology of the canal zone might not have eliminated the sliding problems, especially considering the lesser understanding of rock mechanics at the time of construction. However, some of the instability problems could have been anticipated and reduced, and the costs could certainly have been projected more accurately. As it was, the canal was not completed until 1914, more than thirty years after it was begun.

At that, it has continued to evolve. Parts were widened in the mid-1900s, and a new phase of widening was begun in 1992. A new set of locks is now under construction, scheduled for completion in 2015. The geology remains a challenge, as seen in figure 20.20B; better understanding of it, coupled with better equipment, makes modern excavation more manageable.

Boston's "Big Dig"

Like many cities, Boston experienced increasing traffic congestion over the last half of the twentieth century. The traffic load on the six-lane, elevated Central Artery nearly tripled from 1959 to the end of the century, creating estimated costs of half a billion dollars a year in wasted fuel in idling gridlocked cars, late delivery charges, and elevated accident rates. In a bold move, the city embarked on a massive transportation overhaul. The two major components of this Central Artery/Tunnel Project (quickly christened the "Big Dig") were an eight-to-ten-lane underground expressway to replace the Central Artery, and an extension of the Massachusetts Turnpike across Boston Harbor to Logan Airport via a four-lane tunnel under the harbor. The Central Artery, slated for demolition once the new expressway opened, had to be kept open throughout construction, and various underground structures pass beneath trains and subways that had to remain operational. Logistics aside, the local geology was a challenge. The city is underlain by a complex assortment of glacial sediments, volcanic rocks, and weakly metamorphosed mudstone; strong bedrock is over 120 feet deep in places. It is also densely built, with many historic buildings whose foundations could be destabilized by careless construction. And the water table is shallow; but pumping water out from active excavations could destabilize adjacent rock and soil.

To keep the excavation walls stable while digging down to bedrock, the project relied on extensive use of *slurry walls* (figure 20.21). As trenches were dug, a clay-water slurry was poured in to replace the earth removed. The slurry is dense and stable enough to hold up the excavation walls until steel reinforcing beams can be sunk into place and concrete poured in—displacing the slurry—to form the permanent wall. In other places, the ground was purposely frozen to keep it firm during construction (just as is often done in permafrost areas).

A different approach was taken for the tunnel under Boston Harbor. Tunnel segments were completed on land, towed to their intended positions, sunk into place, joined, and then pumped dry. (A similar technique was used for the construction

Figure 20.21

Boston's "Big Dig" has posed some major geological-engineering challenges. Slurry walls kept excavations from collapsing and nearby building foundations from failing.

Photograph courtesy Massachusetts Turnpike Authority.

of the Rotterdam Subway in Holland in 1964; there, the problem was a very shallow water table that made conventional tunnel construction in place problematic.)

The Big Dig hasn't been cheap. The size and complexity of the project have resulted in a cost of over $14 billion. But the objectives were achieved, with minimal disruption of the city's operation along the way, and the project, begun in 1991, was essentially complete in spring 2006, with all roads and tunnels open. (In July of that same year, some bolts holding up tunnel ceiling panels unexpectedly failed, and a motorist was killed. However, investigation indicated that the failure was not geology-related.)

Dams, Failures, and Consequences

When a building cracks and crumbles due to settling or subsidence, the repairs may be costly, but the toll rarely includes lives. Even when a bridge or tunnel collapses, only those few persons on or in it at the time are affected. A catastrophic dam failure, on the other hand, can destroy whole towns and take thousands of lives in a matter of minutes; one dam can impound a

The St. Francis Dam

The St. Francis Dam was built in California, about 70 kilometers north of Los Angeles. The reservoir it impounded was principally intended for a water supply. The dam, 60 meters high and over 150 meters wide, was completed in 1926. The valley walls on one side of the dam were made of coarse sandstones and other sedimentary rocks. On the other side, the rocks were schists, mica-rich metamorphic rocks that tended to break along parallel planes that sloped toward the dam and reservoir. The contact between the two rock types, over which the dam was built, was a fault. The presence of that fault was known—and even mapped—before the dam was built. The construction itself omitted several features that would have minimized water seepage under the dam. Cracks in the structure were noted as the reservoir began to fill.

On 12 March 1928, the dam abruptly failed. The contents of the reservoir swept down San Francisquito Canyon. Huge chunks of the concrete dam were carried up to a mile downstream. Damages were estimated at $10 million (figure 20.23A); the official death toll was 430.

Possible reasons for the failure became clear from subsequent laboratory tests, and the fault zone did not actually appear to have been a significant factor. The rocks of the valley walls had seemed to be sufficiently strong when dry. But the sedimentary rocks on one side of the dam were rich in gypsum, which is very soluble. When a fist-sized sample of one of the sedimentary rocks was placed in water, it bubbled out air, soaked up water, and disintegrated, in less than an hour, into a heap of sand and clay. This was hardly an appropriate rock type with which to contain a large reservoir of water! Reinvestigation of the site later also revealed a huge old landslide complex, a mile and a half long, within the schists. Recent studies and modeling now suggest that the filling of the reservoir reactivated the slide—somewhat as at Vaiont, Italy—which put pressure on the dam. Hydrostatic pressure from water seepage underneath it may have tended to buoy up the dam, stressing it further. Apparently, the west side of the dam, destabilized and damaged by softening of the gypsum-rich rocks, failed first, and a rapid, muddy washout began. This reduced the support for the rest of the system. As the water began to pour out, it further eroded the base of the dam. The east side of the dam then collapsed. Somewhat astonishingly, the central section of the dam remained standing (figure 20.23B).

As with the Vaiont Dam case, a little more careful investigation in advance could have averted disaster. In the case of the St. Francis Dam, it seems that testing the reaction of the wall rocks to immersion in water and looking hard enough at the slopes to recognize the old landslide complex should have told the engineers everything they needed to know about site suitability!

Other Examples and Construction Issues

Dams, more dramatically than most other structures, thus underscore the need for careful application of the principles of engineering geology before and during construction. They may also underscore the limits of our understanding. The Baldwin Hills reservoir was built near Los Angeles, California, between 1947

A

B

Figure 20.22

Some 21 million acre-feet of water are impounded in Lake Mead behind Hoover Dam. (A) Close-up of the dam. Note road and vehicles for scale. (B) Regional overview gives a better appreciation of the volume of water impounded in the reservoir. Dam is at lower left (arrow); reservoir includes all the water visible above it and at right in photograph.

Photograph (A) by Lynn Betts, courtesy USDA Natural Resources Conservation Service.

tremendous volume of water (figure 20.22). The motivation for intelligent and careful design and siting of dams is thus particularly strong. Unfortunately, past practice has frequently fallen short of the ideal. In chapter 8, we examined the Vaiont Reservoir disaster. In this section, we will look briefly at another well-documented dam disaster. Neither need have occurred; careful geologic investigation beforehand would have shown that neither site was suitable.

A

B

Figure 20.23

Failure of the St. Francis Dam in California: (A) Twisted railroad tracks testify to the force of the floodwaters. (B) The dam after its collapse. Note the waterline on the surrounding hills, marking the water level in the reservoir before failure.

Photographs by H. T. Stearns, USGS Photo Library, Denver, CO.

and 1951. The geologic setting was thoroughly studied and (so it was believed) adequately taken into account. The reservoir was underlain by an active fault zone; the design incorporated a seismic safety factor double that used by engineers in other earthquake-prone areas. Drainage was provided to prevent saturation of the foundations. The design was conservative, and provision was made for monitoring after construction. Nevertheless, on 14 December 1963, a differential slip of more than 10 centimeters occurred along a fault, water began to scour its way out, and two hours later, the dam was abruptly breached. Clearly, the designers did not understand the geology as well as they thought.

Figure 20.24

Sag ponds and reservoirs abound along the San Andreas Fault; here, San Andreas and Crystal Springs reservoir.

Photograph by R. E. Wallace, USGS Photo Library, Denver, CO.

Geology and geography may, in fact, lead to the building of dams on faults. The zone of weakness created by a major fault zone may become a topographic low along which streams flow and lakes accumulate. This is a natural site, then, for a dam and reservoir, but damming such a stream necessarily involves placing a dam on or very close to the fault. The San Andreas is only one of many faults marked, in part, by the presence of natural sag ponds and artificial reservoirs (figure 20.24).

The majority of the approximately 50,000 dams built in the United States over the last hundred years were not built on faults, but they may nevertheless suffer from inadequate design. All dams are designed to survive any imaginable flood load, typically through the provision of a *spillway* that allows water to flow over or around the dam unimpeded when the water level in the reservoir becomes too high (figure 20.25). The worst-case scenario for which a spillway is constructed is commonly a

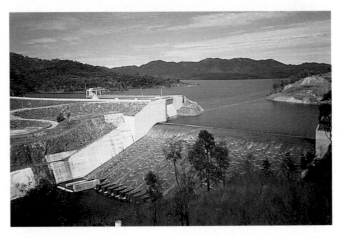

Figure 20.25

Awoonga High Dam in Australia illustrates modern construction methods. View of dam and reservoir; spillway in foreground.

Three Gorges Dam (It's Not Only About Safety)

High-population developing nations such as China have large energy needs. Historically, China has relied on fossil fuels—especially coal—and the implications for greenhouse-gas emissions if per-capita energy consumption rises sharply in this nation of a billion-plus are sobering. Even now, particulate pollution from coal burning is a serious local health problem. Before energy became an issue, flooding was a major concern on the Yangtze River, the world's third largest. It has been estimated that in the twentieth century alone, half a million people died in floods of the Yangtze.

As long ago as 1919, damming the Yangtze was proposed as a means to control the flooding and enhance navigation. Structurally, a logical place was the Three Gorges area, a section of the river edged with soaring cliffs up to 1500 meters (about a mile) high and surrounded by high mountains. In 1953, Mao Zedong advocated construction of a large dam in the Three Gorges area, with a view also to development of a major hydropower facility. Finally, in 1992, the Chinese government announced a commitment to proceed, and

construction began in 1993, on a site within the Three Gorges area that offers solid granite bedrock for miles around.

The Three Gorges Dam (figure 1) is the largest in the world, 185 meters tall, and nearly 2.3 kilometers (about a mile and a half) long. With 26 hydropower generators, the complex has a collective generating capacity of about 18,200 megawatts, making it also the world's largest hydropower project—and six more generating units are planned for installation in one of the banks beside the dam. (For comparison, a typical large nuclear-power plant has a capacity of about 1000 megawatts.) The complex could generate over 10% of China's total energy, and reduce its burning of coal by an estimated 50 million tons per year. Damage and loss of life from flooding should be greatly reduced. Impoundment of water upriver will make the Yangtze navigable by much larger ships farther upstream than before, and a system of locks to get the ships past the dam is part of the project.

As with all large projects of this kind, however, there are negative consequences, and there has been vocal opposition to the

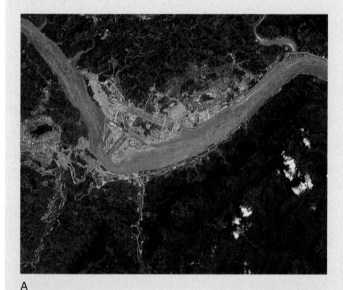

A

B

Figure 1

Three Gorges Dam site. (A) With the dam only partly complete, the sediment-laden Yangtze (gray) still flows past the site (left to right in this image); clear channel is probably an independent lock system for boats. (B) The dam was completed in May 2006. Most of the sediment is now trapped far upstream where the river enters the still-growing reservoir.

(A) Image by Jesse Allen, Earth Observatory, using ASTER data made available by NASA/GSFC/MITI/ERSDAC/JAROS, and U.S./Japan ASTER Science Team.

(B) Image courtesy Earth Sciences and Image Analysis Laboratory, NASA Johnson Space Center.

Three Gorges Dam since long before construction began. The reservoir behind the dam will ultimately stretch 662 km (about 414 miles) long, farther than the distance between Denver, CO and Salt Lake City, UT. It will submerge approximately 125,000 acres of prime farmland, covering 1000 square kilometers (440 square miles) overall. Cities along the banks in the area the reservoir occupies have to be abandoned and their residents moved elsewhere; an estimated 1.3 to 1.9 million people are ultimately being thus displaced. Cultural and archaeological features and sites that cannot be moved—1300 excavated sites, and an estimated 8000 unexcavated sites—are being lost. The scenic qualities of the Three Gorges area will be significantly and permanently altered. Ecologists fear the impact of habitat loss, especially on rare or endangered species that live in the Yangtze—for example, the Chinese river dolphin and Chinese paddlefish, which may already have vanished; the Yangtze soft-shell turtle, Chinese sturgeon, Siberian crane, and others.

Basic safety concerns are natural with a project of this scale. Beyond the structural integrity of the construction, one can wonder about the effect of sedimentation as the muddy Yangtze is impounded: in 1975, the sedimentation-related failure of the Banqiao Dam, which triggered 61 other dam failures, led to 26,000 deaths. Reservoir-induced seismicity has already been observed, whether due to reactivation of old faults or to collapse of karst caves in limestone wallrocks of the reservoir, with quakes up to magnitude 3.4 having been noted since the reservoir began to fill in 2003. A larger concern is the possible effect on two major fault zones in the region, which historically have experienced quakes over magnitude 6, and the dam's ability to withstand such quakes: It is built to withstand earthquakes as large as magnitude 7, but China has certainly experienced larger quakes than that.

Landslides are a major issue. Historic slides in weak, dipping sedimentary rocks are common along the Yangtze. Even the relocation of villages has triggered some new sliding. Within a month of the 2003 start of reservoir filling, old landslides at Shuping and Qianjiangping were reactivated. Many known old slides are being monitored, which enabled scientists to warn residents as slip of the Qianjiangping slide began, but even so, 14 lives were lost, with 10 people missing, and 4 factories and over 100 homes were destroyed by the huge, 24-million-cubic-meter slide. In 2007, reactivation of the Xemaomian slide forced evacuation of a village. Monitoring and bank-strengthening efforts (figure 2) continue. The filling of the reservoir is not the only potential trigger, either. For flood-control purposes, the level of the reservoir, filled to 175 meters above sea level during the winter, will be lowered to 145 meters during the rainy spring/summer season. Geoscientists are concerned that such an annual wetting/drying cycle will further destabilize slopes in the affected zone.

Figure 2

Reinforcing slide-prone slopes along the Yangtze is one strategy for reducing landslide risks.

Photograph by Lynn Highland, U.S. Geological Survey

The planned annual fluctuation in reservoir level may have further consequences. The land exposed when the level is lowered will be attractive as farmland. However, when the level is raised and the land again submerged in the winter, fertilizers, herbicides, and insecticides will be dissolved and washed into the reservoir, polluting the water. Experts are now debating whether farming should, in fact, be allowed in this reservoir border zone.

Preliminary studies indicate additional early environmental impacts of the Three Gorges Dam project. The Yangtze River Basin is the largest in South Asia, and thousands of dams within it had already considerably reduced the sediment load reaching the East China Sea. Within two years after the Three Gorges Dam first began impounding water (2003), that sediment load was cut again by almost half. It is now not much over 100 million tons per year, compared to nearly 500 million tons per year in the mid-twentieth century. Observed results include increased channel erosion and erosion of the Yangtze delta. Future coastal erosion may threaten such important areas as Shanghai. Further, the large and growing reservoir behind the dam, with its broad surface subject to evaporation, is affecting local weather—the evaporation process cooling the air and altering circulation/wind patterns, with demonstrable changes in the distribution of rainfall not just locally but regionally. Monitoring the longer-term impacts should prove interesting.

A

B

Figure 20.26

(A) Cross-sectional view of failed earthen dam at Taum Sauk upper reservoir. Earthen dams without spillways are especially prone to washout. (B) Rushing floodwater from the upper reservoir scoured down to bedrock, carrying away forest and soil.

Photographs from Federal Energy Regulatory Commission.

1000-year flood. Still, as noted in chapter 6, it may be difficult to project just what volume of water that (or any other) large, infrequent flood involves, and changing land-use patterns may be altering the flood-frequency curves also. Dam failures have occurred, too, from dam washout as the sides or face of the spillway were scoured away by floodwaters. In 1981, the U.S. Army Corps of Engineers surveyed dams in this country. They identified 8639 of these dams as ones whose failure would cause serious property damage or loss of life. Of that number, one-third were identified as unsafe, and in over 80% of these cases, the problem was an inadequate or poorly designed spillway.

Thus, dam construction becomes increasingly complex and sophisticated, aided by computers, laboratory tests, field studies, and scale-model experiments, and closer monitoring also increases safety. Still, failures continue. A quite recent one occurred at the Taum Sauk hydroelectric project in Missouri. This project, completed in 1963, is a "pumped storage" system, with an upper and lower reservoir. During peak-demand times, turbines generate electricity as water flows from the upper to the lower reservoir. At night or when demand is light, water is pumped back up from the lower to the upper reservoir. On the night of 14 December 2005, the dam on the upper reservoir abruptly failed (figure 20.26). Investigation suggested that faulty instrumentation/monitoring of the water level in the up-

per reservoir caused it to be overfilled, overtopping and then washing out the dam (which had no spillway). Approximately a billion gallons of water swept downslope. Fortunately, the lower reservoir is designed to hold most of the volume of the upper; there was limited overflow from the lower reservoir, its dam held, and there were no fatalities. The upper reservoir has since been rebuilt, its design improved, and the project is back in operation.

Global concern now is focusing increasingly on the rapid proliferation of dams in some developing countries. In 2000, the World Commission on Dams issued a comprehensive review of large dams worldwide. For this purpose, "large dams" were defined as those over 15 meters high and those 5 to 15 meters high with reservoirs over 3 million cubic meters. Altogether, there are more than 45,000 such dams. Nearly half of these (22,000) are in China, and most of China's large dams have been built since 1950, which means an average of about one large dam *per day* built since that time. The reasons historically have been primarily flood control and irrigation, with increased hydropower-generating capacity a growing motive. In addition to such issues as displacement of people and flooding of habitat and cultural resources (figure 20.27), the World Commission noted serious safety concerns that have emerged as so many large dams have been built so quickly.

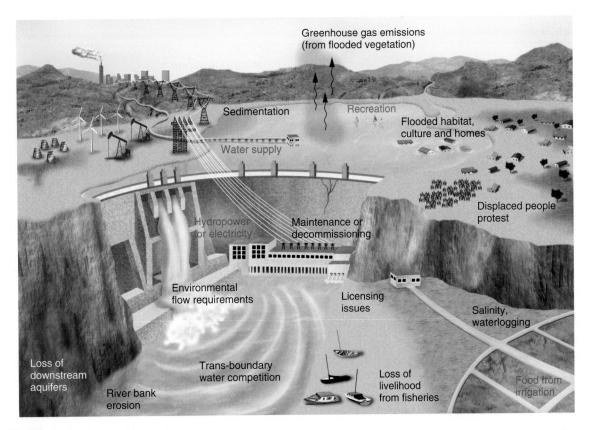

Figure 20.27

Benefits and issues involved in dam construction are many. Safety is an additional, very basic one, which can arise even with new dams, especially if design and construction are hurried.

After World Commission on Dams report, 2000.

Summary

Engineering geology is concerned with making structures as safe and stable as possible, given various kinds of geologic hazards and potential problems. Land-use planning encompasses the same concerns as engineering geology, combined with additional geologic and nongeologic considerations. The aim of the land-use planner is to make the best possible use of limited land, taking all these factors into account. Using the same land for several purposes, either simultaneously or sequentially, is one approach to conserving the land resource. The success of both geological-engineering and land-use-planning efforts is heavily dependent on the accuracy and completeness of the data available to the individuals carrying out these tasks. As with pollution control, relative costs and benefits may have to be weighed in deciding how or whether to undertake a particular construction project.

Key Terms and Concepts

geographic information   multiple use 478   sequential use 479
   system 481

Exercises

Questions for Review

1. Describe the concepts of *multiple use* and *sequential use,* and give two examples of each.

2. Maps of geologic or other factors are useful tools in land-use planning, but their usefulness can be limited by practical problems. Describe two such possible limitations.

3. How can the computer's ability to process large volumes of data assist in the planning process? How can programs be adjusted for different planning objectives?

4. Cite at least ten geologic considerations that might be important in siting an apartment building. How many of the same considerations would be relevant to siting its parking lot?

5. What is permafrost, and why is construction made more difficult by its presence?

6. How is climate change affecting recovery of oil from Alaska's North Slope?

7. How can engineers often minimize problems posed by unstable clays in near-surface rock and soil layers?

8. Dams are not infrequently built over faults. Why?

9. Identify and explain any five concerns or negative consequences associated with large dam projects, such as Three Gorges.

Exploring Further

1. Most city or county planning offices have long-range plans for development in undeveloped areas. Seek out such plans for your area, if available, and investigate what kinds of considerations (geologic or otherwise) went into those plans.

2. Make a walking tour of your neighborhood or of another area to look for signs of careless or thoughtless engineering practice—buildings showing severe cracking or other structural damage from ground failure beneath, serious erosion of steep slopes, large puddles accumulated after rain because roads or buildings dam surface runoff, and so on. Is any particular problem especially common?

3. Go to NASA's Earthobservatory "Image of the Day" site (see NetNotes). What image is featured, and how was it acquired? Were any special remote-sensing techniques involved? Explore other images from the archive to see how satellites with specialized detectors help scientists investigate and monitor the Earth.

4. Check the history/progress/current status of a major project such as the Big Dig, Three Gorges Dam, the Keystone XL pipeline, or the Qinghai-Tibet Railway. Or explore a particular aspect of Three Gorges impacts, such as the history of landslides or seismicity since reservoir filling began in 2003.

5. The online National Atlas site allows you to create customized maps of all kinds of data. Explore at nationalatlas.gov/. Or do the same for maps from the U.S. Geological Survey at nationalmap.gov/.

Introduction

Much of the understanding of geologic processes, including the rates at which they occur, and therefore the kinds of impacts they may have on human activities, is made possible through the development of various means of "telling time" in geologic systems. In this appendix, we will explore several of these methods. A final section examines some applications of geologic age determinations to the study of process rates.

Relative Dating

Arranging Events in Order

Before any ways of establishing numerical ages of rocks or geologic events were known, it was sometimes at least possible to place a sequence of events in the proper order. Among the earliest recognized efforts in this direction were those of Nicolaus Steno. In 1669, he set forth two very basic principles that could be applied to sedimentary rocks. The first, the **Principle of Superposition** (figure A.1), pointed out that, in an undisturbed pile of sediments or sedimentary rocks, those on the bottom were deposited first, followed in succession by the layers above them, ending with the youngest on top. (Today, this idea may seem so obvious as not to be worth stating, but at the time it represented a real step forward in thinking logically about rocks.) The second, the **Principle of Original Horizontality,** was based on the observation that sediments are deposited in approximately horizontal, flat-lying layers. Therefore, if one comes upon sedimentary rocks in which the layers are folded or dipping steeply, they must have been displaced or deformed after deposition and solidification into rock (figure A.2).

In later centuries, these ideas were extended to work igneous rocks into such sequences (figure A.3). If an igneous

A

B

Figure A.2

(A) A tilted sequence of sedimentary rocks in the Canadian Rocky Mountains. Recall also figure 8.26C. (B) These folded marbles and slates in Kings Canyon National Park, California, were originally flat-lying beds of limestone and shale.

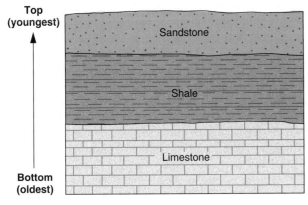

Figure A.1

The Principle of Superposition: The rocks on the bottom of a sequence of undisturbed sedimentary rocks were deposited first, and the depositional ages become younger higher in the sequence.

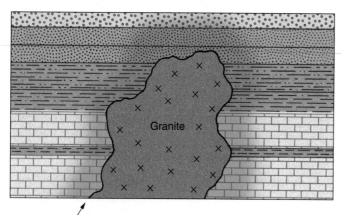

Rocks adjacent to intruding magma may also be metamorphosed by its heat.

A

B

Figure A.3

(A) An igneous rock crosscutting a sedimentary sequence or other rock must postdate the rocks it cuts across. (B) Here, veins of granite cut through a schist; the finer-grained granite at bottom partially crosscuts, and is thus younger than, the coarser.

rock cuts across layers of sedimentary rocks, the sedimentary rocks must have been there first, the igneous rock introduced later. This is sometimes called the *Principle (or Law) of Crosscutting Relationships,* and in fact it applies whether the rocks that are crosscut are sedimentary or not, though the relationships may be easier to recognize if they are. Often, there is a further clue to the correct sequence: The hotter igneous rock may have "baked," or metamorphosed, preexisting rocks immediately adjacent to it, so again the igneous rock must have come second. If an igneous rock contains pieces of other rock types (figure A.4), those pieces must have been picked up as solid chunks by the invading magma, so the rocks making up the inclusions must predate the host rock *(Principle of Inclusion).*

Figure A.4

The darker rock is a schist that was invaded by the magma that formed the (lighter-colored) granite, which picked up pieces of schist. The schist is thus older than the granite.

Such geologic common sense can be applied in many ways. If a strongly metamorphosed sedimentary rock is overlain by a completely unmetamorphosed one, for instance, the metamorphism must have occurred after the first sedimentary rock formed but before the second. Quite complex sequences of geologic events can be unraveled by taking into account such principles. See figure A.5 for examples.

Correlation

Fossils play a role in the determination of relative ages, too. The concept that fossils could be the remains of older life-forms dates back at least to the ancient Greeks, but for some time it fell out of favor. The idea was seriously revived in the 1700s, and around the year 1800, William Smith put forth the **Law of Faunal Succession.** The basic principle is that, through time, life-forms change (figure A.6); old ones disappear from and new ones appear in the fossil record, but the same form is never exactly duplicated at two different times in history. (The same concept could be applied to other contexts, such as car styles in a junkyard with layer upon layer of debris: Car models come and go, and while stylistic features from an earlier era may be revived in later models, each model, as a whole, corresponds to a unique period of manufacture.) This principle, in turn, implies that when one finds exactly the same type of fossil preserved in two rocks, even if the rocks are quite different compositionally and geographically widely separated, they should be the same age. Smith's law thus allowed age *correlation* between rock units exposed in different places (figure A.7). A limitation on its usefulness is that it can be applied only to rocks in which fossils are preserved, which are almost exclusively sedimentary rocks.

A

B

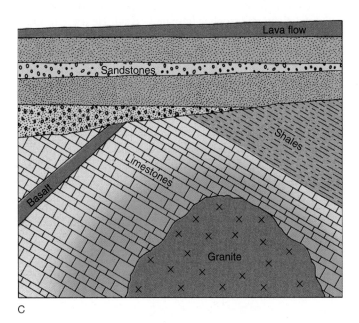

C

Figure A.5

(A) The Principle of Superposition and Principle of Original Horizontality together indicate that the sequence of events shown here in sedimentary rocks of the Grand Canyon was: deposition of the lower layers (dark red and gray); tilting and erosion; deposition of the still-flat-lying upper layers (green, cream, and light red). (B) The coarse gray granite is younger than the darker metasedimentary rocks (Principle of Inclusion) but older than the light quartz/feldspar vein (Crosscutting Relationships). (C) Sorting out relative ages of rock in an outcrop: Of the sedimentary rocks, the limestones must be the oldest (Principle of Superposition), followed by the shales. The granite intrusion and basalt must both be younger than the limestone they crosscut; the granite has also metamorphosed the surrounding limestone. It is not possible to tell whether the igneous rocks predate or postdate the shales (because they are not crosscut or metamorphosed by the igneous rocks), or to determine whether the sedimentary rocks were tilted before or after the igneous rocks were emplaced. After the limestones and shales were tilted and the basalt injected, they were eroded, and then the sandstones were deposited on top. Finally, the lava flow covered the entire sequence.

(A) © The McGraw-Hill Companies, Inc./Doug Sherman, photographer.

The foregoing ideas were all useful in clarifying age relationships among rock units. They did not, however, help answer questions like: How old is this granite? How long did it take to deposit this limestone? How recently, and over how long a period, did this apparently extinct volcano erupt? Has this fault been active in modern times?

Uniformitarianism

Geologic processes were the focus of another early worker—the physician, farmer, and part-time geologist James Hutton. Hutton is widely credited with developing and popularizing the concept of **uniformitarianism.** Uniformitarianism is sometimes described briefly by the phrase "the present is the key to the past." What Hutton meant by this is that because the same physical laws have operated throughout the earth's history, by studying present, active geologic processes and their products, geologists can infer much about how rocks were formed and changed in the past. Today, a volcanic eruption and the hardening of flowing magma into basaltic rock suggest the source of an ancient basalt flow even if geologists cannot now find the volcano from which it erupted. A broad, white sandy beach or a windswept desert might offer some useful insights about the origins of sandstones. Retreating ice sheets in Greenland and Antarctica leave moraines behind, which help geologists to understand the source and significance of the moraines found in the midwestern United States, Canada, and elsewhere.

Uniformitarianism has sometimes been misunderstood to mean that the *rates* of geologic processes have been the same through time as well, that by observing processes like erosion, sedimentation, and seafloor spreading as they now occur, one can know the rate at which they occurred in the past. Unfortunately,

Figure A.6

Ammonites are a now-extinct group of organisms related to such modern marine creatures as the chambered nautilus, squid, and octopus. This display at the Smithsonian Institution shows some of the variety of their fossil forms; inset is a cutaway view of one species.

this is not true, and was not really part of uniformitarianism as originally conceived. Volcanism may have involved the same types of physical and chemical processes throughout geologic history, but it might have been much more active early in the earth's history, before the earth cooled appreciably, than it is now; rates of chemical sedimentation, as well as the proportions of different minerals in the sediments, could have been very different in the past, when the temperature and chemistry of the oceans were also different; and so on. In short, even uniformitarianism

does not allow numerical answers to questions about geologic process rates in the past.

How Old Is the Earth?

Geologists and nongeologists alike have been fascinated for centuries with the very basic question of the earth's age. Many have attempted to answer it, but with little success until the twentieth century.

Early Efforts

One of the earliest widely publicized determinations of the age of the earth was the seventeenth-century work of Archbishop Ussher of Ireland. He painstakingly and literally counted up the generations in the Bible and arrived at a date of 4004 B.C. for the formation of the earth. This very young age was impossible for most geologists to accept; the complex geology of the modern earth seemed to require far longer to develop.

Even less satisfactory from that point of view was the estimate of the philosopher Immanuel Kant. He tried to find a maximum possible age for the earth, assuming that the sun had always shone down on earth and that the sun's tremendous energy output was due to the burning of some sort of conventional fuel. But a mass of fuel the size of the sun would burn up in only 1000 years, given the rate at which the energy is released, plainly an impossible result in view of several millennia of recorded human history. Kant, of course, knew nothing of the nuclear fusion that actually powers the sun.

Nineteenth-Century Views

About 1800, Georges L.L. de Buffon attacked the problem from another angle. He assumed that the earth was initially molten,

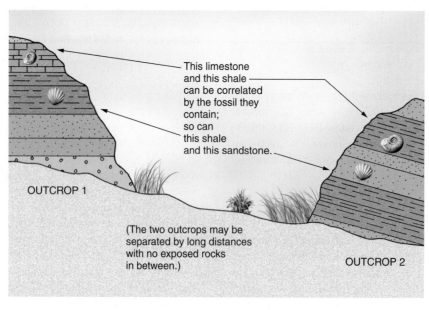

This limestone and this shale can be correlated by the fossil they contain; so can this shale and this sandstone.

OUTCROP 1

(The two outcrops may be separated by long distances with no exposed rocks in between.)

OUTCROP 2

Figure A.7

Similarity of fossils suggests similarity of ages even in different and quite widely separated rocks.

modeled it as a ball of iron, and calculated how long it would take this quantity of iron to cool to present surface temperatures: 75,000 years. To most geologists, this still was not nearly long enough.

Around 1850, physicist Hermann L.F. von Helmholtz took the approach of supposing that the sun's luminosity was due to infall of particles into its center, converting gravitational potential energy to heat and light. This gave a maximum age for the sun (and presumably earth) of 20 to 40 million years. Again, his assumptions were wrong, so his answer was also.

In the late 1800s, Lord William T. Kelvin reworked Buffon's calculations, modeling the earth more realistically in terms of rock properties rather than those of metallic iron. Interestingly, he, too, arrived at an age of 20 to 40 million years for the earth. What he did not take into account, because natural radioactivity was then unknown, was that some heat is continually being *added* to the earth's interior through radioactive decay, so it has taken longer to cool down to its present temperature regime.

The calculations went on, and something was wrong with each. In 1893, U.S. Geological Survey geologist Charles D. Walcott tried to compute the total thickness of the sedimentary rock record throughout geologic history and, dividing by typical modern sedimentation rates, to estimate how long such a pile of sediments would have taken to accumulate. His answer was 75 million years. Walcott was hampered in his efforts by several factors, including the gaps in many sedimentary rock sequences (periods during which no sedimentation occurred or some sediments were eroded away), and the reality that there is no one spot on earth where sediments have always been accumulating. He also had no good idea of the total thickness of sediments deposited in the time before organisms capable of preservation as fossils became widespread, which turns out to be most of earth's history.

In 1899, physicist John Joly published calculations based on the salinity of the oceans. Taking the total dissolved load of salts delivered to the seas by rivers, and assuming that the ocean was initially pure water, he determined that it would take about 100 million years for the present concentrations of salts to be reached. Aside from Joly's assumption that rates of weathering and salt input into the oceans throughout earth's history were constant, he did not consider that the buildup of salts is slowed by the removal of some of the dissolved material—for example, as chemical sediments.

So the debate continued, imaginative and frequently heated, until the discovery of radioactivity provided a much more powerful and accurate tool with which to undertake a solution.

Radiometric Dating

Henri Becquerel did not set out to solve any geologic problems nor even to discover radioactivity. He was interested in a curious property of some uranium salts: If exposed to light, the salts continued to emit light for a while afterward even in a dark room (phosphoresce). He had put a vial of uranium salts away in a drawer on top of some photographic plates well wrapped in black paper. When he next examined the photographic plates, they were fogged with a faint image of the vial, as if they had somehow been exposed to light right through the paper. The uranium was emitting something that could pass through light-opaque materials. We now know that uranium is one of several substances that are naturally radioactive, that undergo spontaneous decay. Once the phenomenon of radioactive decay was reasonably well understood, physicists and geoscientists began to realize that it could be a very useful tool for investigating earth's history.

Radioactive Decay and Dating

As noted in chapter 16, one key property of any particular radioisotope is that it decays at a constant, characteristic rate, with a distinct half-life, which can be determined in the laboratory. One can then, in principle, use the relative amounts of a decaying isotope (parent) and the product isotope into which it decays (daughter) to find the age of the sample; see figure A.8. Suppose that parent isotope A decays to daughter B with a half-life of 1 million years. In a rock that contained some A and no B when it formed, A will gradually decay and B will accumulate. After 1 million years (one half-life of A), half of the atoms of A initially present will have decayed to yield an equal number of atoms of B. After another half-life, half of the remaining half will have decayed, so three-fourths of the initial number of atoms of A will have been converted to B and one-fourth will remain; the ratio of B to A will be 3:1. After another million years, there will be seven atoms of B for every atom of A, and so on. This is **exponential decay,** decreasing abundance over time in a pattern that is just the opposite of the exponential growth of populations discussed in chapter 1. If a sample contains 24,000 atoms of B and 8000 atoms of A, then, assuming no B was present when the rock formed and noting the 3:1 ratio of B to A, we would conclude that the sample was two half-lives of A (in this case, 2 million years) old.

The foregoing is, of course, an oversimplification of the dating of natural geologic samples. Many samples contain some of the daughter isotope as well as the parent at the time of formation, and correction for this must be made. Many other samples have not remained chemically closed throughout their histories; they have gained or lost atoms of the parent or daughter isotope of interest, which would make the apparent date incorrect. In the case of samples with a complex history—an igneous rock that has been metamorphosed once or twice and also weathered, for example—it can be difficult both to obtain a date and to decide which event, if any, is being dated! Fortunately, there are ways to recognize, and in many cases correct for, disturbances in isotopic systems. For a somewhat more thorough discussion of the complexities of dating of geologic systems, refer to pertinent texts in the online reference list.

Choice of an Isotopic System

Several conditions must be satisfied by an isotopic system to be used for dating geologic materials. The parent isotope chosen must be abundant enough in the sample for its quantity to be measurable. Either the daughter isotope must not normally be incorporated into the sample initially, or there must be a means

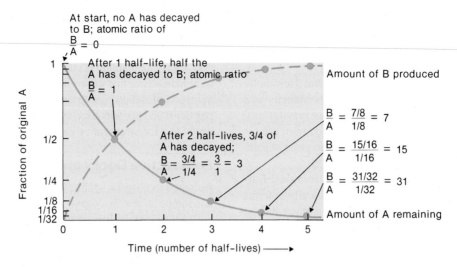

Figure A.8

Decay of radioactive parent A proceeds with complementary accumulation of daughter B. All ratios represent relative numbers of atoms.

of correcting for the amount initially present. The half-life of the parent must be appropriate to the age of the event being dated: long enough that some parent atoms are still present but short enough that some appreciable decay and accumulation of daughter atoms have occurred. A radioisotope with a half-life of ten days would be of no use in dating a million-year-old rock; the parent isotope would have decayed away completely long ago, and there would be no way to tell how long ago. Conversely, a radioisotope with a half-life of 10 billion years would be useless for dating a fresh lava flow because the atoms of the daughter isotope that would have accumulated in the rock would be too few to be measurable. Only about half a dozen isotopes are widely used in dating geologic samples. They include both uranium-235 and uranium-238 (with half-lives of 700 million years and 4.5 billion years, respectively), potassium-40 (1.3 billion years), rubidium-87 (49 billion years), and carbon-14 (5730 years, meaning that carbon-14 is useful only for quite young samples, such as remains from prehistoric civilizations).

Radiometric and Relative Ages Combined

Radiometric dates (dates determined using radioisotopic methods) are sometimes imprecisely called "absolute" ages to distinguish them from the relative ages determined as described earlier. For various technical reasons, accurate radiometric dates cannot be determined for many rocks and fossils, so "absolute" and relative dating methods are often used in conjunction (figure A.9). Undateable sedimentary rocks crosscut by an igneous intrusion must at least be older than the igneous rock, which may be dateable; a fossil found in a rock sandwiched between two dateable lava flows has its age bracketed by the ages of the flows; and so on.

Radiometric dating has proven an extremely powerful tool for studying the earth's history. Because the earth is geologically still very active, no rocks have been preserved unchanged since its formation. However, the dating of meteorites and moon rocks, coupled with geochemical evidence that the whole solar system formed at the same time, has led to the determination of

an age of about 4.5 billion years for the earth. The oldest samples from the continents are close to 4 billion years old. Sea floor is much more readily destroyed/recycled by plate tectonics; the oldest seafloor samples recovered are only about 200 million years old.

The Geologic Time Scale

Particularly in the days before radiometric dating, it was necessary to establish some subdivisions of earth history by which particular intervals of time could be indicated. This initially was done using those rocks with abundant fossil remains. The resultant time scale was subsequently refined with the aid of radiometric dates.

The Law of Faunal Succession and the appearance and disappearance of particular fossils in the sedimentary rock record were first used to mark the boundaries of the time units. The principal divisions were the *eras*—**Paleozoic, Mesozoic,** and **Cenozoic,** meaning, respectively, "ancient life," "intermediate life," and "recent life." The eras were subdivided into *periods* and the periods into *epochs*. The names of these smaller time units were assigned in various ways. Often, they were named for the location of the *type section*, the place where rocks of that age are well exposed and that time unit was first defined. For example, the Jurassic period is named for the Jura Mountains of France. Other periods are named on the basis of some characteristic of the type rocks. The Cretaceous period derives its name from the Latin *creta* ("chalk"), for the chalky strata of that age in southern England and northern Europe; rocks of Carboniferous age commonly include coal beds. All of the relatively unfossiliferous rocks that seemed to be older (lower in the sequence) than the Cambrian period were lumped together as pre-Cambrian, later formally named **Precambrian.**

With the advent of radiometric dating, numbers could be attached to the units and boundaries of the time scale. It became apparent that, indeed, geologic history spanned long periods of time and that the most detailed part of the scale, the **Phanerozoic**

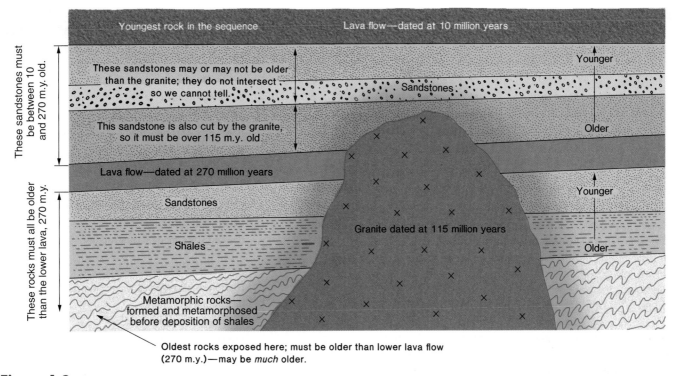

These sandstones must be between 10 and 270 m.y. old.

Youngest rock in the sequence   Lava flow—dated at 10 million years

These sandstones may or may not be older than the granite; they do not intersect so we cannot tell.   Sandstones   Younger

This sandstone is also cut by the granite, so it must be over 115 m.y. old.   Older

Lava flow—dated at 270 million years

These rocks must all be older than the lower lava, 270 m.y.

Sandstones   Younger

Granite dated at 115 million years

Shales   Older

Metamorphic rocks—formed and metamorphosed before deposition of shales

Oldest rocks exposed here; must be older than lower lava flow (270 m.y.)—may be *much* older.

Figure A.9

"Absolute" and relative dating together can put constraints on ages of units not dateable directly.

(Cambrian and later), was by far the shortest, comprising less than 15% of earth history. The approximate time framework of the Phanerozoic is shown in table A.1.

The Precambrian has continued to pose something of a problem. A unit that spans 4 billion years of time seems to demand subdivision—but, in the virtual absence of fossils, on what basis? The principal division that has been made splits the Precambrian into two nearly equal halves—the *Archean* ("ancient") and *Proterozoic* ("pre-life"). The Archean spans the time from the earth's formation to 2500 million years ago, the Proterozoic from 2500 million years ago to the start of the Cambrian (570 million years ago). Universal agreement has not yet been reached on how to subdivide these 2-billion-year blocks of time further.

At the other end of the time scale, a debate has recently developed about whether human impacts on the planet have become so significant that a new name should be assigned for the time from the present going forward—a new epoch, perhaps. The name "Anthropocene" has been proposed, from the Greek *anthropos,* for "human being." A key point of disagreement is whether humans' effects will, over the long march of geologic time, really leave a distinct and substantial mark in the geologic record. This discussion is ongoing.

Geologic Process Rates

The rates at which geologic processes occur can be estimated in a variety of ways, many of which rely on radiometric dating techniques. For example, the rate of seafloor spreading away from a particular spreading ridge can be found by dividing the age of a sample of sea floor into its present distance from the ridge. If a 10-million-year-old seafloor sample is now 400 kilometers from the ridge, then, on average, it has been moved away from the ridge at a rate of

$$\frac{400 \text{ km}}{10,000,000 \text{ yr}} \times \frac{1000 \text{ m}}{1 \text{ km}} \times \frac{100 \text{ cm}}{1 \text{ m}} = \frac{4 \text{ cm}}{\text{yr}}$$

The minimum rate of uplift of rocks in a mountain range might be estimated from the age of marine sedimentary rocks in the mountains, once deposited under water, now high above sea level. Such uplift rates are typically 1 centimeter per year or less, often much less. Beaches formed on Scandinavian coastlines during the last Ice Age have been rising since the ice sheets melted and their mass was removed from the land. From the beach deposits' ages and present elevation, uplift rates can be approximated. Typical rates of this postglacial rebound are on the order of 1 centimeter per year. Radiometric dating has shown that a small volcano may stay active for over 100,000 years, a major volcanic center for 1 to 10 million years. To build a large mountain range may take 100 million years.

Rates of continental erosion due to weathering can be deduced from the loads of major rivers draining the continents, dividing the volume of rock those loads represent by the surface area of the corresponding drainage basin(s). The North American continent is being leveled by erosion at an average rate of 0.03 millimeters per year, or about a tenth of an inch per century. The larger the area over which such

Table B.1 A Brief Mineral Identification Key

Mineral	Formula	Color	Hardness	Other Characteristics
amphibole (e.g., hornblende)	$(Na,Ca)_2(Mg,Fe,Al)_5Si_8O_{22}(OH)_2$	green, blue, brown, black	5 to 6	often forms needlelike crystals; two good cleavages forming 120-degree angle
apatite	$Ca_5(PO_4)_3(F,Cl,OH)$	usually yellowish	5	crystals hexagonal in cross section
azurite	$Cu_3(CO_3)_2(OH)_2$	vivid blue	3½ to 4	often associated with malachite
barite	$BaSO_4$	colorless	3 to 3½	high specific gravity, 4.5 (denser than most silicates)
beryl	$Be_3Al_2Si_6O_{18}$	aqua to green	7½ to 8	usually found in pegmatites
biotite (a mica)	$K(Mg,Fe)_3AlSi_3O_{10}(OH)_2$	black	5½	excellent cleavage into thin sheets
bornite	Cu_5FeS_4	iridescent blue, purple	3	metallic luster
calcite	$CaCO_3$	variable; colorless if pure	3	effervesces (fizzes) in weak acid; cleaves into rhombohedra
chalcopyrite	$CuFeS_2$	brassy yellow	3½ to 4	
chlorite	$(Mg,Fe)_3(Si,Al)_4O_{10}(OH)_2$	light green	2 to 2½	cleaves into small flakes
cinnabar	HgS	red	2½	earthy luster; may show silver flecks
corundum	Al_2O_3	variable; colorless in pure form	9	most readily identified by its hardness
covellite	CuS	blue	1½ to 2	metallic luster
dolomite	$CaMg(CO_3)_2$	white or pink	3½ to 4	powdered mineral effervesces in acid
epidote	$Ca_2FeAl_2Si_3O_{12}(OH)$	green	6 to 7	glassy luster
fluorite	CaF_2	variable; often green or purple	4	cleaves into octahedral fragments; may fluoresce in ultraviolet light
galena	PbS	silver-gray	2½	metallic luster; cleaves into cubes
garnet	$(Ca,Mg,Fe)_3(Fe,Al)_2Si_3O_{12}$	variable; often dark red	7	glassy luster
graphite	C	dark gray	1 to 2	streaks like pencil lead
gypsum	$CaSO_4 \cdot 2H_2O$	colorless	2	
halite	$NaCl$	colorless	2½	salty taste; cleaves into cubes
hematite	Fe_2O_3	red or dark gray	5½ to 6½	red-brown streak regardless of color
kaolinite	$Al_2Si_2O_5(OH)_4$	white	2	earthy luster
kyanite	Al_2SiO_5	blue	5 to 7	found in high-pressure metamorphic rocks; often forms bladelike crystals
limonite	$Fe_4O_3(OH)_6$	yellow-brown	2 to 3	earthy luster; yellow-brown streak
magnetite	Fe_3O_4	black	6	strongly magnetic
malachite	$Cu_2CO_3(OH)_2$	green	3½ to 4	often forms in concentric rings of light and dark green
molybdenite	MoS_2	dark gray	1 to 1½	cleaves into flakes; more metallic luster than graphite
muscovite (a mica)	$KAl_3Si_3O_{10}(OH)_2$	colorless	2 to 2½	excellent cleavage into thin sheets
olivine	$(Fe,Mg)_2SiO_4$	yellow-green	6½ to 7	glassy luster
phlogopite (a mica)	$KMg_3AlSi_3O_{10}(OH)_2$	brown	2½ to 3	closely resembles biotite, but less black
plagioclase feldspar	$(Na,Ca)(Al,Si)_2Si_2O_8$	white to gray	6	may show fine parallel striations on cleavage surfaces
potassium feldspar	$KAlSi_3O_8$	white; often stained pink or aqua	6	two good cleavages forming a 90-degree angle; no striations
pyrite	FeS_2	yellow	6 to 6½	metallic luster; black streak
pyroxene (e.g., augite)	$(Na,Ca,Mg,Fe,Al)_2Si_2O_6$	usually green or black	5 to 7	two good cleavages forming a 90-degree angle
quartz	SiO_2	variable; commonly colorless or white	7	glassy luster; conchoidal fracture

Mineral	Formula	Color	Hardness	Other Characteristics
serpentine	$Mg_3Si_2O_5(OH)_4$	green to yellow	3 to 5	waxy or silky luster; may be fibrous (asbestos)
sillimanite	Al_2SiO_5	white	6 to 7	occurs only in metamorphic rocks; often forms needlelike crystals
sphalerite	ZnS	yellow-brown	3½ to 4	glassy luster
staurolite	$Fe_2Al_9Si_4O_{20}(OH)_2$	brown	7 to 7½	found in metamorphic rocks; elongated crystals may have crosslike form
sulfur	S	yellow	1½ to 2½	sulfurous odor
sylvite	KCl	colorless	2	cleaves into cubes; salty taste, but more bitter than halite
talc	$Mg_3Si_4O_{10}(OH)_2$	white to green	1	greasy or slippery to the touch
tourmaline	(Na,Ca)-$(Li,Mg,Al)(Al,Fe,Mn)_6(BO_3)_3Si_6O_{18}(OH)_4$	black, red, green	7 to 7½	elongated crystals, triangular in cross section; conchoidal fracture

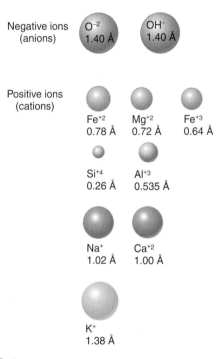

Figure B.1

Sizes of common ions in the crust with radius of each given in Ångstroms (Å). (1 Å = 10^{-10} meters = .000000004 in) Ions of similar size and charge can substitute for each other in minerals' crystal structures. Iron and magnesium frequently do so in the ferromagnesians. Some Al^{+3} substitutes for Si^{+4} in many silicates; Na^+ and Ca^{+2} can replace each other in some feldspars; and so on.

5. More-cohesive, fine-grained sedimentary rocks may be distinguished from fine-grained volcanics because sedimentary rocks are generally softer and more likely to show a tendency to break along bedding planes. Phenocrysts, of course, indicate a (porphyritic) volcanic rock.

6. Foliated metamorphic rocks are distinguished by their foliation (schistosity, compositional banding). Also, rocks containing abundant mica, garnet, or amphibole are commonly metamorphic rocks.

7. Nonfoliated metamorphic rocks, like quartzite and marble, resemble their sedimentary parents but are harder, denser, and more compact. They may also have a shiny or glittery appearance on broken surfaces, due to recrystallization during metamorphism.

Once a preliminary determination of category (igneous/ sedimentary/metamorphic) has been made, table B.2 can be used in conjunction with the appropriate text chapter to identify the rock type. (Keep in mind, however, that the key and text focus on relatively common rock types.)

Table B.2	A Key to Aid In Rock Identification

Igneous Rocks

I. Extremely coarse-grained: Rock is a pegmatite. (Most pegmatites are granitic, with or without exotic minerals.)

II. Phaneritic (coarse enough that all grains are visible to the naked eye).

    A. Significant quartz visible; only minor mafic minerals: *granite*.

    B. No obvious quartz; feldspar (light-colored) and mafic minerals (dark) in similar amounts: *diorite*.

    C. No quartz; rock consists mostly of mafic minerals: *gabbro*.

    D. No visible quartz or feldspar: Rock is ultramafic.

III. Porphyritic with fine-grained groundmass: Go to Part IV to describe groundmass (using phenocryst compositions to assist); adjective "porphyritic" will preface rock name.

IV. Aphanitic (grains too fine to distinguish easily with the naked eye).

    A. Quartz is visible or rock is light in color (white, cream, pink): probably *rhyolite*.

    B. No visible quartz; medium tone (commonly gray or green); if phenocrysts are present, commonly plagioclase, pyroxene, or amphibole: probably *andesite*.

    C. Rock is dark, commonly black; any phenocrysts are olivine or pyroxene: *basalt*

    D. Rock is glassy and massive: *obsidian* (regardless of composition).

    E. Rock consists of gritty mineral grains, ash and glass shards: *ignimbrite* (welded tuff).

Sedimentary Rocks

I. Rock consists of visible shell fragments or of oolites: *limestone*.

II. Rock consists of interlocking grains with texture somewhat like that of igneous rock and is light in color: probable chemical sedimentary rock.

    A. Tastes like table salt: *halite*.

    B. No marked taste; hardness of 2 (if grains are large enough to scratch); does not effervesce: *gypsum*.

    C. Effervesces in weak HCl: limestone (calcite).

    D. Effervesces weakly in HCl, only if scratched: *dolomite*.

III. Rock consists of grains apparently cemented or compacted together: probable clastic sedimentary rock.

    A. Coarse grains (several millimeters or more in diameter), perhaps with a finer matrix: *conglomerate* if the grains are rounded, *breccia* if they are angular.

    B. Sand-sized grains; gritty feel: sandstone. If predominantly quartz grains, *quartz sandstone*; if roughly equal proportions of quartz and feldspar, *arkose*; if many rock fragments, and perhaps a fine-grained matrix, *greywacke*.

    C. Grains too fine to see readily with the naked eye: *mudstone*. If rock shows lamination, and a tendency to part along parallel planes, *shale*.

IV. Relatively dense; compact, dark, no visible grains; massive texture, conchoidal fracture: *chert* (silica).

Metamorphic Rocks

I. Nonfoliated; compact texture with interlocking grains: identified by predominant mineral(s).

    A. If quartz-rich, perhaps with a sugary appearance: *quartzite*.

    B. If calcite or dolomite (identified by effervescence, hardness): *marble*.

    C. Rock consists predominantly of amphiboles: *amphibolite*.

II. Foliated: classified mainly by texture.

    A. Very fine-grained; pronounced rock cleavage along parallel planes, to resemble flagstones: *slate*.

    B. Fine-grained; slatelike, but with glossy cleavage surfaces: *phyllite*.

    C. Coarser grains; obvious foliation, commonly defined by prominent mica flakes, sometimes by elongated crystals like amphiboles: *schist*.

    D. Compositional or textural banding, especially with alternating light (quartz, feldspar) and dark (ferromagnesian) bands: *gneiss*.

Glossary

A

ablation The loss of glacier ice by melting or evaporation.

abrasion Erosion by wind-transported sediment or by the scraping of rock fragments frozen in glacial ice.

absorption field *See* leaching field.

accreted terrane A terrane that has moved from somewhere else to its present relative position on a continent.

acid rain Rain that is more acidic (has lower pH) than normal precipitation.

active margin A continental margin at which there is significant volcanic and earthquake activity; commonly, a convergent plate margin.

active volcano A volcano with a record of eruption within recent history.

aerobic decomposition Decomposition using or consuming oxygen.

aftershocks Earthquakes that follow the main shock when a fault has slipped; of magnitude equal to or lower than the main shock.

A horizon Usually top zone in soil profile, consisting of mix of mineral and organic material.

albedo The fraction of incoming radiation that is reflected back by a surface.

algal bloom An overly exuberant growth of algae in eutrophic water.

alluvial fan A wedge-shaped sediment deposit left where a tributary flows into a more slowly flowing stream, or where a mountain stream flows into a desert.

alpine (valley) glacier A glacier occupying a valley in mountainous terrain.

anaerobic decomposition Decomposition that occurs without using, or in the absence of, oxygen.

andesite Volcanic rock intermediate in composition between basalt and rhyolite.

angle of repose The maximum slope angle at which a given unconsolidated material is stable.

anion An ion with a net negative charge.

anthracite The hardest of naturally occurring coals.

anthropogenic Produced by human activity.

aquifer Rock that is sufficiently porous and permeable to be useful as a source of water.

aquitard Rock of low permeability, through which water flows very slowly.

arête Sharp-spined ridge created by erosion by valley glaciers flowing along either side of the ridge.

artesian system A confined aquifer system in which ground water can rise above its aquifer under its own pressure.

asthenosphere Partially molten, "weak" zone within the upper mantle immediately below the lithosphere.

atom The smallest particle into which a chemical element can be subdivided.

atomic mass number The sum of the number of protons and the number of neutrons in an atomic nucleus.

atomic number The number of protons in an atomic nucleus; characteristic of a particular element.

B

banded iron formation A sedimentary rock consisting of alternating iron-rich and iron-poor bands, found in Precambrian rocks, that may serve as an ore of iron.

barrier islands Long, low, narrow islands parallel to a coastline that protect the coastline somewhat from wave action.

basalt A volcanic rock rich in ferromagnesian minerals; relatively low in silica.

base level The lowest elevation to which a stream can cut down; for most streams, this is the level of the body of water into which they flow, such as another stream, lake, or ocean.

beach Gently sloping shoreline area washed over by waves.

beach face That part of the beach along the water that is regularly washed by waves.

bed load Material moved along a stream bed by flowing water.

bedrock geology The geology as it would appear with the overlying soil and vegetative cover stripped away.

Benioff zone A dipping plane of progressively deeper earthquake foci at a convergent plate boundary associated with a subducting plate.

B horizon Soil layer found below the soil's A and E horizons; also known as the zone of accumulation.

biochemical oxygen demand (BOD) Quantity of oxygen required for aerobic decomposition of organic matter in a system.

biodiesel Fuel derived from vegetable oil or animal fat that can be used in some diesel-powered vehicles, or a blend of regular diesel with such fuels.

biofuels Energy derived from living organisms or from organic matter (biomass).

biogas Methane derived from decaying organic matter.

biomagnification Process through which the concentration of a harmful substance, such as a heavy metal, increases in organisms as it moves up a food chain.

bioremediation Reduction of a pollution hazard by use of organisms—for example, breakdown of toxic organic compounds by microorganisms or fixation of toxic elements in insoluble or immobile forms.

biosphere The sum of all living things on earth.

bitumen The dark, heavy, viscous petroleum found in oil sands.

bituminous A form of coal that is softer than anthracite but harder than lignite.

BOD *See* biochemical oxygen demand.

body waves Seismic waves that pass through the earth's interior; includes P waves and S waves.

braided stream A stream with multiple channels that divide and rejoin.

breeder reactor A reactor in which new fissionable material is produced in quantity at the same time as energy is generated.

brittle Describes materials that tend to rupture before appreciable plastic deformation has occurred.

C

caldera A large, bowl-shaped summit depression in a volcano; may be formed by explosion or collapse.

calving The formation of icebergs by the breakup of a glacier flowing out over water.

capacity (stream) The load that a stream can carry.

cap-and-trade A strategy for reducing pollutant emissions by setting an overall limit on a group (as for SO_2 emissions from all electric utilities) but allowing units within the group to buy or sell emissions allowances, providing a financial incentive to units to reduce emissions.

carbon capture and storage *See* carbon sequestration.

carbonate Nonsilicate mineral containing carbonate groups (CO_3), carbon and oxygen in the proportions of one atom of carbon to three atoms of oxygen.

carbon sequestration Isolation of carbon in some reservoir, from which it does not contribute to atmospheric CO_2.

carrying capacity The ability of a system (or the whole earth) to sustain its population in reasonably healthy and comfortable conditions.

catalyst A substance that promotes chemical reactions.

cation An ion with a net positive charge.

Cenozoic Geologic era spanning the time from 66 million years ago to the present.

chain reaction (nuclear) The process during which fission of one nucleus triggers fission of others, which, in turn, induces fission in others, and so on.

channelization The modification of a stream channel, such as deepening or straightening of the channel, usually with the objective of reducing flood hazards.

chemical sediment Sediment formed at low temperature by direct precipitation from solution.

chemical weathering The breakdown of minerals by chemical reaction with water, with other chemicals dissolved in water, or with gases in the air.

C horizon Soil layer found directly below the soil's B horizon; consists of coarsely broken bedrock.

cinder cone A volcano built of cinders and other pyroclastics piled up around the volcanic vent.

cirque Bowl-shaped depression formed at the head of an alpine glacier.

clastic Broken or fragmented; describes sediments or sedimentary rocks that are formed from fragments of preexisting rocks or minerals.

cleavage (mineral) The tendency of a mineral to break preferentially along planes in certain directions in the crystal structure.

coal A solid, carbon-rich fuel formed from the remains of land plants through the effects of heat and pressure in the earth's crust.

coal-bed methane Methane associated with, and extracted from, coal deposits.

common but differentiated responsibility of states The concept that all nations share responsibility for protecting the global environment, but individual nations' responsibilities differ with their differing contributions to the problems and resources available to address them.

composite volcano *See* stratovolcano.

composting A method of handling (nontoxic) organic waste matter by which it is converted into a beneficial soil additive through controlled decay.

compound A chemical combination of two or more elements, in specific proportions, having a distinctive set of resultant physical properties.

compressive stress Stress tending to compress an object.

concentration factor (ore) The concentration of a metal in a given ore deposit divided by its average concentration in the continental crust.

conditional resources *See* subeconomic resources.

cone of depression A broadly conical depression of the water table or potentiometric surface caused by pumped groundwater withdrawal.

confined aquifer An aquifer overlain by an aquitard or aquiclude.

confining pressure Pressure that is uniform in all directions around a rock or other body.

constructive plate boundary A plate boundary at which new lithosphere is created; e.g., a seafloor spreading ridge.

contact metamorphism Local metamorphism adjacent to a cooling magma body.

contaminant plume A tongue of contaminant-rich water extending away from a point source of groundwater pollution in the direction of groundwater flow.

continental drift The concept that the continents have moved about over the earth's surface.

continental glacier A large glacier covering extensive land area; also known as an *ice cap* or *ice sheet;* may be several kilometers thick.

contour interval The difference in elevation between successive contour lines on a map.

contour line A line joining points of equal elevation on a map.

contour plowing Plowing so that rows run horizontally around the curve of a slope, rather than up and downslope.

convection cells Circulating masses of material driven by temperature differences (hot material rises, then moves laterally, cools, sinks, and is reheated to rise again).

convergent plate boundary Plate boundary at which lithospheric plates are moving toward each other; for example, a subduction zone or continental collision zone.

core The innermost zone of the earth; composed largely of iron.

core meltdown A possible nuclear reactor accident resulting from loss of core coolant and subsequent overheating.

covalent bonding Bonding involving sharing of electrons between atoms.

creep (fault) The slow, gradual slip along a fault zone without major, damaging earthquakes.

creep (rock or soil) Slow, gradual downslope movement of unstable surficial materials (as contrasted with more abrupt landslides).

crest The maximum stage reached during a flood event.

cross section An interpretation of geology and structure in the third (vertical) dimension commonly based on rock exposures and attitudes at the surface; drawn in a particular vertical plane.

crust The outermost compositional zone of the earth; composed predominantly of relatively low-density silicate minerals.

crystalline Describes materials possessing a regular, repeating internal arrangement of atoms.

cumulative reserves The total reserves, including materials already consumed or exploited.

Curie temperature The temperature above which a magnetic material loses its magnetism; different for each such material.

cut bank Steep stream bank being eroded by lateral migration of meanders.

D

Darcy's Law Relationship of groundwater flow rate between two points to the difference in hydraulic head between them.

debris avalanche A mixed flow of rock, soil, vegetation, or other materials.

debris flow Another term for *debris avalanche;* especially, used for a water-saturated one.

decommissioning (nuclear plant) The shutdown of a nuclear reactor at the end of its safe, useful life; includes the disposal of radioactive parts.

deep-focus earthquakes Those with focal depths over 100 km.

deep-well disposal A method for the disposal of toxic liquid waste by injection into deep, porous, permeable strata confined by impermeable rocks.

deflation The wholesale removal of loose sediment by wind erosion.

delta A fan-shaped deposit of sediment formed at a stream's mouth.

desert A barren region incapable of supporting appreciable life.

desertification The process by which marginally habitable arid lands are converted to desert; typically accelerated by human activities.

desert pavement A desert surface produced by the combined effects of wind erosion and overland surface-water runoff, in which the larger rocks are exposed by the selective removal of fine sediment; these rocks, in turn, protect finer material below from erosion.

destructive plate boundary A plate boundary at which lithosphere is consumed; e.g., a subduction zone.

differentiation (earth) Process by which zones of different composition developed from an initially homogeneous earth.

dip-slip fault A fault with predominantly vertical displacement.

discharge (stream) The amount of water flowing past a given point per unit time.

dissolved load Sum of dissolved material transported by a stream.

divergent plate boundary A boundary along which lithospheric plates are moving apart; for example, seafloor spreading ridges and continental rift zones.

divide Topographic high separating drainage basins of different streams; also, plane in a groundwater system from which ground water flows away in different directions on either side.

dormant volcano A volcano with no recent eruptive history but that still looks relatively fresh and unweathered; may become active again in future.

dose-response curve A graph illustrating the relative benefit or harm from a trace element or other substance as a function of the dosage received or amount consumed by a person or organism.

doubling time The length of time required for a population to double in size.

downstream flood A flood affecting a large area of drainage basin or a large stream system; typically caused by prolonged rain or rapid regional snowmelt.

drainage basin The region from which surface water drains into a particular stream.

dredge spoils Sediment dredged from waterways to improve navigation or to increase water capacity.

drift Sediment transported and deposited by a glacier; *see also* till, outwash.

drowned valley Along a coastline, a stream valley that is partially flooded by seawater as a consequence of land sinking and/or sea level rising.

ductile Describes material that undergoes extensive plastic deformation without rupturing.

dune A low mound or ridge of sediment (usually sand) deposited by wind.

dust bowl Any region severely affected by drought and wind erosion; when capitalized (Dust Bowl), refers specifically to a large region of the central United States so affected in the early 1930s.

E

E85 A liquid fuel consisting of 85% ethanol, 15% gasoline.

earthquake Ground displacement and energy release associated with the sudden motion of rocks along a fault.

earthquake cycle The concept that there is a periodic quality about the occurrence of major earthquakes on a given fault zone, with repeated cycles of stress buildup, rupture, and relaxation of stress through smaller aftershocks.

ecliptic plane The plane in space in which the Sun and the orbit of the Earth (and of other planets) lie.

EEZ *See* Exclusive Economic Zone.

E horizon Layer typically found between A and B soil horizons; also known as zone of leaching.

EIS *See* environmental impact statement.

elastic deformation Deformation proportional to applied stress, from which the affected material will return to its original size and shape when the stress is removed.

elastic limit The stress above which a material will cease to deform elastically.

elastic rebound Phenomenon whereby stressed rocks snap back elastically after an earthquake to their pre-stress condition.

electron A subatomic particle with an electrical charge of −1; generally found orbiting an atomic nucleus.

end moraine A ridge of till accumulated at the end of a glacier.

enhanced recovery Any method used to increase the amount of oil or gas recovered from a petroleum reservoir.

environmental geology The study of the interactions between humans and their geologic environment.

environmental impact statement (EIS) An analysis of the environmental impacts to be anticipated from a proposed action and its alternatives; mandated by the National Environmental Policy Act and legislation patterned after it.

eolian Deposited or shaped by wind action.

epicenter The point on the earth's surface directly above the focus of an earthquake.

equilibrium line Line on the surface of a glacier at which accumulation just equals ablation.

estuary A body of water along a coastline that contains a mix of fresh and salt water.

eutrophication The development of high nutrient levels (especially, high concentrations of nitrates and phosphates) in water; may lead to algal bloom.

evaporite A sedimentary mineral deposit formed when shallow or inland seas dry up; also, the minerals commonly deposited in such an environment.

evapotranspiration Movement of water vapor from earth's surface into air by a combination of evaporation from rocks and transpiration through plants.

Exclusive Economic Zone (EEZ) A zone extending to 200 miles offshore from a nation's coast, within which the 1982 Law of the Sea Treaty recognizes that nation's exclusive right to resource exploitation.

exponential decay Breakdown of a fixed percentage or fraction of a substance per unit time.

exponential growth Growth characterized by a constant percentage increase per unit time.

extinct volcano A volcano that has no recent eruptive history and appears very weathered in appearance; not expected to erupt again.

F

fall (rock) Mass wasting by free-fall of material not always in contact with the ground underneath.

fault Planar break in rock along which one side has moved relative to the other.

ferromagnesian A term describing silicates containing significant amounts of iron and/or magnesium; these minerals are usually dark-colored.

firn Dense, coarsely crystalline snow partially converted to ice.

fission (atomic) The process by which atomic nuclei are split into smaller fragments.

fissure eruption The eruption of lava from a crack in the lithosphere, rather than from a central vent.

flood Condition in which stream stage is above channel bank height.

flood-frequency curve A graph of stream stage or discharge as a function of recurrence interval (or annual probability of occurrence).

floodplain A flat region or valley floor surrounding a stream channel, formed by meandering and sediment deposition, into which the stream overflows during flooding.

flow Mass wasting in which materials move in chaotic fashion.

focus The point of first break on a fault during an earthquake.

foliation Parallel alignment of linear or platy minerals in a rock.

fossil fuels Hydrocarbon fuels formed from organic matter.

fracking Hydraulic fracturing used to increase permeability of rocks such as shale.

fusion (nuclear) The process by which atomic nuclei combine to produce larger nuclei.

G

gasification Any process by which coal is converted to a gaseous hydrocarbon fuel.

geographic information system (GIS) A computer-based system for storing, manipulating, and analyzing data associated with particular geographic locations.

geomedicine Study of geographic patterns of health and disease and the relationships between health and substances in the geological environment.

geopressurized natural gas Gas dissolved in deep pore waters.

geopressurized zones Deep aquifers under unusually high pressure, exceeding normal hydrostatic (fluid) pressure.

geothermal energy Energy derived from the internal heat of the earth; its use usually requires a near-surface heat source, such as young igneous rock, and nearby circulating subsurface water.

geothermal gradient The rate of increase of temperature with depth in the earth.

GIS *See* geographic information system.

glacier A mass of ice that moves or flows over land under its own weight.

glass Solid (especially silicate) lacking a regular internal crystal structure.

gouge Finely pulverized rock along a fault plane.

gradient The slope (steepness) of a stream channel along its length.

granite Coarse-grained plutonic igneous rock, typically rich in quartz and feldspars.

greenhouse effect The warming of the atmosphere due to trapping of infrared rays by atmospheric gases, especially as due to the increased concentration of carbon dioxide derived from the burning of fossil fuels.

ground water Water in the zone of saturation, below the water table.

H

half-life The length of time required for half of an initial quantity of a radioactive isotope to decay.

halide Compound of one or more metals plus a halogen element (fluorine, chlorine, iodine, or bromine).

hardness (mineral) The ability of a mineral to resist scratching.

hard water Water containing substantial quantities of dissolved calcium, magnesium, and/or iron.

harmonic tremors A distinctive type of seismic signal, rhythmic and continuous, often associated with magma movement in or beneath a volcano.

heavy metals A group of dense metals, including mercury, lead, cadmium, plutonium, and others, that share the characteristic of being accumulative in organisms and tending to become increasingly concentrated in organisms higher up a food chain.

high-level (radioactive) waste Waste sufficiently radioactive to require special handling in disposal.

horn A peak formed by headwall erosion by several alpine glaciers diverging from the same topographic high.

hot-dry-rock Geothermal resource area in which geothermal gradients are high but indigenous ground water is lacking.

hot spot An isolated center of volcanic activity; often not associated with a plate boundary.

hydraulic gradient The difference in hydraulic head between two points, divided by the distance between them.

hydraulic head Potential energy of water above a given point, reflected in the height of the water surface (above ground), water table in an unconfined aquifer, or potentiometric surface in a confined aquifer.

hydrocarbon Any compound consisting of carbon and hydrogen.

hydrocompaction The process by which a sediment settles, cracks, and becomes compacted when wetted.

hydrograph A graph of stream stage or discharge against time.

hydrologic cycle The cycle through which water in the hydrosphere moves; includes such processes as evaporation, precipitation, and surface and groundwater runoff.

hydrosphere All water at and near the earth's surface that is not chemically bound in rocks.

hydrothermal ores The ores deposited by circulating warm fluids in the earth's crust.

hypothesis A conceptual model or explanation for a set of data, measurements, or observations.

hypothetical resources That quantity of a resource material expected to be found in areas in which like deposits are known to exist.

I

ice age Any period of extensive continental glaciation; when capitalized (Ice Age), the phrase refers to the most recent such episode, from 2 million to 10,000 years ago.

igneous rock A rock formed or crystallized from a magma.

infiltration The process by which surface water sinks into the ground.

infrared Electromagnetic radiation just to the long-wavelength side of the visible light spectrum; heat radiation.

intensity (earthquake) A measure of the damaging effects of an earthquake at a particular spot; commonly reported on the Modified Mercalli Scale.

ion Atom that has gained or lost electrons, so it has a net electrical charge.

ionic bonding Bonding due to attraction between oppositely charged ions.

island arc Chain of volcanic islands formed parallel to a subduction zone, on the overriding plate.

isotopes Atoms of a given chemical element having the same atomic number but different atomic mass numbers.

K

karst Terrain characterized by abundant formation of underground solution cavities and sinkholes; commonly underlain by limestone.

kerogen A waxy solid hydrocarbon in oil shale.

kimberlites Igneous rocks that occur as pipelike intrusive bodies that probably originated in the mantle.

L

lahar A volcanic mudflow deposit formed from hot ash and water, the latter often derived from melting snow on a snow-capped or glaciated volcano.

landslide A general term applied to a rapid mass-wasting event.

laterite An extreme variety of pedalfer soil that is highly leached; common in tropical climates.

lava Magma that flows out at the earth's surface.

lava dome A compact, bulbous, steep-sided structure built of very viscous, silicic lava emitted from a central pipe or vent.

Law of Faunal Succession The concept that life-forms change through time and that therefore each given fossil organism corresponds to only one interval of time.

leachate Water containing dissolved chemicals; applied particularly to fluids escaping from waste disposal sites.

leaching The removal of elements or compounds by dissolution.

leaching field A network of porous pipes and surrounding soil from which septic-tank effluent is slowly released.

levees Raised banks along a stream channel that tend to contain the water during high-discharge events.

lignite The softest of the coals.

liquefaction (coal) Any process by which coal is converted into a liquid hydrocarbon fuel.

liquefaction (soil) A quicksand condition arising in wet soil shaken by seismic waves; soil loses its strength as particles lose contact with each other.

lithification Conversion of unconsolidated sediment into cohesive rock.

lithosphere The solid, outermost zone of the earth, including the crust and a portion of the upper mantle; approximately 50 kilometers thick under the oceans and commonly more than 100 kilometers thick beneath the continents.

littoral drift Sand movement along the length of a beach.

load (stream) The total quantity of material transported by a stream: sum of bed load, suspended load, and dissolved load.

locked fault A section of fault along which friction prevents creep in response to stress.

loess Wind-deposited sediment composed of fine particles (typically 0.01 to 0.06 millimeters in diameter).

longitudinal profile Diagram of elevation of a stream bed along its length.

longshore current Net movement of water parallel to a coastline, arising when waves and currents approach the shore at an oblique angle.

low-level (radioactive) waste Wastes that are sufficiently low in radioactivity that they can be released safely into the environment or disposed of with minimal precautions.

M

magma A naturally occurring silicate melt, which may also contain mineral crystals, dissolved water, or gases.

magmatic deposits (ore) Those deposits associated with magma emplacement or crystallization.

magnitude (earthquake) Measure of earthquake size; often reported using the Richter magnitude scale.

manganese nodules Lumps of manganese and iron oxides and hydroxides, with other metals, found on the sea floor.

mantle The zone of the earth's interior between crust and core; rich in ferromagnesian silicates.

map unit A distinct, identifiable rock unit used in preparing a geologic map.

mass movement *See* mass wasting.

mass wasting The downslope movement of material due to gravity; also known as mass movement.

meanders The curves or bends in a stream channel.

mechanical weathering The physical breakup of rock or mineral grains by surface processes.

Mesozoic Geologic era from 245 to 66 million years ago.

metamorphic rock Rock that is changed in form (deformed and/or recrystallized) through the effects of heat and/or pressure.

methane hydrate A crystalline solid of natural gas and water molecules, found in arctic regions and marine sediments.

Milankovitch cycles Cyclic variations in the amount of sunlight reaching a given latitude, caused by variations in earth's orbit and tilt on its axis.

milling Erosion by the grinding action of sand-laden waves on a coast.

mineral A naturally occurring, inorganic, solid element or compound with a definite composition or range in composition, usually having a regular internal crystal structure.

moment magnitude A measure of earthquake magnitude directly proportional to energy release, calculated from the area of fault plane broken, amount of fault displacement, and shear strength of the rock; represented by M_w.

moraine Landform made of till.

mouth (stream) The point where a stream ends; where it reaches its base level.

mrem One-thousandth of a rem.

multiple barrier concept Waste-disposal approach that involves several mechanisms or materials for isolating the waste from the environment; often used in the context of high-level radioactive wastes.

multiple use The land-use practice in which a given piece of land is used for two or more purposes simultaneously.

N

native element Nonsilicate mineral consisting of a single chemical element.

natural gas Gaseous hydrocarbons, especially methane (CH_4).

neutron An electrically neutral subatomic particle with a mass approximately equal to one atomic mass unit; generally found within an atomic nucleus.

nonpoint source A diffuse source of pollutants, such as runoff from farmland or drainage from a strip-mine.

nonrenewable Not being replenished or formed at any significant rate on a human timescale.

normal fault A dip-slip fault where the block above the fault moves down relative to the block below it; indicates tensional stress.

nucleus The center of an atom, containing protons and neutrons.

O

ocean thermal energy conversion (OTEC) Power generation making use of the temperature difference between deep, cold seawater and warmer near-surface water.

O horizon Top layer of soil, consisting wholly of organic matter; not always present.

oil Any of various liquid hydrocarbon compounds.

oil sand Sedimentary rock, usually sandstone, containing thick, heavy, tarlike hydrocarbon, bitumen.

oil shale A sedimentary rock containing the waxy solid hydrocarbon kerogen.

ore A rock in which a valuable or useful metal occurs at a concentration sufficiently high to make it economically practical to mine.

organic sediments Carbon-rich sediments derived from the remains of living organisms.

OTEC *See* ocean thermal energy conversion.

outwash Glacial sediment moved and redeposited by meltwater.

oxbows Old meanders now cut off or abandoned by a stream.

oxide Nonsilicate mineral containing oxygen combined with one or more metals.

oxygen sag curve A graph depicting oxygen depletion followed by reoxygenation in a stream system below a source of organic waste matter; caused by aerobic decay of the organic matter.

ozone hole Area over which the ozone layer is thinner (containing a lower ozone concentration) than over surrounding areas; commonly develops annually over Antarctica.

ozone layer An ozone-rich layer within the stratosphere, between about 15 and 35 km above earth's surface; absorbs potentially harmful ultraviolet radiation.

P

paleomagnetism The "fossil magnetism" preserved in rocks formed in the past.

Paleozoic Geologic era from 570 to 245 million years ago.

Pangaea The name given to the ancient supercontinent that existed some 200 million years ago.

particulates (pollution) Solid particles suspended in air; includes soot, ash, and dust.

passive margin A geologically quiet continental margin, lacking significant volcanic or seismic activity.

pathogenic Disease-causing; often applied to harmful microorganisms.

pedalfer A moderately leached soil rich in residual iron and aluminum oxide minerals.

pedocal A soil in which calcium carbonate and other readily soluble minerals are retained; characteristic of drier climates.

pegmatite A very coarsely crystalline igneous rock.

percolation Movement of subsurface water through rock or soil under its own pressure.

periodic table The regular arrangement of chemical elements in a chart that reflects patterns of chemical behavior related to the electronic structure of atoms.

permafrost A condition found in cold climates, wherein ground remains frozen year-round at some depth below the surface.

permafrost table The top of the permanently frozen (permafrost) zone.

permeability A measure of how readily fluid can flow through a rock, sediment, or soil.

petroleum Liquid hydrocarbons derived from organic matter and used as fuel.

Phanerozoic The time from 570 million years ago to the present.

photovoltaic cells (solar cells) Devices that convert solar radiation directly to electricity.

phreatic eruption A violent, explosive volcanic eruption like a steam-boiler explosion, occurring when subsurface water is heated and converted to steam by hot magma underground.

phreatic zone *See* saturated zone.

pH scale Scale for reporting acidic or alkaline quality of a liquid.

physical weathering *See* mechanical weathering.

placers The ores concentrated by stream or wave action on the basis of mineral densities and/or resistance to weathering.

plastic deformation Permanent strain in material stressed beyond the elastic limit; the material will not return to its original dimensions when the stress is removed.

plate tectonics The theory that holds that the rigid lithosphere is broken up into a series of movable plates.

plutonic Describes an igneous rock crystallized well below the earth's surface; typically coarse-grained.

point bar A sedimentary feature built in a stream channel, on the inside of a meander, or anywhere the water slows.

point source A single, concentrated, identifiable source of pollutants, such as a sewer outfall or factory smokestack.

polar-wander curve A plot of apparent magnetic pole positions at various times in the past relative to a continent, assuming the continent's position to have been fixed on the earth.

pore pressure Pressure of fluid filling cracks and pores in rock or soil.

porosity Proportion of void space in rock, sediment, or soil.

potentiometric surface A feature analogous to a water table but applied to confined aquifers; indicates the height to which the water's pressure would raise the water if the water were unconfined.

Precambrian The time from the formation of the earth to 570 million years ago (the start of the Cambrian period).

precautionary principle A concept in international law/diplomacy under which nations' activities may be restricted if there is reasonable likelihood that those activities may be harmful; significant damage (as to the environment) need not already have occurred.

precession Cyclic "wobble" of earth's rotational axis, changing its orientation in space.

precursor phenomena Phenomena that precede an earthquake, volcanic eruption, or other natural event, which may be used to predict the upcoming event.

Principle of Original Horizontality The concept that sediments are deposited in approximately horizontal, flat-lying layers.

Principle of Superposition The concept that, in an undisturbed pile of sedimentary rocks, those on the bottom were deposited first and those on top, last.

Prior Appropriation The principle of surface-water law by which users of water from a given source have priority rights to it on the basis of relative time of first use.

proton Subatomic particle with a charge of +1 and a mass of approximately one atomic mass unit; generally found within an atomic nucleus.

pump-and-treat Approach to groundwater purification whereby the water is extracted prior to treatment.

P waves Compressional seismic body waves.

pyroclastic flow Denser-than-air flow of hot gas and ash from a volcano.

pyroclastics The hot fragments of rock and magma emitted during an explosive volcanic eruption.

Q

quick clay The sediment formed from glacial rock flour deposited in a marine setting, weakened by subsequent flushing with fresh pore water.

R

rain shadow A dry zone landward of a mountain range, which is caused by loss of moisture from air passing over the mountains.

rare-earth elements Elements with atomic numbers from 57 through 71.

recharge The processes of infiltration and migration by which ground water is replenished.

recrystallization Atomic rearrangement to form new crystals in a rock while it remains in the solid state; often, the grain sizes in the rock after recrystallization are coarser than in the rock originally.

recurrence interval The average length of time between floods of a given size along a particular stream.

recycling The recovery and reprocessing or reuse of materials already used previously in construction, manufacturing, and so on; a method of conserving resources by minimizing the need for the development of new resources.

regional metamorphism Metamorphism on a large scale, involving increased heat and pressure; often associated with mountain building.

regolith Surficial sediment deposit formed in place, whether or not capable of supporting life. (*See also* soil.)

rem Short for "radiation equivalent man"; a unit of radiation exposure that takes into account the type and energy of radiation involved and the effect of that particular kind of radiation on the human body.

remote sensing Investigation without direct contact, as by using aerial or satellite photography, radar, and so on.

renewable Capable of replenishment or regeneration on a human timescale.

reprocessing (nuclear fuel) The treatment of used fuel elements to extract plutonium or other fissionable material to make new fuel elements.

reserves That quantity of a (resource) material that has been found and is recoverable economically with existing technology.

residence time The average length of time a substance persists in a system; may also be defined as (capacity)/(rate of influx).

resources Those things important or necessary to human life and civilization; also, the total quantity of a given (resource) commodity on earth, discovered and undiscovered, that might ultimately become part of the reserves.

retention pond A large basin designed to catch surface runoff to prevent its flow directly into a stream.

reverse fault A dip-slip fault in which the block above the fault is pushed up and over the lower block; indicates compressional stress.

rhyolite Silica-rich volcanic rock; the volcanic compositional equivalent of granite.

Richter magnitude scale A logarithmic scale used for reporting the size of earthquakes; Richter magnitude (local magnitude, M_L) is proportional to amount of ground shaking (displacement) at the epicenter and is computed on the basis of the amplitude (height) of the largest seismic wave on a seismogram.

Riparian Doctrine The principle of surface-water law by which all landowners bordering a body of water have equal rights to it.

rock A solid, cohesive aggregate of one or more minerals.

rock cycle The concept that rocks are continually subject to change and that any rock may be transformed into another type of rock through an appropriate geologic process.

rockfall *See* fall (rock).

rock flour A fine sediment of pulverized rock produced by glacial erosion.

rupture Breakage or failure of material under stress.

S

saltation The process by which particles are moved in short jumps over the ground surface or stream bed by wind or water.

saltwater intrusion A process by which salt water replaces fresh ground water when the fresh water is being used more rapidly than it is being recharged; especially common in coastal areas.

sanitary landfill A disposal site for solid or contained liquid waste; in simplest form, a dump site at which wastes are covered with layers of earth daily or more often.

saturated zone The region of rock or soil in which pore spaces are completely filled with liquid; also known as the phreatic zone.

scale (map) The ratio of a unit of length on the map to the corresponding actual horizontal distance represented.

scientific method Means of discovering scientific principles by formulating hypotheses, making predictions from them, and testing the predictions.

scrubbers (electrostatic precipitators) Devices that remove pollutant gases and particulates from exhaust gases of power or manufacturing plants.

seafloor spreading The process by which new lithosphere is created at spreading ridges as plates of oceanic lithosphere move apart.

secure landfill A sanitary landfill designed to contain toxic chemical wastes; typically includes one or more impermeable liners and often is monitored by nearby wells.

sediment Surface accumulation of loose, unconsolidated mineral or rock particles.

sedimentary rock A rock formed from sediments at low temperature.

seismic gap A section of an active fault along which few earthquakes are occurring, in contrast to adjacent fault segments; presumably, a locked section of fault.

seismic shadow zone An effect of the earth's liquid outer core, which blocks S waves and thus casts a "seismic shadow" on the opposite side of the earth from the site of a major earthquake.

seismic tomography Technique using velocity variations of seismic body waves to map regions of relatively higher and lower temperature/density within the earth.

seismic waves The form in which most energy is released during earthquakes; divided into body waves and surface waves.

seismograph An instrument used to measure ground motion caused by seismic waves (and other disturbances).

sensitive clay A material similar in behavior to quick clay, but derived from different materials, such as volcanic ash.

septic system A sewage-disposal device involving the slow dispersal of wastes into soil for natural aerobic breakdown.

sequential use The land-use principle by which the same land is used for two or more different purposes, one after another.

settling pond A pond in which sediment-laden surface runoff is impounded to allow sediment to settle out before the water is released.

shale gas Natural gas occurring in shale typically of low permeability.

shallow-focus earthquakes Those with focal depths less than 100 km.

shearing stress Stress that tends to cause different parts of an object to slide past each other across a plane; with respect to mass movements, stress tending to pull material downslope.

shear strength The ability of a material to resist shearing stress.

shield volcano A low, flat, gently sloping volcano built from many flows of fluid, low-viscosity basaltic lava.

silicate Mineral containing silicon and oxygen and usually one or more additional elements.

sinkhole A circular depression in the ground surface commonly caused by collapse into an underground cavern formed by solution.

slide A form of mass wasting in which a relatively coherent mass of material moves downslope along a well-defined surface.

slip face The downwind side of a dune, on which material tends to assume the slope of the angle of repose of the sediment of the dune.

slump A slide moved only a short distance, often with a rotational component to the movement.

soil The accumulation of unconsolidated rock and mineral fragments and organic matter formed in place at the earth's surface; capable of supporting life.

soil moisture The water in the soil in the zone of aeration, or vadose zone.

solifluction Creep or very slow flow in saturated soil.

sorting A measure of the uniformity of a sediment with respect to particle size and/or density.

source (stream) The place where the stream originates.

source separation The sorting of (waste) material by type prior to collection, usually to facilitate the recycling of individual materials or to ready material for a particular disposal strategy, such as incineration.

speculative resources That quantity of a resource material that may be found in areas where similar deposits are not already known to exist.

spoil banks Piles of waste rock and soil left behind by surface mining, especially strip-mining.

stage (stream) The height (elevation) of a stream surface at a given point along the stream's length; usually expressed as elevation above sea level.

strain Deformation resulting from the application of stress.

stratovolcano A composite volcanic cone built of interlayered lava flows and pyroclastic materials.

stream A body of flowing water confined within a channel.

stress A force applied to an object.

striations Parallel grooves in a rock surface cut when a glacier containing rock debris flows over that rock.

strike-slip fault A fault with predominantly horizontal displacement.

strip mining The surface-mining technique in which the cover rock or soil is removed from a near-surface mineral or fuel deposit, and the deposit is then extracted by open excavation, in a series of parallel strips.

subduction zone A convergent plate boundary at which oceanic lithosphere is being pushed beneath another plate (continental or oceanic) into the asthenosphere.

subeconomic resources Those resources already found that cannot be exploited economically with existing technology; also known as conditional resources.

subsurface water The water in the hydrosphere below the ground surface.

sulfate Nonsilicate mineral containing sulfate (SO_4) groups, each made up of one atom of sulfur and four atoms of oxygen.

sulfide Nonsilicate mineral containing sulfur but no oxygen.

surface processes The geologic processes acting principally at the earth's surface; they involve the effects of wind, water, ice, and gravity.

surface water Liquid water above the ground surface.

surface waves The seismic waves that travel along the earth's surface.

surge (storm) Localized increase in water level of an ocean or large lake; caused by extreme low pressure and high winds associated with major storms.

suspended load (stream) The material that is light or fine enough to be moved along suspended in the stream, supported by the flowing water.

sustainable development Development that causes neither serious environmental damage nor such acute resource depletion as to imperil future development or quality of life.

S waves Shear seismic body waves; do not propagate through liquids.

T

tailings The piles of crushed waste rock created as a by-product of mineral processing.

talus Accumulated debris from rockfalls and rockslides.

tar sand *See* oil sand.

tectonics The study of large-scale movement and deformation of the earth's crust.

tensile stress Stress tending to pull an object apart.

terminal moraine The end moraine marking the farthest advance of a glacier.

terrane A geologically distinct region that can be distinguished from adjacent regions of different geology and/or history; commonly bounded by faults.

theory A generally accepted explanation for a set of data or observations; its validity has usually been tested by the scientific method.

thermal inversion The condition in which air temperature increases, rather than decreases, with increasing altitude; the overlying layer of warmer air may then trap warm, rising, pollutant-laden gases.

thermohaline circulation Major ocean circulation pattern, driven by winds and by differences in temperature and salinity of water masses.

thrust fault A reverse fault with very shallowly dipping fault plane.

till Poorly sorted sediment deposited by melting glacial ice.

topsoil The topmost portion of soil, richest in organic matter.

trace elements Those elements present in a system at very low concentrations, typically 100 ppm or less.

transform fault A fault between offset segments of a spreading ridge, along which two plates move horizontally in opposite directions.

transuranic Describes elements with atomic numbers higher than that of uranium, none of which occur naturally on earth, and all of which are radioactive.

triple junction Point at which three plates (and three plate boundaries) meet.

tsunami A seismic sea wave, generated by a major earthquake in or near an ocean basin; sometimes incorrectly called a "tidal wave."

turbidity current A denser-than-water suspension of sediment in water that flows rapidly downslope; develops in ocean basins as a result of submarine landslides or earthquakes.

U

ultraviolet Electromagnetic radiation just to the short-wavelength side of the visible light spectrum; biologically hazardous.

unconfined aquifer An aquifer not overlain directly by an aquitard.

uniformitarianism The concept that the same basic physical laws have operated throughout the earth's history, and therefore, by studying present geologic processes and their products, we can infer much about geologic processes operative in the past.

unsaturated zone A partly saturated region of rock or soil, above the water table; also known as the vadose zone.

upstream flood Flood affecting only localized sections of a stream system; caused by such events as an intense, local cloudburst or a dam failure; typically brief in duration.

V

vadose zone *See* unsaturated zone.

ventifact Literally, "wind-made" rock, shaped by wind erosion.

viscosity Resistance to flow.

volcano Structure built of lava and/or pyroclastics erupted from a central pipe or vent.

volcanic Describes any igneous rock formed at or near the earth's surface.

Volcanic Explosivity Index A scale for reporting the size of an explosive volcanic eruption.

W

watershed Synonym for *drainage basin*.

water table The top of the zone of saturation.

wave base Depth at which the movement of water molecules in waves becomes negligible.

wave-cut platform Steplike surface cut in rock by wave action at sea or lake level.

wave refraction The deflection of waves as they approach shore.

weathering The physical and/or chemical breakdown of rocks and minerals.

well sorted Describes sediments displaying uniform particle size and/or density.

Z

zone of accumulation *See* B horizon.

zone of leaching *See* E horizon.

Clossopteris, 42
CO, 431
CO_2, 105, 212, 219, 220, 226, 428–429, 429–430
Coal, 325–330
Coal-bed methane, 316
Coal-mine fatalities, 329
Coal-mine fires, 329
Coal-to-liquids technology, 326
Coastal zones and processes, 146–166
    barrier islands, 159–161
    beach erosion, 155–158
    cliff erosion, 158–159
    construction/reconstruction, 164–165
    drowned valley, 153
    estuaries, 161, 164
    hurricanes, 152, 162–163
    littoral drift, 149
    longshore current, 149
    sea-level change, 151–154
    sediment transport and deposition, 149
    storms, 149–151
    wave-cut platforms, 153
    wave refraction, 158
    waves and tides, 147–149
College Fjord, Alaska, 200–201
Colorado River Basin, 249–251
Colorado River Compact, 249
Columbia Plateau, 93
Commercial carbon-sequestration facilities, 450
Commission on the Limits of the Continental Shelf, 458
Common but differentiated responsibility of states, 464
Composite volcano, 96
Compound, 23
Comprehensive Environmental Response, Compensation, and Liability Act, 384
Compressive stress, 48
Concentrate-and-contain approach, 376
Concentration factor, 281, 282
Conditional resources, 281
Cone of depression, 234, 235
Confined aquifer, 232–233
Confining pressure, 49
Conglomerate, 35, 37
Conservation
    mineral and rock resources, 298–301
    oil and gas, 319–320
    water resources, 253–254
Conservation Reserve Program (USDA), 274

Construction controls, 467–468
Constructive plate boundaries, 52
Contact metamorphism, 36
Contaminant plume, 420
Continental crust, 5
Continental drift, 8, 43
Continental fissure eruptions, 93
Continental glacier, 193
Continental margins, 147
Continental shelf, 456–458
Continued operation, 75
Contour plowing, 272, 273
Contour strip cropping, 273
Convection cells, 57
Convention on the Law of the Sea, 456–458
Convention on the Regulation of Antarctic Mineral Resource Activities (CRAMRA), 458
Convergent plate boundary, 52–55
Conversion losses, 359
Cook Inlet, Alaska, 50
Copper, 286, 288
Coral bleaching, 222
Coral reef, 275
Core, 3, 5
Core meltdown, 338
Correlation, A–2
Correlative rights, 454
Corundum, 26, 27, B–2
Cost-benefit analysis, 466–467
Council on Environmental Quality, 470
Covalent bonding, 23
Covellite, B–2
Cracking, 310
CRAMRA, 458
Crater Lake, 116, 199
Creep, 64, 173, 185
Crest (stream), 129
Cretaceous period, A–6, A–8
Crime, 278
Crinoids, 37
Crocidolite, 32
Crossbeds, 205, 206
Crosscutting relationships, A–2, A–3
Crust, 3, 5, 280
Crystal skulls, 40
Crystalline, 24
Crystalline silicon, 25
Cumulative reserves, 281
Curie, 242
Curie temperature, 43
Cut bank, 126
Cyanide, 305

D

Dallas Central Wastewater Treatment Plant, 382
Dam failures, 493–498
Dam Safety Bill (1972), 468
Dams, 139–142
Darcy's law, 233, 234
Day After Tomorrow, The (film), 166
DDT, 409, 410, 411–412
Dead zone, 404
Death Valley, 204
Debris avalanche, 175
Debris flow, 175
Decommissioning, 339
Decontamination, 416, 421
Deep time, 7
Deep-well disposal, 378–379
Deepwater Horizon oil spill, 9, 307, 323–324
Deflation, 204
Deforestation, 487
Degradation, 209
Delta, 125
Dengue fever, 221
Denver Coliseum, 479
Desalination, 255–256
Desert, 206–209
Desert pavement, 204
Desertification, 192, 208–210
Destructive plate boundaries, 53
Diamond, 24, 27, 282–284
Diapers, 377
Dichloropentadiene, 422
Dieldrin, 412
Digitized maps, 482
Diisomethylphosphonate, 422
"Dillo Dirt," 382
Dilute-and-disperse approach, 376
Dinosaurs, 6
Dioxin, 409
Dip-slip fault, 65
Discharge, 122, 124, 231
Dishes, 377
Disposal wells, 378–379
Disruption of natural systems, 16–17
Dissolved load, 122
Distillation, 255, 256
Divergent plate boundary, 52
Diversion channels, 136–137
Doctrine of prior appropriation, 453
Dolomite, 30, B–2
Domestic wastewater, 401
Donora, Pennsylvania (1948), 427, 445

Insect-reduction strategies, 415–416
Insecticides, 413, 415
Intensity (earthquake), 70, 71
Interbasin water transfer, 254–255
Interglacials, 198
International environmental initiatives,
    463–466
International resource disputes, 456–458
International Seabed Authority, 456
International water disputes, 249
Ion, 22
Ionic bonding, 23
Iron oxides, 31
Irrigated acreage, 15
Irrigation, 248, 254, 276
Island arc, 55
Isolated silicate structure, 29
Isotopes, 22, A–5 to A–6
Itai-itai, 406

J

Japan earthquake (2011), 79, 81, 341
Johnson, David, 90
Joly, John, A–5
Joshua Tree National Park, 191
Jovian planets, 4
Juan de Fuca ridge, 457
Jupiter, 3, 4
Jurassic period, A–6, A–8

K

Kansai Airport (Osaka), 479
Kant, Immanuel, A–4
Kaolinite, B–2
Karst, 239, 240, 241
Kelvin, William T., A–5
Kerogen, 330
Kesterson Wildlife Refuge, 276
Keystone XL pipeline, 473, 475
Kilauea, 10, 34, 91, 102, 103, 107,
    110, 119
Kimberlite, 284
Kings Canyon National Park, A–1
Kingston Fossil Plant, 328
Kobe earthquake (1995), 81
Koehler, R. W., 445
Konya Dam, 142
Krakatoa, 106
Kyanite, B–2
Kyoto Protocol, 464–466

L

La Niña, 218
LaBrea Tar Pits, 320
Laguna Beach landslide (2005), 170

Lahar, 101–104
Lake Chad, 252–253
Lake Mead, 139, 229, 494
Lake Monoun, 105
Lake Nyos, 105
Laki, 105
Land, 15, 16
Land reclamation, 302–304,
    330, 455
Land surface temperatures, 224
Land-use planning and engineering
    geology, 476–500
    Boston's "Big Dig," 493
    case histories, 490–493, 494
    dam failures, 493–498
    factors to consider, 486–490
    federal land-use policies, 480
    Leaning Tower of Pisa, 491
    maps, 480–486
    multiple use, 478
    Panama Canal, 491–493
    satellite imagery, 486
    sequential use, 479–480
    St. Francis Dam, 494, 495
    testing and scale modeling, 490
    why land-use planning?, 477–478
Landfill, 366–369, 378
Landfill-derived gas, 360
Landsat satellites, 296
Landslide, 106–107, 167, 173, 183, 470.
    See also Mass movements
Landslide monitoring site, 187
Landslide potential (United States), 168
Landslide warnings, 185–187
Lassen Peak, 111
Laterite, 266
Lateritic soil, 266–267
Laurasia, 48
Lava, 34, 97–99
Lava dome, 96, 99
Law. *See* Environmental law and policy
Law of crosscutting relationships,
    A–2, A–3
Law of faunal succession,
    A–2, A–6
Law of the sea, 456–458
Leachate, 366
Leachate problems, 367
Leaching, 262
Leaching field, 379, 380
Lead, 288, 435–438
Leadville, Colorado, 423, 424
Leaning Tower of Pisa, 491
Legal liability of polluter, 458
Legislation/legal issues. *See*
    Environmental law and policy

Levee, 137–139
Libby, Montana, 32–33
Life on the Mississippi (Twain), A–8
Lignite, 325
Limestone, 36, 37
Limonite, B–2
Liquefaction, 75–76, 326
Liquid petroleum and gas, 310
Lithification, 34
Lithosphere, 50, 51
Lithospheric plates, 51
Littoral drift, 149
Llano de Caldera, Atacama Province,
    Chile, 203
Load, 122
Loam, 263
Lodestone, 31
Loess, 205–206, 207
Loma Prieta earthquake (1989),
    73, 74, 80
London, England (1952), 427
London Protocol, 465
Lone Star Geyser, 349
Long Valley caldera, 113–117
Longitudinal profile, 124
Longshore current, 149
Los Angeles Aqueduct, 254
Los Angeles smog, 433
Love Canal, 384, 385
Love wave, 68
Low-level waste, 387
Low-temperature ore-forming processes,
    286–288
Lung disease, 32
Luster, 27, B–1

M

Mad Hatter *(Alice in Wonderland),* 406
Madagascar, 267
Mafic, 91
Magma, 34, 91–93
Magmatic deposits, 282–284
Magnetic reversals, 44
Magnetite, 31, B–2
Magnitude (earthquake), 68–70
Malachite, B–2
Malaria, 221
Mammoth Lakes, 113
Man-made disasters, 9
Manganese nodules, 298, 299
Mantle, 3, 5
Mantoloking, NJ, 160
Maps, 480–486. *See also* United States
Marble, 36, 38
Marine mineral resources, 298
Mars, 3, 4